Mouse Phenotypes

Generation and Analysis of Mutants

SECOND EDITION

A LABORATORY MANUAL

ALSO FROM COLD SPRING HARBOR LABORATORY PRESS

RELATED TITLES

Aging: The Longevity Dividend

Mammalian Development: Networks, Switches, and Morphogenetic Processes

Regeneration

OTHER LABORATORY MANUALS

Budding Yeast: A Laboratory Manual

CRISPR–Cas: A Laboratory Manual

Emerging Model Organisms: A Laboratory Manual, Volume 1

Imaging in Developmental Biology: A Laboratory Manual

Manipulating the Mouse Embryo: A Laboratory Manual, Fourth Edition

Molecular Cloning: A Laboratory Manual, Fourth Edition

Molecular Neuroscience: A Laboratory Manual

Mouse Models of Cancer: A Laboratory Manual

RNA: A Laboratory Manual

Xenopus: A Laboratory Manual

WEBSITE

www.cshprotocols.org

Mouse Phenotypes

Generation and Analysis of Mutants

SECOND EDITION

A LABORATORY MANUAL

Virginia E. Papaioannou
Columbia University Medical Center

Richard R. Behringer
University of Texas MD Anderson Cancer Center

COLD SPRING HARBOR LABORATORY PRESS
Cold Spring Harbor, New York • www.cshlpress.org

MOUSE PHENOTYPES: GENERATION AND ANALYSIS OF MUTANTS, Second Edition
A LABORATORY MANUAL

Publisher	John Inglis	Permissions Coordinator	Carol Brown
Executive Editor	Alejandro Montenegro-Montero	Production Editor	Kathleen Bubbeo
Acquisition Editor	Maria Smit	Production Manager	Denise Weiss
Managing Editors	Maria Smit and Christin Munkittrick	Director of Product Development & Marketing	Wayne Manos
Developmental Editor	Judy Cuddihy		
Senior Project Manager	Inez Sialiano		

Cover art: Cover design by Richard Behringer and Virginia Papaioannou with images kindly provided by colleagues. Images include (1) mouse blastocyst in situ hybridization chain reaction for *Nanog* (green), *Gata6* (red), and DAPI (blue), courtesy of Marta Portela, Mohamed Gatie, and Kat Hadjantonakis, Memorial Sloan Kettering Cancer Center; (2) midgestation mouse embryo immunofluorescent staining, showing blood vessels (magenta) and nerves (green), courtesy of Evan Bardot and Kat Hadjantonakis, Memorial Sloan Kettering Cancer Center; (3) early gestation monozygotic twin embryos antibody stained for GATA6 and CDX2 expression, with Phalloidin marking F-actin and epiblast/EPC pseudocolored using DAPI nuclear stain, courtesy of Shifaan Thowfeequ and Shankar Srinivas, University of Oxford; (4) DAPI-stained late-gestation mouse embryo courtesy of Liang Liang and Paul Trainor, Stowers Institute for Medical Research (some images pseudo-colored by the authors).

Title page art: Courtesy of Phil Soriano.

Library of Congress Cataloging-in-Publication Data

Names: Papaioannou, Virginia E., author. | Behringer, Richard R., author.
Title: Mouse phenotypes : generation and analysis of mutants / Virginia E. Papaioannou, Columbia University Medical Center, Richard R. Behringer, University of Texas MD Anderson Cancer Center.
Description: Second edition. | Cold Spring Harbor, New York : Cold Spring Harbor Laboratory Press, [2023] | "A laboratory manual"--Cover. | Includes bibliographical references and index. |
Summary: "The laboratory mouse is an important model in biomedical research. This volume provides the concepts and tools needed to generate mutant mice and analyze their phenotypes"-- Provided by publisher.
Identifiers: LCCN 2023003303 (print) | LCCN 2023003304 (ebook) | ISBN 9781621824183 (paperback) | ISBN 9781621824190 (epub)
Subjects: LCSH: Mice–Genetics–Laboratory manuals. | Genotype–Laboratory manuals. | Animal mutation–Laboratory manuals.
Classification: LCC QH470.M52 P373 2023 (print) | LCC QH470.M52 (ebook) | DDC 599.35–dc23/eng/20230317
LC record available at https://lccn.loc.gov/2023003303
LC ebook record available at https://lccn.loc.gov/2023003304

To all my former students, and to the future:
Moses, Franklin, Ulysses, and Calliope.

–VEP

❧

To all of my past and present lab members.

–RRB

Contents

Preface

The concept for this manual grew out of our experiences as instructors in the early to mid 1990s for the Cold Spring Harbor Laboratory course on Molecular Embryology of the Mouse, or "The Mouse Course," as it was commonly known. That course, which continues to this day as the course "Mouse Development, Stem Cells, and Cancer," has been, for 40 years, the premier place to learn techniques associated with the use of the mouse in developmental, physiological, and genetic studies, and in particular for making and analyzing genetic alterations through transgenesis and gene targeting.

We found, however, that after three intense weeks of learning the language of mouse embryology and being instructed in sophisticated, cutting-edge techniques of gene manipulation by experts in the field, the students still had questions at the end of the course about how to put it all together in order to analyze a mutant. Their consternation when faced with a complex phenotype resulting from a change brought about by gene manipulation or spontaneous mutation was our biggest challenge as teachers, and there was little in the way of reference books we could point to for guidance. Our original handbook (see below), and now this up-to-date laboratory manual, is our attempt to provide a strategic guide, a road map through the entire procedure of analyzing the phenotype of a mutant mouse strain, starting with the planning stages of obtaining or producing a mutation by mutagenesis experiments, by homologous recombination gene targeting in ES cells, or by gene editing by CRISPR–Cas technology, to the complete analysis of complex phenotypes that might have their effects at any stage of development.

This edition, now in the form of a laboratory manual, has been adapted, expanded, and updated from our book, *Mouse Phenotypes: A Handbook of Mutation Analysis* (Papaioannou and Behringer 2004). The manual provides the concepts and tools needed to make genetic alterations in the mouse and to analyze the resulting mutant phenotypes, with an emphasis on analyzing mutations that affect embryonic development. Advice on strategy and practical hints are provided, to help make the analysis of mutant phenotypes a more efficient and productive enterprise. The book takes the form of a branching pathway of analysis as illustrated in Chapter 1. By answering simple questions about the particular mutation or phenotype, the reader is directed to the relevant chapters of the book, skipping over chapters that do not apply. Specific examples of mutations, protocols, and situations are fully referenced, often based on our own experiences, both in the text and in Appendix 2. Whatever your starting point and whether you are a novice or an experienced embryologist, this book will guide you through a fundamental phenotypic analysis to study gene function in the context of the organism as a whole.

We would like to thank Cold Spring Harbor Laboratory Press for the opportunity to update and expand our original book with a second edition. In the years since its first publication, new technologies have brought genetic manipulations within the reach of virtually any laboratory, rendering the principles of phenotypic analysis even more relevant. We again thank our students in the Cold Spring Harbor course on Molecular Embryology of the Mouse, 1995 and 1996, for the inspiration for this book as well as our co-instructors on the course, Peter Koopman, Terry Magnuson, and Andras Nagy, and course assistants, Debbie Chapman, Yuji Mishina, Jenny Nichols, and Bill Shawlot.

Opening artwork: Engraving by Albrecht Durer, *Adam and Eve* (detail), Metropolitan Museum of Art, Fletcher Fund, 1919.

We thank all of our colleagues and students who generously supplied helpful comments or images to illustrate the book—namely, Naiche Adler, Peter Akinwunmi, Matt Anderson, Ripla Arora, Evan Bardot, Debbie Chapman, Chun-Ming Chen, Jichao Chen, Zhoufeng Chen, Amalene Cooper-Morgan, Frank Costantini, Benoit de Crombrugghe, Sally Cross, Mary Dickinson, Guy Eakin, Shannon Erhardt, Elana Ernstoff, Laurel Fohn, Yasuhide Furuta, Marina Gertsenstein, Saadi Ghatan, Jeremy Gibson-Brown, Sarah Goldin, Bin Gu, Deborah L. Guris, Kat Hadjantonakis, Zach Harrelson, Logan Hsu, Akira Imamoto, Soazik Jamin, Loydie Jerome-Majewska, Alexandra Joyner, Monica Justice, Robert Kelly, Akio Kobayashi, Agata Kurowski, Kin Ming Kwan, Irina Larina, Mark Lewandoski, Liang Liang, Gigi Lozano, Lisette Maddison, Yuji Mishina, Sonja Nowotschin, Lauryl Nutter, George Adebayo, Dmitry Ovchinnikov, Jan Parker-Thornburg, Ayan Ray, Jaime Rivera, Andy Salinger, Tom Saunders, Reena Shakya, Phil Soriano, Shankar Srinivas, Paul Trainor, Dan Turnbull, Aya Wada, and Jun Wang.

Thanks are due also to the staff at Cold Spring Harbor Laboratory Press: John Inglis, Richard Sever, Alejandro Montenegro-Montero, Christin Munkittrick, Maria Smit, Inez Sialiano, Kathleen Bubbeo, Denise Weiss, and Carol Brown, with special thanks to Judy Cuddihy, our enthusiastic and ever-patient Editor. And finally, we thank the members of our labs, past and present, who are the able practitioners of this craft of phenotypic analysis.

Virginia E. Papaioannou
Richard R. Behringer

Reference

Papaioannou VE, Behringer RR. 2004. *Mouse phenotypes: a handbook of mutation analysis.* Cold Spring Harbor Laboratory Press, Cold Spring Harbor, NY.

CHAPTER 1

Strategies for the Production and Phenotypic Analysis of Mutations in the Mouse

The laboratory mouse is one of the primary model organisms for human biology and genetics. The production and phenotypic analysis of mutations in the mouse has made great strides with targeted mutagenesis in embryonic stem (ES) cells and clustered regularly interspaced short palindromic repeats (CRISPR)–Cas gene editing in preimplantation embryos. Mutations in many genes are now available or can be easily produced. For anyone planning to use mutational analysis for a research program, this chapter introduces an overall strategy for obtaining or generating a mutation and systematically analyzing the phenotype.

INTRODUCTION

The laboratory mouse, *Mus musculus*, is one of the principal mammalian species used in biomedical research for basic biological discoveries and as a model organism for human biology and genetics. One of the central resources for understanding gene function using the mouse system is a mutant, and interpretation of mutant phenotypes is of prime importance for understanding how alterations in the genome result in complex changes in phenotype. The development of gene targeting and random gene trapping in embryonic stem (ES) cells, along with ethylnitrosourea (ENU) mutagenesis, revolutionized mouse genetics and resulted in the generation of thousands of mutations. Now, with clustered regularly interspaced short palindromic repeats (CRISPR)–Cas gene editing, the generation of new mutations in the mouse is easier than ever (Seruggia and Montoliu 2014).

Suppose you are interested in using the mouse to perform a genetic analysis of a biological process or to create a model of a human disease. You may have decided that generating a custom mutation in a specific gene would provide you with material to investigate the biological role of that gene and its interactions with other genes in the living organism. You can now search databases for previously generated and available mutations in your gene of interest in ES cells or mice (http://www.informatics.jax.org and www.findmice.org). The majority of these mutations are archived as frozen ES cells, sperm, or embryos and can be revived using standard assisted reproduction technologies. Many mutations are also maintained "on the shelf" as live mice at commercial sources, public repositories, and individual laboratories. If an appropriate mutation for your gene of interest does not exist, then you can choose to generate a mutation in the mouse de novo, but where do you start?

Current technologies allow you to precisely alter the mouse genome in almost any manner of your choosing. But where do you begin in designing the most useful or versatile mutant allele for your

Opening artwork: Detail from *The Garden of Earthly Delights*, Bosch, Hieronymus (c.1450–1516) Credit: Prado, Madrid, Spain/ Bridgeman Images.

© 2024 Cold Spring Harbor Laboratory Press
Cite this chapter as *Cold Spring Harb Protoc*; doi:10.1101/pdb.over107955

studies? Do you choose to produce the mutation by homologous recombination and selection in ES cells or would it be quicker and easier to use CRISPR–Cas-mediated gene editing directly in preimplantation embryos? For your purposes, is a null allele, a conditional null allele, a single-base-pair change, or some other type of allele most useful? Should you include a reporter gene in the allele?

Looking ahead to when you have obtained or produced a mutant allele, and it is in the context of the whole organism, the mouse, the phenotype may not be as straightforward as you had hoped or expected. Where do you start with the analysis of the mutant phenotype? How do you approach abnormalities that occur at various developmental stages or in organs you have not considered since anatomy class to describe the phenotype and begin to understand the molecular and developmental consequences of the genetic change? If the mutation causes defects during embryonic development, what can you learn and how might you circumvent early abnormalities to study the phenotype in adult animals? How can you make the best use of the genetic resource you have developed and uncover causes of complex phenotypes? These are fundamental questions to answer when planning a project that will use valuable resources and typically occupy years of your time.

During the summers of 1995 and 1996, we were the Directors of the Molecular Embryology of the Mouse course at the Cold Spring Harbor Laboratory on Long Island, New York. This annual 3-week course was initiated in 1983 by Brigid Hogan soon after the first transgenic mice were generated by pronuclear injection of zygotes. A group of 10 to 14 students (graduate students, postdocs, and faculty) heard cutting-edge lectures from international experts on mouse genetics and embryology. They were also in the laboratory each day, learning how to isolate mouse embryos at different stages of development, inject zygotes and blastocysts, and perform embryo transfers into surrogate mothers. In the early years of the course, students acquired these skills to generate transgenic mice by zygote microinjection and learned about mouse embryonic development. By the mid-1990s, gene targeting by homologous recombination in ES cells to generate null alleles in mice was becoming a standard protocol and conditional knockouts were being developed. The mouse course expanded to encompass training in this technology as well.

Over the years the course has constantly evolved, incorporating the latest technologies and approaches to generate genetically engineered mice and analyze mutant phenotypes. Many of the students in the course during those two summers that we directed already had a mouse mutant that had been created using gene targeting in ES cells, but they were unsure how to analyze mutant phenotypes, especially an unexpected one like embryonic lethality. Based on our extensive experience in mouse genetics and embryology, we guided the students on how to approach the study of a mutant phenotype. Our experience during the course indicated a need for a practical manual to guide researchers on how to characterize mutant phenotypes in a methodical way. This led us to write the book *Mouse Phenotypes: A Handbook of Mutation Analysis* (Papaioannou and Behringer 2005) and now this new, revised edition. Our book created a strategy for making the best choices in the design of an experimental program and provided guidance in the basic principles of mutant production and analysis to eventually answer the question: How and when does a mutant animal differ from normal? The emphasis was, and still is in this new edition, on detecting phenotypic effects during embryogenesis, when pleiotropic effects of many mutations will first be detected.

EXISTING MUTATIONS

Since the first targeted mutations made in ES cells were recovered in the germline of mice at the end of the 1980s, more than half of all the genes in the mouse genome have been mutated through the combined efforts of individuals in the scientific community and coordinated, high-throughput mutagenesis programs. The level of phenotypic analysis of these mutations is variable, with many mutants analyzed in great detail by individual investigators with specific interests, whereas others have received less attention. The International Mouse Phenotyping Consortium (IMPC; https://www .mousephenotype.org) is an international consortium with the stated goal of phenotyping knockout

Cite this chapter as *Cold Spring Harb Protoc*; doi:10.1101/pdb.over107955

mouse lines for every protein-coding gene in the genome (Lloyd 2011; Bradley et al. 2012). Using primarily targeted mutations that have been produced in a single genetic background through the International Knockout Mouse Consortium (IKMC), the IMPC has developed high-throughput phenotyping platforms using a series of tests, both live and terminal, with standardized protocols for adult and embryonic phenotypes. Phenotyping data is available for mutations in more than 5000 unique genes through the IMPC phenotyping data portal (Koscielny et al. 2014; Birling et al. 2021). This should be a first stop for anyone looking for a particular phenotype or set of phenotypes or the phenotype of a mutation in a particular gene. Depending on the phenotype, the mutation could then be studied at greater depth to reveal the function of the gene with additional tests, or by producing additional alleles (e.g., conditional null or point mutations).

STRATEGY FOR PRODUCING AND PHENOTYPING A MOUSE MUTATION

If, on the other hand, your gene of interest has not been previously mutated or phenotypically characterized, we provide step-by-step guidance for efficient production and analysis of mutant phenotypes regardless of their nature and regardless of when in the life cycle they become evident (see http:/cshprotocols.cshlp.org/site/mousephenotypes). By many estimates, around one-third of homozygous null mutations result in embryonic lethality (Birling et al. 2021), so even if you were not intentionally targeting a gene that regulates development, there is a likelihood that you will end up with an embryonic phenotype. The chapters in this book provide an in-depth strategy for analysis of phenotypes that appear throughout embryonic and postnatal development.

At the outset, learn all you can about the gene or genes you are interested in studying. Know the expression pattern of your gene, either from gene expression databases or by doing expression studies on your own. Equipped with this information, you are ready to begin. The overall strategy of this book takes the form of a logical branching decision pathway (Fig. 1) initiating at three alternative starting points of (1) producing a mutation by homologous recombination or CRISPR–Cas gene editing in ES cells, (2) gene editing by CRISPR–Cas directly in preimplantation mouse embryos, or (3) obtaining an existing mutant from a laboratory or a repository. By starting at the planning stages and answering simple questions about the goals of the project and your resources, you will be directed to the most relevant starting points for analysis, skipping over those chapters that do not apply to your project. Troubleshooting and helpful hints based on our many years of experience will help you avoid costly and time-consuming mistakes.

An important result of this systematic approach is the identification of the earliest stage at which a mutant phenotype is present. Phenotypes found at later stages of development may represent secondary or tertiary consequences of earlier abnormalities. Outcomes of each step in the procedure are examined to determine whether apparent failures could result from impatience, inexperience, or technical problems or whether they are indications of an actual phenotype. Without a systematic approach it very easy to miss important aspects of the phenotype.

Lethality at different embryonic stages might benefit from further analysis of embryo, organ, tissue, or cell developmental potential, covered in later chapters in this book. Chimeras have been used extensively as tools for producing germline mutations via ES cells, but they can also be used in various ways to investigate the phenotypic effects of mutations. They may be useful for circumventing a lethal effect by allowing mutant cells to survive to a later stage of development in a chimeric combination with wild-type cells and, additionally, they provide a method for testing the potential of mutant cells.

A null allele that completely removes the function of a gene may reveal the earliest phenotype of a mutation, but could result in early lethality, precluding the study of later stages. Even in the case of viable or subviable mutations, the accumulation of secondary and tertiary effects that could obscure primary effects occurring later in development or late in life will complicate the analysis. With the extensive pleiotropy of gene effects evident for so many genes in the mouse genome, the full range of temporal and tissue-specific effects of a gene can be much more fully explored if the gene in question is

Getting to a phenotype

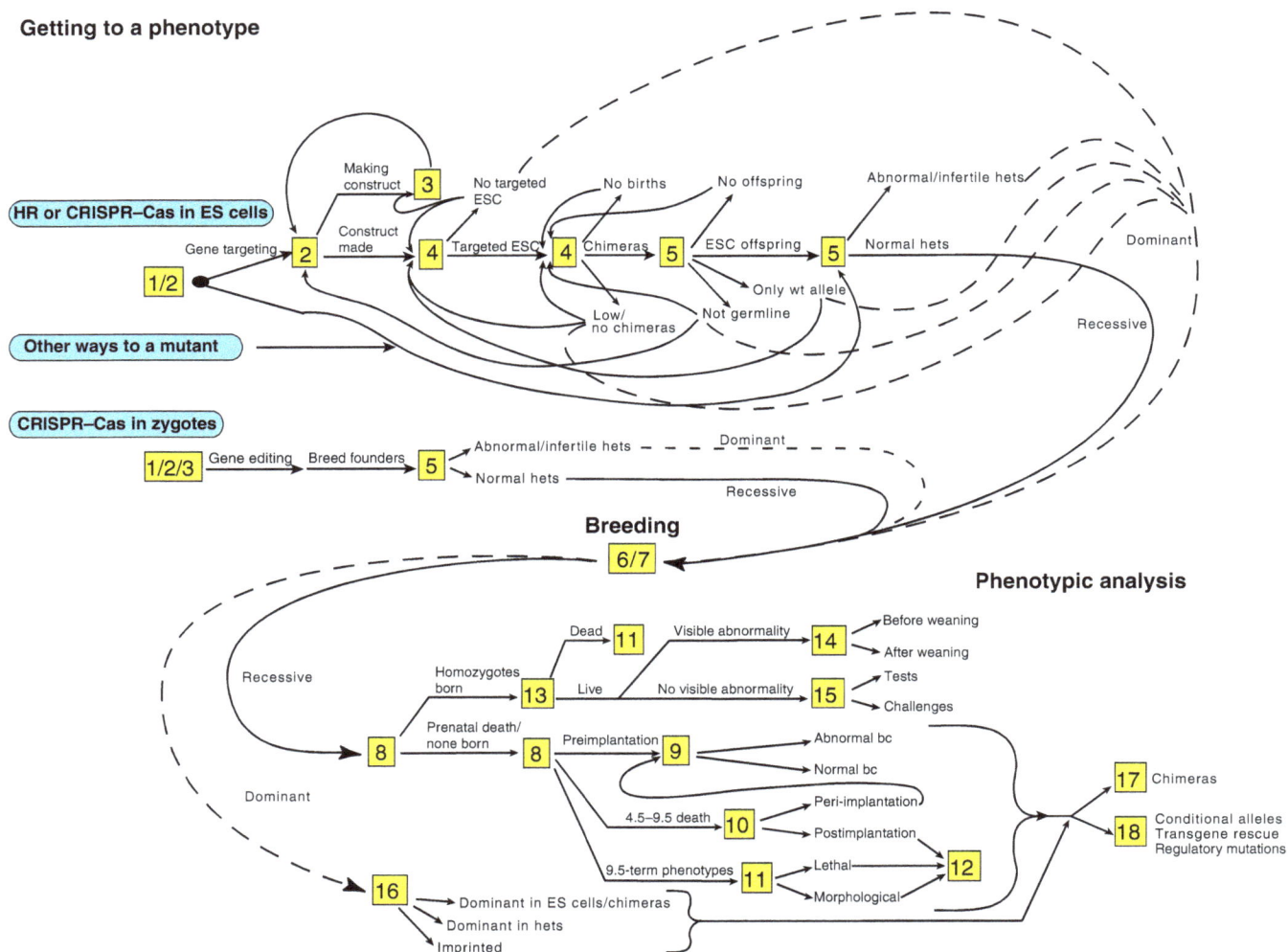

HR or CRISPR–Cas in ES cells

Gene targeting — 1/2 — 2 — Making construct — 3 — No targeted ESC

Construct made — 4 — Targeted ESC — 4 — Chimeras — 5 — ESC offspring — 5 — Normal hets — Dominant

No births — No offspring — Abnormal/infertile hets — Dominant

Low/no chimeras — Not germline — Only wt allele — Recessive

Other ways to a mutant

CRISPR–Cas in zygotes

Gene editing — 1/2/3 — Breed founders — 5 — Abnormal/infertile hets — Dominant

Normal hets — Recessive

Breeding

6/7

Phenotypic analysis

Recessive

Homozygotes born — 13 — Dead — 11 — Visible abnormality — 14 — Before weaning / After weaning

Live — No visible abnormality — 15 — Tests / Challenges

8 — Prenatal death/none born — 8 — Preimplantation — 9 — Abnormal bc / Normal bc

Dominant

4.5–9.5 death — 10 — Peri-implantation / Postimplantation

9.5–term phenotypes — 11 — Lethal / Morphological — 12

16 — Dominant in ES cells/chimeras — Dominant in hets — Imprinted

17 — Chimeras

18 — Conditional alleles / Transgene rescue / Regulatory mutations

FIGURE 1. Flow diagram with a branching decision pathway for obtaining or generating mouse mutations and their phenotypic analysis. From any of the three starting points (indicated in blue boxes) of homologous recombination (HR) or CRISPR–Cas in embryonic stem (ES) cells, CRISPR–Cas gene editing in zygotes, or other ways for obtaining a mutant, the path taken in the branching pathway depends on the outcome at each step of the procedure, taking into account technical and biological issues that can present stumbling blocks along the way. The boxed numbers (in yellow) refer to specific Chapters that present details of each step in the procedure as follows:

Generation of mutants: (1) Chapter 1: *Strategies for the Production and Phenotypic Analysis of Mutations in the Mouse* (Papaioannou and Behringer 2023a) / (2) Chapter 2: *Obtaining or Generating Gene Mutations in Mice* (Papaioannou and Behringer 2023b) / (3) Chapter 3: *Mouse Gene-Targeting Strategies for Maximum Ease and Versatility* (Papaioannou and Behringer 2023c) / (4) Chapter 4: *Embryonic Stem Cell Gene Targeting and Chimera Production in Mice* (Papaioannou and Behringer 2023d) / (5) Chapter 5: *Recovering a Targeted Mutation in Mice from Embryonic Stem Cell Chimeras or CRISPR–Cas Founders* (Papaioannou and Behringer 2023e)

Breeding: (6) Chapter 6: *Strategies for Maintaining Mouse Mutations* (Papaioannou and Behringer 2023f) / (7) Chapter 7: *Special Breeding Techniques for Use in Mouse Mutation Analysis* (Papaioannou and Behringer 2023g)

Phenotypic analysis of embryonic, postnatal, and adult phenotypes, of either dominant (dashed lines) or recessive (solid lines) mutations with emphasis on developmental phenotypes: (8) Chapter 8: *Phenotypic Analysis: Assessing Timing of Recessive Prenatal Lethality in Mice* (Papaioannou and Behringer 2023h) / (9) Chapter 9: *Phenotypic Analysis of Preimplantation Lethality in Mice* (Papaioannou and Behringer 2023i) / (10) Chapter 10: *Phenotypic Analysis of Periimplantation to Mid-Gestation Lethality in Mice* (Papaioannou and Behringer 2023j) / (11) Chapter 11: *Analysis of Mid- to Late-Gestation Phenotypes in Mice* (Papaioannou and Behringer 2023k) / (12) Chapter 12: *Uncovering Phenotypes in Mutant Mice by Determining Embryo, Organ, Tissue, and Cell Developmental Potential* (Papaioannou and Behringer 2023l) / (13) Chapter 13: *Phenotypic Analysis of Perinatal Lethality in Mice* (Papaioannou and Behringer 2023m) / (14) Chapter 14: *Analysis of Postnatal Mutant Phenotypes in Mice* (Papaioannou and Behringer 2023n) / (15) Chapter 15: *The "No Phenotype" Challenge in Analyzing Mutant Mice* (Papaioannou and Behringer 2023o) / (16) Chapter 16: *Phenotypic Analysis of Dominant Mutant Effects in Mice* (Papaioannou and Behringer 2023p) / (17) Chapter 17: *Getting around an Early Lethal Phenotype in Mice with Chimeras* (Papaioannou and Behringer 2023q) / (18) Chapter 18: *Tissue- and/or Temporal-Specific Mutations in Mice Using Conditional Alleles* (Papaioannou and Behringer 2023r)

mutated at the time and in the tissue of the investigator's choosing. If your initial investigation involved a simple loss-of-function allele, now might be the time to consider a conditional null allele. Conditional alleles are discussed during the design phase of gene targeting experiments, and a large proportion of the available existing mutants are available as conditional-ready alleles (Brown and Moore 2012). (18) Chapter 18: Tissue- and/or Temporal-Specific Mutations in Mice Using Conditional Alleles (Papaioannou and Behringer 2023r) provides a detailed guide for the use of conditional null alleles, including the importance of testing not just the functionality of the allele but also the tissue specificity or inducibility of recombinase mouse lines used with conditional alleles. Conditional alleles allow you to look at phenotypes at any time and in any tissue, limited only by the availability of suitable recombinase mouse lines.

Whether or not you are an expert in mouse development, the simple procedures detailed in this guide will allow you to make an intelligent assessment of the phenotype resulting from a mutation and to determine the timing and likely causes of specific developmental failures. Simple protocols are provided that will identify contributing factors and references are given for detailed protocols, including many contained in the mouser's "bible" *Manipulating the Mouse Embryo: A Laboratory Manual* (Behringer et al. 2014). At each step of the way, references from the literature are provided that give actual examples of how different procedures have been applied to specific mutations, many of them drawn from our own experience. From there it is up to you to either continue with an in-depth analysis to tease out all the pleiotropic effects of the gene or to find appropriate expert collaborators. Either way, you will certainly gain a greater appreciation for the beauty and complexity of the developing organism and you may even learn to get over the fear of finding a lethal effect and learn to love an embryonic phenotype.

REFERENCES

Behringer RR, Gertsenstein M, Nagy KV, Nagy A. 2014. *Manipulating the mouse embryo: a laboratory manual*. Cold Spring Harbor Laboratory Press, Cold Spring Harbor, NY.

Birling MC, Yoshiki A, Adams DJ, Ayabe S, Beaudet AL, Bottomley J, Bradley A, Brown SDM, Bürger A, Bushell W, et al. 2021. A resource of targeted mutant mouse lines for 5,061 genes. *Nat Genet* **53**: 416–419. doi:10.1038/s41588-021-00825-y

Bradley A, Anastassiadis K, Ayadi A, Battey JF, Bell C, Birling MC, Bottomley J, Brown SD, Bürger A, Bult CJ, et al. 2012. The mammalian gene function resource: the International Knockout Mouse Consortium. *Mamm Genome* **23**: 580–586. doi:10.1007/s00335-012-9422-2

Brown SD, Moore MW. 2012. The International Mouse Phenotyping Consortium: past and future perspectives on mouse phenotyping. *Mamm Genome* **23**: 632–640. doi:10.1007/s00335-012-9427-x

Koscielny G, Yaikhom G, Iyer V, Meehan TF, Morgan H, Atienza-Herrero J, Blake A, Chen CK, Easty R, Di Fenza A, et al. 2014. The International Mouse Phenotyping Consortium Web Portal, a unified point of access for knockout mice and related phenotyping data. *Nucl Acids Res* **42**: D802–D809. doi:10.1093/nar/gkt977

Lloyd KC. 2011. A knockout mouse resource for the biomedical research community. *Ann NY Acad Sci* **1245**: 24–26. doi:10.1111/j.1749-6632.2011.06311.x

Papaioannou VE, Behringer RR. 2005. *Mouse phenotypes: a handbook of mutation analysis*. Cold Spring Harbor Laboratory Press, Cold Spring Harbor, NY.

Papaioannou VE, Behringer RR. 2023a. Strategies for the production and phenotypic analysis of mutations in the mouse. *Cold Spring Harb Protoc* doi:10.1101/pdb.over107955

Papaioannou VE, Behringer RR. 2023b. Obtaining or generating gene mutations in mice. *Cold Spring Harb Protoc* doi:10.1011/pdb.over107956

Papaioannou VE, Behringer RR. 2023c. Mouse gene targeting strategies for maximum ease and versatility. *Cold Spring Harb Protoc* doi:10.1101/pdb.over107957

Papaioannou VE, Behringer RR. 2023d. Embryonic stem cell gene targeting and chimera production in mice. *Cold Spring Harb Protoc* doi:10.1101/pdb.over107958

Papaioannou VE, Behringer RR. 2023e. Recovering a targeted mutation in mice from embryonic stem cell chimeras or CRISPR–Cas founders. *Cold Spring Harb Protoc* doi:10.1101/pdb.over107959

Papaioannou VE, Behringer RR. 2023f. Strategies for maintaining mouse mutations. *Cold Spring Harb Protoc* doi:10.1101/pdb.over107960

Papaioannou VE, Behringer RR. 2023g. Special breeding techniques for use in mouse mutation analysis. *Cold Spring Harb Protoc* doi:10.1101/pdb.over107961

Papaioannou VE, Behringer RR. 2023h. Phenotypic.analysis: assessing timing of recessive prenatal lethality in mice. *Cold Spring Harb Protoc* doi:10.1011/pdb.over107970

Papaioannou VE, Behringer RR. 2023i. Phenotypic analysis of preimplantation lethality in mice. *Cold Spring Harb Protoc* doi:10.1011/pdb.over107971

Papaioannou VE, Behringer RR. 2023j. Phenotypic analysis of periimplantation to mid-gestation lethality in mice. *Cold Spring Harb Protoc* doi:10.1011/pdb.over107972

Papaioannou VE, Behringer RR. 2023k. Analysis of mid- to late-gestation phenotypes in mice. *Cold Spring Harb Protoc* doi:10.1011/pdb.over107973

Papaioannou VE, Behringer RR. 2023l. Uncovering phenotypes in mutant mice by determining embryo, organ, tissue, and cell developmental potential. *Cold Spring Harb Protoc* doi:10.1011/pdb.over107974

Papaioannou VE, Behringer RR. 2023m. Phenotypic analysis of perinatal lethality in mice. *Cold Spring Harb Protoc* doi:10.1011/pdb.over107975

Papaioannou VE, Behringer RR. 2023n. Phenotypic analysis of postnatal mutant phenotypes in mice. *Cold Spring Harb Protoc* doi:10.1101/pdb.over107976

Papaioannou VE, Behringer RR. 2023o. The "no phenotype" challenge in analyzing mutant mice. *Cold Spring Harb Protoc* doi:10.1101/pdb.over107977

Papaioannou VE, Behringer RR. 2023p. Phenotypic analysis of dominant mutant effects in mice. *Cold Spring Harb Protoc* doi:10.1101/pdb.over107978

Papaioannou VE, Behringer RR. 2023q. Getting around an early lethal phenotype in mice with chimeras. *Cold Spring Harb Protoc* doi:10.1101/pdb.over107979

Papaioannou VE, Behringer RR. 2023r. Tissue- and/or temporal-specific mutations in mice using conditional alleles. *Cold Spring Harb Protoc* doi:10.1011/pdb.over107980

Seruggia D, Montoliu L. 2014. The new CRISPR–Cas system: RNA-guided genome engineering to efficiently produce any desired genetic alteration in animals. *Transgenic Res* **23:** 707–716. doi:10.1007/s11248-014-9823-y

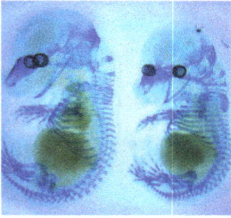

CHAPTER 2

Obtaining or Generating Gene Mutations in Mice

The starting point in a mutational analysis of gene function is obtaining or producing a mutant. Here different methods of obtaining mouse mutants are discussed, including screening for spontaneous mutants, screening for mutants following chemical or X-ray mutagenesis, and producing mutations through targeted manipulation of the genome. Manipulation of the genome can be random, as in different types of insertional mutagenesis. Alternatively, targeted manipulation such as gene targeting using homologous recombination in embryonic stem (ES) cells or gene editing by CRISPR–Cas can be used to produce custom mutations in a specific gene. The basic methods are outlined, and the advantages and disadvantages of homologous recombination and CRISPR–Cas gene editing are discussed. Resources for obtaining mutations that already exist are provided. If, for your planned study, no suitable mutations are available, there is advice about what you should know about your gene of interest before embarking on a gene targeting experiment.

INTRODUCTION

OBTAIN OR GENERATE A MOUSE MUTATION

- EXISTING MUTATION ⟶ Chapter 6

- HR OR CRISPR–Cas IN ES CELLS ⟶ Chapter 3

- CRISPR–Cas IN ZYGOTES ⟶ Chapter 3

A variety of ways are available to obtain mutations in mice. One way is simply to wait for spontaneous mutations to arise. But this can be a tedious pastime, and the mutations are unpredictable to say the least. Thus, various other methods have been devised either to speed up the process or to make it more predictable. The early chapters in this book concentrate on mutations produced by gene targeting either (1) by homologous recombination in embryonic stem (ES) cells followed by germline transmission from ES cell chimeras, or (2) by CRISPR–Cas-mediated gene editing directly in preimplantation embryos. We will discuss the advantages and disadvantages of each approach. However, the method of phenotypic analysis presented in later chapters in this book is by no means limited to this

Opening artwork: Jerome and Papaioannou. 2001. *Nat Genet* **27:** 286–291 ©Nature Publishing Group.

© 2024 Cold Spring Harbor Laboratory Press
Cite this chapter as *Cold Spring Harb Protoc*; doi:10.1101/pdb.over107956

category of mutations created by targeted mutagenesis, but is equally applicable to mutants generated or obtained in other ways.

VARIOUS WAYS TO OBTAIN OR GENERATE MUTANTS

Spontaneous Mutations

The spontaneous mutation rate in mice is one gamete in approximately 100,000 for a specific gene and, of course, the identification of spontaneous mutations is biased toward mutant phenotypes that can be detected visually (e.g., coat color, gross morphology, or behavior). Attempting a genetic screen for spontaneous mutations in mice is thus logistically impractical. However, in addition to targeted mutagenesis, there are many other ways to induce mutations in mice. These methods can be grouped into broad categories, including chemically induced mutations, radiation-induced deletions, insertional mutations, and transgene expression. With these methods, mutations in specific genes of interest can be either selected or directed.

Chemical Mutagenesis Using Ethylnitrosourea

Chemical mutagens can greatly increase the mutation frequency, making genetic screens in mice highly feasible. Numerous groups around the world have performed small- and large-scale chemical mutagenesis screens in mice to isolate dominant and recessive mutations (e.g., Kasarskis et al. 1998; Hrabe de Angelis et al. 2000; Kile et al. 2003). The most powerful chemical mutagen in mice is ethylnitrosourea (ENU). ENU induces an average per-locus mutation frequency of 1.5×10^{-3}. A mutation in a single gene of choice can be recovered on average in 1 in 600–700 gametes screened. To date, every ENU-induced mutation that has been sequenced in the mouse is a point mutation, and, because ENU causes point mutations, it defines single-gene function. ENU creates nonsense, missense, splice, and regulatory mutations that can affect gene, RNA, and protein function; thus, null, neomorph, antimorph, hypomorph, and hypermorph mutations can be isolated. The advantages of ENU mutagenesis are the rich variety of alleles generated, the possibility of identifying mutations by phenotype during the screen, and the fact that many mutations can be generated and isolated using simple methods. The primary disadvantages are the labor and mouse space required to perform a screen and the effort required to map and identify the molecular lesions.

Typically, adult males of a particular inbred strain are injected intraperitoneally with a precise dose of ENU to mutagenize spermatogonial stem cells. ENU treatment causes a depletion of mature germ cells, resulting in transient sterility. The spermatogonial stem cells are spared but their DNA acquires numerous point mutations throughout the genome depending on the dose of ENU administered. The spermatogonial stem cells then undergo spermatogenesis to produce sperm. The ENU-injected males usually regain fertility 11–17 wk post-ENU treatment and are then bred using different breeding schemes designed to generate pedigrees for phenotypic analysis to isolate mutations. A well-designed phenotypic screen of the progeny will identify mutations affecting specific structures or processes of interest (Cordes 2005). Mice with fully penetrant dominant mutations that are viable and fertile are easy to genotype because the dominant phenotype can be used to identify the mutants. However, mice heterozygous for recessive ENU-induced mutations can initially be difficult to genotype because the gene and chromosomal location are usually unknown. In these cases, one must perform test crosses to identify carriers (Fig. 1). Genetic mapping eventually leads to the identification of the chromosomal location of the mutation. Subsequently, linked polymorphic markers can be used to track the mutation.

Advances in genome sequencing and bioinformatic tools (e.g., exome sequencing) have greatly facilitated the identification of ENU-induced mutations (Buchovecky et al. 2013; Geister et al. 2018). Ultimately, the ENU mutation is identified, and, if you are fortunate, it may cause a restriction enzyme polymorphism that can be exploited in a polymerase chain reaction (PCR) genotyping strategy. Alternatively, if there is no restriction enzyme polymorphism, single-stranded conformation polymorphism or other strategies can be used for genotyping.

Cite this chapter as *Cold Spring Harb Protoc*; doi:10.1101/pdb.over107956

A

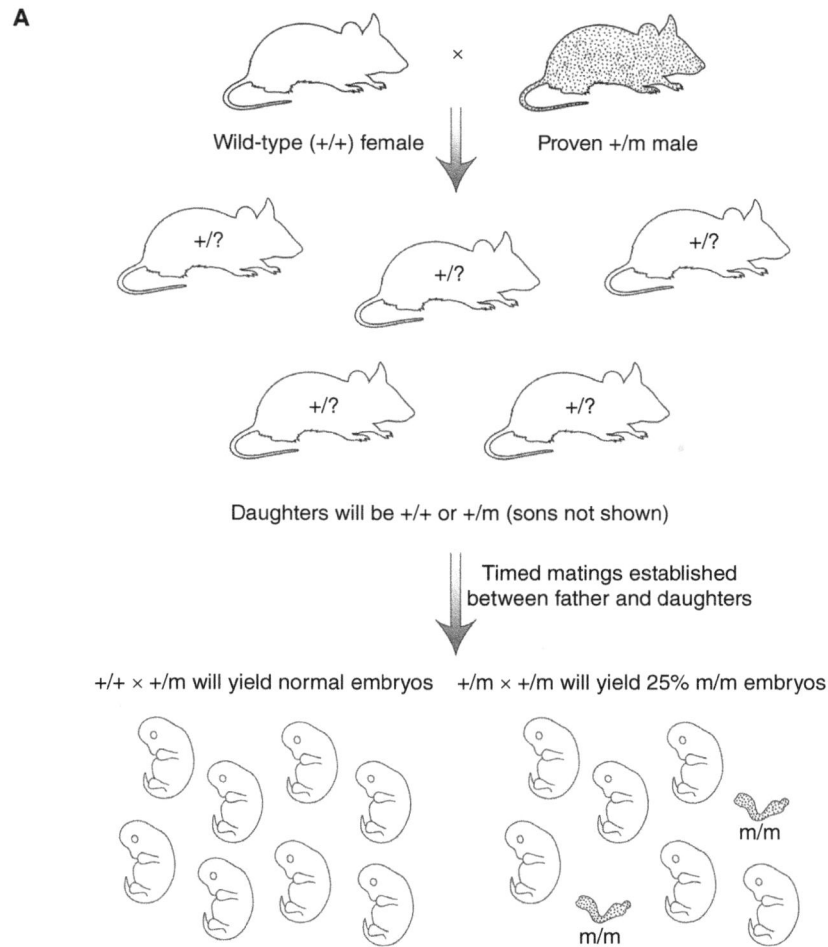

Wild-type (+/+) female   ×   Proven +/m male

+/?   +/?   +/?

+/?   +/?

Daughters will be +/+ or +/m (sons not shown)

Timed matings established
between father and daughters

+/+ × +/m will yield normal embryos   +/m × +/m will yield 25% m/m embryos

m/m

m/m

FIGURE 1. Test crosses for recessive ethylnitrosourea (ENU)-induced mutations. Spontaneous and ENU-induced mutations may not be molecularly defined. Therefore, genetic crosses must be performed to identify mice that carry the mutation. Two situations are shown. (A) A male known to be heterozygous for a recessive embryonic lethal mutation is bred with wild-type females to generate progeny. The daughters from this cross will be +/+ or +/m. To obtain homozygous mutant embryos for analysis, timed matings are established between the male carrier and his daughters for dissection at time points chosen to yield the mutant phenotype. Approximately 50% of the time the cross will be between a +/+ daughter and the +/m father that will yield +/+ and +/m embryos that will have a wild-type phenotype, and 50% of the time the cross will be between a +/m daughter and the +/m father that will yield +/+, +/m, and m/m embryos. Therefore, ~25% of the embryos from this cross should show the mutant phenotype. If new heterozygous males are needed, then the sons from a known carrier are test crossed with their siblings to distinguish +/+ and +/m males. (B) A male known to be heterozygous for a recessive viable but sterile mutation is bred with wild-type females to generate progeny. The offspring from this cross will be +/+ or +/m. To obtain homozygous mutants for analysis, matings are established at random between the siblings. There is a 1 in 4 chance that you have set up a mating between two +/m mice. Once progeny are born they can be examined for the mutant phenotype. Those matings yielding homozygous mutants provide the genetic evidence that both of the parents are heterozygous carriers. (Modified, with permission, from Papaioannou and Behringer 2005, © Cold Spring Harbor Laboratory Press.) (*Figure continued on following page.*)

As a community resource, ENU-mutagenized sperm and corresponding genomic DNA have been cryopreserved from mice (Sakuraba et al. 2005) and a database of random ENU-induced point mutations in specific genes with mutation–phenotype associations has been created (https:// mutagenetix.utsouthwestern.edu) (Wang et al. 2018). In vitro fertilization using the cryopreserved sperm can be used to recover the allele of interest in mice. ES cells have also been treated with ENU to create random mutations (Chen et al. 2000). ES cells containing desired mutations can be used to generate mutant mice via germline transmission through chimeras (Vivian et al. 2002).

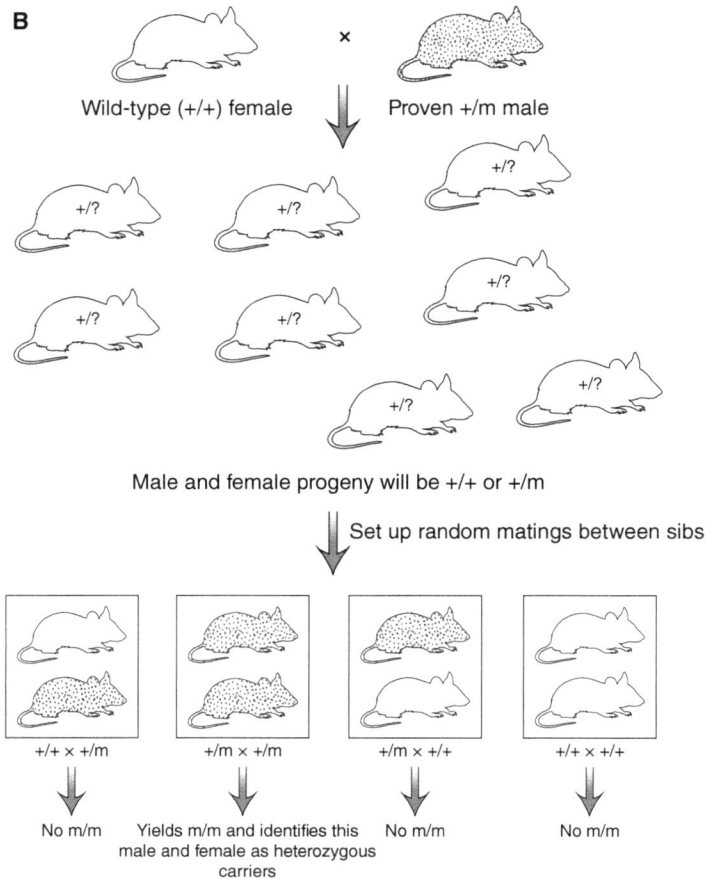

FIGURE 1. *Continued from previous page.*

X-Ray Mutagenesis

Exposing mice to X-rays causes large genetic deletions in germ cells that can be transmitted to progeny. The mutagenized mice are bred to mice with appropriate genetic markers to identify the deletion alleles. Classically, this approach has been used to generate deletion complexes, which usually include multiple loci. One well-characterized deletion complex is the albino deletion complex, a series of overlapping deletions that include the albino (*Tyr*) locus (Holdener and Magnuson 1994). These deletions were identified by test crosses with mice that are homozygous for the recessive albino allele (*Tyr^c*). The presence of white mice in the progeny indicated that the albino locus was deleted in the germ cells of the X-ray-treated parent. By combining various deletions, one can define loci that affect diverse developmental processes. These genetic resources tend to be used primarily for gene discovery rather than as primary tools for understanding the function of one particular gene.

X-rays have also been used in ES cells to create a series of cM-sized deletions at a specific locus (Thomas et al. 1998). This approach can be used to assess large regions of the genome and also regions of haploinsufficiency.

Insertional Mutagenesis

When a piece of exogenous DNA integrates into the genome, it can physically disrupt an endogenous gene or alter its expression at the level of transcription or RNA splicing. This is called an insertional mutation. Insertional mutations can result in various types of alleles but most frequently result in a

Cite this chapter as *Cold Spring Harb Protoc*; doi:10.1101/pdb.over107956

loss of function. The exogenous DNA can be delivered into cells by viruses or as purified DNA molecules. Below we discuss various types of insertional mutations in the mouse.

Viruses

Viruses can be used to infect preimplantation-stage mouse embryos or ES cells to cause insertional mutations (Jaenisch et al. 1983; Robertson et al. 1986). The viral sequence can then be used as a molecular tag to follow proviral genome inserts in pedigrees and ultimately to clone the mutated genes. Viruses can also be modified for specific mutagenesis screens (e.g., gene traps; Friedrich and Soriano 1991).

Transgenes

Insertional mutagenesis is one of the by-products of generating transgenic mice by pronuclear injection of DNA into zygotes. It is estimated that ~5%–10% of random integrations of foreign DNA into the genome fortuitously disrupt an endogenous gene. The transgene can serve as a molecular tag to clone the disrupted gene(s). One of the problems with transgene insertional mutations is that they can be associated with large deletions, duplications, inversions, and translocations. It is thus not always clear if the mutant phenotype is caused by one or multiple gene disruptions. If chromosomal rearrangements are present at the site of transgene integration, molecular cloning of the gene responsible for the mutant phenotype can be challenging. Another potential complication is that the expression of the transgene itself may confuse the analysis of the insertional mutant phenotype.

Gene Traps

Gene traps are DNA or viral vectors typically introduced into ES cells by electroporation or retroviral infection to generate insertional mutations that disrupt the function of endogenous genes (Robertson et al. 1986; Gossler et al. 1989; Friedrich and Soriano 1991; Stanford et al. 2001). Gene trap vectors are engineered to include reporters like *lacZ*, which in principle allow you to follow the expression pattern of the trapped locus. ES cell lines with gene trap insertions can be prescreened for reporter expression patterns or to identify the trapped genes. Ultimately, mice generated from ES cell lines with gene traps can then be screened for mutant phenotypes. Because gene traps are essentially random insertional mutations, the resulting mutant alleles can have diverse activity. It is easy to genotype heterozygous mice for gene trap insertions by virtue of the reporter activity, but identification of homozygous mutant mice is challenging until the locus is cloned. Libraries of ES cell lines with gene trap insertions and associated sequence tags have been generated and are available from the International Gene Trap Consortium (https://igtc.org/), which has integrated the publicly available gene trap lines from a variety of different large-scale screens (Nord et al. 2006; Guan et al. 2010). These resources can facilitate the generation of a mutant for a specific gene by allowing you to acquire the ES cell clone and generate mice from ES cell chimeras.

Transposons

Transposons are another class of insertional mutagens that can be used in mice to generate mutant phenotypes (Horie et al. 2003). *Sleeping Beauty* (*SB*) is a Tc1/mariner-type transposon resurrected from salmonid fish and developed as a transposon mutagenesis system in mice. The SB transposase specifically recognizes *SB* transposon inverted repeats. The *SB* transposon is initially introduced into the mouse genome as a transgene either by pronuclear injection or through transfection of ES cells. Transgenic male mice carrying the *SB* transgene are then bred with transgenic mice that express the SB transposase in the germline. The resulting double transgenic mice are then bred with wild-type mice, and mobilization of the transposable element is monitored in the next generation. The SB transposase causes the transposition of the *SB* transposon through a precise cut-and-paste mechanism to insert at TA dinucleotides. This transposition can occur *in cis* locally on the same chromosome or *in trans* to other chromosomes.

The *piggyBac* (*PB*) transposon system has also been adapted for use in mice (Ding et al. 2005). *PB* transposons insert at TTAA tetranucleotides, which are duplicated upon insertion. One advantage of transposon mutagenesis systems is that the mice create additional mutations simply by breeding.

Transgenic Expression

Finally, transgenic mice generated by various methods are frequently created to express a foreign gene product to elicit a mutant phenotype (e.g., *Gh*; Palmiter et al. 1982; Schilit et al. 2016). These are typically gain-of-function experiments for the intentional misexpression or overexpression of a gene in a specific tissue or many tissues and are different from transgene-induced insertional mutations, mentioned in the previous section, that randomly disrupt endogenous genes. By design, a tissue-specific or broadly expressed promoter may have been used to direct the expression of the gene. If a tissue-specific regulatory sequence was used, then one already knows which tissues to examine first for a phenotype in the transgenic mouse. However, sometimes the resulting phenotypes are not as simple as expected and a more comprehensive analysis will be necessary.

OBTAINING AN EXISTING MUTANT

There are numerous repositories that archive mutations either as live mice or cryopreserved preimplantation embryos or sperm. Therefore, there may already be a mutation in your gene of interest and no need to generate a new mutant. One of the first places to check is Mouse Genome Informatics (MGI; www.informatics.jax.org). Search for your gene using the official gene name, then on the "Gene Detail" page check under "Mutations, Alleles and Phenotypes" to learn what alleles have been reported. Then look under "Find Mice (IMSR)," which takes you to the International Mouse Strain Resource (www.findmice.org). There you will see if live mice, embryos, and/or sperm are available from specific repositories located around the world with links to references and ordering information.

In addition to the gene-trap screens in ES cells mentioned in the previous section, there have also been extensive international efforts to produce mutations in every gene in the mouse genome, making them available through publicly funded repositories. The National Institutes of Health (NIH)-funded Knockout Mouse Project (KOMP) initiated this ambitious mutagenesis project using ES cell technologies to produce knockout or conditional mutations. The European Conditional Mouse Mutagenesis Program (EUCOMM) and the North American Conditional Mouse Mutagenesis project (NorCOMM), using a combination of gene targeting and gene trapping in ES cells, also had the goal of mutating large numbers of genes. In 2007 these three programs, along with the Texas A&M Institute for Genomic Medicine (TIGM), joined forces to minimize overlap and share resources, creating the International Knockout Mouse Consortium (IKMC). By 2012, more than 17,000 different genes had been mutated and multiple mouse lines had been produced (Lloyd 2011; Bradley et al. 2012).

Subsequently, a high-throughput functional genomics project coordinated by the International Mouse Phenotyping Consortium (IMPC; https://mousephenotype.org) has validated and phenotyped a large number of mouse lines (Brown and Moore 2012). By 2021, the number of targeted knockout (null) mutations in mice made by the scientific community and the IMPC totaled more than 11,000 unique genes, over half the mouse genome. Many of these mutations include reporters and a large number are conditional-ready alleles. Through the IMPC, more than 5000 mutant mouse lines, mostly on the C57BL/6N strain background, have been expanded for phenotyping using standardized protocols (Birling et al. 2021). The adult phenotyping pipeline includes a battery of behavioral, metabolic, physiological, and morphometric tests, both live and terminal. For embryonic lethal mutations, which account for more than one-third of mutations, an embryonic pipeline includes assessment of the time of death and micro-computed tomography (microCT) 3D imaging at key developmental stages. Overall, 75% of lines tested have at least one phenotype and many show pleiotropy. Nearly 20% show sexually dimorphic phenotypes. Thus, you may find mutant ES cells, live mice, and/or cryopreserved preimplantation embryos and sperm available from public repositories, many with extensive phenotypic data.

Cite this chapter as *Cold Spring Harb Protoc*; doi:10.1101/pdb.over107956

It is very costly for a repository to keep mutant mice "on the shelf" as live animals. Thus, most strains are cryopreserved as preimplantation embryos or sperm. Once you purchase a mutant from a repository, the mutant will likely have to be recovered from the cryoarchive. This is costly and time-consuming; typically, you will receive mice in 2–4 mo. Alternatively, you or your institution's mouse core facility could purchase frozen preimplantation embryos for thawing and embryo transfer or frozen sperm for in vitro fertilization (IVF) and embryo transfer (Nagy et al. 2006; Takeo and Nakagata 2018). This can also be costly but mice could be born after ∼3 wk.

Another option to obtain a mutant mouse if it has been produced but is not archived in a repository is to do a literature search and contact the laboratory that generated the mutant. That information is also available in MGI. If they no longer have the mutant, they may have provided it to other laboratories that can be contacted. You can also search for recent papers that used the mutant and contact those laboratories. Another approach would be to use social media to ask if others have the mice. If a mutation is not available in your gene of interest or an existing mutation is not of the type that you want, you have the option of making a custom mutation.

GENE TARGETING FOR A CUSTOM MUTATION

Homologous Recombination in ES Cells or CRISPR–Cas Gene Editing?

Over the past three decades, the methods used to produce specific mutations in mice through homologous recombination in ES cells have been honed and improved such that they have become highly accurate, efficient, and almost routine, well within the reach of many laboratories or done as a service through companies or institutional core facilities. The pitfalls and limitations of the technology are well-researched and well-understood. In the past few years, CRISPR–Cas technology has exploded onto the scene, promising faster, simpler, and virtually unlimited possibilities for gene manipulation (Doudna and Mali 2016) directly in preimplantation embryos (Williams et al. 2016), although the technology can also be applied to stem cells in vitro (Smith et al. 2016). The technology is rapidly developing and improvements continue to refine the procedures.

So which method should you choose for your project? The answer to that will depend on your having a full understanding of the capabilities and limitations of the different methods and recognizing your own requirements and resources (Table 1). You may even find that a combination of the methods is the most suitable for your needs. For example, with CRISPR–Cas zygote gene editing, what you gain in time by directly editing the zygotic genome must be counterbalanced by the limitations on the size of inserts for knock-in alleles and the need to outcross founders because of mosaicism. If you choose to do the gene editing in ES cells in vitro, you will have the opportunity to select the desired allele whether you use homologous recombination or CRISPR–Cas editing. The trade-off here is that CRISPR–Cas produces more targeted clones than homologous recombination but homologous recombination has a higher fidelity, resulting in greater efficiency (Rezza et al. 2019). Your own resources may also be an important factor. Cost, availability of services, technical support, and efficiency all need to be considered. If your laboratory or institution has one or the other technology up and running with high efficiency, that may be what you should go with.

Gene Targeting by Homologous Recombination in ES Cells—the Basics

The complex procedures used for targeted gene mutation in the mouse by homologous recombination in ES cells can be broken down into five basic gene targeting steps.

1. Build a targeting construct specific to your favorite gene (Chapter 3: Mouse Gene Targeting Strategies for Maximum Ease and Versatility [Papaioannou and Behringer 2023a]).

2. Produce a mutation in ES cells by electroporating the construct that will target a change to one allele of this gene following homologous recombination (Chapter 4: Embryonic Stem Cell Gene Targeting and Chimera Production in Mice [Papaioannou and Behringer 2023b]).

TABLE 1. Comparison of different gene-editing technologies

Gene manipulation method	Requires ES cell culture	Positive selection used	Negative selection used	Requires chimera production	Type of alleles generated	Knock-ins with small (<2 kb) inserts	Knock-ins with large (>2 kb) inserts	# alleles targeted per manipulation	# alleles transmitted from founder
Homologous recombination in ES cells[a]	Yes	Yes	Yes	Yes	Almost any	Yes	Yes	1	1
CRISPR–Cas editing in zygotes	No	No	No	No	Almost any	Yes	Currently challenging	1 or 2	Multiple

(ES) Embryonic stem.
[a]Includes CRISPR–Cas gene editing in ES cells.

3. Isolate clonal lines of correctly targeted ES cells and grow up a sufficient number for further experiments, freezing some for future use (Chapter 4: Embryonic Stem Cell Gene Targeting and Chimera Production in Mice [Papaioannou and Behringer 2023b]).

4. Produce ES cell chimeras by injecting the correctly targeted ES cells into mouse blastocysts or by doing morula aggregation (Chapter 4: Embryonic Stem Cell Gene Targeting and Chimera Production in Mice [Papaioannou and Behringer 2023b]).

5. Breed the male chimeras to recover offspring carrying the targeted mutation through germline transmission of the targeted allele (Chapter 5: Recovering a Targeted Mutation in Mice from Embryonic Stem Cell Chimeras or CRISPR–Cas Founders [Papaioannou and Behringer 2023c]) and breed them to establish a mutant line (Chapter 6: Strategies for Maintaining Mouse Mutations [Papaioannou and Behringer 2023d]).

CRISPR–Cas Gene Editing—the Basics

The procedures to generate targeted gene mutations in the mouse by gene editing depend on the type of allele you want to generate (null, point mutant, or knock-in). This approach can be used to generate most commonly desired mutations directly in preimplantation embryos. However, complex alleles could be generated first in ES cells by CRISPR–Cas gene editing because you could then screen for the desired allele before generating mice using chimeras. The following steps refer to gene editing in preimplantation embryos.

Steps to Make a CRISPR–Cas Null Allele

1. Choose the exon(s) you want to mutate and use online CRISPR–Cas design tool software (https://en.wikipedia.org/wiki/CRISPR/Cas_Tools) to identify a sequence to design an appropriate guide RNA (Chapter 3: Mouse Gene-Targeting Strategies for Maximum Ease and Versatility [Papaioannou and Behringer 2023a]).

2. Inject zygotes in vitro or electroporate zygotes in vitro or in vivo with a single-guide RNA and Cas9 RNA or protein.

3. Transfer in vitro manipulated embryos into foster mothers.

4. Identify founders with insertions or deletions (indels) in your gene of interest.

5. Cross founders to segregate unique indel alleles (Chapter 5: Recovering a Targeted Mutation in Mice from Embryonic Stem Cell Chimeras or CRISPR–Cas Founders [Papaioannou and Behringer 2023c]) and breed them to establish a mutant line (Chapter 6: Strategies for Maintaining Mouse Mutations [Papaioannou and Behringer 2023d]).

Cite this chapter as *Cold Spring Harb Protoc*; doi:10.1101/pdb.over107956

Steps to Make a CRISPR–Cas Point Mutation Allele

1. Choose the exon in which you want to introduce a point mutation and use online CRISPR–Cas design tool software (https://en.wikipedia.org/wiki/CRISPR/Cas_Tools) to identify a sequence to design an appropriate guide RNA (Chapter 3: Mouse Gene Targeting Strategies for Maximum Ease and Versatility [Papaioannou and Behringer 2023a]).

2. Inject zygotes in vitro or electroporate zygotes in vitro or in vivo with a single-guide RNA, Cas9 RNA or protein, and an oligonucleotide homologous to your target sequence containing the desired point mutation.

3. Transfer in vitro manipulated embryos into foster mothers.

4. Identify founders with the desired point mutation.

5. Cross founders to segregate the desired point mutation from other targeted alleles (indels) (Chapter 5: Recovering a Targeted Mutation in Mice from Embryonic Stem Cell Chimeras or CRISPR–Cas Founders [Papaioannou and Behringer 2023c]) and breed them to establish a mutant line (Chapter 6: Strategies for Maintaining Mouse Mutations [Papaioannou and Behringer 2023d]).

Steps to Make a CRISPR–Cas Knock-In Allele

1. Use online CRISPR–Cas design tool software (https://en.wikipedia.org/wiki/CRISPR/Cas_Tools) to identify a sequence to design a guide RNA to induce a double-strand break in the region of the desired knock-in.

2. Build a targeting construct specific to your favorite gene (Chapter 3: Mouse Gene Targeting Strategies for Maximum Ease and Versatility [Papaioannou and Behringer 2023a]).

3. Inject zygotes in vitro or electroporate zygotes in vitro or in vivo with a single-guide RNA (or two guide RNAs to decrease off-target events), Cas9 RNA or protein, and the targeting vector DNA.

4. Transfer in vitro manipulated embryos into foster mothers.

5. Identify founders with the desired knock-in or conditional allele.

6. Cross founders to segregate the desired mutation from other targeted alleles (indels) (Chapter 5: Recovering a Targeted Mutation in Mice from Embryonic Stem Cell Chimeras or CRISPR–Cas Founders [Papaioannou and Behringer 2023c]) and breed them to establish a mutant line (Chapter 6: Strategies for Maintaining Mouse Mutations [Papaioannou and Behringer 2023d]).

PHENOTYPIC ANALYSIS

Whether you start with an existing mutation or have made one by gene targeting or editing, the final step, and ultimate goal, is the phenotypic analysis in heterozygous and/or homozygous mutant mice. The following chapters in this book deal with breeding mice to maintain the mutation and produce offspring for phenotypic analysis (Chapter 6: Strategies for Maintaining Mouse Mutations [Papaioannou and Behringer 2023d]), and special techniques to use in phenotypic analysis (Chapter 7: Special Breeding Techniques for Use in Mouse Mutation Analysis [Papaioannou and Behringer 2023e]). From there, the branching decision pathway (see Fig. 1 in Chapter 1: Strategies for the Production and Phenotypic Analysis of Mutations in the Mouse [Papaioannou and Behringer 2023f]) leads you through a phenotypic analysis by first determining the earliest time of action of the mutation and then applying strategies and techniques suitable for analysis of different stages of embryonic development, postnatal development or adulthood, including dominant and recessive mutations (see the following chapters in this book).

Using chimeric mice as a phenotyping tool is discussed (Chapter 17: Getting around an Early Lethal Phenotype in Mice with Chimeras [Papaioannou and Behringer 2023g]), and finally the pathway leads to the use of conditional alleles (Chapter 18: Tissue- and/or Temporal-Specific Mutations in Mice Using Conditional Alleles [Papaioannou and Behringer 2023h]).

BEFORE YOU BEGIN

Armchair Biology

What Do You Want and What Do You Expect?

Before you embark on generating or obtaining a mutant, it is worth taking a moment to reflect on exactly what you expect from your gene-targeting experiment. Ideally, this goal was determined long before the desired allele was obtained or designed, but it is never too late to reconsider your approach. The decision to do gene targeting should not be taken lightly because it is an expensive and time-consuming procedure. Even if everything goes well and works the first time, it will be about 6 months from the time you have the finished targeting construct ready for electroporation into ES cells until you could possibly see a homozygous mutant animal. In our experience, a year is a realistic estimate and 2 years is not unusual. CRISPR–Cas gene editing should be quicker because it bypasses the ES cell tissue culture steps but still requires considerable time and effort. That being the case, a lot of thought and a few preliminary experiments could save considerable time and trouble.

First, plan the type of mutation that will be most informative and/or versatile. This will depend on the gene and its products, and how it functions. Chapter 4: Embryonic Stem Cell Gene Targeting and Chimera Production in Mice (Papaioannou and Behringer 2023b) provides some pointers for making the most of the information that you have.

Second, try to predict the mutant phenotype. Although vast experience indicates that the only surefire way to determine the effect of a mutation is to make the mutation, this is not a thinking scientist's approach. There are prognosticators that can take some of the guesswork out of the prediction, thus guiding the experimental strategy. Foremost among these is the expression pattern of the gene. Stories circulate about expression patterns that seem to have nothing to do with the mutant phenotype, and mutations that have no overt phenotype despite extensive gene expression. But we maintain that these situations are the exceptions, if they exist at all, and that the absence of a mutant phenotype in an expressing tissue is more likely a reflection of our ignorance about how to look for it.

BOX 1. IMPRINTING AND X INACTIVATION: EFFECT ON EMBRYOS DUE TO PREFERENTIAL X INACTIVATION OF PATERNAL X CHROMOSOME IN EXTRAEMBRYONIC MEMBRANES

As a rule, X inactivation results from random inactivation of one or the other X chromosome so that females end up with a random mix of cells with either the paternally derived X or the maternally derived X inactive. However, there is an exception to the random inactivation rule in the embryo: There is preferential inactivation of the paternally derived X chromosome in the trophectoderm and primitive endoderm, and consequently in their cellular derivatives. This peculiarity of the embryo, which is a form of imprinting, usually has little effect on experimental work. However, if you have produced a mutation in an X-linked gene that is expressed in either of these tissue lineages (i.e., placental structures or yolk sac endoderm), there will be two classes of heterozygous females with respect to gene expression: one class that will be functionally null in the trophectoderm and primitive endoderm lineages because it inherits the mutant allele from the mother and the paternal chromosome is inactivated, and the other class that will be functionally wild-type because it inherits the mutant allele from the father and the maternal chromosome is active (see Fig. 2). For an example of analysis of this type of mutation, see Esx1 (Li and Behringer 1998).

Cite this chapter as Cold Spring Harb Protoc; doi:10.1101/pdb.over107956

Get to Know Your Gene

Armchair biology can get you into the right frame of mind for dealing with the unexpected by anticipating some of the possibilities of a gene disruption. With the current amount of genome information available, it is possible to learn a lot about your gene that might bear on its function. The following is a checklist of what you should know about the gene to help you prepare before you begin.

- *What is the chromosomal location of your gene?* This is an important piece of information for many reasons. Is the gene X-linked? If so, males will be hemizygous for the mutation and effects might show up in XY ES cells, in chimeras, or in males, where there is only a single X chromosome. Heterozygous females, which will have one wild-type allele active in half their cells because of X inactivation, might escape this problem, unless the gene is expressed in the extraembryonic tissues that are subject to imprinting effects (see Box 1; Fig. 2). Alternatively, they might show mosaic

Mutant allele from mother

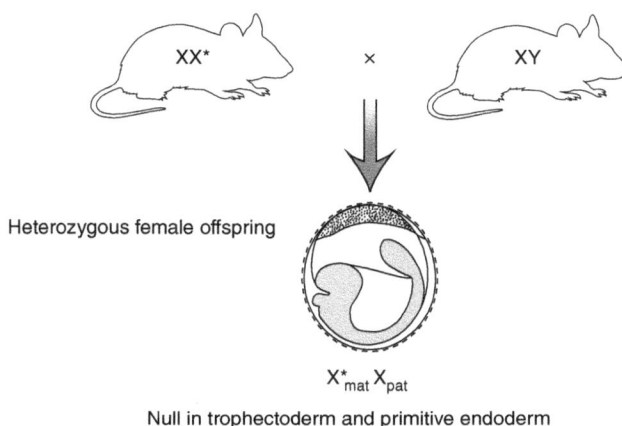

Mutant allele from father

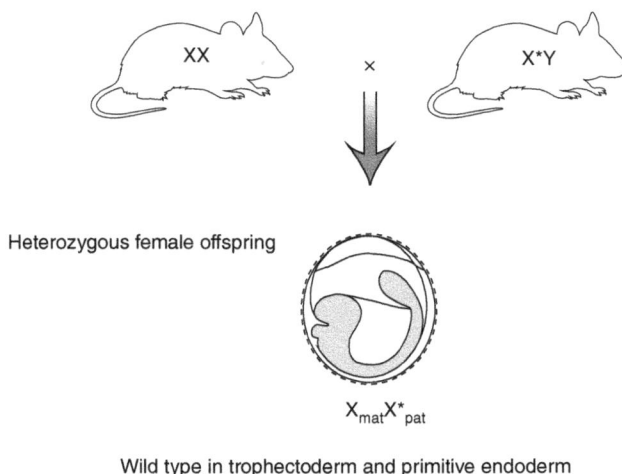

FIGURE 2. Imprinting of an X-linked mutant gene. If an X-linked mutant allele (*) is inherited from the mother (*top cross*), heterozygous female offspring are effectively null for the gene product in the trophectoderm and primitive endoderm lineages because the paternally derived X chromosome (X_{pat}) is preferentially inactivated and the maternally derived chromosome (X^*_{mat}) is active but carries the mutation. If the mutant allele is inherited from the father (*bottom cross*), heterozygous females are wild type for the gene in the extraembryonic tissues because the maternally derived allele is the only one active in the trophectoderm and primitive endoderm lineages. (Modified, with permission, from Papaioannou and Behringer 2005, © Cold Spring Harbor Laboratory Press; embryo reproduced, with permission, from Papaioannou and Hadjantonakis 2003.)

effects reflecting the mosaicism of cells with the mutant allele on the active X chromosome. Is the gene Y chromosome–linked? Similar to X linkage, there will be only one copy of the gene in XY cells.

- *Does the gene fall within an autosomal region subject to imprinting effects?* If so, heterozygous offspring may be differently affected depending on the parent of origin of the mutant allele (Fig. 3).

- *Are there known mutations that map to the chromosomal region where your gene resides?* Check these carefully because if they have not yet been molecularly characterized, they could turn out to be alleles of your gene. To see if this is the case, you might obtain DNA from the known mutants or obtain the mutants themselves.

- *Is your favorite gene a member of a gene family?* If family members overlap in their expression pattern, some overlap of function could lead to redundancy. In this case the areas of unique expression of your gene might indicate areas where a mutant effect is most likely.

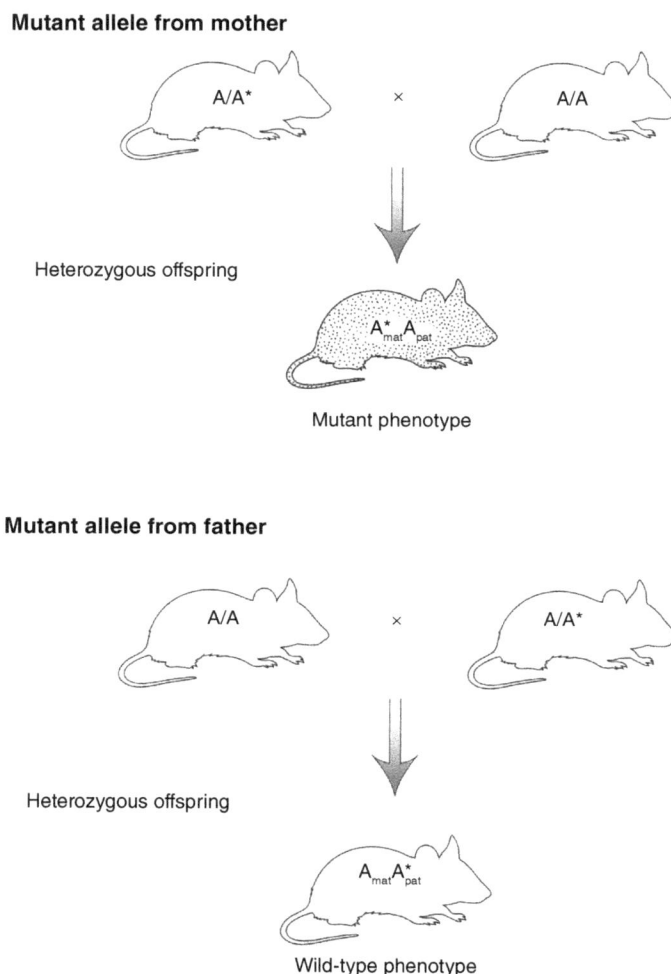

Mutant allele from mother

A/A^* × A/A

Heterozygous offspring

$A^*_{mat}A_{pat}$

Mutant phenotype

Mutant allele from father

A/A × A/A^*

Heterozygous offspring

$A_{mat}A^*_{pat}$

Wild-type phenotype

FIGURE 3. Inheritance of an imprinted, autosomal mutant gene. In this example, the locus is only expressed from the maternal allele. When a mutation (A^*) is inherited from the mother (*top cross*), heterozygous offspring will have one mutant maternal allele (A^*_{mat}) and one inactive paternal allele (A_{pat}) and will thus show a mutant phenotype. When the mutation is inherited from the father (*bottom cross*), the normal, maternally inherited allele will be the active copy and the heterozygous offspring will be phenotypically wild type. It should be noted that the situation would be reversed if the imprinted locus is expressed from the paternal allele. (Reprinted, with permission, from Papaioannou and Behringer 2005, © Cold Spring Harbor Laboratory Press.)

 Cite this chapter as *Cold Spring Harb Protoc*; doi:10.1101/pdb.over107956

- *Is anything known about orthologous or homologous genes in other species?* Look for any existing mutations in your gene in other species or any chromosomal deletions in humans that include this gene. Determine if any human disease loci are mapped to the chromosomal region (www.omim.org). Any of this information could provide a clue as to the expected phenotype.

- *Can you make predictions based on the expression pattern and the nature of the protein?* Expression patterns tell you when and where the action of a mutation could have its origins. Proteins can act within the cells in which they are produced (cell-autonomous) and/or they can act on other cells (cell-non-autonomous).

- *Is anything known about interactions of the gene product with other gene products?* This information may be useful for placing the gene product into known biochemical pathways.

All of these considerations, which can lead to fruitful lines of investigation, are better thought through in advance before embarking on gene targeting. In addition, consider that the mutation may result in a mouse that looks perfectly normal on the surface. Prior knowledge about the gene and gene product could help reveal a subtle or cryptic phenotype.

What Is the Gene Expression Pattern?

A starting point in gene targeting experiments is to define the expression pattern of your chosen gene. It may be your favorite because of its expression in an adult organ of interest or its complicity in a disease process. But remember that the same gene may be used during development or in other tissues and could thus have an effect at an unanticipated time and place. Search the literature for reports of the gene's expression and consider doing a survey of expression in a panel of adult organs and embryonic stages, even if only at the level of reverse transcriptase polymerase chain reaction (RT-PCR) or northern blot analysis (commercial blots are available). Although a detailed expression analysis may not be what you choose to do at this stage, expression in embryos will at least alert you to the possibility of an embryonic phenotype. If you are using a mutation from the KOMP project, there may be detailed expression data available (West et al. 2015). The Gene Expression Database in MGI (GXD; http://www.informatics.jax.org/expression.shtml) is an excellent resource that collates expression data from published papers with emphasis on endogenous gene expression during development. You can check the expressed sequence tag (EST) databases to identify tissues in which transcripts for your favorite gene have been isolated as a guide to expression. There are also numerous transcriptomes available in public databases (e.g., Gene Expression Omnibus [GEO]; www.ncbi.nlm.nih.gov/geo) that can be examined for the expression of your gene of interest.

Equipped with a good understanding of your gene of interest and the different possibilities for making mutations, you are ready to move on.

REFERENCES

Birling MC, Yoshiki A, Adams DJ, Ayabe S, Beaudet AL, Bottomley J, Bradley A, Brown SDM, Bürger A, Bushell W, et al. 2021. A resource of targeted mutant mouse lines for 5,061 genes. *Nat Genet* 53: 416–419.

Bradley A, Anastassiadis K, Ayadi A, Battey JF, Bell C, Birling MC, Bottomley J, Brown SD, Bürger A, Bult CJ, et al. 2012. The mammalian gene function resource: the International Knockout Mouse Consortium. *Mamm Genome* 23: 580–586. doi:10.1007/s00335-012-9422-2

Brown SD, Moore MW. 2012. The International Mouse Phenotyping Consortium: past and future perspectives on mouse phenotyping. *Mamm Genome* 23: 632–640.

Buchovecky CM, Turley SD, Brown HM, Kyle SM, McDonald JG, Liu B, Pieper AA, Huang W, Katz DM, Russell DW, et al. 2013. A suppressor screen in *Mecp2* mutant mice implicates cholesterol metabolism in Rett syndrome. *Nat Genet* 45: 1013–1020. doi:10.1038/ng.2714

Chen Y, Yee D, Dains K, Chatterjee A, Cavalcoli J, Schneider E, Om J, Woychik RP, Magnuson T. 2000. Genotype-based screen for ENU-induced mutations in mouse embryonic stem cells. *Nat Genet* 24: 314–317. doi:10.1038/73557

Cordes SP. 2005. *N*-ethyl-*N*-nitrosourea mutagenesis: boarding the mouse mutant express. *Microbiol Mol Biol Rev* 69: 426–439. doi:10.1128/MMBR.69.3.426-439.2005

Ding S, Wu X, Li G, Han M, Zhuang Y, Xu T. 2005. Efficient transposition of the piggyBac (PB) transposon in mammalian cells and mice. *Cell* 122: 473–483.

Doudna J, Mali P. 2016. *CRISPR–Cas: a laboratory manual.* Cold Spring Harbor Laboratory Press, Cold Spring Harbor, NY.

Friedrich G, Soriano P. 1991. Promoter traps in embryonic stem cells: a genetic screen to identify and mutate developmental genes in mice. *Genes Dev* 5: 1513–1523. doi:10.1101/gad.5.9.1513

Geister KA, Timms AE, Beier DR. 2018. Optimizing genomic methods for mapping and identification of candidate variants in ENU mutagenesis screens using inbred mice. *G3* 8: 401–409. doi:10.1534/g3.117.300292

Gossler A, Joyner AL, Rossant J, Skarnes WC. 1989. Mouse embryonic stem cells and reporter constructs to detect developmentally regulated genes. *Science* 244: 463–465. doi:10.1126/science.2497519

Guan C, Ye C, Yang X, Gao J. 2010. A review of current large-scale mouse knockout efforts. *Genesis* 48: 73–85. doi:10.1002/dvg.20594

Holdener BC, Magnuson T. 1994. A mouse model for human hereditary tyrosinemia I. *Bioessays* 16: 85–87. doi:10.1002/bies.950160203

Horie K, Yusa K, Yae K, Odajima J, Fischer SE, Keng VW, Hayakawa T, Mizuno S, Kondoh G, Ijiri T, et al. 2003. Characterization of *Sleeping Beauty* transposition and its application to genetic screening in mice. *Mol Cell Biol* 23: 9189–9207. doi:10.1128/MCB.23.24.9189-9207.2003

Hrabe de Angelis MH, Flaswinkel H, Fuchs H, Rathkolb B, Soewarto D, Marschall S, Heffner S, Pargent W, Wuensch K, Jung M, et al. 2000. Genome-wide, large-scale production of mutant mice by ENU mutagenesis. *Nat Genet* 25: 444–447. doi:10.1038/78146

Jaenisch R, Harbers K, Schnieke A, Löhler J, Chumakov I, Jähner D, Grotkopp D, Hoffmann E. 1983. Germline integration of Moloney murine leukemia virus at the *Mov13* locus leads to recessive lethal mutation and early embryonic death. *Cell* 32: 209–216. doi:10.1016/0092-8674(83)90511-1

Kasarskis A, Manova K, Anderson KV. 1998. A phenotype-based screen for embryonic lethal mutations in the mouse. *Proc Natl Acad Sci* 95: 7485–7490. doi:10.1073/pnas.95.13.7485

Kile BT, Hentges KE, Clark AT, Nakamura H, Salinger AP, Liu B, Box N, Stockton DW, Johnson RL, Behringer RR, et al. 2003. Functional genetic analysis of mouse chromosome 11. *Nature* 425: 81–86. doi:10.1038/nature01865

Li Y, Behringer RR. 1998. *Esx1* is an X-chromosome-imprinted regulator of placental development and fetal growth. *Nat Genet* 20: 309–311. doi:10.1038/3129

Lloyd KC. 2011. A knockout mouse resource for the biomedical research community. *Ann N Y Acad Sci* 1245: 24–26. doi:10.1111/j.1749-6632.2011.06311.x

Nagy A, Gertsenstein M, Vintersten K, Behringer R. 2006. Oviduct transfer. *Cold Spring Harb Protoc* doi:10.1101/pdb.prot4379

Nord AS, Chang PJ, Conklin BR, Cox AV, Harper CA, Hicks GG, Huang CC, Johns SJ, Kawamoto M, Liu S, et al. 2006. The International Gene Trap Consortium Website: a portal to all publicly available gene trap cell lines in mouse. *Nucl Acids Res* 34: D642–D648.

Palmiter RD, Brinster RL, Hammer RE, Trumbauer ME, Rosenfeld MG, Birnberg NC, Evans RM. 1982. Dramatic growth of mice that develop from eggs microinjected with metallothionein-growth hormone fusion genes. *Nature* 300: 611–615. doi:10.1038/300611a0

Papaioannou VE, Behringer RR. 2005. *Mouse phenotypes: a handbook of mutation analysis.* Cold Spring Harbor Laboratory Press, Cold Spring Harbor, NY.

Papaioannou VE, Behringer RR. 2023a. Mouse gene-targeting strategies for maximum ease and versatility. *Cold Spring Harb Protoc* doi:10.1101/pdb.over107957

Papaioannou VE, Behringer RR. 2023b. Embryonic stem cell gene targeting and chimera production in mice. *Cold Spring Harb Protoc* doi:10.1101/pdb.over107958

Papaioannou VE, Behringer RR. 2023c. Recovering a targeted mutation in mice from embryonic stem cell chimeras or CRISPR–Cas founders. *Cold Spring Harb Protoc* doi:10.1101/pdb.over107959

Papaioannou VE, Behringer RR. 2023d. Strategies for maintaining mouse mutations. *Cold Spring Harb Protoc* doi:10.1101/pdb.over107960

Papaioannou VE, Behringer RR. 2023e. Special breeding techniques for use in mouse mutation analysis. *Cold Spring Harb Protoc* doi:10.1101/pdb.over107961

Papaioannou VE, Behringer RR. 2023f. Strategies for the production and phenotypic analysis of mutations in the mouse. *Cold Spring Harb Protoc* doi:10.1101/pdb.over107955

Papaioannou VE, Behringer RR. 2023g. Getting around an early lethal phenotype in mice with chimeras. *Cold Spring Harb Protoc* doi:10.1101/pdb.over107979

Papaioannou VE, Behringer RR. 2023h. Tissue- and/or temporal-specific mutations in mice using conditional alleles. *Cold Spring Harb Protoc* doi:10.1101/pdb.over107980

Papaioannou VE, Hadjantonakis A-K. 2003. Stem cells from early mammalian embryos: common themes and significant differences. In *Stem cells handbook* (ed. Sell S), pp. 19–30. Humana Press, Totowa, NJ.

Rezza A, Jacquet C, Le Pillouer A, Lafarguette F, Ruptier C, Billandon M, Isnard Petit P, Trouttet S, Thiam K, Fraichard A, et al. 2019. Unexpected genomic rearrangements at targeted loci associated with CRISPR/Cas9-mediated knock-in. *Sci Rep* 9: 3486. doi:10.1038/s41598-019-40181-w

Robertson E, Bradley A, Kuehn M, Evans M. 1986. Germ-line transmission of genes introduced into cultured pluripotential cells by retroviral vector. *Nature* 323: 445–448. doi:10.1038/323445a0

Sakuraba Y, Sezutsu H, Takahasi KR, Tsuchihashi K, Ichikawa R, Fujimoto N, Kaneko S, Nakai Y, Uchiyama M, Goda N, et al. 2005. Molecular characterization of ENU mouse mutagenesis and archives. *Biochem Biophys Res Commun* 336: 609–616. doi:10.1016/j.bbrc.2005.08.134

Schilit SLP, Ohtsuka M, Quadros RM, Gurumurthy CB. 2016. Pronuclear injection-based targeted transgenesis. *Curr Protoc Hum Genet* 91: 15 10 11–15 10 28. doi:10.1002/cphg.23

Smith C, Ye Z, Cheng L. 2016. Genome editing in human pluripotent stem cells. *Cold Spring Harb Protoc* doi:10.1101/pdb.top086819

Stanford WL, Cohn JB, Cordes SP. 2001. Gene-trap mutagenesis: past, present and beyond. *Nat Rev Genet* 2: 756–768. doi:10.1038/35093548

Takeo T, Nakagata N. 2018. In vitro fertilization in mice. *Cold Spring Harb Protoc* doi:10.1101/pdb.prot094524

Thomas JW, LaMantia C, Magnuson T. 1998. X-ray-induced mutations in mouse embryonic stem cells. *Proc Natl Acad Sci* 95: 1114–1119. doi:10.1073/pnas.95.3.1114

Vivian JL, Chen Y, Yee D, Schneider E, Magnuson T. 2002. An allelic series of mutations in *Smad2* and *Smad4* identified in a genotype-based screen of N-ethyl-N-nitrosourea-mutagenized mouse embryonic stem cells. *Proc Natl Acad Sci* 99: 15542–15547. doi:10.1073/pnas.242474199

Wang T, Bu CH, Hildebrand S, Jia G, Siggs OM, Lyon S, Pratt D, Scott L, Russell J, Ludwig S, et al. 2018. Probability of phenotypically detectable protein damage by ENU-induced mutations in the Mutagenetix database. *Nat Commun* 9: 441.

West DB, Pasumarthi RK, Baridon B, Djan E, Trainor A, Griffey SM, Engelhard EK, Rapp J, Li B, de Jong PJ, et al. 2015. A *lacZ* reporter gene expression atlas for 313 adult KOMP mutant mouse lines. *Genome Res* 25: 598–607.

Williams A, Henao-Mejia J, Flavell RA. 2016. Editing the mouse genome using the CRISPR–Cas9 system. *Cold Spring Harb Protoc* doi:10.1101/pdb.top087536

Cite this chapter as *Cold Spring Harb Protoc*; doi:10.1101/pdb.over107956

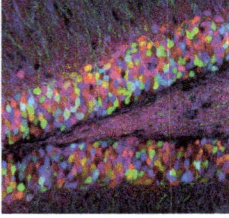

CHAPTER 3

Mouse Gene-Targeting Strategies for Maximum Ease and Versatility

Well-planned strategies are an essential prerequisite for any mutational analysis involving gene targeting. Consideration of the advantages or disadvantages of different methods will aid in the production of a final product that is both technically feasible and versatile. Strategies for gene-targeting experiments in the mouse are discussed, including the rationale behind some of the common elements of gene-targeting vectors, such as homologous DNA and the use of different site-specific recombinases. We detail positive and negative selection as well as screening strategies for homologous recombination events in embryonic stem (ES) cells. For the planning stages of making different types of alleles, we first consider general strategies and then provide details specific to either homologous recombination in ES cells or making alleles by gene editing with CRISPR–Cas in preimplantation embryos. The types of alleles considered are null or knockout alleles, reporter gene knock-in alleles, point mutations, and conditional null alleles.

INTRODUCTION

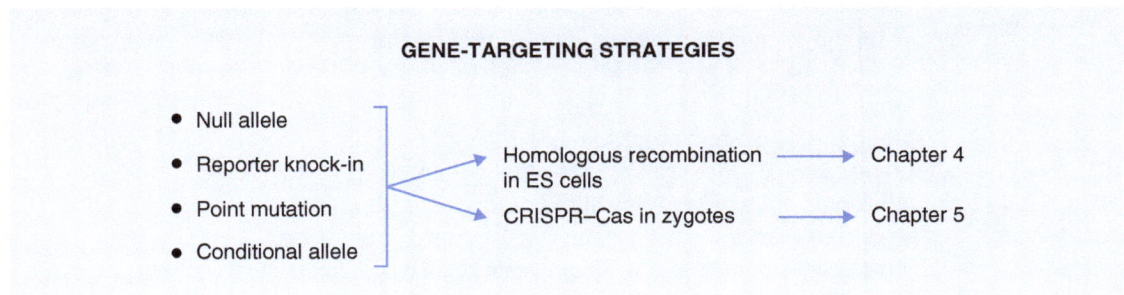

GENE-TARGETING STRATEGIES

- Null allele
- Reporter knock-in
- Point mutation
- Conditional allele

→ Homologous recombination in ES cells ——→ Chapter 4

CRISPR–Cas in zygotes ——→ Chapter 5

The molecular techniques used to build a gene-targeting construct (or vector) are standard (see Chapter 10 in Behringer et al. 2014), and the goal of producing a mutation in a specific gene is conceptually straightforward. However, many aspects of the design of a simple knockout strategy or a more complex knock-in strategy involving a gene-targeting construct will have a bearing on the success of the procedure and on the versatility of the resulting mutation. With little additional work, you may be able to build features into an initial gene-targeting vector that will save you from having to construct a second vector as the project unfolds. Whether you are using homologous recombination in

Opening artwork: Weissman TA, et al. 2011. *Cold Spring Harb Protoc* doi:10.1101/pdb.top114

© 2024 Cold Spring Harbor Laboratory Press
Cite this chapter as *Cold Spring Harb Protoc*; doi:10.1101/pdb.over107957

BOX 1. GENE CHECKLIST BEFORE CONSTRUCTING TARGETED MUTANT ALLELES

Answers to the following fundamental questions about the locus in question will facilitate the best design for your desired mutation.

- How many coding and noncoding exons are present?
- Where is the start of transcription?
- Which exon contains the translation start codon?
- Which exon contains the translation stop codon?
- How many kilobases does the genomic region encoding the open reading frame span?
- Is there evidence for alternative splicing? If so, which exons are alternatively spliced?
- Are protein isoforms generated by the locus?
- Can you easily delete the entire locus, that is, can the entire locus be deleted by removing <20 kb?
- If you delete a specific exon(s), can the remaining exons splice together in-frame?
- If the gene (genomic region) is relatively large (>20 kb), is there a group of contiguous exons that reside within a small (<20-kb) genomic region that could be deleted?
- Do you have the sequence of the locus from the strain of mice to be used or from which the ES cells were derived?
- Have you performed a sequence comparison of nonexonic sequences among human, mouse, and other mammals to identify the location of conserved sequences that might regulate expression?

embryonic stem (ES) cells or using CRISPR–Cas technology to create a mutation either directly in preimplantation embryos or in ES cells, the strategies discussed in this overview will help you anticipate future needs and provide you with information to evaluate the trade-offs inherent in designing different kinds of alleles or using alternative technologies.

Before you begin, make sure that you have thorough and detailed information about the locus you intend to target (Box 1) to facilitate the best design for the mutation. For homologous recombination in ES cells, the first two steps for gene targeting (see Chapter 2: Obtaining or Generating Gene Mutations in Mice [Papaioannou and Behringer 2023a]) are to build a gene-targeting construct specific to your gene of interest and to produce a mutation in ES cells in vitro. For CRISPR–Cas manipulations, the first step is to identify a sequence to generate a guide RNA to target the gene and then produce a null mutation directly in preimplantation embryos. In this chapter, we will present the rationale for different aspects of gene-targeting vector design and discuss strategies for making different types of mutant alleles. Once a targeting strategy and technology has been chosen, you can move on to either Chapter 4: Embryonic Stem Cell Gene Targeting and Chimera Production in Mice (Papaioannou and Behringer 2023b) or Chapter 5: Recovering a Targeted Mutation in Mice from Embryonic Stem Cell Chimeras or CRISPR–Cas Founders (Papaioannou and Behringer 2023c).

TOOLKIT FOR THE DESIGN OF GENE-TARGETING VECTORS

Homologous DNA

Certain details should be considered when generating any type of gene-targeting vector. One of the most important is to use genomic DNA that is isogenic with the ES cell line that will be used for the gene-targeting experiments. In other words, both the genomic DNA used to generate the gene-targeting vector and the ES cells should be derived from the same inbred strain/substrain of mouse. This is also true for CRISPR–Cas knock-in manipulations in embryos: Genomic DNA sequences used in the gene-targeting vector must be identical to the strain of preimplantation mouse embryo manipulated. A complete sequence match between the targeting vector homology and the

Cite this chapter as *Cold Spring Harb Protoc*; doi:10.1101/pdb.over107957

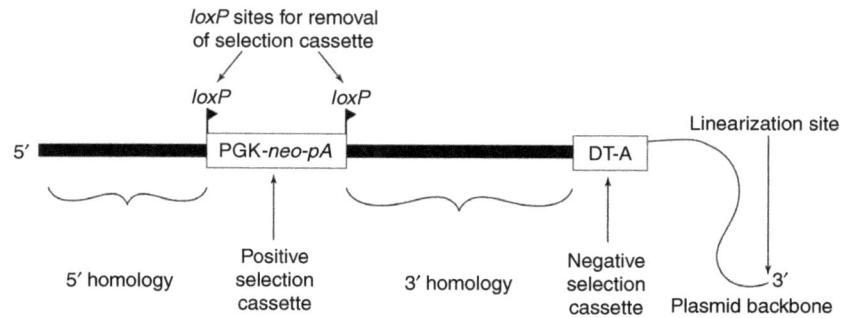

FIGURE 1. Elements of an idealized targeting construct for homologous recombination in embryonic stem (ES) cells. The targeting constructs used for different targeting strategies vary in detail, but the usual components necessary for successful gene targeting include (1) homologous, isogenic DNA totaling 5–8 kb and evenly split between upstream and downstream arms of the targeting construct; (2) a positive selectable marker, such as *neo*, driven by a promoter active in ES cells, such as PGK, and surrounded by recombinase sites, such as *loxP* sites, to facilitate its eventual removal from the targeted allele; (3) a negative selectable marker, such as DT-A, to screen against random integration events, located outside the region of homology; and (4) a linearization site in the plasmid backbone located some distance from the negative selectable marker to protect it from degradation of the ends after linearization. The other essential elements (not shown) are 5′ and 3′ probes to the endogenous locus that are external to the regions of homology in the construct. These are used in Southern analysis of restriction enzyme digests to diagnose 5′ and 3′ crossover events to recognize a correctly targeted allele. Alternatively, a polymerase chain reaction (PCR)-based strategy can be used.

locus to be mutated (i.e., isogenic DNA) facilitates obtaining the maximum frequency of homologous recombination or homology-directed repair. Conversely, potential sequence mismatches resulting from the use of DNA from a different mouse strain can reduce the frequency of homologous recombination or homology-directed repair or even prevent it altogether. Therefore, it is essential to match the specific strain and substrain genomic library used to generate your gene-targeting construct with the ES cell line or embryos you plan to target or, alternatively, to amplify the arms of homology from the ES cells or mouse strain by polymerase chain reaction (PCR) using high-fidelity polymerases.

For homologous recombination in ES cells, the total amount of sequence homology to be used in the gene-targeting vector should be between 5 and 8 kb, which represents a balance between increased chances of homologous recombination and ease of vector construction: Smaller amounts of homology can reduce the targeting frequency and larger amounts can complicate vector construction as well as the identification of homologous recombinants. In general, the homologous sequence should be split between the upstream and downstream arms of homology in the vector (Fig. 1). One arm of homology can be as short as 1 kb in length, with the other arm making up the balance of the total recommended homology of 5–8 kb. The arms of homology in the targeting vector should flank any genomic sequences you wish to delete or modify. If you are using CRISPR–Cas in ES cells, the length of homology used in a knock-in targeting vector depends upon the molecular strategy used (see below).

Site-Specific Recombinases

DNA recombinases that recognize specific sequences are routinely used in genetic strategies to generate tissue-specific and/or conditional mutations and other types of alleles. The primary system currently used is the Cre/*loxP* system, but the Flp/*FRT* and Dre/*rox* systems, which work similarly, are also used (Anastassiadis et al. 2009; Han et al. 2021).

Cre is a DNA recombinase that specifically recognizes 34-base pair (bp) sequences called *loxP* sites. The central 8-bp core of the *loxP* site is flanked by inverted repeats and specifies an orientation to the 34-bp sequence. *loxP* sites can be placed on either side of a segment of DNA. The length of the intervening DNA can be relatively small (<1 kb) or very large (in the range of cMs). The intervening DNA is called "floxed" (flanked by *loxP* sites). Because the *loxP* sites have an orientation, they can

BOX 2. REMOVING A SELECTION CASSETTE FROM ES CELLS IN VITRO

Removing the selection cassette from your targeted ES cell clones in vitro involves transiently expressing the appropriate recombinase (usually Cre or Flp), and then isolating clones and genotyping for the removal of the cassette (Anastassiadis et al. 2013). Electroporate circular recombinase expression plasmids into the ES cell clones, plate at clonal density, and culture without selection. Pick and then genotype the ES cell clones. Usually, ~10%–20% of the clones will have had the cassette removed by the transient recombinase expression. In the past, some investigators have used a floxed *neo*-HSV *tk* dual cassette so that the loss of *tk* can be selected for by culture in gancyclovir or FIAU. This results in nearly all of the surviving ES cell clones having the floxed cassette removed. If you do this, however, remember that HSV *tk* expression causes male sterility (see Box 3) and male chimeras made with ES cells that retain a *neo*-HSV *tk* will likely be sterile.

Triple *loxP* strategies have been used in the past to generate conditional null alleles. In this strategy, two *loxP* sites flank a selection cassette (e.g., *neo*) and the third *loxP* site is added to flank important exons. However, depending on the efficiency and timing of Cre excision following transient exposure to Cre recombinase, three different alleles could be produced requiring some effort to sort out. Nowadays, most labs use a dual Cre/Flp strategy to flox important exons and flrt the selection cassette. Thus, expression of Flp will remove the selection cassette, leaving the important exons floxed.

One caveat of the in vitro approach for removing a selection cassette is that it requires a second electroporation and cell cloning step that could compromise obtaining germline transmission of the targeted allele. In addition, it is necessary to choose the targeted clone(s) to electroporate with the recombinase vector. Often, only one clone is chosen for this purpose, which can be a big mistake because that clone has probably not been screened for germline transmission ability after the initial targeting. An alternative is to remove the selection cassette in vivo, after mice have been produced (see Deletion of Selection Cassettes In Vivo in Chapter 6: Strategies for Maintaining Mouse Mutations [Papaioannou and Behringer 2023d]).

flank a piece of intervening DNA in either direct orientation or inverted orientation. Cre will recognize *loxP* sites flanking a piece of DNA and catalyze the deletion of the intervening DNA if the flanking *loxP* sites are in direct orientation or it will invert the intervening DNA if the *loxP* sites are in inverted orientation. A similar situation occurs with Flp recombinase and *FRT* sites, and Dre recombinase and *rox* sites. DNA sequences flanked by *FRT* sites have been called "flrted."

For ES cell manipulations, selectable marker expression cassettes are floxed or flrted (Fig. 1) so that they can be removed from the targeted locus by Cre or Flp, respectively. In this way, extraneous DNA sequences that might complicate the analysis of the mutant are removed from the targeted locus either in vitro, before chimera production (Box 2), or in vivo, once mutant mice have been produced (see below).

Positive Selection for Vector Incorporation in ES Cells

When using ES cells for gene targeting, positive selection of ES cell clones that have integrated the gene-targeting vector is usually achieved using a neomycin phosphotransferase (*neo*) gene expression cassette that confers resistance to the drug G418, although other selectable markers can be used. However, the presence of a selectable marker gene with its associated promoter (e.g., the *Pgk* promoter) at targeted loci has been shown in some cases to influence the expression of neighboring genes (e.g., *Myf5*; Kaul et al. 2000). Therefore, it is a good idea to flank the positive selectable marker with *loxP* sites to provide the option of removing the selectable marker later with Cre recombinase (Fig. 1).

Once homologous recombination has been achieved, the floxed selectable marker can be deleted either in the targeted ES cells or later in mice. Removal is accomplished in ES cells by transient in vitro Cre expression before advancing to the next step of making chimeras (Box 2). If mice carrying the targeted allele have been produced, the floxed selectable marker can be deleted by pronuclear injection of a Cre expression plasmid or RNA into zygotes carrying the floxed gene, by culturing zygotes in membrane-permeable CRE protein (Ryder et al. 2014), or by breeding mice carrying the floxed gene with Cre transgenic mice (see Chapter 6: Strategies for Maintaining Mouse Mutations [Papaioannou and Behringer 2023d]). In any case, using a floxed allele strategy provides a simple way to generate two different alleles for the targeted gene, one with the selectable marker and one without. Having two

Cite this chapter as *Cold Spring Harb Protoc*; doi:10.1101/pdb.over107957

different alleles could be useful for chimera studies later in the analysis of a phenotype (see Chapter 17: Getting around an Early Lethal Phenotype in Mice with Chimeras [Papaioannou and Behringer 2023e]).

An added benefit of the floxed selectable marker strategy is that it provides the opportunity to make homozygous mutant ES cell lines without ever making a mouse. By deleting the selectable marker from the targeted locus by Cre expression in heterozygous ES cells, you can reuse the original targeting vector and the same selective drug a second time to target the remaining wild-type allele, a strategy called "marker recycling" (Abuin and Bradley 1996). Other recombinase/recombinase target site combinations (e.g., Flp/*FRT*, Dre/*rox*) can also be used to remove selectable markers.

Negative Selection against Random Integration Events in ES Cells

Negative selection against ES cell clones that have incorporated the gene-targeting construct by random insertion can be accomplished using a herpes simplex virus (HSV) thymidine kinase (*tk*) gene outside the region of homology in the construct. Electroporated ES cells are then cultured with ganciclovir or FIAU (1,2-deoxy-2-fluoro-1-β-D-arabinofuranosyl-5-iodouracil). After homologous recombination, the ends of the gene-targeting vector are lost, including the plasmid backbone and *tk* expression cassette. Thus, cells that contain the *tk* gene resulting from random integration of the vector will be killed by the drug selection. Note, however that retention of *tk* can result in male sterility (Box 3). Negative selection can also be achieved using a diphtheria toxin A-chain (DT-A) gene (Fig. 1) that does not require the addition of a drug for selection, but rather kills all cells in which it is incorporated and expressed.

Screening for Homologous Recombination Events in ES Cells or Mice

Even after positive-negative selection, many of the ES cell colonies that survive drug selection will not be homologous recombinants but instead will contain random integration events in which the negative selectable marker was fortuitously lost or damaged. Therefore, all resistant colonies must still be screened using PCR or Southern blot analysis to identify ES cell clones with correct gene targeting. Southern blot analysis requires the use of a diagnostic restriction enzyme digest with a probe, called the external probe, that is not contained within the regions of homology included in the gene-targeting vector. A unique restriction enzyme site is usually introduced along with the positive selection marker during construction of the gene-targeting vector for ease of screening by Southern blotting. The diagnostic restriction enzyme site should be located outside of any floxed selectable marker so that, after Cre expression and deletion of the marker gene, the diagnostic restriction enzyme site remains. Using this strategy, a diagnostic digest and hybridization with an external probe will yield a smaller DNA fragment for the mutant allele compared to the wild-type allele. This is desirable for technical reasons because if the mutant band were larger, it could be hard to distinguish from a partial restriction enzyme digest on a Southern blot.

BOX 3. HSV *tk* AND MALE STERILITY

HSV *tk* expression is toxic to the germ cells of male transgenic mice and causes dominant sterility. This was observed in transgenic mice expressing HSV *tk* in the kidney (Al-Shawi et al. 1988) and in transgenic mice that fortuitously carried an HSV *tk* sequence as part of a larger transgene construct (Braun et al. 1990). Subsequent studies revealed that the coding region of HSV *tk* contains a promoter that is transcriptionally active in the differentiating germ cells of male transgenic mice. Fortunately, HSV *tk* is not active during female gametogenesis, and HSV *tk*-containing transgenes can be transmitted through the female germline. HSV *tk* is widely used in in vitro gene-targeting strategies, and if it is retained at the targeted locus, it can cause unanticipated difficulties in achieving germline transmission of the mutant allele from male chimeras. To get around this problem, the male germ cell promoter of HSV *tk* has been mutated to express a *tk* protein that retains enzyme activity for negative selection but is not expressed in male germ cells and is therefore compatible with male fertility (Salomon et al. 1995).

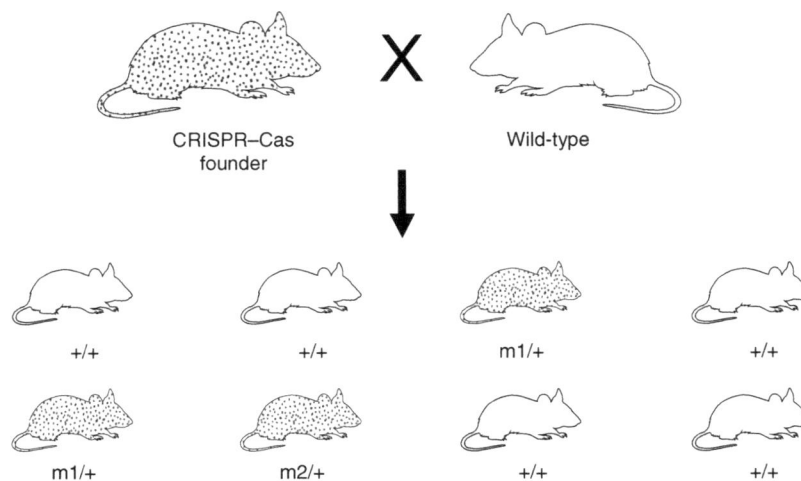

FIGURE 2. Segregation of CRISPR–Cas-induced alleles from founders. A CRISPR–Cas mosaic founder male or female (stippled) is bred to wild-type mice. The progeny that are heterozygous for different mutant alleles (mutation 1, m1 and mutation 2, m2, *stippled*) are identified by polymerase chain reaction (PCR) followed by DNA sequencing and can then be used to generate individual pedigrees.

To characterize the targeted allele fully—and this is essential—both 5′ and 3′ ends of the altered allele must be examined thoroughly to determine the structure of the gene-targeting events that have occurred through both the 5′ and 3′ arms of homology. When using Southern blot analysis, if for some reason both 5′ and 3′ external probes cannot be identified, the targeting event can be identified initially using one external probe, and the allele can be further investigated by additional analysis using a probe within the region of targeting vector homology (e.g., an internal probe such as *neo*).

For CRISPR–Cas manipulations of preimplantation embryos, the resulting mice are screened to identify founders that carry modifications in the locus of interest. Because of variation in the activity of CRISPR–Cas, these founders could carry multiple different alleles. In addition, the founders could be mosaic depending on the timing of CRISPR–Cas activity relative to cleavage (Mizuno et al. 2014; Yen et al. 2014). Once founders are identified they are bred to wild-type mice of the strain appropriate for your studies to segregate alleles in their progeny and to segregate away any off-target events (Fig. 2). The progeny are analyzed to molecularly define the various alleles generated, and individuals carrying the desired alleles are then used to create pedigrees/lines for study.

DIFFERENT TYPES OF ALLELES

General Strategy for Making a Null Allele

The first mutation most investigators aim to make when they initiate the study of a particular gene is a complete loss of function—that is, a null allele. Null alleles are very useful because the phenotype of the homozygous null mouse provides information about the earliest essential role of the product(s) expressed from the locus during development. In addition, null alleles can be used in combination with other alleles for more complex studies. Fortunately, a null allele is one of the simplest mutations to generate, both conceptually and technically, by gene targeting in ES cells or CRISPR–Cas manipulations in preimplantation embryos.

For small genes (<20 kb), the most straightforward strategy is to delete all protein-coding exons. This guarantees the generation of a null allele because protein products cannot be synthesized from the targeted locus. Thus, once the deletion of the DNA sequences has been verified, it is not necessary to analyze mRNA or protein for the targeted gene in homozygous mutants, because there will be none expressed. If the coding sequences are not removed in their entirety, mRNA and protein could still potentially be expressed from the targeted locus, and characterization of variant transcripts generated from the targeted allele will be necessary. This characterization can be time-consuming and difficult to interpret. Partial protein products that could potentially be synthesized from a targeted locus may be problematic if they retain some function. Antibodies for the gene product may not be available to determine if a partial protein product is being expressed, further complicating the interpretation.

An argument against designing or selecting a deletion that removes all coding exons would be the presence of regulatory elements for neighboring genes located within the introns (e.g., *Shh*; Lettice et al. 2003; West et al. 2016). This possibility should be kept in mind, especially if other genes are located nearby. Sequence comparisons of genomic DNA between mouse and other mammals will reveal if there are conserved sequences within the introns that might indicate the presence of such regulatory elements, but they will not necessarily reveal the identity of the gene being regulated.

Generating null alleles for a larger gene requires a little more thought because standard gene-targeting strategies are not as efficient for generating large (>20-kb) deletions; thus, more complex strategies are necessary. With the deletion of all protein-coding exons of a large gene, it is even more likely that regulatory elements of adjacent genes, or even all or part of another gene, could be located within the introns of the gene being deleted. Sequence comparisons of genomic DNA between mouse and other vertebrates might reveal conserved sequences within introns, and searches of transcriptome databases could indicate if another gene is within or overlaps with the gene to be deleted.

With larger genes, the strategy most likely to result in a null allele is the creation of a deletion of up to 20 kb that removes the initial protein-coding exons, including the exon containing the translation initiation site. Alternatively, the translation initiation exon can be maintained but downstream exons deleted such that abnormal transcripts produced are removed by nonsense-mediated decay. Another possibility is that the transcription initiation site(s), if known, could be deleted. If a deletion is designed to remove a subset of internal protein-coding exons, it is important to check that the exons that remain after the deletion cannot splice in-frame to generate a partial protein product. If they can do so, reconsider the position of the deletion or introduce a frameshift mutation. In fact, null alleles produced by any method should be thoroughly evaluated to ensure the elimination of residual proteins produced through such mechanisms as alternative splicing, alternative promoter usage, or translation initiation from non-AUG start codons, all of which have been documented in gene targeting (Hosur et al. 2020).

Making Null Alleles by Homologous Recombination in ES Cells

The foregoing considerations about the final mutation design are relevant regardless of the method used to produce the mutation. Homologous recombination in ES cells allows good control for producing the correct allele before making mutant mice. Deletions for mutating either small (Fig. 3) or large genes (Fig. 4) by homologous recombination in ES cells are created most simply by using a replacement gene-targeting strategy: A selectable marker replaces coding exons and a positive–negative selection scheme is used to enrich for ES cell clones that have undergone homologous recombination. The primary features of a replacement vector are a plasmid backbone containing a positive selection marker positioned between two regions of homology and a negative selectable marker next to one of the arms of homology (see Fig. 1). It is not essential to use a negative selection marker, although its inclusion will increase the ratio of targeted to randomly integrated clones, thus reducing the workload during screening of drug-resistant ES cell clones. Linearized targeting vectors are usually introduced into ES cells by electroporation, followed by culture in selective medium, and subsequent identification of homologous recombinant ES cell clones (see Chapter 11 in Behringer et al. 2014).

Making Null Alleles with CRISPR–Cas

Null mutations made by CRISPR–Cas manipulations of preimplantation embryos are usually generated by creating small insertions/deletions (indels) (Fig. 5). The size of these indels varies depending on the amount of DNA inserted or removed by nonhomologous end joining repair after the CRISPR–Cas-induced double-strand break. If CRISPR–Cas acts in the zygote, two different alleles

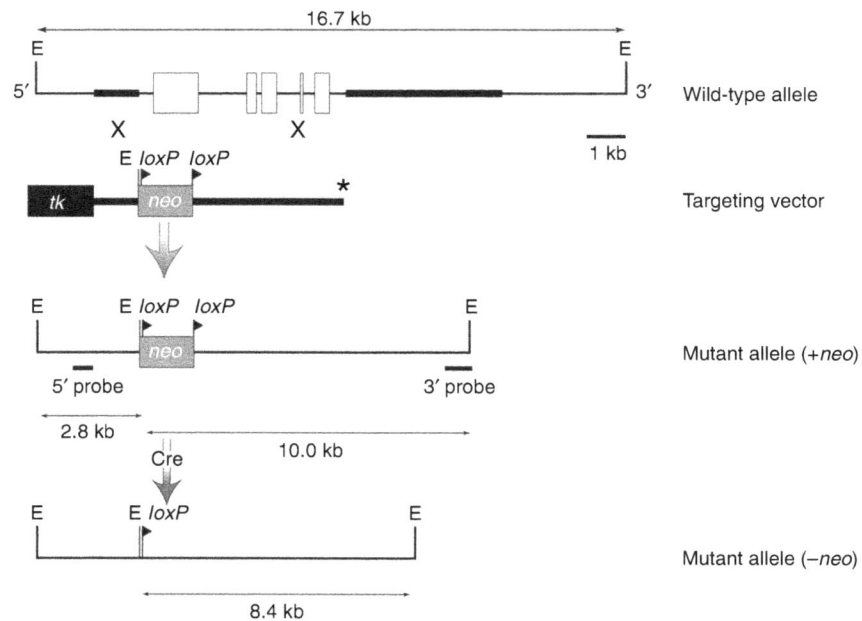

FIGURE 3. Deletion of all protein-coding exons for a small gene by homologous recombination in ES cells. All of the protein-coding exons of the wild-type allele (*open boxes*) are replaced by a floxed neomycin (*neo*) cassette to ensure the generation of a null allele. In this case, the total amount of homology in the targeting vector (5.5 kb) is divided as a 1.2-kb 5′ arm and a 4.3-kb 3′ arm (*thick lines*). The *neo* cassette includes an *Eco*RI (E) site to identify homologous recombinants by Southern analysis. An HSV thymidine kinase (*tk*) cassette is placed adjacent to the 5′ arm of homology for negative selection. The site of vector linearization is indicated by the asterisk (*). In this case, 5′ and 3′ external probes are used to identify the targeted mutant allele. The sizes of the wild-type and mutant bands for Southern analysis are indicated. The floxed *neo* cassette can be removed by Cre. (+*neo*) Allele retaining the *neo* cassette; (−*neo*) allele in which the *neo* cassette has been deleted.

might be generated; four may be generated if the cut/repair occurs after DNA replication. If the CRISPR–Cas components act on the targeted allele after cell division, then each blastomere could have different alleles such that the resulting founder mouse would be mosaic for two or more alleles (Yen et al. 2014; Clark et al. 2020).

A guide RNA is identified to target a specific sequence within the target gene, typically an exon (Chapter 2: Obtaining or Generating Gene Mutations in Mice [Papaioannou and Behringer 2023a]). It is also possible to use CRISPR–Cas to create larger defined deletions by introducing two guide RNAs that flank the region to be deleted (Fujii et al. 2013; Boroviak et al. 2016). The guide RNA(s) and Cas (RNA, plasmid, or protein) are introduced into preimplantation embryos by microinjection or electroporation (see Chapter 7 in Behringer et al. 2014; Hashimoto et al. 2016). The resulting pups born from these manipulations are then screened by PCR for mutations in the locus of interest. It is likely the founders carrying the various CRISPR–Cas-induced mutations will be mosaic and can carry multiple different mutant alleles in their germline that are isolated by breeding (Fig. 2).

Reporter Gene Knock-Ins

On many occasions during phenotypic analysis of a mutant allele, it is useful to have a simple cellular reporter to follow the expression of the gene or to track mutant cells. Introducing a reporter into a specific locus by gene targeting (a so-called knock-in strategy) can be used to achieve the goal of faithful recapitulation of the expression of an endogenous gene by a readily detectable reporter. This method exploits the intact regulatory elements of the endogenous chromosomal locus, at the same time rendering the gene nonfunctional by deletion or alteration of crucial protein-coding exons. Knock-in strategies are also possible that preserve the function of the gene of interest. One of the important steps in designing a knock-in allele is to determine the position within the gene to introduce

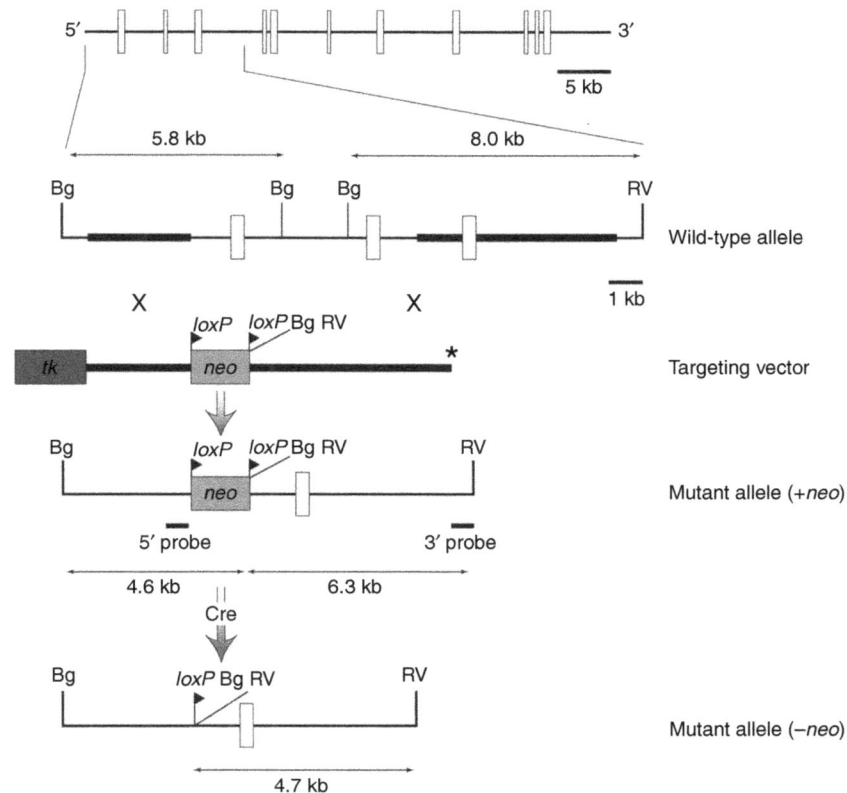

FIGURE 4. Deletion of the initial protein-coding exons for a large gene by homologous recombination in ES cells. The coding exons (*open boxes*) of this gene span 38 kb (*top*). Therefore, the first two coding exons are deleted. This region (6.3 kb) is replaced by a floxed *neo* cassette. A total of 9.0 kb of homology is divided into a 3.0-kb 5′ arm and a 6.0-kb 3′ arm (*thick lines*). The *neo* cassette includes *Bgl*I (Bg) and *Eco*RV (RV) sites to identify homologous recombinants by Southern analysis. An HSV *tk* cassette is placed on the 5′ arm of homology. The site of vector linearization is indicated by the asterisk (*). A 3′ external probe is used for the initial identification of homologous recombinants. Once these clones are identified, they are expanded and can be analyzed with an internal probe to characterize the structure of the 5′ recombination event. The sizes of the wild-type and mutant bands for Southern analysis are indicated. The floxed *neo* cassette can be removed by Cre. (+*neo*) Allele retaining the *neo* cassette; (*neo*) allele in which the *neo* cassette has been deleted.

the reporter. The reporter gene can be engineered (1) as an ATG fusion with the endogenous transcript, (2) as a fusion transcript by generating a bicistronic transcript using an internal ribosome entry site (IRES) sequence or T2A peptide cleavage sequence, (3) by creating a chimeric transcript using a splice acceptor sequence, or (4) by a combination of both an IRES and splice acceptor.

Depending on the strategy, insertion of the reporter gene sequence can be within protein coding or untranslated exons or within introns. A conservative approach to obtain a faithful reporter is to make a simple insertion without deleting any sequence. This will avoid unintentional removal of regulatory elements in a deleted sequence. However, you might consider a small deletion of the protein-coding sequences within an exon to increase the chances of generating a null allele. Introduction of a reporter into untranslated sequences could potentially yield wild-type, hypomorphic (destabilization of the transcript), hypermorphic (stabilization of the transcript), or null alleles, depending on the specific situation (e.g., *Vegfa*; Miquerol et al. 2000).

There are some notes of caution.

• In spite of your best intentions and design efforts, knock-ins will not necessarily reflect the endogenous expression pattern of the targeted gene. This is because the inserted reporter could act as a dominant insertional mutation by altering gene transcription. We just do not know enough about the regulation of genes to be able to avoid this in all cases. Thus, validation of the reporter

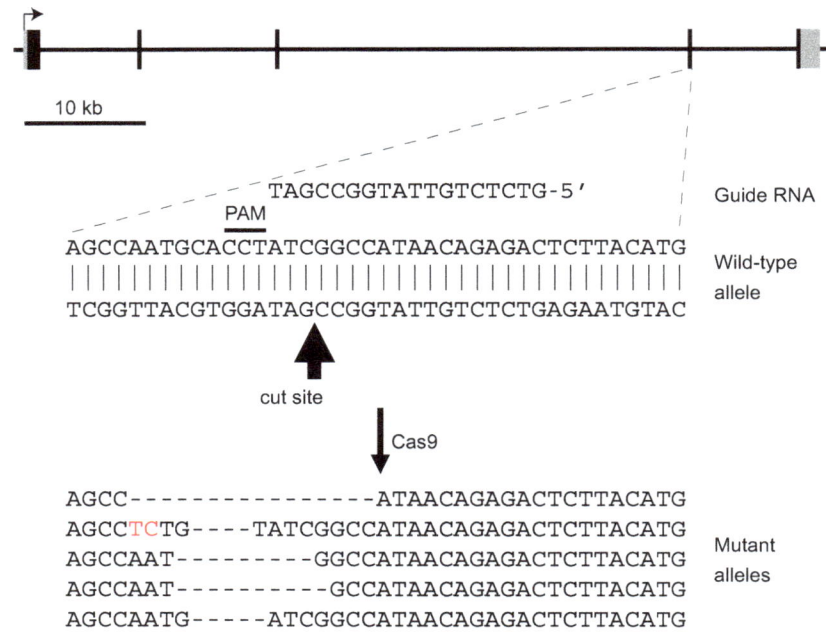

FIGURE 5. Generation of null alleles using CRISPR–Cas. The genomic structure of the *Tyr* locus and sequence from exon 4 is shown. A guide RNA adjacent to a PAM (protospacer-adjacent motif) sequence is identified to make a double-strand cut at a predicted site in exon 4 (thick arrow). Guide RNA and Cas9 (DNA, RNA, or protein) are introduced into zygotes by injection or electroporation. Various possible insertion or deletion (indel) alleles created by CRISPR–Cas are shown below. (Red sequence) insertion; (dashes) deletions; (black boxes) protein coding exons; (gray boxes) untranslated regions; (arrow on exon 1) transcription start site. (Adapted from Fig. 1 in Yen et al. 2014.)

knock-in requires comparisons between endogenous and reporter expression. Once validated for specific stages and tissues, the knock-in can be used with confidence to report the expression of the targeted locus during mutant analysis.

- The activity of a reporter protein or its relative stability may result in a slightly different expression pattern compared with that of the endogenous mRNA or protein expression pattern. For example, X-Gal staining reveals the enzymatic activity of the protein β-galactosidase, which may have different degradation kinetics from the endogenous protein of the mutated gene. If antibodies to the endogenous protein and the marker proteins are available, the extent of this discrepancy can be determined by comparing the two.

Knock-in reporters are useful for following the mutant cells because they mark cells that transcribe the targeted locus. This can be very helpful for learning whether the expressing cells are still present in a mutant animal and, if so, how their behavior differs from wild-type cells. In addition, knock-in reporters can be used in combination with null alleles to determine whether the locus is regulated by negative feedback mechanisms. Assuming that the knock-in allele is also a null allele, the expression of the reporter in knock-in/wild-type heterozygotes is compared with expression in knock-in/null heterozygotes, both of which have a single knock-in reporter allele. If expression of the reporter is more extensive in the knock-in/null heterozygotes, this would suggest a negative feedback regulation of the gene by its own gene product.

Knock-In Alleles by Homologous Recombination in ES Cells

For homologous recombination in ES cells, knock-in alleles are usually generated using a replacement gene-targeting strategy (Fig. 6). The reporter is followed directly by a floxed positive selection marker gene, such as *neo*, that can be removed later by Cre recombinase, either in vitro or in vivo. As mentioned above, alternative recombinase systems could also be used. The 5′ and 3′ arms of homology

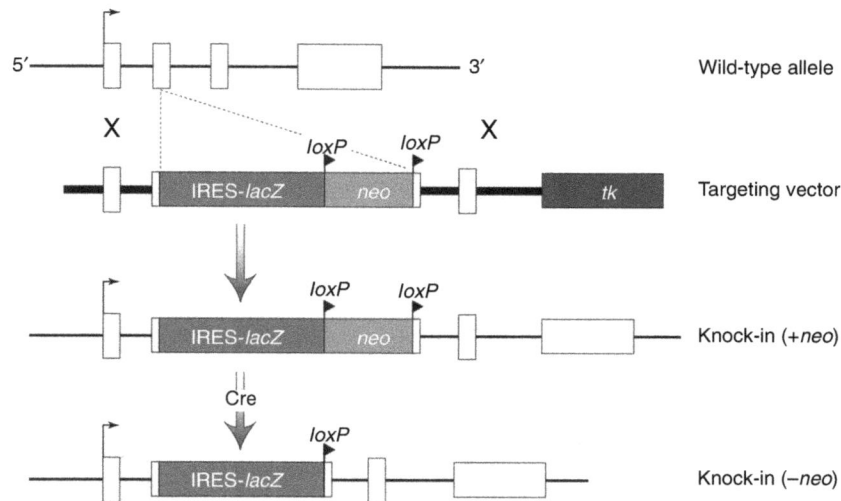

FIGURE 6. Reporter knock-in by homologous recombination in ES cells. In this example, a *lacZ* reporter with a poly(A) is introduced into the second exon of a gene followed by a floxed *neo* expression cassette that can subsequently be removed either in targeted cells or in animals by Cre recombinase. An internal ribosome entry site (IRES) is used to bypass the requirement of having *lacZ* in frame with the coding region of the endogenous gene. An HSV thymidine kinase (*tk*) cassette is placed adjacent to the 3′ arm of homology for negative selection. The floxed *neo* cassette can be removed by Cre. Gene sequences can be deleted when making a knock-in, but one must be aware that regulatory sequences might be removed. Instead of a reporter, the coding region of any gene of interest can be knocked in. Many times, knock-ins also knock out endogenous gene function. (*Thick lines*) Regions of homology; (+*neo*) allele retaining the *neo* cassette; (−*neo*) allele in which the *neo* cassette has been deleted.

flank the reporter-*neo* sequences and a negative selection marker can also be added outside the region of homology. Strategies to identify homologous recombinants by Southern analysis or PCR are the same as for generating a standard null allele.

Knock-In Alleles with CRISPR–Cas

Multiple methods to generate knock-in alleles using CRISPR–Cas have been developed (Gu et al. 2018; Miura et al. 2018; Abe et al. 2020). The feasibility and efficiency of these methods depends on the length of DNA to be introduced into the target locus and perhaps the target locus itself. Relatively new approaches to introduce exogenous sequences into a target locus have exploited the stages of preimplantation development that have higher recombination activity. One fundamental difference for generating knock-in alleles by CRISPR–Cas manipulations in preimplantation embryos versus using ES cells is that there is no selectable marker within the gene-targeting vector.

In one approach, used for introducing shorter sequences, long single-stranded DNA (lssDNA) with ~50–100 nt of homology for a target locus on each end, can be used to knock in up to ~1.5–2.0 kb of heterologous sequence (Fig. 7A; Quadros et al. 2017). Conveniently, lssDNA of this length can be synthesized commercially. Advances in high-fidelity synthesis of lssDNA should facilitate the introduction of longer sequences into a target locus by this approach.

A second approach, used to knock in longer sequences, is to generate a standard double-stranded DNA (dsDNA) gene-targeting vector with 0.5- to 2.0-kb homology flanking the sequence to be introduced into a target locus (Fig. 7B). In one study, two-cell-stage embryos were injected with dsDNA gene-targeting vectors when chromatin is relatively open and there is likely higher homologous recombination activity during an extended G_2 phase of the cell cycle (Gu et al. 2018). In addition, targeting efficiencies were further increased when the dsDNA gene-targeting vectors were biotinylated and a streptavidin Cas9 fusion protein was used to localize the repair template to the target locus. In another study, zygotes were injected, coinciding with the S phase of the cell cycle (Abe et al. 2020). Both methods result in the generation of knock-in alleles at reasonably high frequencies.

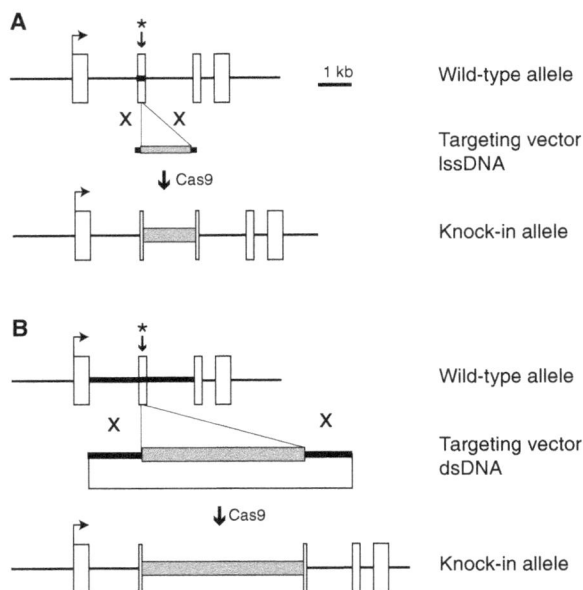

FIGURE 7. Strategies to generate CRISPR–Cas knock-in alleles. Guide RNAs with associated PAM sequences are identified to generate a double-strand break in the exon of choice. After knock-in, the guide RNA and PAM sequences are separated by the insert, preventing recutting of the edited locus by CRISPR–Cas. Asterisks (*) indicate location of guide RNA and associated PAM sequence. (A) Knock-in of an insert (gray) up to ∼2.0 kb. A long single-stranded (lss) DNA vector with 50–100 nt arms of homology (thick lines, 5′ and 3′ of the insert) is used. (B) Knock-in of an insert (gray) >2.0 kb. A double-stranded (ds) DNA vector with 0.5- to 2.0-kb arms of homology (*thick lines*, 5′ and 3′ of the insert) is used.

Producing Point Mutations or Small Changes

Another popular and versatile type of mutation to engineer is a point mutation. These can be used to create a stop codon that will truncate a protein product, alter transcription factor binding sites for transcriptional regulation studies, alter a specific amino acid by a missense mutation to create a variant protein, or perhaps mimic a mutation identified by human genetic studies as a disease-causing mutation. Many other types of alterations, such as splice mutations and untranslated region (UTR) modifications, are also possible with point mutations or other small changes.

Strategies for Point Mutations by Homologous Recombination in ES Cells

There are three general methods for creating point mutations or small changes by gene targeting in ES cells, each with advantages and potential disadvantages. These include (1) the Cre/loxP strategy (Arango et al. 1999), (2) the double-replacement strategy (Moore et al. 1995), and (3) the hit-and-run strategy (Hasty et al. 1991). Currently, however, the more straightforward and preferred way to generate this type of mutation is by CRISPR–Cas manipulations of preimplantation embryos.

Strategies for Point Mutations Using CRISPR–Cas

CRISPR–Cas manipulations in preimplantation embryos can be used to generate point mutations or to insert small sequence modifications (e.g., epitope tags, loxP sites). A guide RNA is identified to target a specific sequence within your favorite gene. The cut site should preferably be within 15 bp of the intended point mutation/epitope tag for highest efficiency of targeting. An oligonucleotide ∼200 nt in length with 60–100 nt of homology flanking the intended point mutation/epitope tag is synthesized. This oligonucleotide also includes silent mutations in the PAM (protospacer-adjacent motif) site (or the spacer sequence) that prevent recutting of the edited locus by CRISPR–Cas. The oligonucleotide can also be designed to create a new restriction enzyme site to aid in the genotyping of the mutant mice (Fig. 8). Alternatively, there are PCR methods that will directly detect point mutations (e.g., Hindson et al. 2011). The guide RNA, the oligonucleotide that contains your desired mutation in a segment of homology, and Cas (RNA, plasmid, or protein) are introduced into preimplantation embryos by microinjection or electroporation. The resulting pups born from these manipulations are then screened by PCR for mutations in the locus of interest to identify founders. As with CRISPR-induced indels, it is likely the founders will carry various CRISPR–Cas-induced

Cite this chapter as *Cold Spring Harb Protoc*; doi:10.1101/pdb.over107957

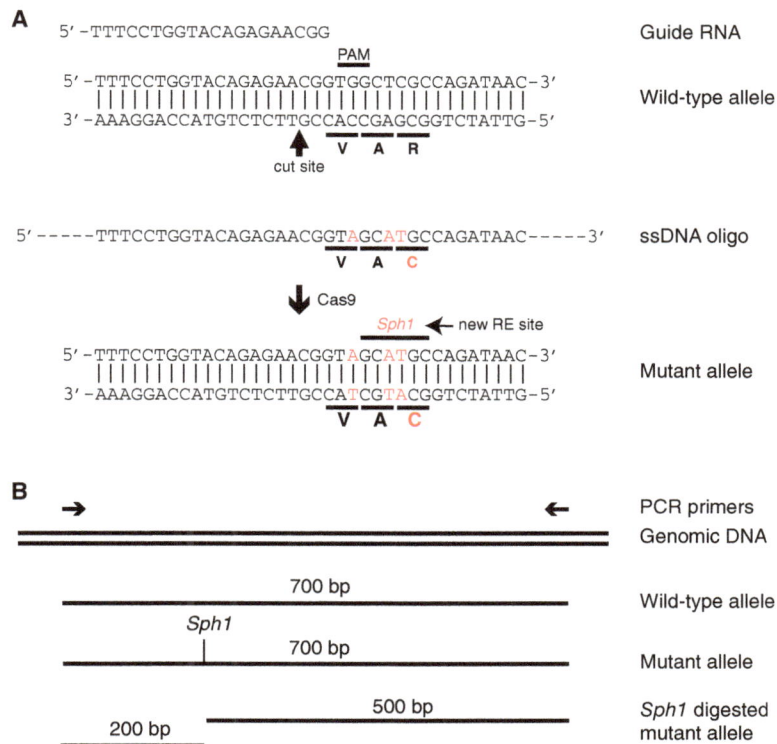

FIGURE 8. (A) Strategy to generate a CRISPR–Cas-induced point mutation. An arginine (R) to cysteine (C) missense mutation is introduced using a ∼100-nt single-stranded DNA oligonucleotide (ssDNA oligo) with the desired sequence changes (red text). Publicly available programs are used to identify candidate guide RNAs with associated PAM sequences to the target sequence. CRISPR–Cas cuts 3–4 bp 5′ from the PAM site. A nucleotide change in the ssDNA oligo will mutate the PAM site to prevent recutting of the allele after the point mutation has been created. In this example, another nucleotide change in the ssDNA oligo creates a new restriction enzyme (RE) site that is exploited for PCR genotyping. (B) PCR genotyping strategy to identify mice carrying the point mutation. Primers will amplify a wild-type sequence (∼700 bp). If the mutations described in A are present, then digestion with *Sph1* will cut the mutant allele into ∼200- and ∼500-bp DNA fragments that can be visualized on an agarose gel. (A) alanine, (V) valine.

mutations in their germline, including indels as well as the desired point mutation. Once progeny from the founders are generated, it is important to confirm the sequence of the intended point mutation and surrounding sequences. The same CRISPR–Cas strategy can be used in ES cells where screening for the desired mutation can be done in ES cell clones before making chimeras.

Conditional Null Alleles

Many genes are expressed during embryogenesis and in adult tissues and are therefore likely to have multiple roles in different tissues at different stages of development. Targeted mutations such as null alleles may cause early embryonic lethality, precluding the study of gene function at later stages of development or after birth. Conditional genetic strategies provide methods to bypass early lethality to study complex gene function. If you anticipate that your gene might fall into this category, and especially if you are mainly interested in the role of the gene in a specific tissue or organ, consider making a conditional null allele or obtaining one from the International Mouse Phenotyping Consortium (IMPC) (https://www.mousephenotype.org/).

Conditional null alleles are usually generated by targeting two *loxP* sites in the same orientation into noncoding regions (e.g., into introns or gene flanking regions) that flank critical protein-coding exons (Fig. 9). Once mice with a conditional allele have been produced, you may then use Cre recombinase at the time and in the tissue that you choose to act on these *loxP* sites to cause the time- or tissue-specific deletion of the floxed exons. Conditional alleles must be thoroughly charac-

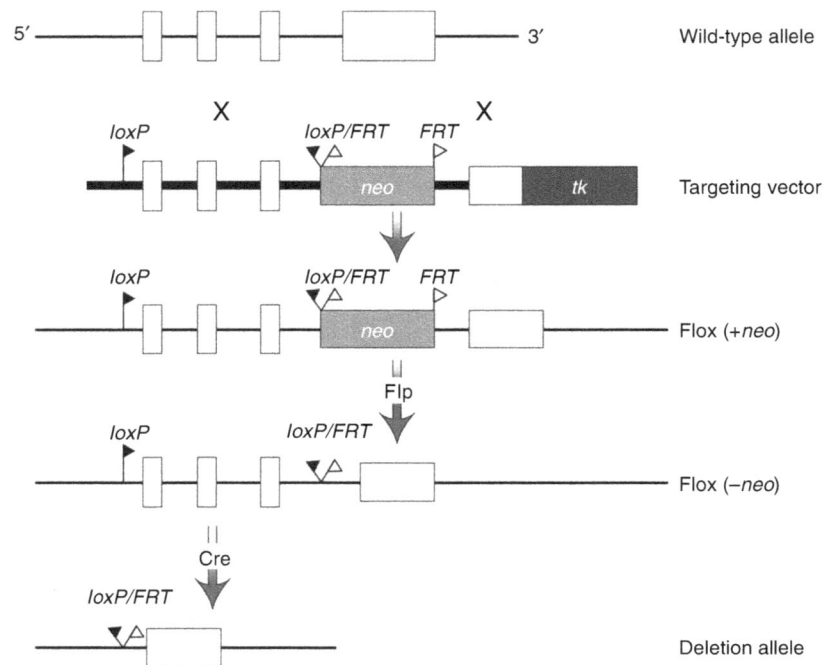

FIGURE 9. Generation of a conditional null allele by homologous recombination in ES cells. In this strategy, a replacement vector is generated in which an *FRT*-flanked *neo* expression cassette is introduced into an intron. *loxP* sites are engineered to flank exons. It is very useful to engineer a restriction site 5′ of the 5′ *loxP* site to identify targeting events by Southern analysis that have included both *loxP* sites. An HSV *tk* cassette is placed adjacent to the 3′ arm of homology. After homologous recombination, a conditional allele is generated that has *neo* in an intron. This allele may behave as wild-type, a hypomorph, or even a null. *neo* can be removed by Flp expression, yielding a conditional allele that will most likely behave as wild-type unless the *loxP* sites have unintentionally disrupted an important regulatory sequence. Cre expression will delete the intervening exons to generate a loss-of-function allele. (Thick lines) Regions of homology; (Flox [+*neo*]) floxed allele that contains the *neo* cassette; (Flox [−*neo*]) floxed allele without the *neo* cassette.

terized to determine that they function as predicted (see Chapter 18: Tissue- and/or Temporal-Specific Mutations in Mice Using Conditional Alleles [Papaioannou and Behringer 2023f]).

A conditional null allele with a linked reporter can also be engineered to monitor the excision of floxed exons by Cre recombinase using a reporter gene such as *lacZ* or *GFP* (Fig. 10). This type of allele allows you to visualize recombination events (i.e., knockout of the gene of interest) with cellular resolution. This can be particularly useful if you do not have an antibody to the protein encoded by the gene of interest or if the cells that express the gene are scattered throughout organs, thus making RNA in situ analysis at the cellular level challenging.

There are a number of ways to design a conditional allele with a linked reporter. Basically, critical exons are flanked by *loxP* sites, a promoterless reporter gene is inserted downstream from the coding region of the gene of interest, and, if using ES cells, a selectable marker that subsequently can be excised in vitro or in vivo is incorporated for gene targeting. Ideally, the conditional allele should behave like a wild-type allele and not express the reporter. After Cre-mediated excision of the floxed exons, the allele should become null and express the reporter under the control of the endogenous regulatory sequences. In contrast to the "standard" conditional alleles in which only *loxP* sites remain in the locus after removal of the selection cassette, there is a greater likelihood of an introduced reporter altering the transcriptional regulation of the gene in question. In addition, just like any other knock-in, the introduced reporter may or may not mimic the expression of the endogenous locus.

It is thus important to characterize thoroughly conditional alleles that use reporters, assaying for normal gene activity before Cre-mediated excision and validating the reporter expression pattern after Cre excision. To validate the reporter expression pattern, generate mice heterozygous for the recom-

Cite this chapter as *Cold Spring Harb Protoc*; doi:10.1101/pdb.over107957

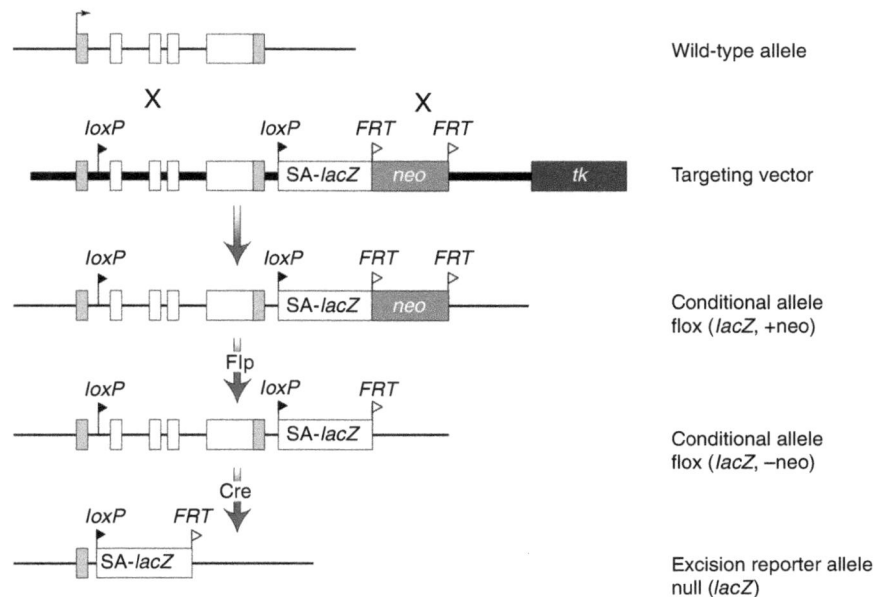

FIGURE 10. Generation of a conditional null allele with a linked reporter by homologous recombination in ES cells. In this strategy, a replacement vector is generated in which a splice acceptor (SA)-*lacZ* pA cassette followed by an *FRT*-flanked *neo* expression cassette is introduced 3′ of the gene of interest. *loxP* sites are engineered to flank exons. It is very useful to engineer a restriction site 5′ of the 5′ *loxP* site to identify targeting events by Southern analysis that have included both *loxP* sites. An HSV *tk* cassette has been placed adjacent to the 3′ arm of homology. After homologous recombination, a conditional allele is generated that has the *lacZ* and *neo* expression cassettes 3′ of the locus. This allele may behave as wild-type, a hypomorph, or even a null. The *lacZ* reporter should be silent because it is not transcribed and does not have a basal promoter. *neo* can be removed by Flp expression in vitro or in vivo, yielding a conditional null allele with a silent reporter. Cre expression will delete the intervening exons to generate a loss-of-function allele and a chimeric transcript containing exon 1 and *lacZ*, leading to β-galactosidase expression. (*Boxes*) Exons; (*shaded boxes*) untranslated regions; (Flox [*lacZ*, +*neo*]) floxed allele that contains the SA-*lacZ* and *neo* cassettes; (Flox [*lacZ*, −*neo*]) floxed allele that contains the SA-*lacZ* cassette without the *neo* cassette; (Null [*lacZ*]) recombined allele with the SA-*lacZ* cassette.

bined allele and compare the reporter expression pattern with that of the endogenous gene during development. This information will define the potential limitations of your reporter. In a number of instances, this type of conditional allele with a linked reporter of excision has acted as a hypomorphic allele before Cre-mediated excision (e.g., mice homozygous for the conditional allele or mice heterozygous for the conditional allele and a null allele of the gene show a phenotype) (e.g., *Fgf8*; Meyers et al. 1998). This would limit the usefulness of the conditional allele, but many experiments are still possible depending on the stage that you want to analyze and when the mice are compromised. The utility of this type of allele in contrast to standard conditional alleles is that recombination can be measured directly and with cellular resolution. So, consider your experiments and decide on the type of allele that best suits your needs.

Conditional Null Alleles by Homologous Recombination in ES Cells

There are many gene-targeting strategies to make conditional alleles by homologous recombination in ES cells, but all require the use of a positive selectable marker, which may either remain in the conditional allele or be removed, for example, by using a second type of recombinase system, such as Flp recombinase (Figs. 9 and 10). In many cases, a conditional allele with the selectable marker in place can function as a wild-type allele. However, this will have to be determined for each individual case because the presence of the selectable marker and its promoter could potentially generate a null or hypomorphic allele as described in the previous section. If the presence of the selectable marker results in a decrease in the level of gene expression, this hypomorphic allele could provide a useful genetic alteration for further analysis.

To ensure that the recombinase sites will function as planned in the targeted allele, it is worth testing for excision of the floxed and flrted exons in vitro in ES cells before mice are made. Demonstrate that the specific recombinase sites flanking the selection cassette function to delete the selection cassette in vitro (Box 2). In addition, show that the recombinase sites flanking critical exons function to delete those exons in vitro. If you are waiting until later to remove the selection cassette in vivo by breeding, take the time now to express the appropriate recombinase(s) in vitro to ensure the deletion can be produced. If one of the recombinase sites is damaged, then you will be generating mice with an allele that will not function properly, wasting a lot of time and effort.

Mice with the conditional allele should be phenotypically identical to wild-type mice and thus easily maintained as homozygous stocks. If this is not the case, you have a hypomorphic or null allele, which was not intended but could still be useful. Once you have a conditional allele in mice, introduction of Cre recombinase by breeding with *cre*-transgenic mice will result in Cre-mediated excision of the floxed coding exons, producing a deletion allele at a time and tissue of your choosing. For breeding strategies to make use of a conditional allele for time- or tissue-specific gene ablation, see Chapter 18: Tissue- and/or Temporal-Specific Mutations in Mice Using Conditional Alleles (Papaioannou and Behringer 2023f).

Conditional Null Alleles Using CRISPR–Cas

CRISPR–Cas can be used to generate conditional null alleles by directly modifying zygotes (Clark et al. 2020). Multiple approaches have been developed. One method is independent of the distance between the placement of the two *loxP* sites and the others are limited by this distance.

The introduction of two *loxP* sites using single-stranded oligonucleotides is independent of the distance between the two sites (Fig. 11A). *LoxP* sites are 34 bp in length. Thus, they can be introduced into the genome using a guide RNA, Cas9, and an oligonucleotide that includes the *loxP* sequence surrounded by target gene homology (see Producing Point Mutations or Small Changes, above). You can attempt to place both *loxP* sites directly into zygotes simultaneously, but obtaining both *loxP* sites flanking the intended exon(s) on the same chromosome can be inefficient (Yang et al. 2013; Gurumurthy et al. 2019). It is more likely that you will obtain mice containing one or the other of the *loxP* sites targeted in the correct position (Fig. 11B). If male founders with a single correctly targeted *loxP* site are generated, then you can wait until they are sexually mature, isolate and cryopreserve their sperm, and subsequently perform in vitro fertilization (IVF) of oocytes obtained by superovulation of wild-type females (Behringer et al. 2018; Takeo and Nakagata 2018a,b), and repeat the CRISPR–Cas injection or electroporation to insert the second *loxP* site. You could do this through natural matings to preserve the male founder, but the oocyte yield from plugged females even if superovulated will be lower and the process could take much longer. Although you must go through an extra generation to obtain the floxed allele using this sequential targeting approach, it appears to work relatively efficiently to generate the desired conditional null allele. If you perform this sequential targeting approach, there are a few caveats to consider.

First, the founder male with one *loxP* site correctly inserted at the target locus should be genotyped very carefully. It is possible that CRISPR–Cas induced an indel at the introduction site for the second *loxP* site, which would prevent its insertion. Thus, you must select founder males that have a wild-type sequence at the location for insertion of the second *loxP* site or redesign the guide to target the indel sequence. Second, you can use oligonucleotides to introduce both *loxP* sites during the second round. This increases the chances of obtaining *loxP* sites in *cis* on the same chromosome, but genotyping will be required to determine if the *loxP* sites are in *cis* or *trans*.

There are two methods of making a conditional allele that are dependent on the distance between the two *loxP* sites.

Cite this chapter as *Cold Spring Harb Protoc*; doi:10.1101/pdb.over107957

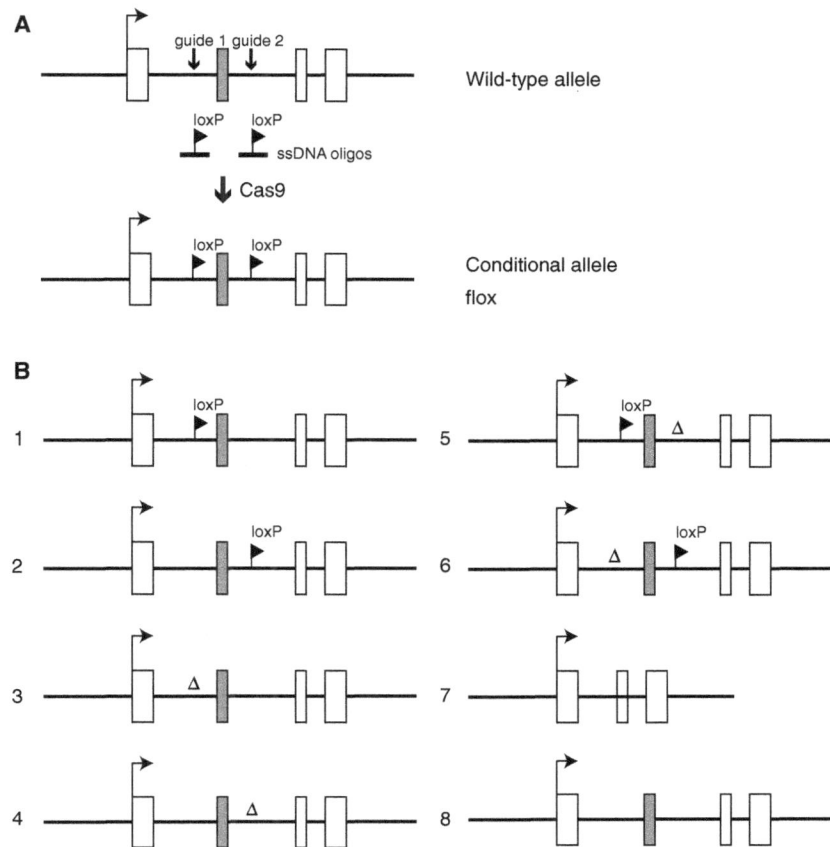

FIGURE 11. Generation of a conditional null allele using CRISPR–Cas and oligonucleotides. (*A*) Wild-type and conditional (flox) alleles are shown. (Open boxes) Exons; (arrow on exon 1) start of transcription. Two guide RNAs recognize the locations for introduction of *loxP* sites to flank a critical exon (shaded). Single-stranded DNA oligonucleotides of ~100 nt with the *loxP* sites centrally located are synthesized. Guide RNAs, ssDNA oligos, and Cas9 (DNA, RNA, or protein) are introduced into zygotes by injection or electroporation. If successful, the exon will be flanked by *loxP* sites. (*B*) Other possible outcomes from the experiment in *A*: (1 and 2) only one *loxP* site is introduced; these alleles could be used for a subsequent round of targeting to insert the second *loxP* site; (3 and 4) indels (Δ) are created at one site or the other; (5 and 6) a *loxP* site is correctly inserted at one site but the other site has an indel (Δ); (7) exon 2 is deleted; (8) the locus remains wild-type.

1. If the two sites are to be placed within 2.0 kb of each other, then a lssDNA with ~50–100 bp of homology for the target sequence on each end can be used to introduce both *loxP* sites flanking the desired exon(s) (Fig. 12A; Miura et al. 2018).

2. If the two *loxP* sites are to be placed farther than 2.0 kb apart, then a dsDNA targeting vector can be used to knock in both *loxP* sites at once (Fig. 12B; Yao et al. 2018). When you generate the dsDNA targeting vector, the *loxP* sites are placed such that the PAM site and guide RNA target sequence are split so that CRISPR–Cas does not cut the targeting vector or recut the edited locus.

It is likely that new methods or variations of current methods will yield ever more efficient approaches to generate conditional null alleles by directly modifying zygotes using CRISPR–Cas. So, keep an eye on the literature! Once a targeting strategy and technology has been chosen, you can move on to either Chapter 4: Embryonic Stem Cell Gene Targeting and Chimera Production in Mice (Papaioannou and Behringer 2023b) or Chapter 5: Recovering a Targeted Mutation in Mice from Embryonic Stem Cell Chimeras or CRISPR–Cas Founders (Papaioannou and Behringer 2023c).

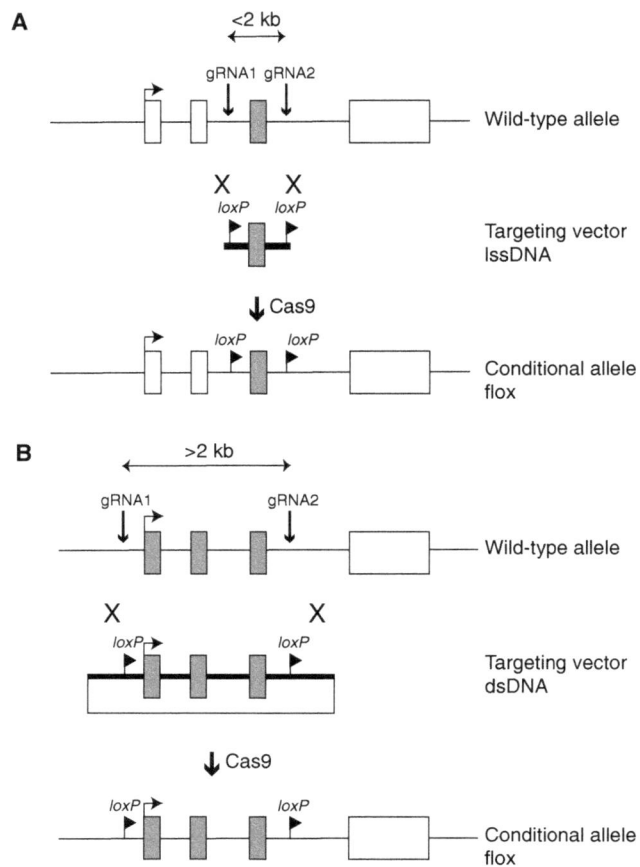

FIGURE 12. Generation of a conditional null allele using CRISPR–Cas and long single-stranded DNA (lssDNA) or double-stranded DNA (dsDNA) targeting vectors. Wild-type and conditional alleles (flox) are shown. (Open boxes) exons; (arrow on exon 1) start of transcription. (A) If the placement of *loxP* sites is <2 kb apart, then a lssDNA is synthesized incorporating *loxP* sites that will flank a critical exon (*shaded*); 50–100 nt of homology is included at each end of the lssDNA. Guide RNA, lssDNA, and Cas9 (DNA, RNA, or protein) are introduced into zygotes by injection or electroporation. (B) If the placement of *loxP* sites is >2 kb apart, then a dsDNA gene-targeting vector is created incorporating *loxP* sites that will flank critical exons (shaded); 0.5–2.0 kb of homology is included at each end of the targeting vector. The PAM sequence is mutated in the dsDNA target vector so that it will not be cut by CRISPR–Cas before or after gene editing. Guide RNA, dsDNA targeting vector, and Cas9 (DNA, RNA, or protein) are introduced into zygotes by injection or electroporation.

ACKNOWLEDGMENTS

We thank Yasuhide Furuta, Lisette Maddison, Lauryl Nutter, Janice Parker-Thornburg, and Thomas Saunders for helpful comments.

REFERENCES

Abe T, Inoue KI, Furuta Y, Kiyonari H. 2020. Pronuclear microinjection during S-phase increases the efficiency of CRISPR-Cas9-assisted knockin of large DNA donors in mouse zygotes. *Cell Rep* 31: 107653. doi:10.1016/j.celrep.2020.107653

Abuin A, Bradley A. 1996. Recycling selectable markers in mouse embryonic stem cells. *Mol Cell Biol* 4: 1851–1856. doi:10.1128/MCB.16.4.1851

Al-Shawi R, Burke J, Jones CT, Simons JP, Bishop JO. 1988. A Mup promoter-thymidine kinase reporter gene shows relaxed tissue-specific expression and confers male sterility upon transgenic mice. *Mol Cell Biol* 8: 4821–4828. doi:10.1128/mcb.8.11.4821-4828.1988

Anastassiadis K, Fu J, Patsch C, Hu S, Weidlich S, Duerschke K, Buchholz F, Edenhofer F, Stewart AF. 2009. Dre recombinase, like Cre, is a highly efficient site-specific recombinase in E. coli, mammalian cells and mice. *Dis Model Mech* 2: 508–515. doi:10.1242/dmm.003087

Anastassiadis K, Schnütgen F, von Melchner H, Stewart AF. 2013. Gene targeting and site-specific recombination in mouse ES cells. *Methods Enzymol* 533: 133–155. doi:10.1016/B978-0-12-420067-8.00009-X

Arango NA, Lovell-Badge R, Behringer RR. 1999. Targeted mutagenesis of the endogenous mouse *Mis* gene promoter: in vivo definition of genetic pathways of vertebrate sexual development. *Cell* **99**: 409–419. doi:10 .1016/S0092-8674(00)81527-5

Behringer RR, Gertsenstein M, Nagy KV, Nagy A. 2014. *Manipulating the mouse embryo: a laboratory manual.* Cold Spring Harbor Laboratory Press, Cold Spring Harbor, NY.

Behringer R, Gertsenstein M, Nagy KV, Nagy A. 2018. Administration of gonadotropins for superovulation in mice. *Cold Spring Harb Protoc* doi:10.1101/pdb.prot092403

Boroviak K, Doe B, Banerjee R, Yang F, Bradley A. 2016. Chromosome engineering in zygotes with CRISPR/Cas9. *Genesis* **54**: 78–85. doi:10 .1002/dvg.22915

Braun RE, Lo D, Pinkert CA, Widera G, Flavell RA, Palmiter RD, Brinster RL. 1990. Infertility in male transgenic mice: disruption of sperm development by HSV-*tk* expression in postmeiotic germ cells. *Biol Reprod* **43**: 684–693. doi:10.1095/biolreprod43.4.684

Clark JF, Dinsmore CJ, Soriano P. 2020. A most formidable arsenal: genetic technologies for building a better mouse. *Genes Dev* **34**: 1256–1286. doi:10.1101/gad.342089.120

Fujii W, Kawasaki K, Sugiura K, Naito K. 2013. Efficient generation of large-scale genome-modified mice using gRNA and CAS9 endonuclease. *Nucleic Acids Res* **41**: e187. doi:10.1093/nar/gkt772

Gu B, Posfai E, Rossant J. 2018. Efficient generation of targeted large insertions by microinjection into two-cell-stage mouse embryos. *Nat Biotechnol* **36**: 632–637. doi:10.1038/nbt.4166

Gurumurthy CB, O'Brien AR, Quadros RM, Adams J, Alcaide P, Ayabe S, Ballard J, Batra SK, Beauchamp MC, Becker KA, et al. 2019. Reproducibility of CRISPR–Cas9 methods for generation of conditional mouse alleles: a multi-center evaluation. *Genome Biol* **20**: 171. doi:10.1186/ s13059-019-1776-2

Han X, Zhang Z, He L, Zhu H, Li Y, Pu W, Han M, Zhao H, Liu K, Li Y, et al. 2021. A suite of new Dre recombinase drivers markedly expands the ability to perform intersectional genetic targeting. *Cell Stem Cell* **28**: 1160–1176.e1167. doi:10.1016/j.stem.2021.01.007

Hashimoto M, Yamashita Y, Takemoto T. 2016. Electroporation of Cas9 protein/sgRNA into early pronuclear zygotes generates non-mosaic mutants in the mouse. *Dev Biol* **418**: 1–9. doi:10.1016/j.ydbio.2016.07.017

Hasty P, Ramírez-Solis R, Krumlauf R, Bradley A. 1991. Introduction of a subtle mutation into the Hox-2.6 locus in embryonic stem cells. *Nature* **350**: 243–246. doi:10.1038/350243a0

Hindson BJ, Ness KD, Masquelier DA, Belgrader P, Heredia NJ, Makarewicz AJ, Bright IJ, Lucero MY, Hiddessen AL, Legler TC, et al. 2011. High-throughput droplet digital PCR system for absolute quantitation of DNA copy number. *Anal Chem* **83**: 8604–8610. doi:10.1021/ ac202028g

Hosur V, Low BE, Li D, Stafford GA, Kohar V, Shultz LD, Wiles MV. 2020. Genes adapt to outsmart gene-targeting strategies in mutant mouse strains by skipping exons to reinitiate transcription and translation. *Genome Biol* **21**: 168. doi:10.1186/s13059-020-02086-0

Kaul A, Koster M, Neuhaus H, Braun T. 2000. Myf-5 revisited: loss of early myotome formation does not lead to a Rib phenotype in homozygous *Myf-5* mutant mice. *Cell* **102**: 17–19. doi:10.1016/S0092-8674(00) 00006-4

Lettice LA, Heaney SJ, Purdie LA, Li L, de Beer P, Oostra BA, Goode D, Elgar G, Hill RE, de Graaff E. 2003. A long-range *Shh* enhancer regulates expression in the developing limb and fin and is associated with preaxial polydactyly. *Hum Mol Genet* **12**: 1725–1735. doi:10.1093/hmg/ddg180

Meyers EN, Lewandoski M, Martin GR. 1998. An *Fgf8* mutant allelic series generated by Cre- and Flp-mediated recombination. *Nat Genet* **18**: 136–141. doi:10.1038/ng0298-136

Miquerol L, Langille BL, Nagy A. 2000. Embryonic development is disrupted by modest increases in vascular endothelial growth factor gene expression. *Development* **127**: 3941–3946. doi:10.1242/dev.127 .18.3941

Miura H, Quadros RM, Gurumurthy CB, Ohtsuka M. 2018. Easi-CRISPR for creating knock-in and conditional knockout mouse models using long ssDNA donors. *Nat Protoc* **13**: 195–215. doi:10.1038/nprot.2017 .153

Mizuno S, Dinh TT, Kato K, Mizuno-Iijima S, Tanimoto Y, Daitoku Y, Hoshino Y, Ikawa M, Takahashi S, Sugiyama F, et al. 2014. Simple generation of albino C57BL/6J mice with G291T mutation in the tyrosinase gene by the CRISPR/Cas9 system. *Mamm Genome* **25**: 327–334. doi:10.1007/s00335-014-9524-0

Moore RC, Redhead NJ, Selfridge J, Hope J, Manson JC, Melton DW. 1995. Double replacement gene targeting for the production of a series of mouse strains with different prion protein gene alterations. *Biotechnology* **13**: 999–1004. doi:10.1038/nbt0995-999

Papaioannou VE, Behringer RR. 2023a. Obtaining or generating gene mutations in mice. *Cold Spring Harb Protoc* doi:10.1011/pdb.over107956

Papaioannou VE, Behringer RR. 2023b. Embryonic stem cell gene targeting and chimera production in mice. *Cold Spring Harb Protoc* doi:10.1011/ pdb.over107958

Papaioannou VE, Behringer RR. 2023c. Recovering a targeted mutation in mice from embryonic stem cell chimeras or CRISPR–Cas founders. *Cold Spring Harb Protoc* doi:10.1011/pdb.over107959

Papaioannou VE, Behringer RR. 2023d. Strategies for maintaining mouse mutations. *Cold Spring Harb Protoc* doi:10.1011/pdb.over107960

Papaioannou VE, Behringer RR. 2023e. Getting around an early lethal phenotype in mice with chimeras. *Cold Spring Harb Protoc* doi:10.1011/pdb .over107979

Papaioannou VE, Behringer RR. 2023f. Tissue- and/or temporal-specific mutations in mice using conditional alleles. *Cold Spring Harb Protoc* doi:10.1011/pdb.over107980

Quadros RM, Miura H, Harms DW, Akatsuka H, Sato T, Aida T, Redder R, Richardson GP, Inagaki Y, Sakai D, et al. 2017. Easi-CRISPR: a robust method for one-step generation of mice carrying conditional and insertion alleles using long ssDNA donors and CRISPR ribonucleoproteins. *Genome Biol* **18**: 92.

Ryder E, Doe B, Gleeson D, Houghton R, Dalvi P, Grau E, Habib B, Miklejewska E, Newman S, Sethi D, et al. 2014. Rapid conversion of EUCOMM/KOMP-CSD alleles in mouse embryos using a cell-permeable Cre recombinase. *Transgenic Res* **23**: 177–185. doi:10.1007/s11248-013-9764-x

Salomon B, Maury S, Loubière L, Caruso M, Onclercq R, Klatzmann D. 1995. A truncated herpes simplex virus thymidine kinase phosphorylates thymidine and nucleoside analogs and does not cause sterility in transgenic mice. *Mol Cell Biol* **15**: 5322–5328. doi:10.1128/MCB.15.10 .5322

Takeo T, Nakagata N. 2018a. In vitro fertilization in mice. *Cold Spring Harb Protoc* doi:10.1101/pdb.prot094524

Takeo T, Nakagata N. 2018b. Mouse sperm cryopreservation using cryoprotectant containing L-glutamine. *Cold Spring Harb Protoc* doi:10.1101/ pdb.prot094516

West DB, Engelhard EK, Adkisson M, Nava AJ, Kirov JV, Cipollone A, Willis B, Rapp J, de Jong PJ, Lloyd KC. 2016. Transcriptome analysis of targeted mouse mutations reveals the topography of local changes in gene expression. *PLoS Genet* **12**: e1005691. doi:10.1371/journal.pgen .1005691

Yang H, Wang H, Shivalila CS, Cheng AW, Shi L, Jaenisch R. 2013. One-step generation of mice carrying reporter and conditional alleles by CRISPR/ Cas-mediated genome engineering. *Cell* **154**: 1370–1379. doi:10.1016/j .cell.2013.08.022

Yao X, Zhang M, Wang X, Ying W, Hu X, Dai P, Meng F, Shi L, Sun Y, Yao N, et al. 2018. Tild-CRISPR allows for efficient and precise gene knockin in mouse and human cells. *Dev Cell* **45**: 526–536.e525. doi:10.1016/j.devcel.2018.04.021

Yen ST, Zhang M, Deng JM, Usman SJ, Smith CN, Parker-Thornburg J, Swinton PG, Martin JF, Behringer RR. 2014. Somatic mosaicism and allele complexity induced by CRISPR/Cas9 RNA injections in mouse zygotes. *Dev Biol* **393**: 3–9. doi:10.1016/j.ydbio .2014.06.017

CHAPTER 4

Embryonic Stem Cell Gene Targeting and Chimera Production in Mice

Producing a custom gene mutation in embryonic stem (ES) cells, whether through homologous recombination or CRISPR–Cas gene editing, is the first step along the way to getting the mutation into live mice. However, there are a number of additional steps along the way, each presenting technical challenges. Here, we provide a guide for troubleshooting when the results are not as expected and to distinguish technical problems from possible biological effects of the mutation. From the isolation of clonal lines of targeted ES cells through the production of ES cell chimeras with the targeted ES cell clone, we discuss common technical problems and their most likely causes and solutions. We also provide guidance for situations where the mutation has a phenotype in the form of a dominant effect on ES cells or in chimeras.

INTRODUCTION

ISOLATE CLONAL ES CELL LINES

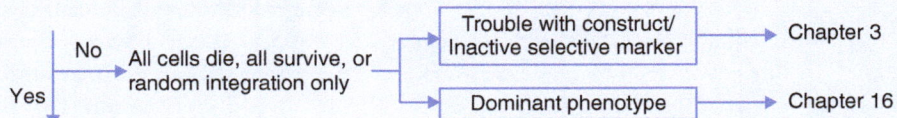

All cells die, all survive, or random integration only
- No → Trouble with construct/Inactive selective marker → Chapter 3
- No → Dominant phenotype → Chapter 16
- Yes ↓

PRODUCE CHIMERAS

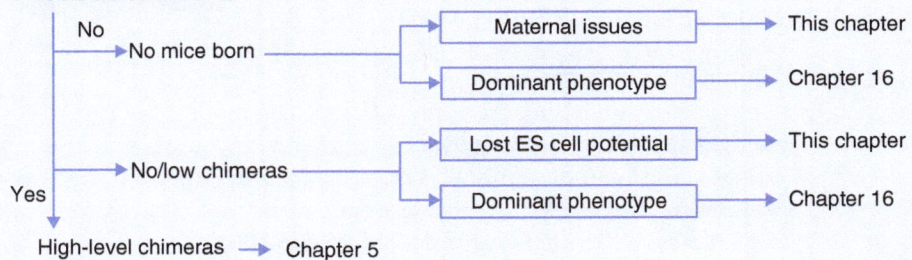

No mice born
- No → Maternal issues → This chapter
- No → Dominant phenotype → Chapter 16

No/low chimeras
- Lost ES cell potential → This chapter
- Dominant phenotype → Chapter 16

- Yes ↓

High-level chimeras → Chapter 5

The means to an end in science are often circuitous as unexpected results open new avenues of investigation. Sometimes these lead to serendipitous discoveries, but just as often they can be an annoyance and distraction if attainment of the intended end point is delayed. This chapter is designed to guide you through the minefield of technical and biological issues that can sidetrack the recognition of

Opening artwork: Adapted from Chapman et al. 2003. *Mech Dev* **120**: 837–847.

BOX 1. TEN MOST COMMON MISTAKES IN ES CELL GENE-TARGETING EXPERIMENTS AND THEIR CONSEQUENCES

1. Starting with 'bad' ES cells, that is, cells that are not germline competent.

2. Forgetting to add targeting vector DNA to the cuvette before electroporation; all ES cells die during selection.

3. Using the wrong G418 concentration; no cells die or all cells die.

4. Using β-mercaptoethanol at too high a concentration; all cells die.

5. Electroporating ES cells that were already drug-resistant; all cells live.

6. Switching the lids of the 96-well plates during the freeze down of the master plate or inverting the orientation of the duplicate 96-well plate used for genotyping, leading to the use of the wrong ES cell clones to generate chimeras.

7. Killing ES cells during freeze down of the master plate; targeted clones can be identified but not retrieved.

8. Using the wrong probe to identify homologous recombinants; no targeted ES cell clones can be identified or nontargeted clones were used to make chimeras.

9. Retaining herpes simplex virus thymidine kinase in ES cells, causing male chimera sterility; no germline transmission of the ES-derived genotypes (see Box 3 in Chapter 3: Mouse Gene-Targeting Strategies for Maximum Ease and Versatility [Papaioannou and Behringer 2023d]).

10. *loxP* site in the targeting vector is not functional; not discovered until after germline transmission and crosses with Cre mice.

a mutant phenotype in a gene-targeting experiment using embryonic stem (ES) cells in mice, whether you have produced the ES cell mutation by homologous recombination or by gene editing using CRISPR–Cas in ES cells. If you are using CRISPR–Cas to target genes directly in embryos, you can skip this chapter and go to Chapter 5: Recovering a Targeted Mutation in Mice from Embryonic Stem Cell Chimeras or CRISPR–Cas Founders (Papaioannou and Behringer 2023a).

Once a targeting vector has been electroporated into ES cells, the next steps are to isolate clonal lines of correctly targeted ES cells and grow up a sufficient number for further experiments and future use, and then to make chimeras with the mutant cells. We assume that you started your gene-targeting experiment with a source of proven germline competent ES cells and that you are following the detailed protocols provided by other sources for their culture, electroporation, and selection (see Chapter 11 in Behringer et al. 2014 and Joyner 1993). In this chapter, we address some of the pitfalls that might be encountered in recovering clonal lines of correctly targeted ES cells and making germline chimeras, including the possibility that you might have early indications of a dominant phenotypic effect, and direct you to the next steps in diagnosing problems or moving ahead with the analysis. Ruling out the common mistakes in gene-targeting experiments will lead to the most efficient way of determining the effects of a mutation and steers you clear of time-consuming technical discoveries that others have made many times before you (Box 1).

ISOLATE CLONAL LINES OF CORRECTLY TARGETED ES CELLS

Careful design of a targeting construct and meticulous technique for ES cell culture, electroporation, and selection should lead to the successful isolation of correctly targeted ES cells that are easily propagated and expanded, allowing you to freeze aliquots as backup and for future use (see Chapter 11 in Behringer et al. 2014). If this is the case, you can skip this section and get on with making chimeras (Produce ES Cell Chimeras with Your Targeted ES Cell Clone, below). If not, there are a number of possible causes presented here in approximate order of decreasing likelihood that you will need to explore.

Cite this chapter as *Cold Spring Harb Protoc*; doi:10.1101/pdb.over107958

Technical Difficulties

If no or very few ES cell colonies survived drug selection, it is worth considering that you may have neglected to add the DNA targeting construct to the cell suspension before electroporation, failed to linearize it, or used the incorrect concentration, thus leaving all cells susceptible to the selective drug. Alternatively, you might have added too high a concentration of the selective drug. If you are using drug-resistant feeder cells, whether or not they die can provide an indication of whether the drug dose is toxic.

Other technical problems that might lead to the death of all cells, such as problems with the electroporator, culture medium, incubators, etc., can easily be ruled out by the simple control of culturing two extra dishes of ES cells—one electroporated, the other not—without the selective drug. These ES cells should grow well, whereas the experimental drug-treated cells should all die, apart from resistant colonies. Try another electroporation with this control, making sure that you add the DNA construct, to determine whether your electroporation and culture conditions are optimal.

If a large number of clones or a solid lawn of ES cells survives drug selection, but few or none have incorporated the targeting construct, you may have neglected to add the selective drug or used an inadequate dose of the drug. Or perhaps you thawed the wrong vial of ES cells for the electroporation and used some that were already drug-resistant. It is always worth determining a "kill curve" for a particular ES cell line and batch of drug so that you know what to expect. Simply determine a dose–response curve for the drug, selecting the lowest dose that kills 100% of exponentially growing cells within 5–7 d.

Trouble with the Targeting Construct

If you are getting drug-resistant colonies with random integrants of your targeting construct, but no homologous recombination at the targeted locus, it could be that you have a less-than-perfect targeting construct or that your locus is hard to target. Before going further, make sure that you are using the correct genotyping probe. Is the band of the correct, predicted size? It is worthwhile to sequence the probe fragment to ensure that it is correct. If you are using polymerase chain reaction (PCR), double-check your strategy. Some loci are notoriously difficult to target but we have yet to hear a convincing explanation for this, or more importantly, how to predict it. Your best bet is to review Chapter 2: Obtaining or Generating Gene Mutations in Mice (Papaioannou and Behringer 2023b) and make sure that you have done everything you can to optimize the chances of homologous recombination, particularly with respect to the use of isogenic DNA and to having sufficiently long arms of homology. If you have any doubt, consider redesigning and rebuilding the targeting construct. However, before you do, electroporate again with the construct that you have and screen for additional clones, which usually takes less time and effort than rebuilding the targeting construct. On the other hand, the low efficiency may be part of a learning curve that will improve with practice, so do not get discouraged.

Inactive Selectable Marker

If ES cells die, but you have ruled out the technical difficulties listed above and drug-resistant feeders are fine, a fairly remote possibility is that the selectable marker in the construct may be inactive, either because of mutations introduced during cloning or the activity of a transcriptional repressor in the region of homology used for building the construct. In the first case, you could sequence the selectable marker you used and replace it with another if it has been mutated. If this is not the problem, consider a different placement of the selectable marker in the construct—for example, use a different region of homology in hopes of avoiding the putative repressor.

Is This a Phenotype?

Finally, there is the possibility that *this is your phenotype!* Perhaps the mutation of a single copy of your favorite gene is sufficient to render the heterozygous ES cells incapable of growing—a dominant effect. Explore this last possibility but continue to work, because we know of no actual case of a gene with this particular heterozygous mutant effect of killing ES cells. But how would you go about testing it if you think that you have that rare case? Go to Chapter 16: Phenotypic Analysis of Dominant Mutant Effects in Mice (Papaioannou and Behringer 2023c) to investigate possible dominant effects. If you can rule out this possibility, try electroporating again and select a large number of clones for further analysis. You simply may not have tested enough clones to recover a rare homologous recombination event. If all else fails, make a new targeting vector incorporating greater or different regions of homology.

It should be noted that X-linked or Y-linked genes are special cases in that only one copy is present in XY ES cells, so the cells are functionally hemizygous for the gene except for genes located within the pseudoautosomal region shared by the X and Y chromosomes (Perry et al. 2001). The possibility that a mutation in the single copy results in an ES cell phenotype is thus more likely since it does not require a dominant effect.

EXPAND AND FREEZE CORRECTLY TARGETED ES CELL CLONES FOR FUTURE USE

How Many Clones Are Enough?

To verify that a phenotype is the result of the targeted mutation you created, it is useful to show that two independently targeted ES cell clones give the same phenotype. However, to be on the safe side, it is worth recovering more than two targeted clones—five to 10 would be reasonable. These cells have been subjected to electroporation, drug treatment, and cloning. Even if you started with a parent ES cell line with excellent germline potential (which you should have tested), the resultant clones could have picked up mutations, chromosomal aberrations, or restrictions of developmental potential during these procedures, so that any single clone might not provide germline transmission through chimeras. Similarly, if you have only one or two clones, they might succumb to some technical glitches along the way, so it is wise to have a few more as backup.

In the event that your targeting efficiency is very low and it is hard to get a reasonable number of targeted clones without massive effort, check the technical aspects of the electroporation procedures because this is often where the problem lies. Check the DNA concentration and purity. Make sure that the DNA is completely dissolved and well mixed with the ES cell suspension before electroporation. In addition, make sure to use the correct cell concentration and electroporation parameters. If the efficiency is still low and you have consulted Chapter 3: Mouse Gene-Targeting Strategies for Maximum Ease and Versatility (Papaioannou and Behringer 2023d) for hints on optimizing the targeting construct, persevere a bit longer. In our experience, a definite learning curve exists and gene-targeting efficiency often improves with electroporation practice.

How Many Cells Are Enough?

As soon as you have identified a targeted clone, culture and expand it, verify by regenotyping that it is indeed targeted, and freeze down a sufficient number of cells for future experiments. Five to 10 vials of each targeted clone should be fine, but you might want to archive these in more than one location to guard against losing them all in a freezer disaster.

PRODUCE ES CELL CHIMERAS WITH YOUR TARGETED ES CELL CLONE

The next step in gene targeting using ES cells, whether you used homologous recombination or CRISPR–Cas gene editing, is to get the mutation into live mice through the production of chimeras that will pass the mutation through their germline. Generating mouse chimeras with correctly targeted ES cells can be done either by injection into blastocysts or by aggregation with or injection into morulae (see Chapter 12 in Behringer et al. 2014). Ideally, the ES cells will make a large contribution to the chimera that will be detectable by a high contribution to the coat color phenotype (i.e., high-level chimeras), provided you made use of a suitable coat color marker in your chimeras. If this is the case and you have obtained high-level chimeras with your targeted ES cells, you can skip the rest of this chapter and go on to Chapter 5: Recovering a Targeted Mutation in Mice from Embryonic Stem Cell Chimeras or CRISPR–Cas Founders (Papaioannou and Behringer 2023a). Here, we provide some pointers in planning the composition of your chimeras to take advantage of coat color markers, and in subsequent sections we troubleshoot possible reasons for a failure to obtain high-level chimeras.

Planning the Composition of Chimeras

The particular strain and substrain of origin of the ES cells you are working with will determine the most suitable host embryo strain to use in making chimeras. The best options will incorporate a coat color genotype that will allow visual identification of the contribution of ES cells to the resulting chimeras so that high-level chimeras can be easily identified and selected for subsequent recovery of the targeted mutation. However, you will also need to consider how you are going to breed the chimeras once they are identified for the next step of recovery of the mutation (see Breeding Schemes for ES Cell Chimeras in Chapter 5: Recovering a Targeted Mutation in Mice from Embryonic Stem Cell Chimeras or CRISPR–Cas Founders [Papaioannou and Behringer 2023a]). Another practical consideration is the cost and ease of obtaining mice to provide the host embryos, so a readily available commercial strain is best. Remember that this strain is simply a temporary host that will deliver the targeted mutation into live mice, without making a permanent genetic contribution to the resulting mutant strain.

Historically, ES cells derived from 129 strain mice were the most commonly used for gene targeting because of the ease of obtaining the ES cells from embryos of this strain (Robertson 1987) and because of their propensity to colonize the germline in chimeric hosts. In terms of genetic purity, however, the 129 strain could hardly be a worse choice as there are a number of different substrains with significant genetic differences. Nonetheless, if your starting point in gene targeting was a 129-derived ES cell line, it is important to know what substrain you are working with. The 129 strain originated in 1928 from crosses of several fancy coat color mice. From these crosses, a variety of 129 substrains were developed either intentionally or simply by random genetic drift following separation of breeding colonies, and various contaminating crosses were made either intentionally or inadvertently. The result today is a range of 129 substrains differing or segregating for a variety of alleles, including coat color genes (Table 1). All of these, however, carry the dominant *white bellied agouti* allele (A^w), which simplifies the choice of embryo host for making chimeras as the chimeras made with any *non agouti* (*a/a*) strain, such as C57BL/6J or C57BL/6N, will be easily identified by agouti coat color patches (Fig. 1A).

In recent years, technical advances have made it possible to obtain germline-competent ES cells from many different strains, any of which might be used for gene targeting (van der Weyden et al. 2011). Of particular note are ES cells derived from the C57BL/6N substrain, which have been used in the high-throughput gene-targeting projects of the International Knockout Mouse Consortium (IKMC; https://www.mousephenotype.org/about-impc/about-ikmc/). If C57BL/6N-derived ES cells were used for gene targeting, there are a number of options for host blastocysts depending

TABLE 1. Derivation of some common 129-embryonic stem (ES) cell lines with the old and new 129 substrain nomenclature[a] and the relevant coat color alleles these ES cell lines carry

| ES cell line | Substrain of embryo of origin | | Relevant coat color alleles in ES cells |
	Old nomenclature	New nomenclature	
AB1	129/SvEvBrd	129S5	A^w
CCE	129/SvEv	129S6	A^w
E14TG2a[b]	129/OlaHsd	129P2	A^w; Tyr^{c-ch}; p
J1	129/SvJae	129S4	A^w
R1	129/Sv × 129/J	129S1 × 129X1	A^w; Tyr^{c-ch}; Tyr^+; p; p^+

[a]http://www.informatics.jax.org/mgihome/nomen/strain_129.shtml
[b]The ES cell–derived component of chimeras made with this cell line will be very pale colored.

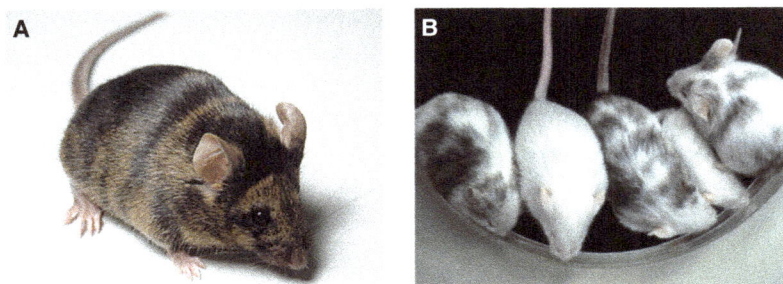

FIGURE 1. Examples of chimera coat colors. (*A*) The case of a 129-derived embryonic stem (ES) cell line combined with C57BL/6J (B6) blastocysts. The chimera has patches of agouti and black hairs by virtue of the *white-bellied agouti* (A^W/A^W) genotype of the ES cells and the *nonagouti* genotype (*a/a*) of the B6 embryos. This animal is somewhat >50% agouti and would be a reasonable candidate for test breeding. (*B*) Chimeras made with 129-ES cells and albino, non-agouti blastocysts (*Tyr^c/Tyr^c*; *a/a*) have white, black, and agouti patches. The white patches are areas populated by blastocyst-derived albino melanocytes, the agouti patches have 129-derived melanocytes and hair follicles (the site of action of the *agouti* gene), and the black patches have 129-derived melanocytes but blastocyst-derived (*a/a*) follicles. These animals range from ~5% to 50% chimerism.

on whether the ES cells have recessive *agouti* alleles (*a/a*, e.g., ES cell line JM8) or whether they contain a dominant *agouti* allele (*A/a*; e.g., ES cell line JM8A3 or JM8A1.N3) (Pettitt et al. 2009). Either will show up as pigmented patches if an albino host is chosen (Fig. 1B). To take advantage of the tried and true (and cost-effective) C57BL/6J strain as a host embryo, the chimeras will be evident by agouti patches only if the ES cell line carries a dominant *agouti* allele. Whatever strain is chosen for the host embryo, be sure to plan ahead and anticipate how you will recognize not just the chimeras, but also how you will test breed the chimeras for detection of germline transmission (see What Strain to Breed Chimeras with? in Chapter 5: Recovering a Targeted Mutation in Mice from Embryonic Stem Cell Chimeras or CRISPR–Cas Founders [Papaioannou and Behringer 2023a]).

No Mice Born Following Transfer of Chimeric Embryos

Technique, Technique

Practice control embryo transfers until you have a good success rate (see Chapter 6 in Behringer et al. 2014). You should be able to achieve an 80%–90% success rate in embryo recovery after embryo transfer before you start with important experimental embryos. Otherwise, you are wasting your time. Try to reduce surgical trauma and, in particular, treat the uterus gently, because rough treatment can cause a pregnancy to fail. Make sure your embryo culture medium is correct (see Chapter 4 in Behringer et al. 2014). In addition, be sure that your pseudopregnant embryo transfer recipients are really pseudopregnant (male mice have been known to plug females that are not in estrus) by checking that they have corpora lutea (CL) in the ovary at the time that you do the embryo transfer (Protocol 4.1: Counting Murine Corpora Lutea to Determine the Number of Oocytes Ovulated [Papaioannou and Behringer 2023e]). If not, you are essentially throwing the embryos away, so use another recipient. Finally, check the light cycle control in your animal room to ensure that it is turning on and off correctly (14 h light/10 h dark is good, but 12/12 is fine, too), as erratic lighting disturbs the reproductive cycle.

Maternal Issues

It is possible, especially if you saw that the recipients were pregnant (i.e., they looked bulgy around the middle 10–12 d after embryo transfer), that pups were born but cannibalized before you had the chance to see them. Be aware of the date that the chimeras are due to be born and search the bedding material for telltale remains on that date (if you find any, even small remains can be genotyped to determine whether the newborns were chimeric). Sacrifice the surrogate female and examine the uterus shortly after the expected due date for signs that she recently gave birth. There will be red parturition sites on the mesometrial side of the flaccid, distended uterus if she gave birth.

If cannibalism seems to be the problem, there are several ways to improve this inadequate maternal behavior.

1. Minimize crowding during late pregnancy. Keep only one or two pregnant mice in the same cage (however, note that local institutional rules may dictate the number of animals per cage).

2. Avoid bedding or cage changes within 1–2 d of the expected birth date, but be sure that the cage is relatively clean. This means that you may have to put a "do not disturb" sign on the cage and take the responsibility of keeping it clean, fed, and watered rather than relying on the schedule of animal husbandry staff. Provide nesting material a few days in advance of the due date. You can use commercial nesting materials or you can place a tissue or paper towel in the cage.

3. No matter how anxious you are to see the babies, do not disturb the mother during or shortly after birth. If you observe the pups strewn around the cage and/or mutilated, then and only then is the time for drastic intervention in the form of cross-fostering onto a lactating female (Protocol 4.2: Cross-Fostering of Newborn Mice to Counteract Bad Parenting [Papaioannou and Behringer 2023f]). If cannibalization is a persistent problem, a last resort is to preempt the birth by delivering the pups by Caesarean section (see Chapter 6 in Behringer et al. 2014) a day before the expected birth date (E18.5) and cross-fostering onto a lactating female.

Phenotype

It is just possible that the mutant ES cells have caused the death of all chimeric embryos if the mutation is dominant (or is X or Y linked) and has a very detrimental effect (e.g., *Gata1*; Pevny et al. 1991). Looking at the uteri of the foster mothers shortly after the expected birth date will indicate whether any postimplantation embryonic development took place, because the implantation sites or resorption sites (places at which embryos implanted but then died and were partially resorbed by the uterus) should still be visible as swellings or blood spots on the uterus.

If you suspect that the chimeras are dying, either before or after implantation, try including in the embryo transfer a few wild-type embryos of a distinct coat color along with the chimeric embryos to determine if death is specific to the chimeras. If all of the embryos fail, it is more likely to be the result of a more general problem as detailed in the previous sections. If nonchimeric embryos develop but the ES cell chimeras do not, you might just have a dominant effect. Go to Chapter 16: Phenotypic Analysis of Dominant Mutant Effects in Mice (Papaioannou and Behringer 2023c).

Pups Are Born but None Are Chimeric or the Level of Chimerism Is Low

Technique

Were the ES cells kept under optimal conditions during the injection experiment? Trypsinized cells should be held on ice in buffered medium in a capped test tube. Some of these trypsinized cells are transferred into the injection or aggregation dish for chimera production. pH and osmolarity must be correctly maintained in the injection or aggregation dish. Watch the color of the medium and use a sufficient volume of medium in the dish to reduce concerns of evaporation. You can check if this manipulation is compatible with ES cell viability by culturing the "leftover" ES cells after a blastocyst injection or morula aggregation experiment under standard ES cell culture conditions to see if they still grow well.

Are you injecting 12 to 15 ES cells, which seems to be the optimal number for most ES cell lines? Are you confident that your blastocyst injection or morula aggregation procedure is effective and that the ES cells are incorporated into the embryos before transfer to the foster mother? If you think that the lack of incorporation of ES cells might be a problem, you could try injecting or aggregating some marked ES cells (e.g., cells with a *lacZ* or a green fluorescent protein reporter gene); culturing the chimeric embryos overnight; and then looking for the cell marker as an indication of ES cell contribution. If marked cells are not present, it is a good idea to reevaluate your injection or aggregation technique as well as your ES cell–handling technique, because the cells might have died or they might never have been incorporated. If the cells are present in good numbers and appear to be incorporated into the embryo, continue working through the other possibilities.

The Targeted ES Cell Clone Has Lost Developmental Potential

It is a good idea to always check that your parental ES cell line is competent to make chimeras and to contribute to the germline before using it for gene targeting. You can do this yourself by making chimeras with the unmanipulated parental ES cell line or you can depend on information supplied by the person or company from whom you received the ES cells in the first place. If you are sure about the competence of your parental ES cells, try another targeted clone. You should have more than one if you followed our recommendations above. If not, go back and isolate more targeted ES cell clones.

One common reason for a restriction in germline potential is that the ES cells have become aneuploid. If this has happened, a subpopulation of euploid cells might still be in the culture and if the clone is a valuable one (i.e., you have only one targeted clone), it might be worth trying to isolate the euploid cells by subcloning and karyotyping cells (see Chapter 8 in Behringer et al. 2014) and then making chimeras with euploid subclones.

Incompatible Strain Combination

It may be that the host embryo strain is not conducive to making chimeras with your ES cell strain. Search the literature to discover what strains have been successfully used with your particular ES cells. If you are using a compatible combination, try again with larger numbers. If not, consider changing to a different strain of mouse for the host embryos.

Possible Coat Color Phenotype

If offspring are all nonchimeric by coat color, but the sex ratio is skewed heavily in favor of males, consider that the mutation you made may have affected the coat color. Some of the offspring may actually be chimeras, hence the skewed sex ratio, but the coat color is not as expected because of the heterozygous mutation—that is, they are cryptic chimeras. This is fairly unlikely, but it has been known to happen (e.g., *Pdgfra*; see Soriano 1997). The phenotypic sex of XX ⟷ XY chimeras will be skewed in favor of males, assuming that you are using an XY ES cell, because of the nature of sex determination. Genotype some of the offspring to be sure that they are not cryptic chimeras. If they are, you will have an interesting dominant phenotype. Go to Chapter 16: Phenotypic Analysis of Dominant Mutant Effects in Mice (Papaioannou and Behringer 2023c). Otherwise, make more chimeras with a different clone.

Could This Be a Phenotype?

Consider the possibility that cells heterozygous for the mutation are incapable of contributing to chimeras because of haploinsufficiency for the gene product, if you made a null mutation, or because of a defective gene product, if your mutation results in a neomorphic or antimorphic protein. If the targeted ES cells grew well, the defect might only become evident as the ES cells differentiate (or try to) in the embryo in vivo. Go to Chapter 16: Phenotypic Analysis of Dominant Mutant Effects in Mice (Papaioannou and Behringer 2023c). Keep in mind that if the gene is X or Y linked, the mutation will affect the only copy of the gene in an XY ES cell line.

REFERENCES

Behringer RR, Gertsenstein M, Nagy KV, Nagy A. 2014. *Manipulating the mouse embryo: a laboratory manual*. Cold Spring Harbor Laboratory Press, Cold Spring Harbor, NY.

Joyner AL. 1993. *Gene targeting: a practical approach*. IRL Press, Oxford, New York, Tokyo.

Papaioannou VE, Behringer RR. 2023a. Recovering a targeted mutation in mice from embryonic stem cell chimeras or CRISPR–Cas founders. *Cold Spring Harb Protoc* doi:10.1101/pdb.over107959

Papaioannou VE, Behringer RR. 2023b. Obtaining or generating gene mutations in mice. *Cold Spring Harb Protoc* doi:10.1011/pdb.over107956

Papaioannou VE, Behringer RR. 2023c. Phenotypic analysis of dominant mutant effects in mice. *Cold Spring Harb Protoc* doi:10.1101/pdb.over107978

Papaioannou VE, Behringer RR. 2023d. Mouse gene-targeting strategies for maximum ease and versatility. *Cold Spring Harb Protoc* doi:10.1101/pdb.over107957

Papaioannou VE, Behringer RR. 2023e. Counting murine corpora lutea to determine the number of oocytes ovulated. *Cold Spring Harb Protoc* doi:10.1101/pdb.prot107986

Papaioannou VE, Behringer RR. 2023f. Cross-fostering of newborn mice to counteract bad parentage. *Cold Spring Harb Protoc* doi:10.1101/pdb.prot107987

Perry J, Palmer S, Gabriel A, Ashworth A. 2001. A short pseudoautosomal region in laboratory mice. *Genome Res* 11: 1826–1832. doi:10.1101/gr.203001

Pettitt SJ, Liang Q, Rairdan XY, Moran JL, Prosser HM, Beier DR, Lloyd KC, Bradley A, Skarnes WC. 2009. Agouti C57BL/6N embryonic stem cells for mouse genetic resources. *Nat Methods* 6: 493–495. doi:10.1038/nmeth.1342

Pevny L, Simon MC, Robertson E, Klein WH, Tsai SF, D'Agati V, Orkin SH, Costantini F. 1991. Erythroid differentiation in chimaeric mice blocked by a targeted mutation in the gene for transcription factor GATA-1. *Nature* 349: 257–260. doi:10.1038/349257a0

Robertson EJ. 1987. Embryo-derived stem cell lines. In *Teratocarcinomas and embryonic stem cells: a practical approach* (ed. Robertson EJ), pp. 71–112. IRL Press, Oxford, Washington, DC.

Soriano P. 1997. The PDGF alpha receptor is required for neural crest cell development and for normal patterning of the somites. *Development* 124: 2691–2700. doi:10.1242/dev.124.14.2691

van der Weyden L, White JK, Adams DJ, Logan DW. 2011. The mouse genetics toolkit: revealing function and mechanism. *Genome Biology* 12: 224. doi:10.1186/gb-2011-12-6-224

Protocol 4.1

Counting Murine Corpora Lutea to Determine the Number of Oocytes Ovulated

One can determine if and how many oocytes have been ovulated in a female mouse by counting the number of corpora lutea (CL) on the ovaries during the process of preimplantation embryo collection. A simple method of harvesting the ovaries and observing with a dissecting microscope and top lighting is provided along with a description of how to recognize CL. As the embryos rarely, if ever, cross the uterotubal junction, this provides a measure of the maximum number of embryos expected to be recovered from each oviduct or uterine horn, a number that can be valuable in analyzing early lethal mutations.

INTRODUCTION

Corpora lutea (CL) are the visible and transient signs of ovulation. At the sites of extrusion of oocytes from the surface of the ovary, the ovarian follicles form endocrine tissues known first as corpora hemorrhagica and later as CL. These structures are essential for the maintenance of pregnancy because they are a major source of progesterone. They are also useful diagnostic aids for two specific experimental situations in mice. One is during embryo transfer to pseudopregnant recipients. If there are no CL present on either ovary upon exposure of the reproductive tract before embryo transfer, the recipient did not ovulate and is therefore not pseudopregnant; she is not a suitable host for embryo transfer. The second situation is in the diagnosis of preimplantation embryonic phenotypes, when the CL can be taken as a measure of the maximum number of embryos that could be expected, provided all ovulated oocytes were fertilized. This protocol details the counting of CL during morula or blastocyst recovery from the oviducts or uterus, which can aid in determining whether there is early preimplantation embryonic lethality.

MATERIALS

It is essential that you consult the appropriate Material Data Safety sheets and your institution's Environmental Health and Safety Office for proper handling of equipment and hazardous materials used in this protocol.

Reagents

Embryo culture medium (Chapter 4, Protocol 1 in Behringer et al. 2014)
Phosphate buffered saline (PBS)
Pregnant females at embryonic day (E) 0.5 to E3.5

Cite this protocol as *Cold Spring Harb Protoc*; doi:10.1101/pdb.prot107986

Equipment

Dissection instruments
Dissecting microscope with top and transmitted light
Petri dishes

METHOD

1. Sacrifice the pregnant female and remove the whole reproductive tract including the ovaries, keeping track of right and left sides.

2. Separate the ovaries from the oviducts and surrounding bursa and place them in PBS in separate drops or Petri dishes, marked right and left, so that you can eventually correlate the CL counts with embryo counts from each oviduct or uterine horn.

3. Flush the embryos separately from the right and left oviducts or uterine horns (depending on the age of the embryos) into fresh drops of culture medium in the usual manner (see Chapter 4, Protocol 1 in Behringer et al. 2014) but without counting the embryos, keeping right- and left-isolated embryos separate.

   It is important to do the CL counts blind for unbiased data. At this point, it might be useful to have another investigator deal with the embryos while you do the CL counts.

4. Under a dissecting microscope with mostly top lighting, look for uniform, raised, pink hemispheres (∼0.75 mm in diameter) on the surface of each ovary. Turn the ovary over to ensure that you find them all.

   The CL are generally rosy pink shortly after ovulation compared to the whiter background ovarian tissue (Fig. 1). If you use a little transmitted light as well, they appear denser and more uniform compared to the relatively translucent remainder of the ovary.

5. Correlate the number of CL with the number of embryos recovered from each side of the reproductive tract.

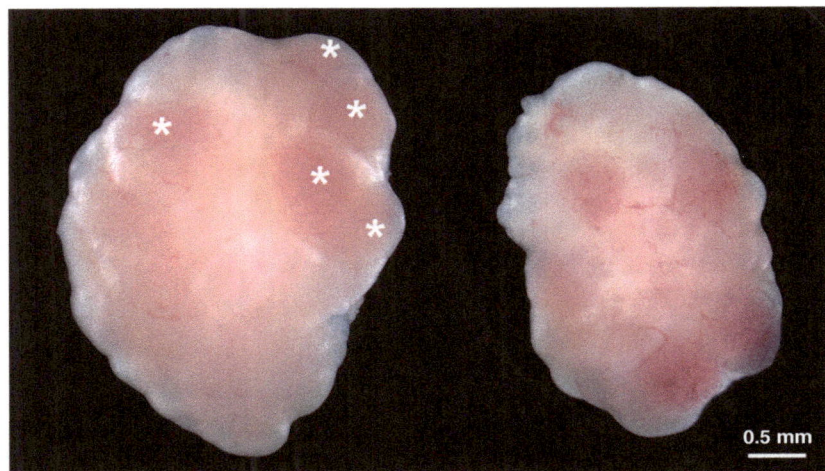

FIGURE 1. Two ovaries from an E2.5 female. Five CL are on the ovary on the *left* (marked with asterisks) and five are on the ovary on the *right* that are not marked. Can you find them?

DISCUSSION

CL from a previous estrous cycle may be present in ovaries, because the CL begin to degenerate after ~10 d in the absence of fertilization. However, these "old" CL will not be anywhere near as prominent as the CL from the recent ovulation cycle. Be on the lookout for ovulated but unfertilized ova, as these will also be represented by CL on the ovaries.

REFERENCES

Behringer RR, Gertsenstein M, Nagy KV, Nagy A. 2014. *Manipulating the mouse embryo: a laboratory manual.* Cold Spring Harbor Laboratory Press, Cold Spring Harbor, NY.

Cross-Fostering of Newborn Mice to Counteract Bad Parenting

Newborn mice can be cross-fostered to a lactating female if the birth mother is not taking care of them properly. The pups are removed from the birth mother, warmed, and placed in a clean cage with a recently postpartum, lactating female. Hints on improving the chances of successful fostering are provided.

INTRODUCTION

Sometimes the maternal instinct in mice is just not what it should be, especially in first-time mothers. A number of factors, such as overcrowding, disturbance, an alteration in the light cycle, temperature, or the introduction of a strange mouse—male or female—around the time of parturition can lead to maternal neglect of offspring, or worse, cannibalization. If you catch the mother in the act, the pups might be scattered around the cage, cold to the touch, and possibly have bite marks. If the pups are particularly valuable, it is worth trying to cross-foster them onto a lactating female.

MATERIALS

It is essential that you consult the appropriate Material Data Safety sheets and your institution's Environmental Health and Safety Office for proper handling of equipment and hazardous materials used in this protocol.

Reagents

Lactating female, 1- to 2-d-postpartum, with litter (in her home cage)

Equipment

Clean mouse cage with fresh bedding
Holding cage
Slide warmer or warming pad (e.g., Deltaphase Isothermal Pad, Braintree Scientific, Inc.)

METHOD

1. Remove the neglected or abandoned pups you wish to foster from their cage and proceed with the following steps.

   *This method is similar to that used following the cesarean delivery of offspring (see Protocol: **Caesarean Section and Fostering** [Nagy et al. 2006] and Behringer et al. 2014).*

2. Warm the pups by placing them in a Petri dish on a warming pad or slide warmer set at 37°C or hold them in your gloved hand until they are warm.

   Do not use a lamp, as this is likely to overheat and dehydrate the pups.

3. Locate a lactating female who has given birth within the past day or two and is successfully nursing her young. She can be any strain, but determine how you will distinguish her pups from the foster pups later, possibly by incorporating a coat color marker (see Box 1. Marking Newborn Pups in Chapter 14: Analysis of Postnatal Mutant Phenotypes in Mice [Papaioannou and Behringer 2023]). If the foster mother's litter is large, decrease the size by removing as many of her pups as you plan to foster. Remove the largest ones to give the foster pups a competitive chance.

4. Remove the foster mother to a holding cage while you try to fool her into thinking that these are all her babies. To do this, take both sets of pups and some soiled bedding from the foster's home cage and mix well. The pups should be at the same body temperature. Just for good measure, we suggest adding a few drops of the foster mother's urine to the mix (simply hand restrain her above the babies and let it drip).

   We know of one investigator who briefly anesthetizes the foster mother supposedly to induce amnesia, but we have not found this to be necessary.

5. When the pups are active and warm and smelling of the foster family's home cage, place them in the nest and reintroduce the foster mother. You will know immediately if she is going to accept the babies. If not, they will be dumped outside the nest in rapid order.

REFERENCES

Behringer RR, Gertsenstein M, Nagy KV, Nagy A. 2014. *Manipulating the mouse embryo: a laboratory manual.* Cold Spring Harbor Laboratory Press, Cold Spring Harbor, NY.

Nagy A, Gertsenstein M, Vintersten K, Behringer R. 2006. Caesarean section and fostering. *Cold Spring Harb Protoc* doi:10.1101/pdb.prot4381

Papaioannou VE, Behringer RB. 2023. Analysis of postnatal mutant phenotypes. *Cold Spring Harb Protoc* doi:10.1101/pdb.over107976

Cite this protocol as *Cold Spring Harb Protoc*; doi:10.1101/pdb.prot107987

CHAPTER 5

Recovering a Targeted Mutation in Mice from Embryonic Stem Cell Chimeras or CRISPR–Cas Founders

Following the production of chimeras from targeted embryonic stem (ES) cells or obtaining founders from CRISPR–Cas gene editing in preimplantation embryos, the desired targeted mutation must be recovered and established in the heterozygous state in a strain or stock of mice for further study. The breeding schemes for ES chimeras and CRISPR–Cas founders differ. For ES cell chimeras, we discuss the relative benefits of breeding from male or female chimeras. We discuss the importance of genetic background and provide practical advice for putting the mutation on inbred or outbred backgrounds or producing a coisogenic strain. For CRISPR–Cas founders, which will most likely be mosaic for different mutations, initial breeding strategies are discussed to maintain a desired genetic background at the same time as producing progeny to segregate different alleles. Strategies for testing the progeny to recognize indels, missense mutations, and knock-in mutations are discussed. In the event that ES cell chimeras or CRISPR–Cas founders produce no offspring or fail to transmit the mutant allele(s), there is a trouble-shooting guide to pinpoint the problem. If heterozygous offspring from the chimeras or founders are normal, fertile, and of both sexes, the analysis of homozygous effects of the mutation can now begin; if not, possible dominant effects are considered.

INTRODUCTION

A primary goal for generating mutations in the mouse is to produce animals that can transmit the mutation to their progeny (i.e., germline transmission). In chimeras, embryonic stem (ES) cells must differentiate into functional germ cells carrying the mutation for this goal to be reached. Similarly, CRISPR–Cas-manipulated preimplantation embryos must also differentiate mutation-carrying gametes. Here we discuss different breeding schemes for ES cell chimeras and CRISPR–Cas founders. From these matings, the most straightforward situation is if male and female offspring heterozygous for the targeted allele are recovered from chimeras or CRISPR–Cas founders and can breed to generate homozygous mutant mice. However, this is not always the case and there are a number of possible reasons for this, which are detailed in this chapter, with an indication of how to proceed depending on which experimental situation you encounter.

Opening artwork: Courtesy of Shuo-Ting Yen and Richard Behringer.

© 2024 Cold Spring Harbor Laboratory Press
Cite this chapter as *Cold Spring Harb Protoc*; doi:10.1101/pdb.over107959

RECOVER MUTANT ALLELE FROM CHIMERAS OR CRISPR–Cas FOUNDERS

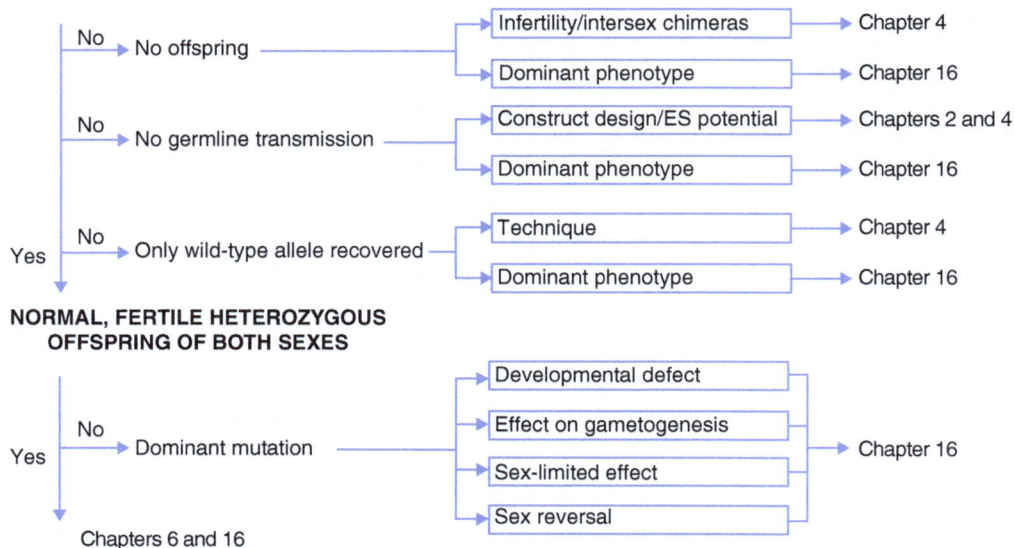

```
          No      No offspring ─────────────┬──► Infertility/intersex chimeras ──► Chapter 4
                                            └──► Dominant phenotype ──────────────► Chapter 16

          No      No germline transmission ─┬──► Construct design/ES potential ───► Chapters 2 and 4
                                            └──► Dominant phenotype ──────────────► Chapter 16

  Yes     No      Only wild-type allele recovered ─┬──► Technique ─────────────────► Chapter 4
                                                   └──► Dominant phenotype ─────────► Chapter 16
```

NORMAL, FERTILE HETEROZYGOUS OFFSPRING OF BOTH SEXES

```
          No      Dominant mutation ────────┬──► Developmental defect
                                            ├──► Effect on gametogenesis
  Yes                                       ├──► Sex-limited effect        ──► Chapter 16
                                            └──► Sex reversal

          Chapters 6 and 16
```

BREEDING SCHEMES FOR ES CELL CHIMERAS

Breed Male or Female Chimeras?

The object of test-breeding ES cell chimeras is to identify the chimeras that transmit the ES cell genotype with the targeted mutation through their germline (so-called germline chimeras or transmitters) and to produce as many ES cell-derived offspring for experiments as possible in the shortest time with the least amount of work. The first decision is whether to breed female and male chimeras, or only the males. Because almost all ES cell lines used for gene targeting are male-derived (XY), the sex ratio of chimeras produced will be skewed in favor of males because XX ⟷ XY chimeras tend to develop as males. The infrequent female chimeras do not generally transmit the ES cell genotype because XY cells rarely, if ever, complete meiosis in an ovary and it is thus not worth test breeding them. However, if you observe that the sex ratio of chimeras is not skewed and that there are an equal number of female chimeras with a high proportion of ES cell–derived coat color (high-level chimeras), it may be worth test breeding them as it might indicate that the Y chromosome has been lost from the ES cells, and in mice, XO cells make functional oocytes.

Breed Chimeras with Which Strain?

The second decision is which strain of mice to use for test breeding the chimeras for germline transmission. There are several options, depending on the strain of origin of the ES cells you used for gene targeting, the strain of the host blastocyst, and whether you want the mutation on an inbred or outbred background. Historically, the first commonly used ES cell lines were of 129 strain origin and the usual scheme was to combine a 129-derived ES cell line (A^w/A^w, homozygous *white bellied agouti* at the agouti locus) with C57BL/6J embryos (*a/a*, homozygous *nonagouti* at the agouti locus), which allowed the easy detection of chimeras based on the appearance of agouti hairs in the chimeras' coats (see Fig. 1 in Chapter 4: Embryonic Stem Cell Gene Targeting and Chimera Production in Mice [Papaioannou and Behringer 2023a]).

Cite this chapter as *Cold Spring Harb Protoc*; doi:10.1101/pdb.over107959

However, with the recognition that C57BL/6J is the best-characterized inbred strain and is the reference strain for mouse genome sequencing, the International Knockout Mouse Consortium (IKMC) made use of newly developed C57BL/6N substrain ES cell lines for most of their high-throughput gene targeting. These ES cells have high germline potential, a characteristic previously found only in 129-derived ES cells. One subline was engineered to repair one allele of the *agouti* gene restoring the agouti coat color to aid in identifying germline chimeras and germline transmission. Thus, the choice of host blastocyst and chimera test breeding strain will depend on which ES cells you started with, so be careful to determine the strain and coat color genotype of the ES cells whether you are producing the mutation yourself or obtaining ES cells from IKMC (see Chapter 2: Obtaining or Generating Gene Mutations in Mice [Papaioannou and Behringer 2023b]). Following are some options and examples for breeding chimeras with different strain combinations to achieve a particular outcome.

Test Breed Chimeras to Produce an F_1 between Two Inbred Strains

If you used 129-derived ES cells and C57BL/6J blastocysts, you can take advantage of the coat color difference between the strains by test breeding with C57BL/6J mice: All pure black (non-agouti) offspring are derived from the host blastocyst C57BL/6J component of the chimera and all agouti offspring (Fig. 1A) are from the ES cell component (*agouti* is dominant over *nonagouti*). It is only necessary to genotype the agouti offspring; they are distinguishable from the black ones by 7 d of age, when the yellowish, agouti hairs begin to show at the back of the neck. Half of the agouti offspring should be heterozygotes for the targeted mutation, and as a bonus, they will all be otherwise genetically uniform, robust (C57BL/6Jx129) F_1 mice, although, of course, further breeding will produce F_2 mice segregating alleles for every locus at which the two strains differ (Fig. 1A). The F_1 mice could also be repeatedly backcrossed to C57BL/6J to produce a congenic strain, for example, genetically identical apart from a differential chromosomal region around the targeted locus (see Protocol 6.1: Backcrossing to Generate a Congenic Mouse Strain [Papaioannou and Behringer 2023c]). An advantage of this breeding scheme is that the black offspring of the test mating, which are all pure C57BL/6J, can be raised and recycled for use in test breeding and/or as blastocyst donors in additional chimera experiments.

Similarly, if you used C57BL/6N-derived ES cells, breeding the chimeras with another inbred strain will result in F_1 mice between the two strains. Depending on the coat color genotype of the ES cells and the host embryo strain used to make the chimeras, you may (Fig. 1B) or may not be able to distinguish all of the ES-derived from host-derived offspring by coat color (Pettitt et al. 2009).

Test Breed Chimeras to Put the Mutation on an Outbred Background

The biggest advantage of this scheme is the increased fecundity associated with outbred mice. The main disadvantage is that your mutation will be on an outbred genetic background, usually not the first choice for genetic studies. Nevertheless, if you use this scheme to obtain larger litters and more test offspring, you can always switch to another breeding scheme once you have identified the germline transmitting chimeras if you want to have the mutation on a defined genetic background.

If you used 129 ES cells, pick an outbred strain, albino or not, that is known to be homozygous *nonagouti* (Black Swiss is one possibility) to distinguish ES cell–derived offspring by their agouti coat. If you use an albino outbred mouse, also be aware that some 129 ES cell lines (notably R1; see Table 1 in Chapter 4: Embryonic Stem Cell Gene Targeting and Chimera Production in Mice [Papaioannou and Behringer 2023a]) segregate for albino alleles that might give you white or dilute coat colors—but at least you know they are derived from the ES cell component of the chimera.

If you used C57BL/6N ES cells, which generally show high levels of chimerism and germline transmission, rather than using this breeding scheme, it is probably advantageous to produce a coisogenic strain first (see next section) and then outcross if you want the mutation on a hybrid or an outbred background for any reason.

A

B

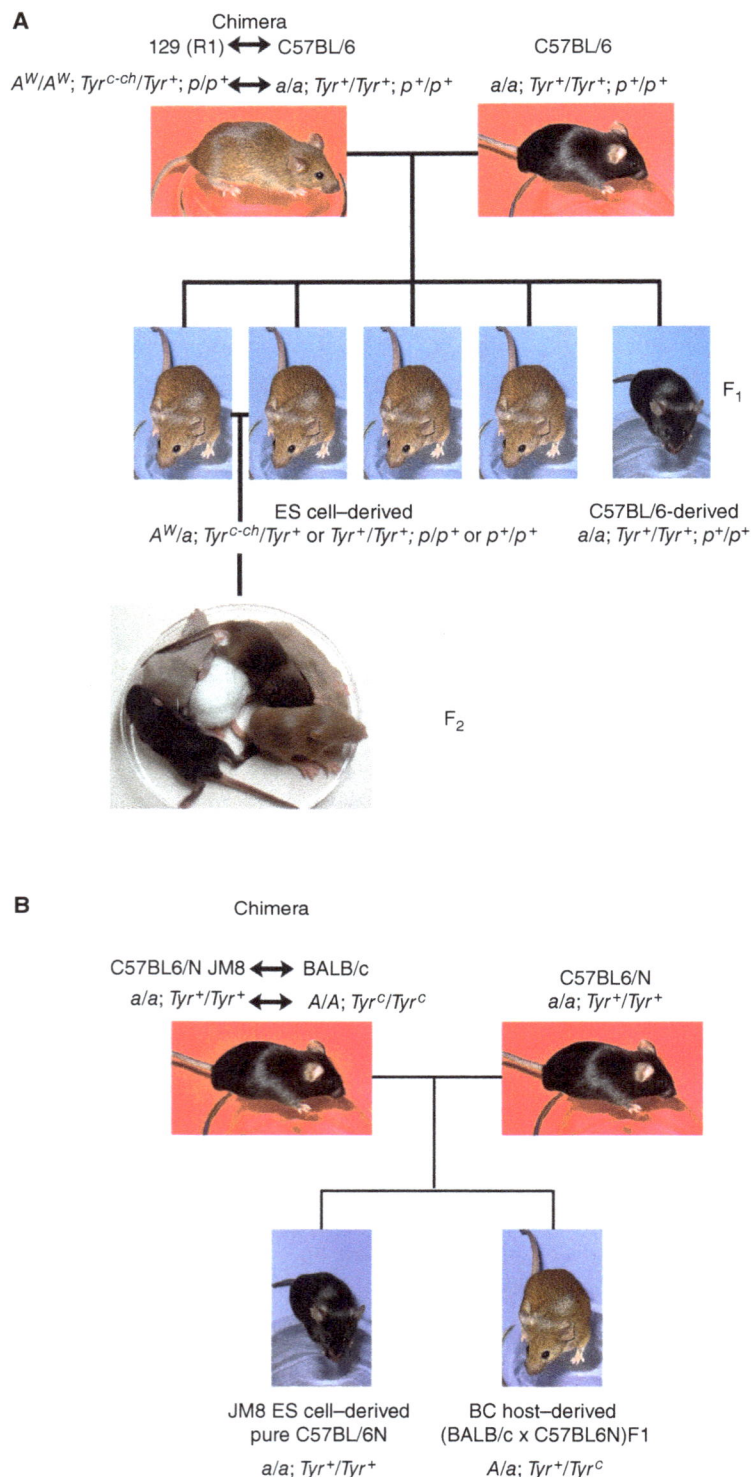

FIGURE 1. Representative examples of coat colors obtained in chimera test-breeding programs with different embryonic stem (ES) cells and host blastocysts. (*A*) In this example, a near-100% chimera made with R1 129-ES cells (indistinguishable from a wild-type agouti mouse) and a C57BL/6J (B6) blastocyst is crossed with a B6 mouse. The F$_1$ offspring are either agouti or black, depending on whether they come from the ES cell–derived or B6-derived germ cells of the chimera, respectively. When heterozygous ES cell–derived offspring are intercrossed, segregating recessive alleles at the linked *albino* (*Tyr*) and *pink-eyed dilution* (*p*) locus from the R1 ES cells, as well as the segregating *agouti* alleles, provide a spectrum of coat colors in the F$_2$ generation ranging from agouti and black through various shades of agouti or non-agouti dilute color. These coat color alleles will segregate independently from the mutant allele, unless, of course the mutant allele is linked to one of the coat color loci. In the case of ES cell lines, such as AB-1, which are not segregating for *albino* or *pink-eyed dilution* alleles, offspring of F$_1$ mice will be either black or agouti, independent of the mutant allele. (*B*) In this example, an apparently 100% chimera (no visible white hairs) made with JM8 C57BL/6N ES cells and a BALB/c blastocyst is crossed with a C57BL6/N mouse. The F$_1$ offspring are black or agouti, depending on whether they come from the ES cell–derived C57BL6/N or the blastocyst-derived BALB/c germ cells of the chimera, respectively. All of the black animals are pure C57BL/6N and half will have the mutant allele. (Mice pictured are coat color examples, not mice of the actual strains indicated.)

Cite this chapter as *Cold Spring Harb Protoc*; doi:10.1101/pdb.over107959

Test Breed Chimeras to Produce a Coisogenic Strain

The main advantage of this scheme is that the resulting ES cell–derived offspring will be coisogenic with the ES cell strain (i.e., differing only at the targeted locus). Thus, there is no need for lengthy repeated backcrossing to have the mutation on an inbred background, which is most often the goal in genetic studies. In the case of 129 ES cell ⟷ C57BL/6J chimeras the disadvantages are that 129 strain mice are not very good breeders, and that you are giving up the possibility of using a coat color marker to distinguish ES cell–derived from C57BL/6J-derived offspring of the chimera in the test cross. Only agouti offspring will be produced and therefore all will have to be genotyped. In this case, this breeding scheme is most useful for producing a 129 coisogenic strain once you have identified a chimera that is a high-level germline transmitter using the F_1 breeding scheme described above. Another caveat is that the substrain of 129 from which the ES cells were derived may not be readily available from commercial sources for test breeding (see Table 1 in Chapter 4: Embryonic Stem Cell Gene Targeting and Chimera Production in Mice [Papaioannou and Behringer 2023a]).

If your chimeras were made with C57BL/6N ES cells, test breeding to produce a coisogenic strain will be done with C57BL/6N mice (not C57BL/6J, which is a different substrain). However, depending on the strain and genotype of the host blastocyst used (see Chapter 4: Embryonic Stem Cell Gene Targeting and Chimera Production in Mice [Papaioannou and Behringer 2023a]) and whether the ES cells were *a/a* (ES cell lines Jm8, JM8.N4, or JM8.F6), or carried the repaired *agouti* allele and were *A/a* (ES cell lines JM8A3 or JM8A1.N3), the resulting black offspring may need to be genotyped to determine that they are pure C57BL/6N rather than host blastocyst–derived (Pettitt et al. 2009).

BREEDING SCHEME FOR CRISPR–Cas FOUNDERS

Most likely, you made your CRISPR–Cas allele in an inbred strain either for convenience or because of some particular characteristic of that strain. Thus, to take advantage of having the mutant allele coisogenic with the strain of the founder, test breeding should be done with the same inbred strain of mice, preferably from male founders for speed and efficiency, although female founders can also be used. Any genetic crosses you wish to make with other strains or mutations can then be performed once you have identified the intended, correctly targeted allele(s) and have established a breeding colony.

Once you have obtained offspring resulting from a CRISPR–Cas-induced genetic modification experiment, be it by zygote injection or electroporation, they will need to be screened to identify which carry the desired mutant allele. This is most efficiently performed by screening tail or ear punch DNA from each founder by polymerase chain reaction (PCR) amplification of the target region. CRISPR–Cas manipulations can result in founders containing multiple alleles (Yen et al. 2014). Thus, the analysis of the PCR products depends on the type of allele you intended to generate. In addition, CRISPR–Cas founders will likely be mosaics with both wild-type and mutant cells depending on the timing of gene editing relative to cleavage of the embryos (Yen et al. 2014). This should be kept in mind during test breeding. It is important to develop at least two independent pedigrees for the desired mutant allele(s) to show that they result in the same mutant phenotype.

Indel Mutations

A region surrounding the desired targeted mutation is amplified by PCR from tail or ear punch DNA of each potential founder. The band generated by PCR amplification is sequenced. Typically, the sequence will be clear until it reaches a region that reads the mutant allele(s). Then the sequence breaks down and is unreadable. It is also possible to perform the Surveyor Assay (Qiu et al. 2004). In this assay, the amplified DNA is denatured and mixed with denatured wild-type DNA of the same region. DNA hybrids are then treated with the CEL I nuclease that cuts at mismatched regions, resulting in smaller DNA fragments. The cut fragments are resolved on an agarose gel. Mice positive for indels at the target locus using either assay would then be bred with wild-type mice of the same strain to produce progeny and segregate individual mutant alleles (see Fig. 2 in Chapter 3: Mouse Gene-Targeting Strategies for Maximum Ease and Versatility [Papaioannou and Behringer 2023d]). The resulting progeny are screened by PCR. DNA sequencing of the DNA made should identify the precise mutation that was recovered. Desired mutant alleles are then maintained in separate pedigrees for subsequent phenotype analysis. Repeated crossing with wild-type mice for several more generations will segregate away most potential off-target mutations.

Missense Mutations

The strategy to generate a missense mutation by CRISPR–Cas includes a silent mutation that creates a novel restriction site (see Chapter 3: Mouse Gene-Targeting Strategies for Maximum Ease and Versatility [Papaioannou and Behringer 2023d]). The band generated by PCR of potential founder tail or ear punch DNA is digested with the restriction enzyme and resolved on an agarose gel. Cutting of the amplified band with the introduced restriction enzyme site identifies mice that may contain the intended allele. These founders are then bred with wild-type mice of the same strain to produce progeny and segregate mutant alleles. The resulting progeny are genotyped by PCR and DNA sequencing to identify mice carrying the intended missense mutation. These mice are used to establish separate pedigrees for phenotype analysis. If you did not incorporate a silent mutation to prevent recutting or create a novel restriction site, you will have to screen your founders as described for indels (above).

Knock-In Mutations

Tail or ear punch DNA of potential founders should be tested with a PCR genotyping strategy to identify mice carrying the targeted allele. Once identified, the founders are bred with wild-type mice of the same strain to produce progeny and segregate mutant alleles. The resulting progeny are then genotyped by PCR and DNA sequencing to identify mice carrying the intended knock-in mutation. These mice are used to establish separate pedigrees for phenotype analysis.

PROBLEMS RECOVERING TARGETED ALLELE(S) FROM CHIMERAS OR CRISPR–Cas FOUNDERS

Usually, the level of coat-color chimerism will reflect the level of germline transmission of the targeted allele in ES cell chimeras, and most CRISPR–Cas founders will pass on the alleles even if they are mosaic. However, there are some instances when this is not so straightforward. The following sections present explanations for some of the other possible outcomes.

ES Cell Chimeras or CRISPR–Cas Founders Produce No Offspring

Unexplained Infertility

This is a catchall phrase and could apply to either your chimeras, CRISPR–Cas founders, or the mice with which you mated them. Changing the test breeding females (selecting young, estrous females) (see Fig. 2 in Box 1, and Behringer et al. 2016) will provide the male chimera or CRISPR–Cas founder with a change of scenery, but do not spend too much time here. It is only if you have very few male chimeras or CRISPR–Cas founders that you should use breeding space on nonproductive test matings; it is much better to go back a step to generate additional chimeras or founders. After all, the goal is to have germline transmission of more than one ES cell clone or multiple CRISPR–Cas-generated alleles in your gene of interest.

BOX 1. SELECTING ESTROUS FEMALES

The estrous cycle in the mouse ranges from 3 to 4 d and ovulation occurs around the midpoint of the dark period of the light cycle. Mating also occurs around the middle of the dark period, but a lot of variation is possible. Many animal rooms are set to a 12 h light/12 h dark cycle, but a 14/10 light/dark cycle tends to make the time of ovulation and mating tighter and you might notice less variability in embryonic stages using this cycle. One way of getting timed pregnancies is simply to leave females with males continuously, but a more efficient method is to select estrous females and place only those in proestrus or metestrus with males (Behringer et al. 2016). In either case, check for vaginal copulation plugs in the morning, because they can fall out or dissolve during the course of the day. These plugs are white or cream colored, solidified components of the ejaculate (Fig. 2) and simply indicate that mating has taken place. Estrous females can be recognized by the appearance of the external vaginal epithelium by visual inspection. Simply lift the tail of the female and gently probe the vagina (the same way you would check for a vaginal plug the next morning), so as not to induce pseudopregnancy. You want to select females with two or more of the following vaginal characteristics: dry but not flaky; pink, not red, white, or blue; swollen, so that the tissue bulges outward; and/or epithelium that folds into corrugations on upper and lower vaginal lips. Alternatively, but less conveniently, the stage of the estrous cycle can be determined by vaginal smear (Gonzalez 2016).

FIGURE 2. Appearance of a vaginal plug on the morning after copulation. The ejaculate from the male forms a solid plug in the vagina that can remain there for 8–24 h. A probe, such as the dental spatula shown here, can be used to gently probe the vagina to detect the plug. (Reprinted, with permission of the Licensor through PLSclear, from Papaioannou and Johnson, 2000. *Gene targeting: a practical approach* [ed. Joyner AL], pp. 133–175, Oxford University Press, Oxford.)

Intersex Chimeras

For ES cell chimeras, a common cause of infertility is that the chimera is neither male nor female (i.e., an intersex). Most ES cells in common usage are XY (i.e., they were originally derived from male embryos), and you have a 50:50 chance that any injected blastocyst will be a female (XX). Usually, XX ⟷ XY chimeras develop as phenotypic males and fortuitously the biology is on the side of the investigator because only the XY (ES-derived) cells make functional sperm. However, possibly because of the level of contribution of the XY ES cells, XX ⟷ XY chimeras sometimes have both male and female reproductive organs or ovotestes and consequently they are infertile. Examination of the gonads will tell you whether this is a possibility; intersex gonads have areas of ovarian and testicular tissue and the uterus, vas deferens, and/or external genitalia may be abnormal. Before sacrificing the chimera, you could do a laparotomy using anesthesia if you wanted to be sure that this is the cause of infertility. Even if an intersex produces copulation plugs, it is highly unlikely that it will be of any further use.

Dominant Sex Reversal

It is possible that mutation of the gene in question could cause a dominant sex reversal even in chimeras or CRISPR–Cas founders. Because most investigators use XY ES cells, the consequence of dominant sex reversal would be the generation of predominantly phenotypically female chimeras that still retain the Y chromosome and would be unable to produce viable oocytes (or sperm). If all of your chimeras are female and infertile, you can determine whether they all have a Y chromosome using a Y-specific PCR primer set (Protocol 5.1: Sex Genotyping Mice by Polymerase Chain Reaction [Papaioannou and Behringer 2023e]) remembering that half of them should have a male component derived from the host embryo. If it turns out that they all have a Y chromosome, go to Chapter 16: Phenotypic Analysis of Dominant Mutant Effects in Mice [Papaioannou and Behringer 2023f]) for the next step as this is a dominant phenotypic effect.

Normally you should obtain both male and female CRISPR–Cas founders. If the sex ratio is heavily skewed toward one sex and a significant proportion of your founders are not fertile, then the mutant allele might be acting dominantly, causing sex reversal. Genotype for the presence or absence of the Y chromosome to determine the genetic sex of the founders and compare that to their phenotypic sex. If it matches, you can rule out sex reversal. If it does not match, you may have discovered a new gene involved in sex development.

Chimeras or CRISPR–Cas Founders Are Fertile but Do Not Transmit the Mutant Allele

Impatience

The absence of germline transmission of the ES cell component from chimeras or CRISPR–Cas founders may simply indicate that you are overanxious and the sample is too small to draw conclusions. From chimeras, whole litters may be derived from one component or the other, possibly because of patches of cells populating a testis tubule, and the composition of litters can change with time. Remember that the number of offspring from any chimera gives you only a sampling of the germline of that animal. If you have a chimera with a 5% contribution of ES cells to the germline, then you can expect only one out of 20 offspring to be of the ES cell genotype; it is all about probability. Generally, the level of chimerism in the coat (assuming that you are using a coat color marker to assess chimerism) (see Box 2) is a guide to the contribution of ES cells to the germline, but chimerism in the coat is no guarantee of chimerism in the germline (or any other tissue, for that matter). Here is where your gambling instincts can come in handy: If you have only low-level contribution of ES cells to chimeras, it may not be worth breeding them very extensively. If you have some chimeras with low-level contribution and some with high-level contribution, limited animal space is better used by concentrating on the game with better odds. In other words, breed only the high-level chimeras. If you have no success with further test breeding, make more chimeras. (See Pups Are Born but None Are Chimeric or the Level of Chimerism Is Low in Chapter 4: Embryonic Stem Cell Gene Targeting and Chimera Production in Mice [Papaioannou and Behringer 2023a] for pointers on improving the level of chimerism.)

CRISPR–Cas founders can be homozygous or heterozygous for the mutant allele in all cells of their body but it is also likely that a proportion of the founders will be mosaic, with some cells carrying the mutant allele and some carrying only wild-type alleles. The ratio of mutation-carrying and wild-type cells will vary between founders (Yen et al. 2014). Therefore, if a small fraction of the cells in a founder carries the mutant allele, it may take a while to recover that allele in progeny. The best remedy for this is to generate multiple founders so that you are not relying on a single founder to produce progeny carrying the mutant allele.

Unintended Consequences of ES Cell Gene Targeting Construct Design

By any chance, is there a *tk* gene in the targeted allele? An active herpes simplex virus thymidine kinase (HSV *tk*) gene is often used for negative selection in gene-targeting experiments. However, HSV *tk* expression is incompatible with spermatogenesis. If you used HSV *tk* in your construct, you must remove it before making chimeras. Either remake the targeting construct by putting in a different

Cite this chapter as *Cold Spring Harb Protoc*; doi:10.1101/pdb.over107959

BOX 2. COMMON COAT COLOR MARKERS

The multicolored litters that can emerge in the second generation of breeding an ES cell–derived targeted mutation often surprise researchers. But it is not for nothing that coat color genetics has been the darling of the mouse fancy for hundreds of years. Do not be worried by the segregation of coat color genes and do not be fooled into thinking it is a phenotype. A look at coat color genes carried by ES cells and a quick review of dominance and epistasis are in order. Although many targeted alleles are now available in C57BL/6N-derived ES cells, historically, the most commonly used ES cells were derived from 129 mice. However, the 129 "strain" is anything but uniform; it has many different substrains with different combinations of coat color alleles among them (see Table 1 in Chapter 4: Embryonic Stem Cell Gene Targeting and Chimera Production in Mice [Papaioannou and Behringer 2023a]). Similarly, one C57BL/6N ES cell line is segregating for a corrected *agouti* allele. If the parental ES cell line used for gene targeting came from mice heterozygous for coat color alleles, these will eventually segregate in the offspring of a germline transmitter to provide you with the entertainment of trying to decipher complex coat colors. When you choose an inbred strain for test matings, you know that the mice are homozygous at all loci, but cross these with a segregating strain and the rainbow unfurls (see Fig. 1A).

Dozens of loci affect coat color, but for practical purposes, only a few need to be considered. One is the *albino* locus. The most recessive allele, *albino* or Tyr^c, results in no pigment at all in the homozygous state and so is epistatic (takes precedence) over all other coat color alleles and essentially hides them (thus, by looking, you cannot tell what other coat color alleles an albino mouse carries). Many random-bred mice are homozygous for *albino*; this genotype (Tyr^c/Tyr^c) results in the standard pink-eyed, white laboratory mouse —but they might be heterozygous for alleles at other coat color loci. Some suppliers indicate known segregating coat color loci in their catalogs. There are also other alleles at the *albino* locus, such as *chinchilla* (Tyr^{c-ch}), which reduces but does not eliminate pigment; thus, the result in a homozygote is a dilution of whatever other colors are present. The "wild type," Tyr^+, is non-albino, resulting in full pigmentation in coat, skin, and eyes. It is fully dominant to all other albino alleles.

A second locus of interest is the *agouti* locus, which affects the distribution of pigment types, yellow or black, throughout the hair and coat. A large allelic series can be found at this locus, with the bottom recessive allele resulting in exclusively black pigment and the top dominant allele resulting in exclusively yellow pigment (in the hair). The most common alleles, *agouti* (A) and *white-bellied agouti* (A^w), result in banding of individual hairs with yellow and black pigments and a belly lighter than the back. These alleles are commonly referred to as "wild type," because variations of these alleles occur in nature in many mammals and result in a brownish, mousy color that is an excellent protective camouflage. This is the "agouti" mouse, whereas the "non-agouti" mouse is typically solid black.

Common 129-derived ES cell lines, such as R1, D3, and AB-1, were derived from 129 mice homozygous for *white-bellied agouti*, or *agouti* alleles at the *agouti* locus. Some of these, however, were heterozygous for *albino* and/or *chinchilla* and also for different alleles at another dilution locus called *pink-eyed dilution* (*p*). The interplay of these alleles with the alleles carried by the strain to which you mate your germline chimera will result in white, dilute, solid color, and agouti mice with black or pink eyes.

selection cassette or, if the HSV *tk* has *loxP* or *FRT* sites around it, go back to the targeted ES cells and remove HSV *tk* by transient transfection with the appropriate recombinase (see Box 2, Removing a Selection Cassette from ES Cells In Vitro, in Chapter 3: Mouse Gene-Targeting Strategies for Maximum Ease and Versatility [Papaioannou and Behringer 2023d]) and then start again making more chimeras. In addition, an HSV *tk* gene that has been modified such that it does not cause male sterility can still be used as a selectable marker. If you have a strong reason for needing HSV *tk* in mice, investigate this altered gene (Cohen et al. 1998).

Lost Potential of ES Cells

The ES cell clone may have lost the potential to contribute to the germline and proceed through spermatogenesis (or oogenesis, if you are using an XX ES cell line) through some effect unrelated to the targeted allele. Although it will be hard to show that this is the cause, and it is probably not worth the effort, you might reach this conclusion if you have produced a large number of fertile, high-level chimeras, none of which transmits the ES cell genotype. This is a signal to move on to another targeted clone. If the same happens with multiple targeted clones, karyotype the clones and the parent cell line to determine if there is a high proportion of cells with aneuploidy. If not, perhaps you have a phenotype; see the next paragraph.

Phenotype

It is possible that the specific mutation you made renders the heterozygous cells in chimeras or CRISPR–Cas founders, or homozygous mutant cells in the case of a CRISPR–Cas founder, incapable of contributing to the germline or completing gametogenesis, even though they make cellular contributions to other tissues in chimeras or CRISPR–Cas founders (*Klhl10*; Yan et al. 2004). This might apply to only a very small number of genes. But if you think this is possible based on the expression pattern of the gene during gametogenesis or on the nature of the protein, for example, as a last resort, go to Chapter 16: Phenotypic Analysis of Dominant Mutant Effects in Mice (Papaioannou and Behringer 2023f) to investigate this possible dominant effect.

Only the Wild-Type Allele Is Recovered from CRISPR–Cas Founders or ES Cell Germline Chimeras

Technical Glitches with CRISPR–Cas Founders or ES Cell Chimeras

If only wild-type progeny are obtained in multiple litters from individual CRISPR–Cas founders, it would be prudent to regenotype your founder. If the founder's genotype is reconfirmed for the targeted mutation, then continue breeding. However, you should have multiple founders breeding so that your study does not rely on a single founder.

With ES cell germline chimeras, it can be very unsettling when they fail to transmit the targeted allele. But take heart and rest assured that if your chimeras transmit the wild-type allele derived from the ES cells at least your parent ES cells and clones have good developmental potential and your blastocyst injection or embryo aggregation skills are good, even if you have not yet reached your goal.

It is unfortunately quite common that the lids of 96-well plates get switched, the labels rub off, the wrong ES clone is recovered, or the Southern blot or PCR genotyping is misinterpreted. Go back over all of your data relating to genotyping the putative targeted ES cell clones and regenotype the clone you injected. If the clone turns out not to be targeted, determine all the possible ways that you might have mixed up a genuinely targeted cell clone with a nontargeted one. If you have been taking careful notes and proceeding systematically, you might still be able to find the genuinely targeted clone in the freezer.

If it seems that the ES clone with which you are working was indeed correctly targeted when it was isolated, there is the remote possibility that a rare wild-type contaminant cell overgrew the targeted cells or that a gene conversion event and subsequent selection for the revertant cells took place during culture. Contamination of a targeted ES cell clone with wild-type cells can be assessed by examining the Southern blot used to identify the homologous recombinants. Both the wild-type and targeted bands should be of comparable intensity. If the intensity of the wild-type band is stronger than the targeted band, then you may have contamination of wild-type cells. One way to prevent overgrowth of contaminating or revertant wild-type cells is to continue ES cell culture in selective medium until shortly before making chimeras; however, this may adversely affect the efficiency of chimera formation.

Finally, you could take a tail or ear punch sample from the chimeras to do a Southern blot to determine if the cells used to make the chimera were targeted. However, the most effective and proactive way to prevent such problems is to regenotype cell clones as you inject or aggregate them, a smart practice that could be made part of the regular chimera-making routine. Simply collect some ES cells at the end of the experiment and run them through your genotyping protocol. If you have been injecting wild-type cells, you can abort the experiment and go back to either isolate new ES cell clones or work with other clones that you know are targeted.

Cite this chapter as *Cold Spring Harb Protoc*; doi:10.1101/pdb.over107959

BOX 3. CIRCUMVENTING THE PROBLEM OF A HAPLOID EFFECT ON SPERM

Having a mutant allele that detrimentally affects haploid gametes is a difficult but not impossible situation for analysis, since you may need to have a steady supply of chimeras or founders to produce the defective gametes for study. There may be some hope of circumventing the problem, however, if the gene affects only spermatogenesis. The mutant allele could then be transmitted through the female germline of CRISPR–Cas founders (Takeda et al. 2016). When working with ES cells, the choices are to:

1. Screen all of your targeted ES cell clones for the Y chromosome in the hope that one of them will have fortuitously lost it and could therefore produce XO female chimeras with a good chance of transmitting the targeted allele;

2. Alternatively subclone the ES cells and screen for loss of the Y;

3. Test breed all of your female chimeras anyway, on the chance that one of them might produce oocytes from the XY ES cells (transmission of XY ES cells from female chimeras has been often reported but it is still unclear whether oocytes are derived from XY cells or whether the Y chromosome is lost somewhere along the line, resulting in XO cells that produce oocytes); or

4. Start over, this time making a conditional allele (see Chapter 3: Mouse Gene-Targeting Strategies for Maximum Ease and Versatility [Papaioannou and Behringer 2023d]).

Intended but Unpredicted Consequence of the Genetic Alteration, in Other Words, a Phenotype

The mutant-bearing spermatids (postmeiotic) or later gametes from the chimera or CRISPR–Cas founder, or possibly the heterozygous embryos resulting from the test mating, might not be surviving because of a lack of the gene product in the gametes or haploinsufficiency in the case of embryos (*Vegfa*; Carmeliet et al. 1996). For further analysis of this kind of phenotype in the haploid gamete or heterozygous embryos, you will be dependent on having a steady supply of good, germline transmitting chimeras or CRISPR–Cas founders. Therefore, after you have established that this is indeed the situation (see Chapter 16: Phenotypic Analysis of Dominant Mutant Effects in Mice [Papaioannou and Behringer 2023f]), you may need to go back and make more chimeras or founders, repeatedly (Box 3), or consider generating a conditional null allele.

Another possibility is a dominant perinatal lethal phenotype that you might have missed if you are genotyping offspring at weaning or assessing germline transmission at 7 d postnatally when agouti

BOX 4. IDENTIFYING MALE AND FEMALE MICE

How do you determine if your mouse is male (has testes) or female (has ovaries)? We use this definition of males and females because some mutants have testes but are genetically XX or have ovaries but are genetically XY. Typically, you assess the phenotype by examining the anatomy of the external genitalia of the mouse. The genetic sex of a mouse can be determined by PCR genotyping for Y-chromosome-linked genes such as *Sry* (see Protocol 5.1: Sex Genotyping Mice by Polymerase Chain Reaction [Papaioannou and Behringer 2023e]).

Male and female mice can be distinguished at birth. The distance between the genital papilla and the anus is longer in males than in females (Fig. 3A–D). Males also have pigmentation in this region unless you are working with an albino strain (Fig. 3C). The simplest way to assess the sex of newborn mice is to sort the pups in the litter, comparing the genital region of each mouse to others in the litter. Usually, you will be able to sort the litter into two groups, males and females.

Once the mice develop their fur, other external anatomical features can be used to distinguish males and females. As the pups age, the difference in the distance between the genital papilla and anus becomes more obvious (Fig. 3E–J). Mammary gland/nipple formation on the ventral side of the body can be used to identify females since male mice do not have nipples because of the effects of androgens during development. Look at the underside of the mouse for two parallel rows of five spots each in females where the nipples have formed (Fig. 3E,F). Non-agouti black female mice, such as C57BL/6, have light-colored fur surrounding the nipples next

Continued

BOX 4. *Continued*

to the black fur, making them very easy to see. This is also true for the commonly used agouti pigmentation alleles. For albino females, the nipples will be pink next to the white fur. Once the mice are sexually mature (~6-wk-old), the vagina will be open in females and it is very easy to distinguish males from females (Fig. 3G–J). If you have any doubts, simply compare the mouse in question with a known male or female.

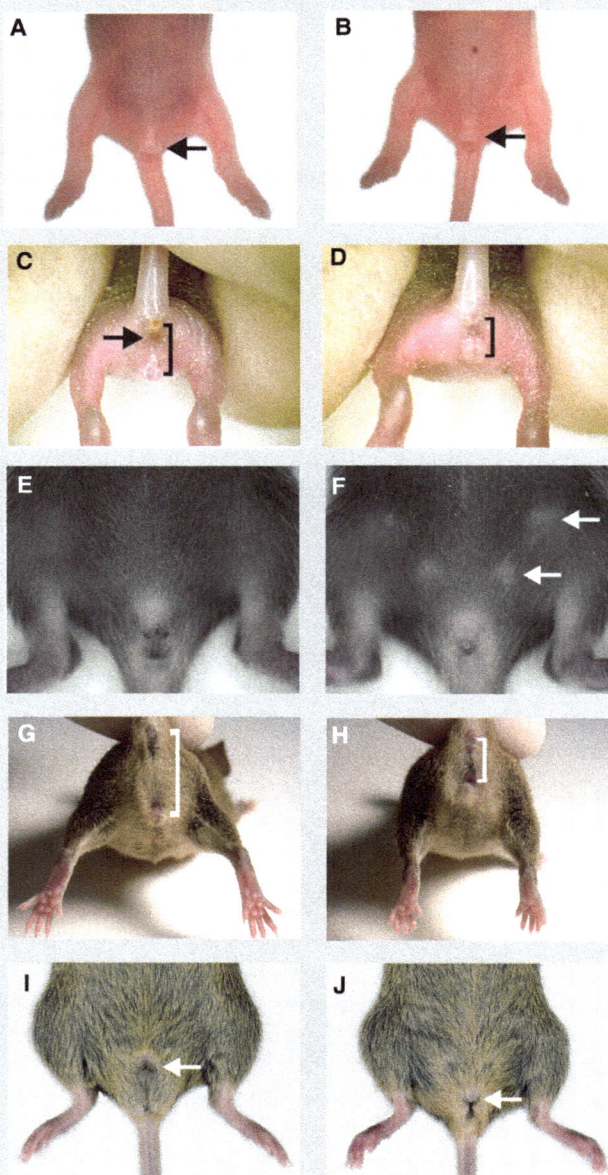

FIGURE 3. External morphology of male and female mice. Ventral view of newborn male (*A*) and female (*B*) mice. Photographs show slight differences in the genital papillae of male and female neonates (arrows); the papilla of males is larger. Rear view of ~2-d-postpartum male (*C*) and female (*D*) mice. The distance between the genital papilla and the anus (brackets) is longer in males than in females. In pigmented strains, the scrotal area of the male neonates has pigment (arrow). Ventral view of ~10-d-old C57BL/6J male (*E*) and female (*F*) mice. At this age, the nipples of the female can be seen because of the lighter surrounding hair (arrows). Rear view of male (*G*) and female (*H*) adults. The genital–anal distance (brackets) is much longer in males compared with females. The vaginal opening is just caudal of the genital papilla. Ventral view of male (*I*) and female (*J*) adults showing the gross differences in their external genitalia. Arrows point to the genital papilla.

hairs become apparent mice. Check for perinatal losses, and if they are present, genotype the dead pups to see if these are the missing heterozygotes. If this is the case, you have a phenotype and a dominant effect (e.g., *Sox9*; Bi et al. 2001). Go to Chapter 16: Phenotypic Analysis of Dominant Mutant Effects in Mice (Papaioannou and Behringer 2023f).

Are Heterozygous Offspring of the Chimeras or CRISPR–Cas Founders Normal, Fertile, and of Both Sexes?

If you have either germline-transmitting chimeras or CRISPR–Cas founders that produce offspring that are heterozygous for the intended targeted allele(s), and are normal, fertile and of both sexes (Box 4), congratulations! You are on your way and can concentrate on analyzing a homozygous mutant phenotype in the next generation. Go on to Chapter 6: Strategies for Maintaining Mouse Mutations (Papaioannou and Behringer 2023g) and eventually to Chapter 7: Special Breeding Techniques for Use in Mouse Mutation Analysis (Papaioannou and Behringer 2023h). As a precaution, genotype the first heterozygous offspring from chimeras by Southern blot analysis with an external probe just in case there was a mix-up between isolation of the targeted ES cell clones and making the chimeras. Once you have confirmed the correctly targeted genotype of the heterozygotes, use a quicker PCR strategy for genotyping mice in subsequent generations. However, if the answer to the question posed in this section is no, congratulations because you have a phenotype (see below).

Dominant Mutation

If the heterozygous mice make it to term, the heterozygous mutation is clearly compatible with embryonic development. However, a dominant phenotype could manifest as abnormalities at birth or at any stage thereafter. Watch for postnatal heterozygote-associated abnormalities and analyze the mice as described in Chapter 14: Analysis of Postnatal Mutant Phenotypes in Mice (Papaioannou and Behringer 2023i) for homozygous mutants.

Dominant Effect on Gametogenesis

If the male and female heterozygotes are infertile, but otherwise morphologically normal, this could be an indication that your mutant has a dominant effect on gametogenesis or some aspect of reproductive behavior (e.g., *Klhl10*; Yan et al. 2004). However, because you got the mutation through the germline of either a male or female chimera or CRISPR–Cas founder in the first place, it has to be something that can be rescued by coexistence with wild-type cells in a chimera or mosaic founder, at least in that sex (male or female, depending on your breeding scheme). Also, there are many reasons, other than a possible effect of your mutation, that a mouse might be infertile and it is best to breed a reasonable number before coming to this conclusion. If infertility is always associated with the mutation in one sex or the other, or both, this is the phenotype, and you can go to Chapter 16: Phenotypic Analysis of Dominant Mutant Effects in Mice (Papaioannou and Behringer 2023f) for help with phenotyping.

Sex-Limited Dominant Effect

If heterozygotes of only one or the other sex have physical, growth, or survival problems postnatally— a sex-limited phenotype—consult an experienced reproductive biologist or endocrinologist to determine how the effect is limited by the sex of the heterozygote. If fertility is the problem and heterozygotes of only one sex are fertile, the mutation may be specific to spermatogenesis or oogenesis, or to some aspect of reproductive behavior, in which case you can maintain the mutant by breeding heterozygotes of the opposite sex (e.g., *ods*; Bishop et al. 2000). We doubt that the majority of investigators will ever reach this point in the diagnosis, because these effects should be rare, but for those who do, consider collaborating with investigators interested in gametogenesis or reproductive behavior if this is not your primary interest.

Dominant Sex Reversal

If the heterozygotes are all female, it is possible that your mutation causes dominant sex reversal, resulting in XY females (e.g., *ods*; Bishop et al. 2000). The heterozygous XY female offspring will probably be sterile but the heterozygous XX females may be fertile. Thus, minimally, you should be able to maintain the mutation through the female germline. Making the mutation homozygous will be a challenge (see Chapter 7: Special Breeding Techniques for Use in Mouse Mutation Analysis [Papaioannou and Behringer 2023h] for breeding tips and Chapter 16: Phenotypic Analysis of Dominant Mutant Effects in Mice [Papaioannou and Behringer 2023f] for further analysis).

REFERENCES

Behringer R, Gertsenstein M, Nagy KV, Nagy A. 2016. Selecting female mice in estrus and checking plugs. *Cold Spring Harb Protoc* doi:10.1101/pdb.prot092387

Bi W, Huang W, Whitworth DJ, Deng JM, Zhang Z, Behringer RR, de Crombrugghe B. 2001. Haploinsufficiency of *Sox9* results in defective cartilage primordia and premature skeletal mineralization. *Proc Natl Acad Sci* 98: 6698–6703. doi:10.1073/pnas.111092198

Bishop CE, Whitworth DJ, Qin Y, Agoulnik AI, Agoulnik IU, Harrison WR, Behringer RR, Overbeek PA. 2000. A transgenic insertion upstream of *Sox9* is associated with dominant XX sex reversal in the mouse. *Nat Genet* 26: 490–494. doi:10.1038/82652

Carmeliet P, Ferreira V, Breier G, Pollefeyt S, Kieckens L, Gertsenstein M, Fahrig M, Vandenhoeck A, Harpal K, Eberhardt C, et al. 1996. Abnormal blood vessel development and lethality in embryos lacking a single VEGF allele. *Nature* 380: 435–439. doi:10.1038/380435a0

Cohen JL, Boyer O, Salomon B, Onclerco R, Depetris D, Lejeune L, Dubus-Bonnet V, Bruel S, Charlotte F, Mattéï MG, et al. 1998. Fertile homozygous transgenic mice expressing a functional truncated herpes simplex thymidine kinase ΔTK gene. *Transgenic Res* 7: 321–330. doi:10.1023/A:1008893206208

Gonzalez G. 2016. Determining the stage of the estrous cycle in female mice by vaginal smear. *Cold Spring Harb Protoc* doi:10.1101/pdb.prot094474

Papaioannou VE, Behringer RR. 2023a. Embryonic stem cell gene targeting and chimera production in mice. *Cold Spring Harb Protoc* doi:10.1101/pdb.over107958

Papaioannou VE, Behringer RR. 2023b. Obtaining or generating gene mutations in mice. *Cold Spring Harb Protoc* doi:10.1101/pdb.over107956

Papaioannou VE, Behringer RR. 2023c. Backcrossing to make a congenic mouse strain. *Cold Spring Harb Protoc* doi:10.1101/pdb.prot108039

Papaioannou VE, Behringer RR. 2023d. Mouse gene-targeting strategies for maximum ease and versatility. *Cold Spring Harb Protoc* doi:10.1101/pdb.over107957

Papaioannou VE, Behringer RR. 2023e. Sex genotyping mice by polymerase chain reaction. *Cold Spring Harb Protoc* doi:10.1101/pdb.prot108062

Papaioannou VE, Behringer RR. 2023f. Phenotypic analyses of dominant mutant effects in mice. *Cold Spring Harb Protoc* doi:10.1101/pdb.over107978

Papaioannou VE, Behringer RR. 2023g. Strategies for maintaining mouse mutations. *Cold Spring Harb Protoc* doi:10.1101/pdb.over107960

Papaioannou VE, Behringer RR. 2023h. Special breeding techniques for use in mouse mutation analysis. *Cold Spring Harb Protoc* doi:10.1101/pdb.over107961

Papaioannou VE, Behringer RR. 2023i. Analysis of postnatal mutant phenotypes in mice. *Cold Spring Harb Protoc* doi:10.1101/pdb.over107976

Pettitt SJ, Liang Q, Rairdan XY, Moran JL, Prosser HM, Beier DR, Lloyd KC, Bradley A, Skarnes WC. 2009. Agouti C57BL/6N embryonic stem cells for mouse genetic resources. *Nat Methods* 6: 493–495. doi:10.1038/nmeth.1342

Qiu P, Shandilya H, D'Alessio JM, O'Connor K, Durocher J, Gerard GF. 2004. Mutation detection using Surveyor nuclease. *Biotechniques* 36: 702–707. doi:10.2144/04364PF01

Takeda N, Yoshinaga K, Furushima K, Takamune K, Li Z, Abe S, Aizawa S, Yamamura K. 2016. Viable offspring obtained from *Prm1*-deficient sperm in mice. *Sci Rep* 6: 27409. doi:10.1038/srep27409

Yan W, Ma L, Burns KH, Matzuk MM. 2004. Haploinsufficiency of kelch-like protein homolog 10 causes infertility in male mice. *Proc Natl Acad Sci* 101: 7793–7798. doi:10.1073/pnas.0308025101

Yen ST, Zhang M, Deng JM, Usman SJ, Smith CN, Parker-Thornburg J, Swinton PG, Martin JF, Behringer RR. 2014. Somatic mosaicism and allele complexity induced by CRISPR/Cas9 RNA injections in mouse zygotes. *Dev Biol* 393: 3–9. doi:10.1016/j.ydbio.2014.06.017

Sex Genotyping Mice by Polymerase Chain Reaction

A simple method to determine the genetic sex of a mouse is to amplify DNA from a male-specific gene by polymerase chain reaction (PCR). This protocol is used to detect the Y-chromosome-specific gene *Sry* in tissue lysates of tail tip or ear punch samples.

INTRODUCTION

To determine if you have a genetic male (XY) or genetic female (XX or XO) mouse, use polymerase chain reaction (PCR) to amplify a sequence, such as *Sry*, that is specific to the Y chromosome. Here we provide a robust PCR assay for *Sry*. Primers and the PCR method for *Sry* and an autosomal, internal control gene (*Rapsn*) are given (Kobayashi et al. 2004).

MATERIALS

It is essential that you consult the appropriate Material Data Safety sheets and your institution's Environmental Health and Safety Office for proper handling of equipment and hazardous materials used in this protocol.

Reagents

PCR Master Mix (2×) (Promega)
Primers
    Sry-forward: 5′-TGACTGGGATGCAGTAGTTC-3′
    Sry-reverse: 5′-TGTGCTAGAGAGAAACCCTG-3′
    Sry amplicon size ∼230 base pairs (bp)
    Rapsn-forward: 5′-AGGACTGGGTGGCTTCCAACTCCCAGACAC-3′
    Rapsn-reverse: 5′-AGCTTCTCATTGCTGCGCGCCAGGTTCAGG-3′
    Rapsn amplicon size ∼590 bp
Tissue lysates (see Behringer et al. 2019)

    Tail tip, ear punch, or other tissue lysate from a known male and female as positive and negative controls, respectively

    Tail tip, ear punch, or other tissue lysate from experimental animals

Equipment

Gel electrophoresis equipment
PCR machine

METHOD

1. Prepare the following mix

	μL per reaction	10 reactions
2× Master Mix (Promega)	10.0	100.0
Water	4.5	45.0
Sry-forward (10 μM)	1.0	10.0
Sry-reverse (10 μM)	1.0	10.0
Rapsn-forward (100 ng/μL)	0.25	2.5
Rapsn-reverse (100 ng/μL)	0.25	2.5
Genomic DNA solution (tissue lysate)	3.0	30.0
Total volume	20.0	200.0

2. Set PCR cycle conditions for 5 min at 95°C; 35 cycles for 30 sec at 95°C, 30 sec at 65°C, 45 sec at 72°C; for 10 min at 72°C; store at 4°C.

3. Visualize the amplified sequences by standard agarose gel electrophoresis and DNA visualization protocols (see Green and Sambrook 2019).

   Genetic females (XX or XO) will have a single ∼590-bp Rapsn band. Genetic males (XY) will have an Sry band at ∼230 bp and a Rapsn band at ∼590 bp.

ACKNOWLEDGMENTS

We thank Akio Kobayashi for details of the *Sry* PCR genotyping assay.

REFERENCES

Behringer R, Gertsenstein M, Nagy KV, Nagy A. 2019. Preparation of polymerase chain reaction template DNA from mouse tail tissue. *Cold Spring Harb Protoc* doi:10.1101/pdb.prot092700

Green MR, Sambrook J. 2019. Protocol: agarose gel electrophoresis. *Cold Spring Harb Protoc* doi:10.1101/pdb.prot100404

Kobayashi A, Shawlot W, Kania A, Behringer RR. 2004. Requirement of *Lim1* for female reproductive tract development. *Development* 131: 539–549. doi:10.1242/dev.00951

CHAPTER 6

Strategies for Maintaining Mouse Mutations

Rules for naming a new mutation are provided. The majority of new mutations are recessive and thus easily maintained in a mouse strain. Considerations on the choice of genetic background are given, depending on how the mutant was produced and how you intend to analyze it. General information on maintaining a mutant colony to perpetuate the mutation and to efficiently produce homozygous mutant mice for analysis is provided. Also discussed are special breeding techniques to delete a selection cassette in vivo, if you produced the mutation in embryonic stem (ES) cells, and to maintain a mutant with a balancer chromosome. In the event of either male or female infertility in the heterozygotes, assisted reproductive techniques may be necessary.

INTRODUCTION

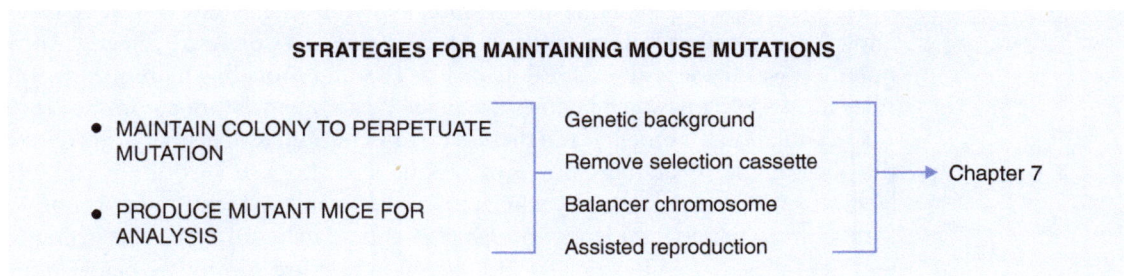

STRATEGIES FOR MAINTAINING MOUSE MUTATIONS

- MAINTAIN COLONY TO PERPETUATE MUTATION

- PRODUCE MUTANT MICE FOR ANALYSIS

Genetic background

Remove selection cassette

Balancer chromosome

Assisted reproduction

→ Chapter 7

Although dominant mutations are certainly possible, by far, the majority of targeted mutations, or spontaneous ones, for that matter, turn out to be recessive. Thus, it is highly likely that successful gene targeting will result in viable, fertile, heterozygous male and female mice. The next step will be to mate heterozygotes to produce homozygous mutant offspring, which will be expected to have a phenotype. But first, let us digress for a moment to consider the best strategies for this all-important mating and for maintaining this hard-won mutant. If you are a veteran mouse breeder with a background in genetics and experience with mutant mice, you can skip this introduction. If you have a dominant phenotype and were sent here from Chapter 16: Phenotypic Analysis of Dominant Mutant Effects in Mice (Papaioannou and Behringer 2023a), look through this introduction for hints on breeding in general and specific tips for the special situation of breeding a dominant infertility mutation. In either

Opening artwork: Printed with permission from ©2023 The Jackson Laboratory.

© 2024 Cold Spring Harbor Laboratory Press
Cite this chapter as *Cold Spring Harb Protoc*; doi:10.1101/pdb.over107960

case, then move on to Chapter 7: Special Breeding Techniques for Use in Mouse Mutation Analysis (Papaioannou and Behringer 2023b) before proceeding with the phenotypic analysis.

Whether you started with an existing mutation or produced one by gene targeting in embryonic stem (ES) cells or using CRISPR–Cas gene editing, you will need to establish a breeding colony to perpetuate the mutation as well as to produce mutants for phenotypic analysis. Some mutations may affect viability or fertility, requiring special breeding techniques. If you produced a new allele, however, the first task is naming that allele.

NAMING A NEW MUTANT ALLELE

The descriptive, sometimes fanciful names given to mouse mutants in the past, such as *varitint waddler*, *kinky waltzer*, or *screw tail*, are gradually giving way to names based less on the phenotype of the mutant and more on primary features of the gene product of the wild-type allele. Many genes have changed names as the genes have been cloned and their protein products discovered. A familiar example is the *albino* mutation originally named to describe an all-white mouse with no pigment; the locus is now known as tyrosinase (*Tyr*) after the enzyme that the gene encodes, and the descriptive mutant allele names such as *albino*, *himalayan*, and *chinchilla* are now identified by superscript symbols (Tyr^c, Tyr^{c-h}, Tyr^{c-ch}).

If the mutant allele you are analyzing was created as a result of random insertion of a transgene (i.e., insertional mutagenesis) or was the result of a gene trap, the mutation will be named as a mutant allele of the gene that was disrupted, and the superscript will contain the symbol "Tg" for transgene or "Gt" for gene trap. The name of the gene or vector content that was inserted is added in parentheses but can be omitted if the mutation is uniquely identified without it. The gene name can be attached to a strain designation to indicate the genetic background on which it is maintained (Fig. 1).

When naming a new mutation in a known gene made by gene targeting, you will be limited to using the accepted gene name and symbol, and your main concern will be in adding the superscript "tm" for targeted mutation made in ES cells (Fig. 1A) or "em" for endonuclease-mediated if the mutation was made with CRISPR–Cas (Fig. 1B), and providing a unique identifier for the allele. The rules for deciding new gene names are made by a nomenclature committee (http://www.informatics .jax.org/mgihome/nomen/), and there is not much room for creativity. However, you may personalize the allele name by adding your own unique laboratory or institution code. If you do not have a laboratory or institution code, you can get one from the Institute for Laboratory Animal Research (https://www.nationalacademies.org/ilar/lab-code-database). Note that some journals may require that new allele names be approved by the nomenclature committee before publication.

Mutant alleles produced in ES cells by the high-throughput efforts of the International Knockout Mouse Consortium (IKMC; https://www.mousephenotype.org/about-impc/about-ikmc/) are of several different types (e.g., gene-trapped alleles, reporter tagged, conditional), some with the capacity of producing several different types of derivative alleles with Cre or Flp recombinase. If you received either mutant ES cells or mice from IKMC, the mutation will already have a name (Fig. 1C), but it is worth studying the special rules for nomenclature (http://www.informatics.jax.org/mgihome/nomen/ IKMCnomen.shtml) to make sure you know what you are dealing with initially and that you can correctly name any derivative alleles you produce.

ANIMAL FACILITIES

By far, the preferred animal facility is a specific pathogen–free (SPF) barrier facility. The specific pathogens that are excluded from such facilities include ecto- and endoparasites, certain bacteria, mycoplasma, and viruses. SPF facilities, rapidly becoming the accepted standard for research institutions, are facilities that you should promote if they are not already in place at the institution in which you intend to maintain your mutant mice.

Cite this chapter as *Cold Spring Harb Protoc*; doi:10.1101/pdb.over107960

A Mode of production: targeted mutation

Sequential number of mutations at this locus made in this lab

C57BL/6Tac-*Tbx6*^{tm1Pa}

Registered code for lab of origin

Inbred strain designation: strain/subline and holder or supplier

Gene symbol for locus

Mode of production: endonuclease-mediated

B C57BL/6N-*Tyr*^{em1Bhr}

"a" denotes conditional-ready allele

C C75BL/6NTac-*Oaf*^{tm1a(EUCOMM)Wtsi}

Large-scale mutagenesis project

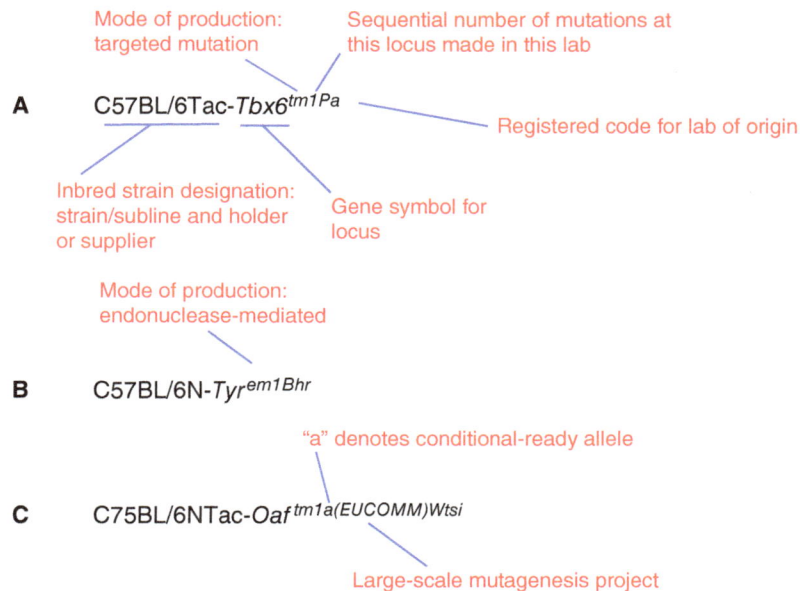

FIGURE 1. Examples of allele names of new mutations made by different methods. The allele name contains the gene symbol of the mutated gene in italics, preceded by the name of the strain of mouse on which the mutation is maintained and followed by a superscript to provide a unique identifier. The superscript contains information on how the allele was made (e.g., [tm] targeted mutagenesis, [em] endonuclease-mediated targeting, [Gt] gene trap), a serial number from the laboratory indicating the sequential number of mutations made in this locus by this laboratory, and a laboratory code for the investigator or institution making the mutation. Additional information may be included for alleles made in large-scale mutagenesis projects. (*A*) This example designates the first mutant allele of the *Tbx6* locus made in the Papaioannou laboratory (Pa) using targeted mutagenesis (tm) in embryonic stem cells, maintained as a congenic strain with the C57BL/6 inbred strain from Taconic Farm (Tac) (Chapman and Papaioannou 1998). (*B*) This example is the first mutation of the *Tyrosinase* (*Tyr*) gene made by the Behringer lab (Bhr) by endonuclease-mediated mutagenesis (em) in the C57BL/6N inbred strain (Yen et al. 2014). (*C*) This example is the first allele of the *out at first homolog* (*Oaf*) gene made by targeted mutagenesis in ES cells by the Wellcome Trust Sanger Institute (Wtsi) for the EUCOMM project, maintained on the C57BL/6NTac strain. The "a" in the superscript indicates that this is a knockout-first, conditional-ready allele. Derivative alleles made by recombinase (Cre or Flp) exposure are indicated by "b" or "c" following recombinase excision of genomic regions. See http://www.informatics.jax.org/mgihome/nomen/IKMCnomen.shtml for full details on nomenclature of derivative alleles and alleles made by other means.

An SPF facility uses procedures and equipment for animal housing, handling, and husbandry that maintain animals in a SPF environment. They also rigorously monitor incoming mice, tumors, cell lines, reagents, equipment, and personnel to prevent the introduction of pathogens into the facility. These procedures may require animal quarantine or rederivation of mice by cesarean section before importation, often at considerable expense and sometimes inconvenience for the investigator. However, these procedures are well worth the trouble in the long run, and investigators should insist on strict adherence to the rules.

The advantages of working with SPF mice should be self-evident, but they bear repeating. Animals infected with pathogens may have increased mortality and impaired breeding performance, necessitating a much larger number of animals for both maintenance and experiments than if they were healthy. The effects of endemic pathogens may be transient or subclinical, but will have effects on the host immune system that could invalidate research data. Variations in the population of mice due to the natural course of infections could lead to variations that confound the phenotypic analysis or that could be mistaken for a phenotype. Mice with poor health status also present a roadblock for collaborations, because they will have to be rederived or quarantined before import into any SPF facility. All in all, healthy mice make better research subjects and are the basis for accurate phenotypic analysis of genetic mutations.

Sometimes, disasters do strike animal facilities. As insurance, send some mutants to a collaborator at a different institution and/or submit them to a mutant repository such as the Mutant Mouse

Resource and Research Center (MMRRC; https://www.jax.org/research-and-faculty/resources/mutant-mouse-resource-research-center), the European Mouse Mutant Archive (EMMA; https://www.infrafrontier.eu/resources-and-services/deposit-mice-emma-repository), or the Center for Animal Resources and Development (CARD; http://card.medic.kumamoto-u.ac.jp/card/english) or other repositories. Alternatively, freeze some sperm or embryos for safekeeping either on your own (for details, see Taft 2017b; Takeo and Nakagata 2018; Taft 2018, which originally appeared in Chapter 16 of Behringer et al. 2014) through your institutional core facility, or with a commercial vendor, and archive them in more than one location.

GENETIC BACKGROUND

Among strains of mice, many differences are readily apparent: color, behavior, size, shape, and disease susceptibility, to name a few. The basis for these strain differences in phenotype is allelic variation in genes, the so-called genetic background. It is against this background that major gene effects—spontaneous or induced single-gene mutations—are analyzed. Because no gene functions in isolation and a phenotype is the product of the gene's action and the action of all other genes affecting the system, it stands to reason that single-gene effects, however large, are subject to variation in phenotype, depending on the background genotype on which they are examined (Box 1).

If your mutation was made directly in zygotes using CRISPR–Cas, the strain was chosen at the outset, perhaps for some special characteristic of that strain or for convenience. Crossing the founders with the same strain will produce mutant mice that are coisogenic with the strain (i.e., genetically identical to the parent inbred strain except for the altered allele) and thus have a uniform genetic background. If you used ES cells for gene targeting, your choice of breeding schemes (see Breeding Schemes for ES Cell Chimeras in Chapter 5: Recovering a Targeted Mutation in Mice from Embryonic Stem Cell Chimeras or CRISPR–Cas Founders [Papaioannou and Behringer 2023c]) will determine whether the first heterozygous mutants you obtain are on either an F_1 background between two inbred strains, a mixed or outbred genetic background, or a uniform, coisogenic background. An F_1 or outbred genetic background may be advantageous for rapidly expanding the colony of mice carrying the mutation because they have the advantage of hybrid vigor. However, the drawback is that the first homozygous mutant mice will inevitably have a mixed genetic background, and there could be genetic

BOX 1. GENETIC BACKGROUND EFFECTS: MODIFIERS OF PHENOTYPE

The phenotype may have variable penetrance (appearing in only some individuals) or variable expressivity —that is, it may vary considerably from one individual to the next. The source of this variability could be either developmental or genetic. Such genetic effects on the phenotype from genes other than the mutated locus are known as genetic background effects and the genes that cause them are genetic modifiers. They can be revealed by the observation that the penetrance or expressivity of the phenotype is different when observed on different genetic backgrounds (i.e., on different strains), and they may be the result of one or many genes in which the two strains have different alleles. Typically, genetic background effects are minor and there is usually not much interest in chasing down exactly what other gene(s) is causing the variation; it could be a single gene with a minor effect or a combination of many genes each with a small effect. It is enough to know that the phenotype is subject to background variation. Occasionally, however, a radically different phenotype may result when a mutation is placed on a different genetic background (e.g., *Egfr*; Threadgill et al. 1995), making it much more enticing to map and identify the specific gene modifier(s) as an important contributor to the phenotype. It could be that an allele at a single locus acts as a major genetic modifier of the phenotype and is therefore of significant biological interest to the study of your favorite gene. Thus, whether or not there seems to be variation in the phenotype of your mutant, you could have a surprise when it is examined on a different genetic background.

Cite this chapter as *Cold Spring Harb Protoc*; doi:10.1101/pdb.over107960

variability affecting the mutant phenotype. In other words, in gaining the advantage of hybrid or outbred vigor, the advantage of a uniform genetic background will be lost in this first analysis.

Even if you choose initially to analyze your mutant on a mixed or outbred background, it is advantageous to transfer your allele to one or more inbred strains as outlined in the following paragraphs. This will allow you to look at the mutant effect on different genetic backgrounds, potentially uncovering genetic modifiers of the targeted gene and possibly reducing phenotypic variability. The choice of strain may be dictated by the expected phenotype; for example, if you are working with a suspected tumor suppressor gene, you might choose an inbred strain with a high susceptibility to tumors. Or the strain might be chosen for convenience, such as commercial availability or availability of genomic sequence. A good choice of inbred strain is C57BL/6J or related substrains such as C57BL/6N, not because these strains are particularly easy to work with, but simply because these mice are the most widely used in research, are available in large numbers from commercial sources, and many mutations are maintained in these strains. In addition, the C57BL/6J genome was the first to be sequenced, assembled, and annotated and is the mouse reference genome for the Mouse Genomes Project (https://www.mousegenomes.org/). One small caveat should be noted: C57BL/6J females have a high incidence of septate vaginas, sometimes approaching 10%. This is apparent as bilateral vaginal openings separated by a membranous septum rather than a single vaginal opening. In our experience, these particular females are not good breeders. It is probably best to cull them when they are identified.

If you used ES cells for gene targeting, then breeding a germline-transmitting chimera to mice of the same strain as the ES cells is the quickest way to place the mutant allele onto an inbred background. In the first generation, you will have a coisogenic line. For example, if you used C57BL/6N ES cells for gene targeting, then breeding chimeras with C57BL/6N mice results in a coisogenic strain. Note, however, that this is a different substrain from the commonly used C57BL/6J strain, with known genetic differences (e.g., a *Crb1* mutation for retinal degeneration) that could affect your experimental results (Mattapallil et al. 2012; Simon et al. 2013). Similarly, if you used a 129-ES cell line, be aware that there are many different 129 substrains (see Table 1 in Chapter 4: Embryonic Stem Cell Gene Targeting and Chimera Production in Mice [Papaioannou and Behringer 2023d]). If mice of the specific substrain from which the ES cells were derived are not available for breeding, then any cross with a different 129 substrain will not produce a true coisogenic inbred strain, but rather a mixed substrain background.

If your ES cells and the test breeders were of different inbred strains (e.g., if you used 129-ES cells and test-bred chimeras to C57BL/6J females), then with the first generation (C57BL/6J × 129) F_1 mice you are already halfway there, genetically speaking, to producing a congenic strain. However, note that a congenic strain is different from a coisogenic strain in that it is identical except for a differential chromosomal region surrounding the mutant allele. From the F_1 generation, it is a simple matter of backcrossing heterozygous animals to C57BL/6J mice with forced heterozygosity at your mutant locus for nine generations (Protocol 6.1: Backcrossing to Generate a Congenic Mouse Strain [Papaioannou and Behringer 2023e]).

MAINTAINING A MUTANT COLONY: THE BASICS

It is useful to think about the breeding of your mutant mouse colony as having two components: (1) maintenance of a production colony to perpetuate the mutation and (2) production of mutant mice or embryos for phenotypic analysis.

Maintaining a Production Colony to Perpetuate the Mutation

Your object is to keep the mutation that you have produced in a live animal as a permanent resource for yourself and the scientific community. Unless the mutation is severely detrimental to the health or reproductive capacity of the heterozygous mice (see Assisted Reproduction to Deal with Infertility, below), this is usually quite straightforward using basic husbandry techniques as approved by your local Institutional Animal Care and Use Committee (IACUC) for housing, genotyping, and marking individual mice (for details, see Chapter 3 in Behringer et al. 2014). We routinely take the tip of the tail for genotyping either at 7–10 d postnatally or at the time of weaning. However, you can also isolate DNA from an ear punch or toe clip used for marking individual mice. The advantage of the earlier date, especially if space is limited, is that genotypes will be known by the time of weaning and unneeded mice can be culled. Waiting until weaning requires one less visit to the mouse room, but all mice will have to be weaned and stored until genotypes are known. If the mutation is dominant or semidominant, the genotyping task may be easier, provided that the phenotype is fully penetrant (i.e., all of the heterozygotes show the effect). If you can distinguish every heterozygote by gross morphology or behavior, you can dispense with the genotyping step and simply breed the animals with a mutant phenotype to phenotypically (and thus genotypically) wild-type mice.

When maintaining a colony on a mixed or outbred genetic background, beware of inbreeding depression. If you start with a small number of heterozygous mice and intercross them in a small closed colony with forced heterozygosity (i.e., selecting heterozygotes for breeding), you will inbreed at a fairly rapid pace. What was originally a vigorous breeding colony may quickly lose reproductive potential and threaten to die out. Inbreeding depression can actually be more of a problem when you start with an outbred population than when you work only with inbred mice. This is because inbreeding depression is the result of uncovering detrimental recessive alleles that segregate in a population. In the case of an outbred stock, the effects will be particularly severe between the second and eighth generations of inbreeding as the deleterious alleles become homozygous. In the case of starting with a cross between inbred strains, all of the really deleterious alleles will already have been eliminated from the population during the initial derivation of the inbred strains.

This inbreeding depression has nothing (necessarily) to do with your mutation, but effects of the mutation could exacerbate the problem. The solution is quite simple: Bring in new blood. This can be accomplished by (1) producing additional vigorous hybrid heterozygotes by continuing to breed the germline chimeras, if you used ES cells, or (2) outcrossing the heterozygotes to an inbred strain or even an outbred stock of mice.

Remember that for expansion of a production colony, crossing a heterozygote with a wild-type mouse—useful if the heterozygotes are in demand for experiments—produces exactly the same proportion (50%) of heterozygous mice as crossing two heterozygotes together. Furthermore, with a heterozygous × wild-type cross, the genotyping is simplified because you need only distinguish between two genotypes. In the event that homozygous mutants are viable and fertile, these mice can be mated with wild type, and 100% of the offspring will be heterozygotes (with no genotyping necessary). However, do not be tempted to keep the production colony going by breeding only homozygous mutants, because you always want to keep the wild-type allele in the colony for producing wild-type littermates with the same or similar genetic background for experimental controls.

Maintaining a mutation on an inbred background should not create problems, although you will almost certainly need to maintain a larger number of breeding pairs than those of mixed background to produce a similar number of offspring. If you produced a coisogenic strain, the reproductive characteristics of the colony should be the same as those of the original inbred strain, and you do not have to worry about any further inbreeding depression since deleterious alleles will have been eliminated from the strain during its original derivation. The mutation is maintained by forced heterozygosity and the strain is called a segregating inbred strain. If you are backcrossing your mutation onto a different inbred strain (see Protocol 6.1: Backcrossing to Generate a Congenic Mouse Strain [Papaioannou and Behringer 2023e]), however, you will have hybrid vigor in the early generations that gradually disappears in the later generations as the congenic strain approaches genetic uniformity with the other inbred strain. Be careful of this loss of productivity and setup more mating pairs as the reproductive performance declines.

Your goal in the production colony is to produce a constant supply of heterozygous and wild-type animals in sufficient numbers to replenish breeders as they age or their fertility declines and to supply plenty of heterozygous animals for the mutant analysis. In the case of a prenatal homozygous lethal mutation, a heterozygous female will have to be sacrificed for each litter studied, so the number required can be considerable. Do not allow your stocks to get "stale." This entails culling mice that get too old for breeding and replacing them with young mice. Maintaining a colony usually involves keeping a certain number of "stock" cages of genotyped male and female mice for future use. Keep a watchful eye on these cages and replace stock mice, as they age, with weanlings.

As you expand your colony of mice carrying the mutation, you will likely be breeding heterozygous males with wild-type females and heterozygous females with wild-type males. If you notice a mutant phenotype in the resulting heterozygotes, but only from one of these types of crosses, then you have evidence for an autosomally imprinted mutant phenotype (see Special Considerations for Imprinted Genes in Chapter 16: Phenotypic Analysis of Dominant Mutant Effects in Mice [Papaioannou and Behringer 2023a]).

Production of Mutant Mice or Embryos for Analysis

From your actively breeding mutant-maintenance colony, production of mutants for analysis is straightforward. If male and female heterozygotes are fertile, mate them together: One-quarter of the offspring will be homozygous mutants; one-half will be heterozygotes, which might have a phenotype if the mutation is dominant or semidominant; and the remaining quarter will be wild type and can serve as controls. Set up some heterozygous males in individual cages to serve as studs in sufficient numbers to supply your experimental needs. Once their genotype and fertility have been confirmed by the production of homozygous mutant offspring, these males become priceless. Leave them in their home cage, mate them by sequentially introducing heterozygous females that you can select for estrus, and remove the females when they have mated (i.e., when a vaginal plug has been detected [see Box 1 and Figure 2 in Chapter 5: Recovering a Targeted Mutation in Mice from Embryonic Stem Cell Chimeras or CRISPR–Cas Founders (Papaioannou and Behringer 2023c)]). This scheme will produce a steady supply of pregnant females for the analysis of offspring or embryos. As a standard safeguard to ensure that the female is really a heterozygote, take a tissue sample to regenotype each of the mothers you use to generate embryos or offspring for experiments. That way, your Mendelian proportions will not be skewed by inclusion of litters from incorrectly genotyped mothers. This is especially important early on when you are still determining when your mutant phenotype shows up and whether it is variable in presentation.

MAINTAINING A MUTANT COLONY: BEYOND THE BASICS

Deletion of Selection Cassettes in Vivo

If you generated a targeted mutation in ES cells, you used a drug-resistance expression cassette in the gene-targeting strategy to isolate DNA transformants and identify homologous recombinants. These cassettes usually have their own promoters, which can potentially influence the expression of neighboring genes and/or the expression of your targeted locus. Therefore, in most cases, it is desirable to remove the selection cassette from the targeted locus. This is relatively simple using the Cre/*loxP* or Flp/*FRT* recombinase systems (see Site-Specific Recombinases in Chapter 3: Mouse Gene-Targeting Strategies for Maximum Ease and Versatility [Papaioannou and Behringer 2023f]) and the general principles apply to other recombinase systems as well.

Floxed or flrted cassettes can be removed either in vitro in the targeted ES cells before making chimeras (see Box 2 in Chapter 3: Mouse Gene-Targeting Strategies for Maximum Ease and Versatility [Papaioannou and Behringer 2023f]) or in vivo in ES cell–derived mice by crosses with mice that express the appropriate recombinase. Floxed (flanked by *loxP* sites) or flrted (flanked by *FRT* sites) selection cassettes can be removed in vivo by crosses with transgenic mice that express Cre or Flp recombinase, respectively, in germ cells, blastomeres, the inner cell mass (ICM), or the epiblast before primordial germ cell formation (e.g., *CmvCre* and *Sox2Cre*; Arango et al. 1999; Hayashi et al. 2003).

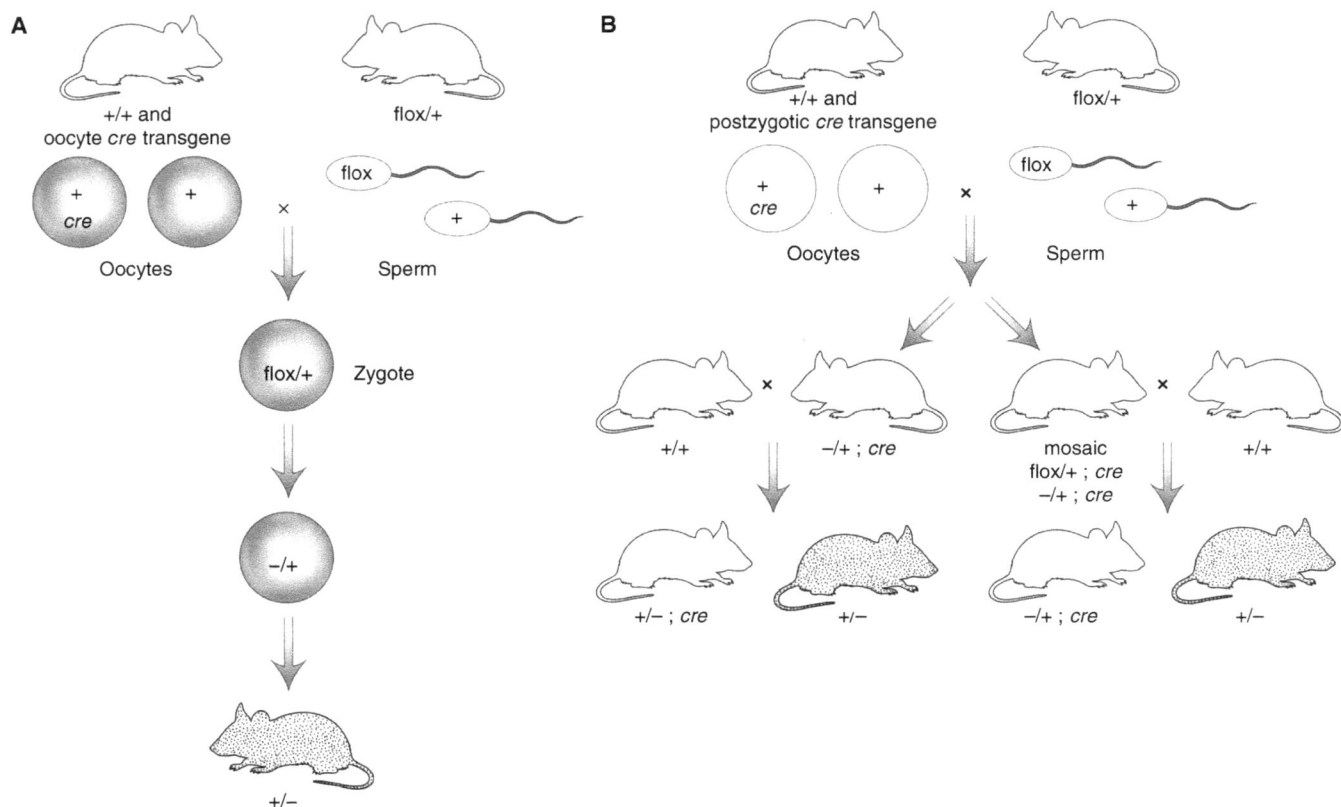

FIGURE 2. Deletion of selection cassettes in vivo. (*A*) Oocyte-Cre deleter strategy. Transgenic mice express Cre during oogenesis, before the final meiotic divisions. Two types of oocytes will be generated: those that carry the *cre* transgene and those lacking the *cre* transgene. However, all of the oocytes will have Cre protein (shading). When sperm from a male carrying a floxed drug-selection cassette fertilize these oocytes, the floxed cassette will be deleted in the zygote even in the absence of the *cre* transgene. Mice heterozygous for the cassette-deleted allele (–) and without the *cre* transgene are then selected to perpetuate the mutation. This strategy is attractive because you delete the drug-selection cassette (or any floxed sequence) and get rid of the *cre* transgene in one step. (*B*) Postzygotic Cre deleter strategy. Transgenic mice express Cre in blastomeres, the inner cell mass (ICM), or the epiblast before germ cell allocation. Two types of oocytes will be generated from these mice: those that carry the *cre* transgene and those lacking the *cre* transgene. In contrast to the oocyte-Cre deleter strategy, no oocytes will have Cre protein. When sperm from a male carrying a floxed drug-selection cassette fertilize these oocytes, the floxed cassette will be deleted postzygotically only if the *cre* transgene is also inherited. Sometimes this results in a somatic and germ cell mosaic (i.e., some cells retain the drug-selection cassette and some delete the floxed sequence). Mice with the floxed sequence deleted can be recovered by an additional cross to wild-type mice. This also serves to segregate the *cre* transgene away from the recombined allele.

These transgenics or knock-ins are sometimes called "deleter" mice. The first of two situations to consider is that mice heterozygous for a floxed (or flrted) selection cassette are crossed with mice heterozygous for a *cre* (or *flp*) transgene that is expressed in developing oocytes before meiotic divisions (Fig. 2A). The deleter female will ovulate two types of oocytes: those that carry the recombinase transgene and those that do not. However, all of the oocytes will have recombinase protein because of the premeiotic expression of the transgene. Thus, when sperm carrying the floxed (or flrted) cassette fertilize the recombinase-carrying oocytes, all of the resulting zygotes (recombinase transgene positive or negative) will have the floxed (or flrted) cassette removed. Genotype and select the progeny that are negative for the deleter transgene or knock-in allele. This strategy can also be used for crosses between male germline ES cell chimeras transmitting a floxed (or flrted) allele and deleter females for removal of the selection cassette upon germline transmission.

In the second situation, mice heterozygous for a floxed (or flrted) selection cassette are crossed with mice heterozygous for a *cre* (or *flp*) transgene that is expressed in blastomeres, the ICM, or the

Cite this chapter as *Cold Spring Harb Protoc*; doi:10.1101/pdb.over107960

epiblast before primordial germ cell formation (Fig. 2B). Because recombinase expression occurs after fertilization, deletion of the floxed (or flrted) selection cassette will occur only in double heterozygous mice that carry both the recombinase transgene and the floxed (or flrted) allele. Therefore, to generate mice carrying the targeted mutation without the selection cassette, you will need to breed the double heterozygotes with wild-type mice to segregate the deleted allele and the recombinase transgene or knock-in allele.

It is important to remember that if you want to remove a floxed or flrted selection cassette and simultaneously maintain your targeted allele on an inbred genetic background, you must have a Cre- or Flp-expressing mouse of the same inbred background, which may be hard to find. Some Cre deleter mouse strains are available that are congenic with C57BL/6 substrains. 129 inbred Cre deleter mice have been reported, but with so many different 129 substrains (see Table 1 in Chapter 4: Embryonic Stem Cell Gene Targeting and Chimera Production in Mice [Papaioannou and Behringer 2023d]), generating a true coisogenic strain will be limited by *cre*-transgenic mouse strain availability. A solution to this problem is to generate inbred zygotes that contain your targeted allele with the floxed or flrted selection cassette and then to microinject pronuclei with a Cre or FRT expression plasmid—a circular form for transient expression without integration. The resulting progeny may have the selection cassette removed in all tissues or they may be mosaic. If they are mosaic, you can backcross to the inbred strain to segregate the alleles. It is also possible to expose two-cell mouse embryos to a commercially available cell membrane-permeable CRE protein in vitro to remove a floxed sequence (Ryder et al. 2014).

Maintaining a Mutation Using a Balancer Chromosome

If you have a recessive lethal mutation, you must genotype heterozygotes at each generation to maintain the line. Balancer chromosomes, which have been engineered in the mouse, contain large (cM) inversions that suppress the recovery of recombination events over the region of the inversion. In addition, the inversion is marked with a visible dominant marker (e.g., coat color) and a linked recessive lethal mutation. Thus, mice heterozygous for a balancer chromosome can be distinguished from wild-type mice by visual inspection for the dominant marker, whereas mice homozygous for the balancer chromosome die as embryos (e.g., *Dph2l1*; Chen and Behringer 2004). To balance a recessive lethal mutation, breed mice heterozygous for your mutation to mice carrying a balancer chromosome with an inversion that spans the region containing your gene mutation (Fig. 3). Then select males and females that are heterozygous for your mutation and the balancer chromosome. Breed these mice together to generate progeny. Progeny homozygous for your gene mutation will die due to the mutation's effects, as will mice homozygous for the balancer chromosome because of the recessive lethal mutation included on it. Therefore, the only mice born from this cross will be mice heterozygous for your gene mutation and heterozygous for the balancer chromosome (i.e., they will have the same genotype as their parents and no genotyping is necessary). A balancer chromosome does not necessarily have to have a linked recessive lethal mutation if the dominant visual marker can be used to distinguish mice heterozygous or homozygous for the balancer.

Balancer chromosomes can also be used to maintain a recessive infertility gene. As with the lethal mutation, mice homozygous for the balancer chromosome will die, and mice heterozygous or homozygous for your mutation and the balancer gene will be distinguishable by the visible marker on the balancer chromosome. No genotyping or test breeding is necessary. One limitation is that relatively few engineered balancer chromosomes have been generated in the mouse covering only a fraction of the genome (Zheng et al. 1999; Nishijima et al. 2003; Chick et al. 2004; Ye et al. 2016).

Assisted Reproduction to Deal with Infertility

It may be that your mutant gene causes dominant infertility or that you simply find yourself in the situation of having the mutant gene in only a few mice that have stopped breeding, even though this might not be part of the mutant phenotype. In either case, to rescue the mutation, a series of increasingly involved steps can be taken to assist reproduction. Of course, if the infertility is

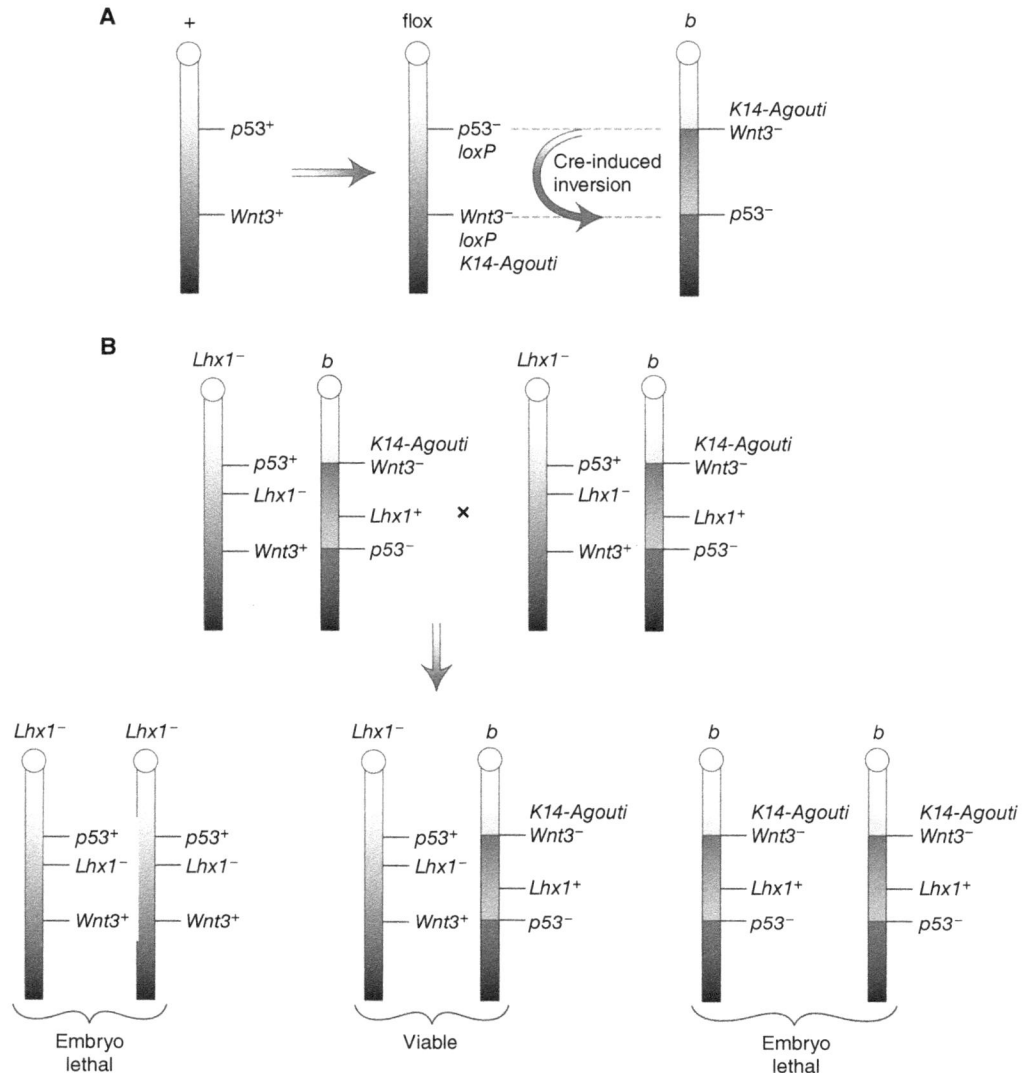

FIGURE 3. Maintaining a recessive lethal or sterile mutation using a balancer chromosome. (*A*) Diagram of a balancer chromosome generated by chromosome engineering on chromosome 11 (Zheng et al. 1999). Two *loxP* sites in reverse orientation were introduced in *cis* into the *p53* and *Wnt3* loci using gene targeting in ES cells (flox chromosome). Cre expression caused the inversion of this 24-cM region. The targeting strategy also disrupted the *p53* and *Wnt3* genes and introduced a *K14-Agouti* transgene into the *Wnt3* locus. The balancer chromosome (*b*) has (1) a 24-cM inversion that suppresses recombination with the wild-type chromosome (+) within the inverted region, (2) a recessive embryonic lethal mutation (i.e., *Wnt3⁻*), and (3) a dominant coat color marker (i.e., agouti pigmentation). Mice heterozygous for the balancer chromosome (*b*/+) can be recognized on both agouti and non-agouti genetic backgrounds because they have lighter ears and tails because of *K14*-directed expression of *Agouti*. Mice homozygous for the balancer chromosome (*b*/*b*) die because they are *Wnt3⁻*/*Wnt3⁻*. (*B*) Breeding scheme to balance a recessive embryonic lethal mutation. *Lhx1* resides on chromosome 11 between *p53* and *Wnt3*. *Lhx1⁻*/*Lhx1⁻* mutants die around mid-gestation. To balance the *Lhx1⁻* allele, *Lhx1⁺*/*Lhx1⁻* mice are first bred to mice carrying the chromosome 11 balancer chromosome (not shown). Agouti progeny with light ears and tails (i.e., carrying the balancer chromosome) are genotyped for the presence of *Lhx1⁻*. Finally, *Lhx1⁻*/*b* males and females are intercrossed (shown). The only progeny that will survive to birth will be those with the same genotype as their parents (*Lhx1⁻*/*b*). Therefore, no genotyping is necessary to maintain this stock. All you have to do is set up any of the progeny from these crosses as breeding pairs. When necessary, the *Lhx1* mutation can be separated from the balancer chromosome by outcrossing to wild-type mice, genotyping the progeny, and selecting *Lhx1* heterozygotes.

Cite this chapter as *Cold Spring Harb Protoc*; doi:10.1101/pdb.over107960

limited to only one sex, breeding heterozygotes of the fertile sex to wild-type mice can perpetuate the mutation. But when you want to examine homozygous mutants, you may need to rely on assisted reproduction to overcome the infertility of the other sex.

Dealing with Male Infertility

If an important mutant male mouse is not breeding even after setting up matings with different 6- to 8-wk-old females, then a series of assisted reproduction methods can be performed to rescue his germline. First, superovulated weanling females can be provided to the male (for details, see Behringer et al. 2018). If a plug is obtained, you can collect the fertilized oocytes and transfer them into pseudopregnant foster mothers to generate progeny (for details, see Nagy et al. 2006). It is possible that you will recover oocytes, but they will not be fertilized, even if a plug is present. If this occurs, attempt a second mating and if this does not yield fertilized oocytes, then in vitro fertilization (IVF) is indicated (for details, see Taft 2017a, which originally appeared in Chapter 15 of Behringer et al. 2014).

For IVF, isolate sperm from the cauda epididymis of the male mutant. Because each male has two epididymides, one can be removed surgically for IVF without sacrificing the animal, leaving the other intact for a second attempt later. Transfer fertilized oocytes that progress to the two-cell stage after IVF into the oviducts of pseudopregnant foster mothers to generate progeny. At each IVF attempt, sperm can be frozen for subsequent manipulations (for details, see Takeo and Nakagata 2018; Taft 2017b).

At this point, if the IVF attempts are unsuccessful, some options are possible. Epididymal or testicular sperm can be used for intracytoplasmic sperm injection (ICSI) into oocytes (e.g., *Tpn1*, *Tpn2*; Zhao et al. 2004; for details, see Ward and Yanagimachi 2018). In addition, if the mouse is extremely important, one could consider animal cloning (Ogura 2017), although this is impractical as a means of perpetuating mutations. If this is a targeted mutation, it would be much simpler to recreate the mouse the way it was obtained through ES cells or CRISPR–Cas in preimplantation embryos. If the infertility is a dominant effect of the mutation, you may never be able to get homozygous mutants by assisted reproduction but the reproductive phenotype can be studied in the heterozygous offspring of the chimeras.

Dealing with Female Infertility

If an important mutant female mouse is not breeding, first exchange the male partner for a fresh, young one (6–8 -wk-old) and wait to see if a pregnancy results. If the female does not become pregnant within 1 mo, inject the appropriate hormones for superovulation and pair with a wild-type male. If you obtain a plug, collect fertilized oocytes for transfer into wild-type recipient females to circumvent a potentially nonreceptive reproductive tract (e.g., *Lif*; Stewart et al. 1992). Save the mutant ovaries for ovary transplantation. Ovary transplantation is a standard methodology used for rescuing the germline of female mice (for details, see Behringer 2017) that could also be used if no plug is obtained. Ovaries or pieces of ovaries can be transplanted into the bursae of wild-type histocompatible hosts where they will ovulate functional oocytes and, if fertilized, produce progeny. If your female mutant is on an inbred background, females of that strain or an F_1 hybrid in which one component is that same inbred strain can be used as hosts. If the mutant is on a B6/129 mixed genetic background, you can use a B6129F1 female as a histocompatible host. If your mutant is on an outbred genetic background, you will have to consider transferring the ovaries into an immune-compromised female recipient (e.g., a *nude* mouse). Donor ovaries can be isolated from adult mice, prepubertal females, or even newborn females. If the ovary transplant fails, you might want to consider animal cloning as a last resort (for details, see Ogura 2017); freeze some tissues (e.g., tail fibroblasts) from the female for this purpose.

Under certain circumstances, other types of artificial reproductive techniques, such as IVF, or ICSI might be appropriate. As with male infertility, a dominant infertility effect might preclude the production of homozygous mutants, but the reproductive phenotype can be studied in the heterozygous mice.

REFERENCES

Arango NA, Lovell-Badge R, Behringer RR. 1999. Targeted mutagenesis of the endogenous mouse *Mis* gene promoter: in vivo definition of genetic pathways of vertebrate sexual development. *Cell* 99: 409–419. doi:10.1016/S0092-8674(00)81527-5

Behringer RR. 2017. Mouse ovary transplantation. *Cold Spring Harb Protoc* doi:10.1101/pdb.prot094458

Behringer RR, Gertsenstein M, Nagy KV, Nagy A. 2014. *Manipulating the mouse embryo: a laboratory manual.* Cold Spring Harbor Laboratory Press, Cold Spring Harbor, NY.

Behringer RR, Gertsenstein M, Nagy KV, Nagy A. 2018. Protocol: administration of gonadotropin for superovulation in mice. *Cold Spring Harb Protoc* doi:10.1101/pdb.prot092403

Center for Animal Resources and Development (CARD). 2008. http://card.medic.kumamoto-u.ac.jp/card/english [Accessed October 6, 2022].

Chapman DL, Papaioannou VE. 1998. Three neural tubes in mouse embryos with mutations in the T-box gene, *Tbx6*. *Nature* B391: 695–697. doi:10.1038/35624

Chen CM, Behringer R. 2004. *Ovca1* regulates cell proliferation, embryonic development, and tumorigenesis. *Genes Dev* 18: 320–332. doi:10.1101/gad.1162204

Chick WS, Mentzer SE, Carpenter DA, Rinchik EM, You Y. 2004. Modification of an existing chromosomal inversion to engineer a balancer for mouse chromosome 15. *Genetics* 167: 889–895. doi:10.1534/genetics.104.026468

European Mouse Mutant Archive (EMMA). 2020. https://www.infrafrontier.eu/resources-and-services/deposit-mice-emma-repository [Accessed October 6, 2022].

Hayashi S, Tenzen T, McMahon AP. 2003. Maternal inheritance of Cre activity in a *Sox2Cre* deleter strain. *Genesis* 37: 51–53. doi:10.1002/gene.10225

International Knockout Mouse Consortium (IKMC). https://www.mousephenotype.org/about-impc/about-ikmc/ [Accessed October 6, 2022].

Mattapallil MJ, Wawrousek EF, Chan CC, Zhao H, Roychoudhury J, Ferguson TA, Caspi RR. 2012. The *Rd8* mutation of the *Crb1* gene is present in vendor lines of C57BL/6N mice and embryonic stem cells, and confounds ocular induced mutant phenotypes. *Invest Ophthalmol Vis Sci* 53: 2921–2927. doi:10.1167/iovs.12-9662

MGI Nomenclature for mutant alleles generated in ES cell lines by the IKMC. 2022. http://www.informatics.jax.org/mgihome/nomen/IKMCnomen.shtml [Accessed October 6, 2022].

Mouse Genome Informatics. 2022. http://www.informatics.jax.org/mgihome/nomen/ [Accessed October 6, 2022].

Mouse Genomes Project. https://www.sanger.ac.uk/data/mouse-genomes-project/ [Accessed October 6, 2022].

Mutant Mouse Resource and Research Center (MMRRC). 2022. https://www.jax.org/research-and-faculty/resources/mutant-mouse-resource-research-center [Accessed October 6, 2022].

Nagy A, Gertsenstein M, Vintersten K, Behringer R. 2006. Oviduct transfer. *Cold Spring Harb Protoc* doi:10.1101/pdb.prot4379

National Academies Lab Code Database. 2022. https://www.nationalacademies.org/ilar/lab-code-database [Accessed October 6, 2022].

Nishijima I, Mills A, Qi Y, Mills M, Bradley A. 2003. Two new balancer chromosomes on mouse chromosome 4 to facilitate functional annotation of human chromosome 1p. *Genesis* 36: 142–148. doi:10.1002/gene.10207

Ogura A. 2017. Cloning mice. *Cold Spring Harb Protoc* doi:10.1101/pdb.prot094425

Papaioannou VE, Behringer RR. 2023a. Phenotypic analysis of dominant mutant effects in mice. *Cold Spring Harb Protoc* doi:10.1101/pdb.over107978

Papaioannou VE, Behringer RR. 2023b. Special breeding techniques for use in mouse mutation analysis. *Cold Spring Harb Protoc* doi:10.1101/pdb.over107961

Papaioannou VE, Behringer RR. 2023c. Recovering a targeted mutation in mice from embryonic stem cell chimeras or CRISPR–Cas founders. *Cold Spring Harb Protoc* doi:10.1101/pdb.over107959

Papaioannou VE, Behringer RR. 2023d. Embryonic stem cell gene targeting and chimera production in mice. *Cold Spring Harb Protoc* doi:10.1101/pdb.over107958

Papaioannou VE, Behringer RR. 2023e. Backcrossing to generate a congenic mouse strain. *Cold Spring Harb Protoc* doi:10.1101/pdb.prot108039

Papaioannou VE, Behringer RR. 2023f. Mouse gene targeting strategies for maximum ease and versatility. *Cold Spring Harb Protoc* doi:10.1101/pdb.over107957

Ryder E, Doe B, Gleeson D, Houghton R, Dalvi P, Grau E, Habib B, Miklejewska E, Newman S, Sethi D, et al. 2014. Rapid conversion of EUCOMM/KOMP-CSD alleles in mouse embryos using a cell-permeable Cre recombinase. *Transgenic Res* 23: 177–185. doi:10.1007/s11248-013-9764-x

Simon MM, Greenaway S, White JK, Fuchs H, Gailus-Durner V, Wells S, Sorg T, Wong K, Bedu E, Cartwright EJ, et al. 2013. A comparative phenotypic and genomic analysis of C57BL/6J and C57BL/6N mouse strains. *Genome Biol* 14: R82. doi:10.1186/gb-2013-14-7-r82

Stewart CL, Kaspar P, Brunet LJ, Bhatt H, Gadi I, Kontgen F, Abbondanzo SJ. 1992. Blastocyst implantation depends on maternal expression of leukaemia inhibitory factor. *Nature* 359: 76–79. doi:10.1038/359076a0

Taft R. 2017a. In vitro fertilization in mice. *Cold Spring Harb Protoc* doi:10.1101/pdb.prot094508

Taft R. 2017b. Mouse sperm cryopreservation using cryoprotectant containing monothioglycerol (MTG). *Cold Spring Harb Protoc* doi:10.1101/pdb.prot094490

Taft R. 2018. Mouse embryo cryopreservation by slow freezing. *Cold Spring Harb Protoc* doi:10.1101/pdb.prot094540

Takeo T, Nakagata N. 2018. Mouse sperm cryopreservation using cryoprotectant containing L-glutamine. *Cold Spring Harb Protoc* doi:10.1101/pdb.prot094516

Threadgill DW, Dlugosz AA, Hansen LA, Tennenbaum T, Lichti U, Yee D, LaMantia C, Mourton T, Herrup K, Harris RC, et al. 1995. Targeted disruption of mouse EGF receptor: effect of genetic background on mutant phenotype. *Science* 269: 230–234. doi:10.1126/science.7618084

Ward MA, Yanagimachi R. 2018. Intracytoplasmic sperm injection in mice. *Cold Spring Harb Protoc* doi:10.1101/pdb.prot094482

Ye Z, Sun L, Li R, Han M, Zhuang Y, Wu X, Xu T. 2016. Generation of a mouse full-length balancer with versatile cassette-shuttling selection strategy. *Int J Biol Sci* 12: 911–916. doi:10.7150/ijbs.15172

Yen ST, Zhang M, Deng JM, Usman SJ, Smith CN, Parker-Thornburg J, Swinton PG, Martin JF, Behringer RR. 2014. Somatic mosaicism and allele complexity induced by CRISPR/Cas9 RNA injections in mouse zygotes. *Dev Biol* 393: 3–9. doi:10.1016/j.ydbio.2014.06.017

Zhao M, Shirley CR, Hayashi S, Marcon L, Mohapatra B, Suganuma R, Behringer RR, Boissonneault G, Yanagimachi R, Meistrich ML. 2004. Transition nuclear proteins are required for normal chromatin condensation and functional sperm development. *Genesis* 38: 200–213. doi:10.1002/gene.20019

Zheng B, Sage M, Cai WW, Thompson DM, Tavsanli BC, Cheah YC, Bradley A. 1999. Engineering a mouse balancer chromosome. *Nat Genet* 22: 375–378. doi:10.1038/11949

Backcrossing to Generate a Congenic Mouse Strain

> Genetic background can have subtle or profound effects on mutant phenotypes, providing additional information regarding the function of the gene. If your mutation is maintained on one genetic background but you wish to analyze it on another, it is a simple matter to transfer the mutation to a recipient strain background by repeated backcrossing (introgression) as detailed in this protocol. The resulting strain is called a congenic strain, defined as a strain carrying the mutation within a segment of chromosome from the donor strain with the remainder of the genome from the recipient strain.

INTRODUCTION

Genetic background, the various alleles at all loci within the genome, can greatly influence the phenotypic expression of a mutation (e.g., *Egfr*; Threadgill et al. 1995). The consequences of this genetic variation are observed in the distinctive characteristics of inbred mouse strains that have been established. For example, C57BL/6J mice have an ~4% frequency of abnormal eyes, SJL/J male mice are very aggressive, and FVB/NJ mice have large litters (Tam and Cheung 2020). These different genetic backgrounds could enhance or suppress the phenotype resulting from a particular mutation in a mouse you have acquired or generated. Thus, it is important to consider on which genetic background you will perform your study. You are not limited to studying the phenotype resulting from a mutation on just one genetic background. A simple way to move a mutation from one genetic background to another, a process called introgression, is to generate a congenic strain, as described in this protocol. A mouse heterozygous for a mutation of interest (donor) is bred to mice of the genetic background of choice (recipient). Heterozygotes are identified among the progeny using standard genotyping procedures and then backcrossed to the chosen inbred strain. This is repeated for at least 10 backcross generations, the formal definition of a congenic strain. This congenic strain contains the mutation within a chromosomal segment from the donor strain with the rest of the genome from the recipient strain. The size of the remaining donor chromosomal segment containing the mutation depends on random recombination events that occur between the donor and recipient genomes during meiosis at each backcross generation.

MATERIALS

It is essential that you consult the appropriate Material Data Safety sheets and your institution's Environmental Health and Safety Office for proper handling of equipment and hazardous materials used in this protocol.

© 2024 Cold Spring Harbor Laboratory Press
Cite this protocol as *Cold Spring Harb Protoc*; doi:10.1101/pdb.prot108039

Reagents

Sexually mature donor mouse carrying the mutation of interest on any genetic background
Sexually mature mice of the inbred strain of the chosen genetic background

METHOD

1. Cross a mouse heterozygous for the mutation (donor) to a mouse of the chosen inbred strain (recipient). This is called an outcross (Fig. 1).

   If the mutation is originally on an inbred background (e.g., a mutation made with CRISPR–Cas gene editing in C57BL/6N zygotes and maintained with C57BL/6N matings), then the offspring of the outcross will be genetically uniform F_1 mice, with half of their genome coming from the C57BL/6N strain and half coming from the other, recipient inbred strain. Incidentally, the genetic background of this F_1 generation can always be recreated after a congenic strain is established by outcrossing mice from the congenic strain to the original C57BL/6N strain.

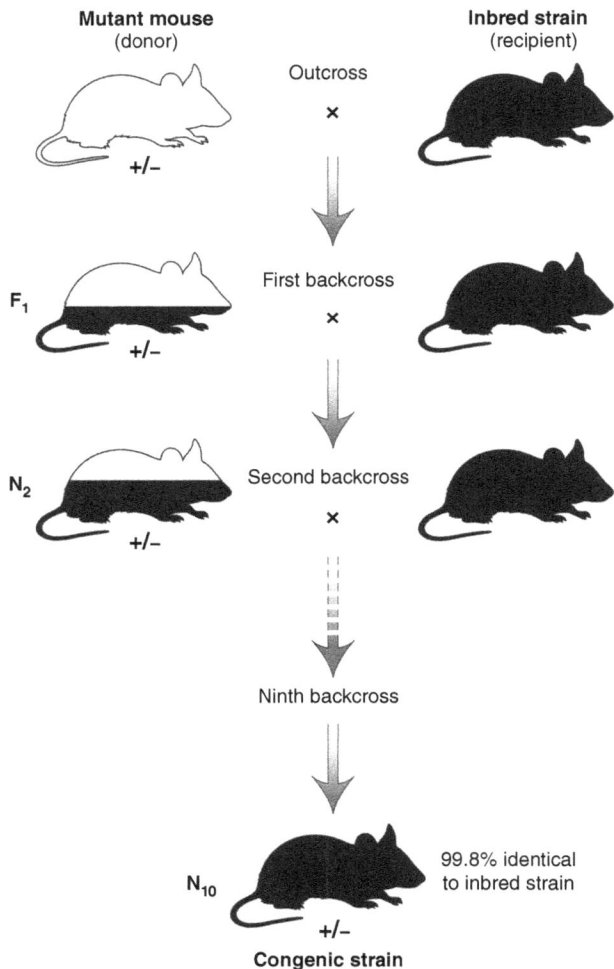

FIGURE 1. Backcross scheme to generate a congenic strain. By crossing a heterozygous mutant (donor) to an inbred strain (recipient) and selecting heterozygous offspring from the F_1 generation to cross back to the inbred strain for the next generation (N_2), a congenic strain that is 99.8% identical to the inbred strain but carries the mutation will be achieved after the ninth backcross.

If the mutation is originally on a mixed genetic background, as might be the case for a spontaneous mutation, the outcross creates a unique background that is 50% from the mixed background and 50% from the recipient inbred background.

2. Genotype the F_1 offspring using standard genotyping methods and select mice heterozygous for the mutation to backcross with mice of the recipient inbred strain. (For details, see Chapter 13 in Behringer et al. 2014 and Behringer et al. 2019.)

3. Genotype progeny at each backcross generation, selecting heterozygotes for the next backcross.

   A total of 10 crosses to an inbred strain define a congenic background. In all the crosses, mice selected for heterozygosity at the mutant locus are backcrossed to mice of the recipient inbred strain. At each generation (called generation N_2, ..., N_n) the genetic diversity is decreased and the mice contain more and more of the recipient inbred genome. At N_{10}, this reaches 99.8% and the mutant strain is considered congenic with the recipient inbred strain (Fig. 1).

4. Keep pedigree records of all matings.

   Although you may want to set up multiple crosses at each generation to ensure fertile matings, the final congenic strain must trace back through a single N_{10} founder before it is expanded into the final breeding colony.

TROUBLESHOOTING

You may notice a decline in fertility as you progress through the generations. That is because you started with an F_1 (or a mixed genetic background) that had hybrid vigor and you are moving toward a genetically uniform inbred background. However, the fertility should approach that of the recipient inbred strain and not get any worse, unless, of course, the mutation has a heterozygous phenotype affecting fertility.

DISCUSSION

When backcrossing, it makes sense to use a male of the recipient inbred strain, because you can use the same mouse for several generations if he is a good breeder. Simply mate him with his heterozygous daughters and granddaughters. If instead you choose to use recipient inbred females, just be sure to use a recipient inbred male for at least one generation to get the Y chromosome of the recipient inbred strain into your congenic line.

Theoretically, after nine backcrosses the N_{10} offspring are genetically identical except for a differential chromosomal region of ~20 cM (~40,000 kb) containing the mutant allele. Because heterozygotes were selected at each generation, the chromosomal region around the mutation was selected, and it will only become smaller by random crossover events during meiosis. It would take another 40 generations of backcrossing to reduce the size of the differential chromosomal segment to <5 cM by random crossover. However, for most practical purposes, accepting 99.8% genetic uniformity is a fairly good alternative. In fact, you might want to monitor the phenotype during earlier generations of backcrossing because an effect might be seen quite early on; by the third backcross generation, the mice are already ~90% identical to the recipient inbred strain. Thus, experiments could be initiated to study the phenotype of the mutation at the third backcross generation while you continue backcrosses to generate the congenic strain. Sometimes one backcross can greatly influence the phenotype of a mutation. In some cases, a severe mutant phenotype can become less severe by an outcross to an outbred stock (e.g., Swiss Webster). The details of the genetic background should always be included in the Methods section of publications.

Congenic mouse lines can be generated more quickly using the "speed congenic" or marker-assisted method. In this method, mice are genotyped for genome-wide polymorphic markers at each backcross generation. The mice that have the highest percentage of markers from the recipient inbred background are then selected for use in the next backcross generation. This selective breeding reduces the number of generations needed to generate a congenic line but requires more work than simply breeding mice (Markel et al. 1997). Also, there is the possibility of decreasing breeding time by making use of superovulation and embryo transfer (Behringer 1998).

REFERENCES

Behringer R. 1998. Supersonic congenics? *Nat Genet* **18**: 108. doi:10.1038/ng0298-108

Behringer RR, Gertsenstein M, Nagy KV, Nagy A. 2014. *Manipulating the mouse embryo: a laboratory manual*. Cold Spring Harbor Laboratory Press, Cold Spring Harbor, NY.

Behringer RR, Gertsenstein M, Nagy KV, Nagy A. 2019. Isolation of high-molecular-weight DNA from mouse tail tips. *Cold Spring Harb Protoc* doi:10.1101/pdb.prot092692

Markel P, Shu P, Ebeling C, Carlson GA, Nagle DL, Smutko JS, Moore KJ. 1997. Theoretical and empirical issues for marker-assisted breeding of congenic mouse strains. *Nat Genet* **17**: 280–284. doi:10.1038/ng1197-280

Tam WY, Cheung KK. 2020. Phenotypic characteristics of commonly used inbred mouse strains. *J Mol Med* **98**: 1215–1234. doi:10.1007/s00109-020-01953-4

Threadgill DW, Dlugosz AA, Hansen LA, Tennenbaum T, Lichti U, Yee D, LaMantia C, Mourton T, Herrup K, Harris RC, et al. 1995. Targeted disruption of mouse EGF receptor: effect of genetic background on mutant phenotype. *Science* **269**: 230–234. doi:10.1126/science.7618084

Cite this protocol as *Cold Spring Harb Protoc*; doi:10.1101/pdb.prot108039

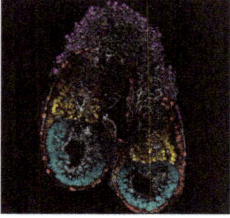

CHAPTER 7

Special Breeding Techniques for Use in Mouse Mutation Analysis

Certain specialized breeding techniques may come in handy during the analysis of a mutation in order to further understanding of the mutation and its interactions with other genes. Different mutant alleles of the gene in question might be available from other sources or mutations with similar phenotypes could potentially be alleles. This could be determined by complementation testing. In the production of a conditional allele, retention of exogenous DNA in the allele could fortuitously disrupt a regulatory element and thus result in a hypomorphic allele, which can be simply tested by breeding. Mutations in different genes frequently affect the same organ, tissue, or cell type through genetic interactions. Common approaches to investigate and interpret genetic interactions are detailed here for gene families, in which there may be redundancy or genetic compensation of different genes, for genes that constitute different components of a biochemical pathway, for genes with overlapping expression patterns, and for unrelated genes that produce similar mutant phenotypes.

INTRODUCTION

SPECIAL BREEDING TECHNIQUES FOR MUTATIONAL ANALYSIS

- COMPLEMENTATION
- HYPOMORPHIC ALLELES
- GENETIC INTERACTIONS   → Chapter 8 or 16
- CHEATING MENDEL

The most exciting part of any mutation analysis is uncovering a mutant phenotype. This can also be the most challenging part of your study as the well-designed and well-characterized mutation in a specific gene is now functioning throughout development in the context of the entire genome. How the genotype leads to a particular phenotype can be a challenge to unravel. At this point in your study of the effects of the mutation, and as you proceed with a systematic phenotypic analysis, there are some special breeding schemes that you might productively use to deepen your understanding of the mutation and its interaction with other genes. The four sections in this chapter center around understanding your allele and its

Opening artwork: Courtesy of Shifaan Thowfeequ and Shankar Srinivas.

© 2024 Cold Spring Harbor Laboratory Press
Cite this chapter as *Cold Spring Harb Protoc*; doi:10.1101/pdb.over107961

interaction with other genes. As you characterize the heterozygous or homozygous mutant phenotype, keep these strategies in mind and apply them when they become appropriate.

COMPLEMENTATION TESTING

Different alleles of the same gene can be very informative about gene function. Be sure to check mutant mouse resources to see if there are any other existing mutations of your gene. Also, once you have a phenotype, be on the lookout for other mutations, either spontaneous or induced in mutagenesis screens, for example, that resemble the phenotype of your mutation. It might be that the mutations are in two different genes but both affect the same developmental pathway, resulting in similar phenotypes (see Testing for Genetic Interactions, below). Alternatively, it may be that they are both mutations in the same gene. This can easily be sorted out if both genes have been cloned, but if one has not been cloned, you can test for allelism of the two mutations in question by what is known as a complementation test (Fig. 1) (e.g., $Tbx6$; Watabe-Rudolph et al. 2002). This test is done by breeding heterozygotes for each mutation together. If the mutations are in separate genes, they will complement one another and the phenotype of all the offspring will be wild type, including the double heterozygotes (assuming both mutations are recessive). If they are alleles of the same gene, the offspring that inherit both alleles (i.e., the compound heterozygotes) have no wild-type allele and will have a phenotype resembling the homozygous phenotypes of the separate mutations. In other words, the mutations do not complement one another and we see a phenotype. The mutations will most likely differ molecularly because of their independent origins, and therefore the compound heterozygotes may not be identical to either parental homozygous type. It certainly will be worth your while to determine the molecular lesion in the second allele to better

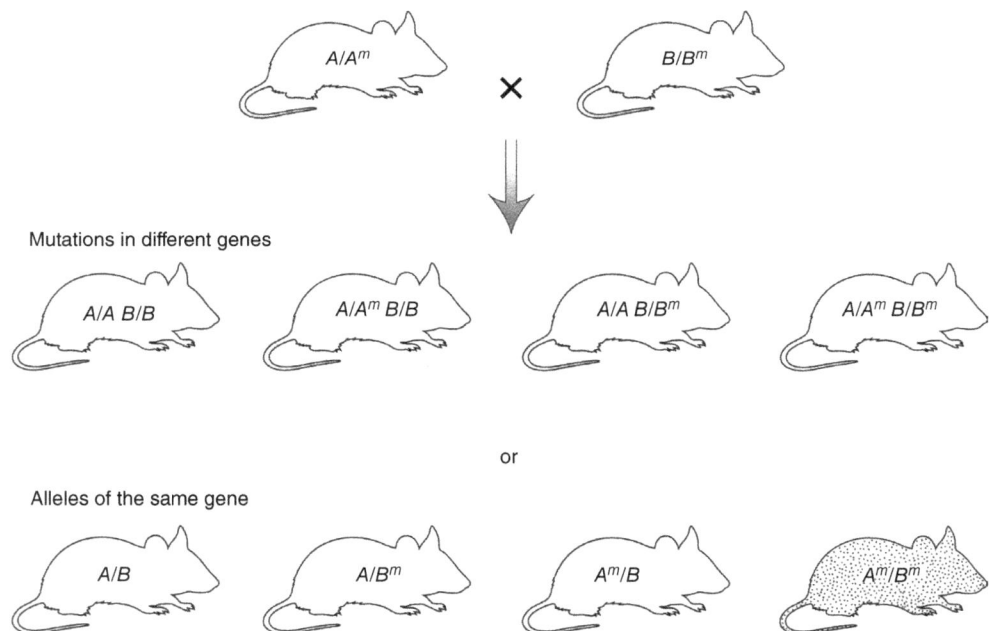

FIGURE 1. Test for allelism between two recessive mutations. Suppose that you produced a mutation (A^m) in gene A and the homozygote has a similar phenotype to a mutation uncovered in an ethyl nitrosurea (ENU) screen. The ENU mutation was called B^m and although the gene has not yet been cloned, it maps to the same chromosomal region as A. To test if B^m is an allele of A, cross A/A^m mice together with B/B^m mice. If the mutations are in different genes, all of the offspring will be wild type in phenotype; if A^m and B^m are alleles of the same-gene, one-quarter of the offspring will be heterozygous for the two alleles (A^m/B^m) and will have a phenotype similar to A^m/A^m or B^m/B^m, depending on the nature of the molecular lesions.

understand the structural/functional relationship of the mutant alleles. With two mutant alleles and a wild-type allele, you are on the way to having an allelic series, which can be very informative about gene function.

TESTING FOR A HYPOMORPHIC ALLELE

Retaining a floxed (flanked by *loxP* sites) or flrted (flanked by *FRT* sites) selection cassette, or even a *loxP* site, within your targeted locus (e.g., within an intron in a conditional allele before deletion of floxed or flrted exons) might fortuitously produce a mutant effect. The promoter of the selection cassette may interfere with the targeted locus, or the cassette or the *loxP* site might have fortuitously disrupted a regulatory element. The resulting allele may have wild-type activity, reduced activity, or no activity (a null). How do you determine which is the case? One way is simply to generate homozygotes for the allele to see if you get a mutant phenotype. If you already have a null allele for this gene, you can compare the two homozygous mutant phenotypes. If they are identical, then it is probable that the presence of the selection cassette or other exogenous sequence has disrupted the expression of the targeted gene to produce a null. If the phenotype is milder than the null phenotype, then you may have a hypomorphic allele. If so, then when you combine the potential hypomorphic allele with the null allele, the resulting phenotype should be more severe than the homozygous hypomorph, but less severe than the homozygous null (e.g., *Tbx6*; Watabe-Rudolph et al. 2002). This can be very useful for understanding how different levels of gene activity influence development.

Another way to assess the activity of a targeted mutation is to combine it with a deficiency that includes the targeted locus. A number of chromosomes with relatively large (cM) deletions are available in mice that can be used for this purpose. Thus, if the phenotype of your targeted allele over the deficiency is less severe than the homozygous null mutant, then your allele may be a hypomorph. Some caution is necessary because the mutation over the deficiency will also be haploid for all of the other genes contained within the deficiency, which could influence the phenotype.

TESTING FOR GENETIC INTERACTIONS

Mutations in many different genes may affect a given tissue. Combining heterozygous and/or homozygous mutations for different genes in the same mouse may reveal informative interactions between the genes. When the mutant phenotype of your gene of interest changes on different genetic backgrounds, it is the result of interactions between your mutation and alleles of genes that differ between the strains used to maintain the mutation (see Box 1 in Chapter 6: Strategies for Maintaining Mouse Mutations [Papaioannou and Behringer 2023a]). Epistasis describes a situation in which the effects of a mutation in one gene essentially mask the effects of a mutation in another gene —for example, a null mutation in an enzyme that catalyzes an early step in a biochemical pathway will be epistatic over all genes that act on later parts of the pathway. On the other hand, a genetic interaction between two mutations would be indicated if mutations in different genes are brought together in a single mouse and the phenotype is different from the mutant phenotypes of the individual genes rather than simply a combination of the two. Genetic interaction is a broad term that does not suggest the mechanism of the interaction. The genes may function in either the same or parallel biochemical pathways. If they function in the same biochemical pathway, then the two gene products may have identical biochemical activities (functional redundancy) or they may be different components of the same biochemical pathway. If two genes function in parallel biochemical pathways, mutation of one gene may compromise a tissue and mutation of the second may further compromise the tissue, producing a more severe abnormality because of the additive effects of both genes. Remember that a biochemical pathway can function within a cell or between cells. Thus, the expression patterns of the genes in question do not necessarily have to be coexpressed to justify a genetic interaction experiment.

It is very easy to breed two different mouse mutants together, but the interpretation of the resulting phenotypes of the double mutants can be difficult. In addition, genetic interactions may show variable penetrance and variable expressivity so generate sufficient data to do statistics (see Box 1 in Chapter 8: Phenotypic Analysis: Assessing Timing of Recessive Prenatal Lethality in Mice [Papaioannou and Behringer 2023b]). Below are some of the more common approaches that you can pursue when investigating a genetic interaction.

Gene Families: Redundancy or Compensation

There is a nearly 30% chance that a gene knockout will produce a homozygous mutant mouse that is viable, fertile, and with no obvious abnormalities (Birling et al. 2021). Why does this seem to occur so frequently? One explanation is that the mammalian genome has multiple copies of related genes that together regulate the same biological processes. Thus, if you have a mutation in only one member of a gene family, you may not reveal an essential role for that particular gene unless you also include mutations in other family members (e.g.; *Myod1*, *Myf5*; Rudnicki et al. 1993). Certainly, if you are pursuing a gene-targeting experiment, search the databases to determine if a related gene(s) exists and if there is any overlap in its expression pattern with your gene. If there is no overlap, then the two genes will not have redundant functions. However, overlap of expression in a particular tissue during development or in the adult may result in tissue that does not show defects unless both genes are mutated because of functional redundancy.

Redundancy can be revealed by generating mice that carry heterozygous or homozygous mutations for each of the related genes (Fig. 2A). Whether you discover new mutant phenotypes in the double heterozygotes (sometimes called transheterozygotes), in mice heterozygous for one gene and homozygous mutant for the other gene, or in double-homozygous mutants will depend on the total levels of the two gene products in a particular tissue at a specific time in development. It is relatively easy to document coexpression (e.g., by in situ hybridization or immunohistochemistry); however, it is more difficult to measure the levels of gene products (mRNA or protein) in specific cells or tissues expressed in specific regions of a developing embryo or in adult tissues. If you can isolate the cells or tissue in question, then real-time reverse transcription–polymerase chain reaction (RT-PCR) or other quantitative assays of mRNA can be used; with sufficient amounts of tissue, a western blot can also be performed. Bulk RNA-seq of the tissue of interest or single-cell RNA-seq can be used to quantify transcripts.

With respect to the effect on the phenotype, functional redundancy may be indistinguishable from a compensatory up-regulation, in which the expression of one gene family member is up-regulated in the absence of another, thus compensating for a single mutant effect (Fig. 2B). As with functional redundancy, a phenotype or a more severe phenotype may only become evident when mutations in more than one family member are combined, but distinguishing redundancy from compensatory up-regulation requires knowing the level of gene expression of the putative compensating gene family member before and after combination with your mutation.

Different Components of a Biochemical Pathway

Many biochemical pathways that typically involve the participation of multiple unrelated proteins have been defined. Thus, you may already know that your gene encodes a component of a biochemical pathway that has been defined in vitro. Are these in vitro findings also true in vivo? One way to test this possibility is to cross your mutant with mice mutant for different components of the biochemical pathway. Simplistically, you might expect the phenotypes of mutations in these genes to be similar, if not identical, if the mutations result in limiting amounts of gene products, and this has been seen many times (see Similar Mutant Phenotypes Caused by Unrelated Genes, below) (Fig. 3). However, it may be more complicated. An in vivo test to determine if the two gene products participate in a common biochemical pathway would be to cross the two mutants together to generate double

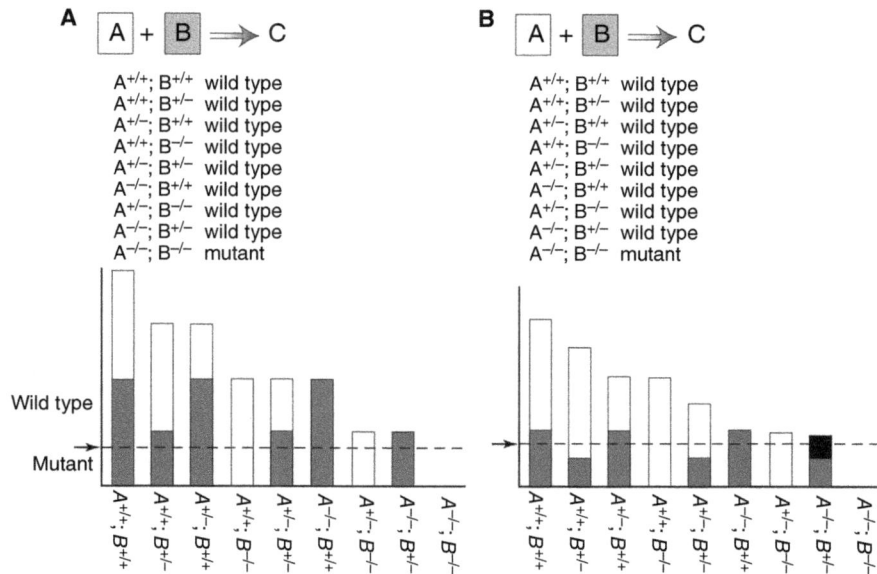

FIGURE 2. Redundancy and compensation. (A) Redundant action of gene family members coexpressed in the same tissue. A (open boxes) and B (shaded boxes) are gene family members that regulate a downstream target C. In this example, the wild-type level of the A protein is equal to the wild-type level of the B protein, and the bars on the histogram indicate the total level of A plus B proteins. The arrow and dotted line in the histogram indicate a hypothetical threshold level of protein that is required in this situation to maintain a wild-type phenotype. When A and B levels are reduced by half (i.e., in the compound heterozygotes) or even by three-quarters (i.e., in the null, heterozygous genotypes), there are still sufficient total levels of A and/or B proteins to maintain a wild-type phenotype. Only when both A and B are eliminated in the double-homozygous genotype does a mutant phenotype develop. (B) Compensation by the up-regulation of a gene family member coexpressed in the same tissue. In this example, the wild-type level of the A protein is higher than the wild-type level of the B protein. The arrow and dotted line in the histogram indicate a hypothetical threshold level of protein that is required in this situation to maintain a wild-type phenotype. One dose of A (i.e., $A^{+/-}$; $B^{-/-}$) is sufficient to maintain a wild-type phenotype, but one dose of B (i.e., $A^{-/-}$; $B^{+/-}$) would not appear to be sufficient. However, in the $A^{-/-}$; $B^{+/-}$ animals, the B locus is somehow sensitive to a deficiency in total A and B proteins and its expression is up-regulated sufficiently (darker shading) to result in a wild-type phenotype.

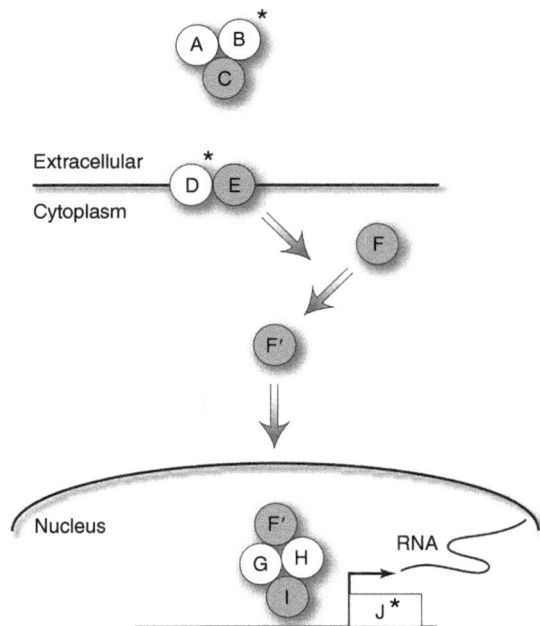

FIGURE 3. Genes in the same biochemical pathway. Extracellular proteins (A–C) can interact with cell membrane proteins (D,E) to transduce signals, perhaps by modifying cytoplasmic proteins (F to F′) that in this example translocate into the nucleus to form a transcriptional complex (F′–I) leading to the expression of a downstream target (J). In vitro studies elucidating this pathway provide a framework for in vivo genetic analyses. Mutations (asterisks) that lead to alterations in individual components of the pathway could independently compromise the pathway leading to a mutant phenotype in animals. In addition, combining mutations for multiple components of the biochemical pathway (i.e., generating double or triple heterozygotes) may reveal a genetic interaction if the gene products of the individual components are at limiting amounts when heterozygous.

heterozygotes. If a new phenotype appears, then you have evidence for a genetic interaction. Because you already had in vitro evidence that the two gene products participated in a biochemical pathway, it would be reasonable to suggest that this was also true in vivo. However, if the two components of the same biochemical pathway were not limiting in the single heterozygotes, then you might not see a new phenotype in the double heterozygotes, and the lack of a new phenotype in double heterozygotes would not preclude the possibility that the two genes functioned in the same biochemical pathway. You could obtain more support for the idea that the two gene products truly acted in the same biochemical pathway if you could examine the expression of a downstream response gene. RNA-seq of the relevant tissues can be used to determine transcriptomes for such an analysis.

Different Genes with Overlapping Expression Patterns

Gene expression studies obtained by microarray and now RNA-seq have yielded tremendous amounts of information available in public databases (https://www.ncbi.nlm.nih.gov/geo/; http://www.informatics.jax.org/expression.shtml). Thus, if you are studying a particular tissue, you will probably know a lot about the expression of other genes in that tissue. Take advantage of this information and the availability of many mouse mutants. Cross your mutant with other mice that have mutations in genes coexpressed in the tissue in which your gene is expressed to look for new mutant phenotypes in double mutants. Be cautious about interpretations by keeping in mind the types of proteins encoded by the two different genes. If no biochemical connection is obvious or you were unable to establish such a connection, then you may simply be compromising the function of a specific tissue in different ways by the additive effects of each mutation, not necessarily revealing a common biochemical pathway. For example, one mutation may slightly alter a tissue without causing an obvious phenotype. A mutation in a different gene may have the same effect. However, when both mutations are brought together, the two subthreshold defects combine to alter the tissue (e.g., proliferation, migration, adhesion) enough to produce a mutant phenotype.

Once you establish a genetic interaction, more biochemical information will be needed to formulate a molecular/biochemical interpretation. However, even if you do not find that the two gene products act in the same biochemical pathway but are probably acting in parallel pathways, the double mutants can be useful since the genetic interaction still results in a mutant tissue. Thus, these tissue-specific alterations can be helpful for understanding the role of a specific tissue during embryonic development or adult tissue function.

Similar Mutant Phenotypes Caused by Unrelated Genes

Once you know the mutant phenotype of your gene, look for other mutations that result in the same or similar phenotypes. If you have mice with identical mutant phenotypes, this suggests that the gene products may participate in the same biochemical pathway (e.g., ligand and receptor or two transcription factors that form a complex) (Fig. 3) (e.g.; *Amh*, *Amhr2*; Behringer et al. 1994; Mishina et al. 1996). Thus, the two genes may be coexpressed in the same cell but could also be expressed in different cells. First, determine the cells and tissues in which the two genes are normally expressed. Then, check the expression of your gene in mice mutant for the other gene and vice versa to investigate a possible epistatic relationship. Generate double heterozygotes. Do the double heterozygotes have a phenotype that is not found in the single heterozygotes? If they do, then you have evidence for a genetic interaction. At this point, you will have to do more experiments to determine if there is a biochemical relationship between the two gene products. As pointed out in Different Components of a Biochemical Pathway, above, a negative result would not provide enough information to draw conclusions.

Mutant Phenotypes Caused by Overexpression of Downstream Genes

A mutation may eliminate a negative regulator of a downstream gene or gene product (i.e., an antagonist). Thus, a mutant phenotype may actually be caused by the overexpression of a downstream gene (Fig. 4) (e.g., *Leftb*, *Cer1*, *Nodal*; Perea-Gomez et al. 2002). This may be indicated by a molecular

Cite this chapter as *Cold Spring Harb Protoc*; doi:10.1101/pdb.over107961

Model   A ⊣ B

          A⁻ ⊣ B↑

Result	Genotype	Phenotype
	$A^{+/+}$; $B^{+/+}$	wild type
	$A^{-/-}$; $B^{+/+}$	mutant phenotype
	$A^{-/-}$; $B^{+/-}$	suppression of mutant phenotype

FIGURE 4. Negative regulator of a downstream gene. If A encodes a negative regulator of B, then a phenotype resulting from the loss of A could be due to the up-regulation (overexpression) of B. This model could be tested by reducing the number of wild-type copies of B (e.g., heterozygosity of B). Suppression of the A mutant phenotype would provide genetic evidence for the model.

marker analysis if you find that one of the markers is up-regulated in a particular tissue that is abnormal. According to this hypothesis, if you reduced the expression of the downstream gene in the mutants, you should suppress the mutant phenotype, indicating a genetic interaction. Therefore, if you suspect such a situation, you can test this idea by crossing your mutant with mice carrying a mutation in the downstream gene, thus reducing the number of wild-type alleles of the downstream gene. This should reduce the amount of downstream over expression and suppress the mutant phenotype.

INCREASING THE FREQUENCY OF HOMOZYGOUS MUTANT MICE

Germ cell–expressing Cre mouse lines can be used in combination with conditional alleles to increase the frequency of homozygous mutant mice from 25% to up to 100%, which could be very useful for

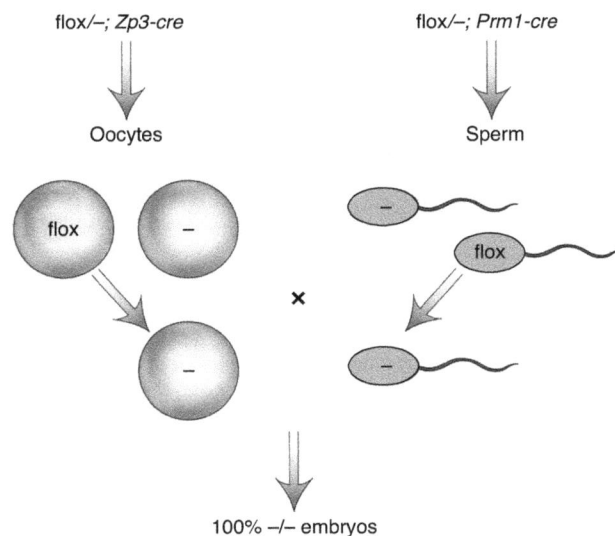

FIGURE 5. Cheating Mendel. Germ cell–specific recombinases can be used in combination with floxed (or flrted) alleles to increase the proportion of homozygous mutant mice or embryos in a litter. For example, *Zp3-cre* female mice express Cre in oocytes (shading) and *Prm1-cre* male mice express Cre in spermatids. Females can be generated that are heterozygous, as in this example, or homozygous for the floxed allele and hemizygous for the oocyte-specific *cre* transgene (flox/–; *Zp3-cre* or flox/flox; *Zp3-cre*), and males can be generated that are heterozygous, as in this example, or homozygous for the floxed allele and hemizygous for the spermatid-specific *cre* transgene (flox/–; *Prm1-cre* or flox/ flox; *Prm1-cre*). Cre expression in both gametes will convert floxed alleles to null (–) alleles. Thus, females produce only null oocytes and males produce only null sperm. From this cross, only homozygous mutant (–/–) mice or embryos can be generated.

analyses such as expression profiling. This strategy has been dubbed (by Gail Martin) "cheating Mendel" because the outcome gives the illusion that Mendel's laws have been broken (Fig. 5) (e.g.; *Sox9*; Chaboissier et al. 2004). Here's how it works: First, obtain transgenic mouse lines that express Cre in the male and female germlines. For example, protamine 1 (*Prm1*)-*cre* transgenic mice express Cre in spermatids and zona pellucida 3 (*Zp3*)-*cre*-transgenic mice express Cre in oocytes. Breed mice with these *cre* transgenes to mice with your conditional allele to obtain progeny that are homozygous for the conditional allele and hemizygous for the germ cell–specific *cre* transgenes. Thus, when the male generates spermatids, Cre expression will convert them from conditional to null. This is facilitated by the syncytial nature of spermatogenesis in which gene products from one nucleus are shared with neighboring sperm cells through a common cytoplasm. In females, the developing oocytes will express Cre before the final meiotic division, loading the oocytes with Cre protein. Thus, the conditional allele will be converted to a null in the resulting oocytes. When these males and females are bred together, they will theoretically generate only homozygous recombined (null) embryos. In practice, the yield of homozygous recombined embryos may be a bit <100%, depending on the efficiency of germ cell Cre activity. You can also manipulate this breeding strategy to generate 50% homozygous mutants if one of the parents is heterozygous for the conditional allele. This system is especially useful in the rare cases of a haploinsufficiency lethal mutation.

REFERENCES

Behringer RR, Finegold MJ, Cate RL. 1994. Mullerian-inhibiting substance function during mammalian sexual development. *Cell* 79: 415–425.

Birling MC, Yoshiki A, Adams DJ, Ayabe S, Beaudet AL, Bottomley J, Bradley A, Brown SDM, Bürger A, Bushell W, et al. 2021. A resource of targeted mutant mouse lines for 5,061 genes. *Nat Genet* 53: 416–419.

Chaboissier M-C, Kobayashi A, Vidal VIP, Lützkendorf S, van de Kant HJG, Wegner M, de Rooij DG, Behringer RR, Schedl A. 2004. Functional analysis of *Sox8* and *Sox9* during sex determination in the mouse. *Development* 131: 1891–1901. doi:10.1242/dev.01087

Mishina Y, Rey R, Finegold MJ, Matzuk MM, Josso N, Cate RL, Behringer RR. 1996. Genetic analysis of the Mullerian-inhibiting substance signal transduction pathway in mammalian sexual differentiation. *Genes Dev* 10: 2577–2587.

Mouse Genome Informatics. 2022. http://www.informatics.jax.org/expression.shtml [Accessed October 6, 2022].

NCBI Gene Expression Omnibus. https://www.ncbi.nlm.nih.gov/geo/ [Accessed October 6, 2022].

Papaioannou VE, Behringer RR. 2023a. Strategies for maintaining mouse mutations. *Cold Spring Harb Protoc* doi:10.1101/pdb.prot107960

Papaioannou VE, Behringer RR. 2023b. Phenotypic analysis: assessing timing of recessive prenatal lethality in mice. *Cold Spring Harb Protoc* doi:10.1101/pdb.over107970

Perea-Gomez A, Vella FD, Shawlot W, Oulad-Abdelghani M, Chazaud C, Meno C, Pfister V, Chen L, Robertson E, Hamada H, et al. 2002. Nodal antagonists in the anterior visceral endoderm prevent the formation of multiple primitive streaks. *Dev Cell* 3: 745–756.

Rudnicki MA, Schnegelsberg PN, Stead RH, Braun T, Arnold HH, Jaenisch R. 1993. MyoD or Myf-5 is required for the formation of skeletal muscle. *Cell* 75: 1351–1359.

Watabe-Rudolph M, Schlautmann N, Papaioannou VE, Gossler A. 2002. The mouse rib-vertebrae mutation is a hypomorphic *Tbx6* allele. *Mech Dev* 119: 251–256.

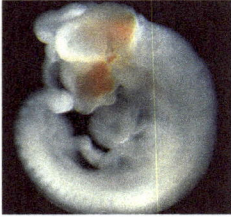

CHAPTER 8

Phenotypic Analysis: Assessing Timing of Recessive Prenatal Lethality in Mice

Once a recessive mutation has been established in a mouse strain in the heterozygous state, the task of phenotypic analysis of the homozygous mutants can begin. This chapter leads you through a sequence of steps to determine whether the homozygous mutants are present at birth or whether the mutation causes prenatal lethality. In the case of a prenatal lethality, the time of death of the mutants, which could occur at any time during pre- or postimplanation development, must be firmly established before further phenotypic analysis. Here, we present a detailed plan to efficiently determine the time of prenatal death of the mutants and provide a guide for developmental landmarks to establish how far they progress during gestation. To determine whether or not homozygous mutants are present or normal at any given time point, it is important to recover a sufficient number of embryos. Examples of a simple Chi square test for Mendelian segregation is provided to establish statistical significance for the genotype/phenotype distribution.

INTRODUCTION

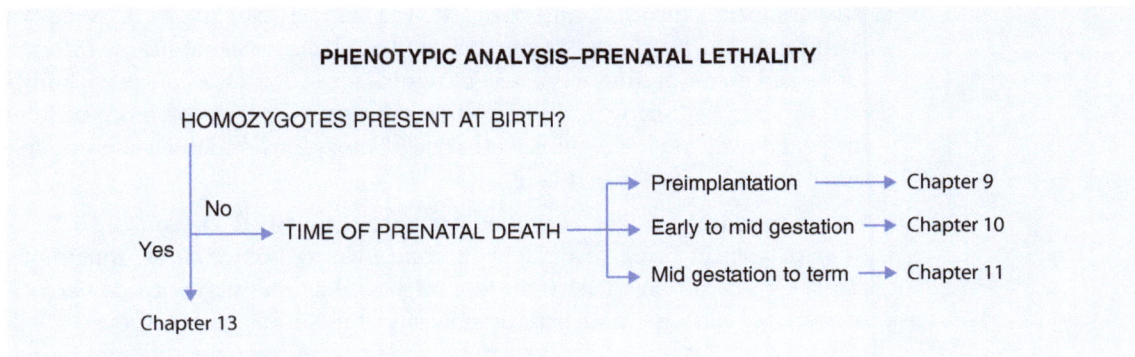

This chapter is the starting point for the analysis of spontaneous or induced mutations in heterozygous or homozygous mutant mice. However you were referred to this chapter, you now have your long-sought phenotype resulting from either a targeted or gene-edited mutation in your favorite gene, a

Opening artwork: Courtesy of Zhou-Feng Chen and Richard Behringer.

Cite this chapter as *Cold Spring Harb Protoc*; doi:10.1101/pdb.over107970

mutation in an unknown gene, or a selected mutation of your choice. The phenotype may be just what you predicted based on gene expression and gene product characteristics, or, if it is a spontaneous mutation, you might have selected it because of the interesting phenotype. But quite possibly, the mutation resulted in a different phenotype than you expected or novel mutant characteristics in addition to those you expected. At first analysis, it may even look like there is no mutant phenotype at all, but this is unlikely, and a closer examination or some manipulations will usually reveal the effects of the mutation.

This chapter presents strategies to pinpoint the timing of a mutant effect. It is written for the situation in which the mutation is recessive and thus the phenotype is being analyzed in the progeny of heterozygous mice that are viable and fertile. However, the analytical methods are applicable to dominant effects as well (see Chapter 16: Phenotypic Analysis of Dominant Mutant Effects in Mice [Papaioannou and Behringer 2023a]). Thus, we begin by examining the progeny of matings between known heterozygous mice.

DETERMINING WHETHER HOMOZYGOUS MUTANTS ARE PRESENT AT BIRTH

The first question is whether the homozygous mutants are present at birth … but how will you know if they are or not? You may have noticed smaller than normal litters, but it is impractical to genotype mice at birth, so most likely the first time you will be certain of missing homozygous mutants is when you genotype the mice between 10 d and weaning. If they are not present then, they could still be born but die sometime between birth and genotyping. As a routine, check females near the expected date of parturition, count newborn offspring, with minimal disruption of the nest, and look for any dead pups, which should be genotyped. At the time of genotyping or weaning of the surviving offspring, see if the number of pups correlates with the number born, or whether postnatal losses and the lack of homozygous mutants indicate that they were born but died before weaning, in which case, you should start with Chapter 13: Phenotypic Analysis of Perinatal Lethality in Mice (Papaioannou and Behringer 2023b) to continue the analysis of postnatal mutants.

There are a number of trivial possibilities that might account for not detecting homozygous mutants at birth.

- The number of progeny analyzed (i.e., the sample size) might be too small to be statistically significant. Do not jump to conclusions based on small numbers. Instead, review Mendelian ratios and do a Chi square (χ^2) test of significance (Box 1). A low probability is still a probability, and you are the one who decides the significance level. Breed more mice to obtain around 30 offspring until you are convinced that no homozygous mutants are born, and check through the other possible causes listed here.

- One of the parents might have been incorrectly genotyped. Do you find a 2:1 ratio of heterozygous to wild-type offspring that might indicate missing homozygous mutants from a heterozygous cross? Or do you find a 1:1 ratio that might indicate a misgenotyped parent, resulting in a cross between a wild type and heterozygote and thus a 1:1 heterozygous to wild-type ratio in the offspring? If this is the case, regenotype the parents or mate different animals. If not, continue checking the other possibilities.

- You are experiencing technical problems with the genotyping assay. Up to this point in generating or obtaining your mutant, you have only needed to distinguish wild-type and heterozygous mice. Double check that your polymerase chain reaction (PCR) or Southern assay can distinguish the homozygous mutants as well and then check every technical aspect of the assay to make sure it is working correctly.

- The nontrivial possibility is that the mutation results in prenatal lethality.

Cite this chapter as *Cold Spring Harb Protoc*; doi:10.1101/pdb.over107970

BOX 1. WORKED EXAMPLES OF A CHI SQUARE TEST OF MENDELIAN SEGREGATION

If you want to know whether the frequency of genotypes you observe is what you expect on the basis of Mendelian segregation, a Chi square (χ^2) test is appropriate. It goes without saying that the larger your sample, the more confidence you can have in the result. The test provides a probability only; you will have to decide the level of significance, although 5% is typically used. After performing the test, you will also see where the discrepancy lies between what you observe and what you expect.

First example

Let us say that you have collected 36 postimplantation embryos from heterozygote x heterozygote crosses, 34 of which you successfully genotyped (the PCR for the other two failed or the samples were lost), and you found the following distribution of genotypes: 10 +/+, 22 +/−, and 2 −/−. In addition, during the dissections, you noted 10 decidual swellings that contained only trophoblast giant cells and could not be genotyped. From the 34 embryos genotyped, you would expect the following distribution of genotypes under Mendel's laws: 8.5 +/+, 17 +/−, and 8.5 −/−, which is the perfect 1:2:1 ratio. Obviously, we are not dealing in half embryos, so this is only the theoretical expectation. Now, compare your results with what you expected using the following formula, which provides a measure of the deviation of each observed value from the expected value for each genotypic class:

$$\frac{(\text{observed} - \text{expected})^2}{\text{expected}}.$$

Add these deviations for each genotypic class and you have the χ^2 statistic:

$$\frac{(10 - 8.5)^2}{8.5} + \frac{(22 - 17)^2}{17} + \frac{(2 - 8.5)^2}{8.5} =$$

$$0.26 \quad + \quad 1.47 \quad + \quad 4.97 \quad = \quad 6.7 = \chi^2.$$

Any elementary statistics text contains tables of this statistic. Using the appropriate degrees of freedom, locate this χ^2 value in the table and read off the probability. The degree of freedom is the number of categories that can vary independently given the number of observations. In other words, with 34 embryos, two categories could vary independently, but the third will then be fixed as the difference between 34 and the sum of the numbers in the other two classes. In this case, with three genotypic categories, there are two degrees of freedom. From the χ^2 table, the χ^2 value corresponds to a probability of between 5% and 2%. This means that the probability of getting this distribution of genotypes out of 34 embryos by chance is <5% (but >2%). If you set the level of significance at 5%, the distribution is significantly different from expected, but if you decide to go with a more stringent level of significance (say, 1%), then it is not significantly different from expected at the 1% level. As we said, the 5% level of significance is the usual, but being so close to the significance level means that you might want to collect a larger sample before drawing final conclusions.

An examination of the individual values contributing to the χ^2 value can also be informative. This example has close to the expected number of +/+ embryos and so the deviation is small, as is the contribution to the χ^2 value (0.26). The number of −/− embryos is quite far off from the expected and the deviation is correspondingly large, making a relatively large contribution to the χ^2 value (4.97), and indicating that embryos are missing from this class. However, the deficiency in +/− embryos also contributes a fairly substantial proportion to the total deviation.

Finally, think about those 10 deciduae containing giant cells and the two embryos that were excluded from the calculations because they were not successfully genotyped. To the extent that technical losses, such as the failure of a PCR reaction or loss of samples, are independent of the genotype of the embryo, we are probably safe to exclude the two lost embryos from the calculations as random losses. However, biological losses as represented by the conceptuses that consisted only of giant cells could indeed be genotype dependent. In fact, this could be a phenotype. For the sake of argument, you might want to include these giant-cell-only deciduae in the −/− class and do another χ^2 test, under the assumption that these are mutant embryos. Now does your distribution fit Mendel's prediction?

Continued

BOX 1. *Continued.*

Second example

You are genotyping offspring of a heterozygous cross at the time of weaning. You do not observe any homozygous mutant offspring and want to know how many you should genotype before concluding that the homozygous mutant animals die sometime before weaning. If you observe a perfect 1:2 ratio of +/+ to +/− animals with no −/−, then by using the χ^2 test, you reach significance at the 5% level with 18 offspring, and at the 1% level with 30 offspring, and you can be 95% or 99% confident, respectively, that the homozygous mutant animals die before weaning. However, these levels of significance will be reached with even fewer offspring if there is a deficiency of heterozygotes, because this will also contribute to the χ^2 value. If you see a deficiency of +/− animals, it may mean that something additional is operating, such as a heterozygous effect on viability.

ASSESSING TIME OF DEATH OF PRENATAL LETHAL MUTANTS

Narrowing down the time of death of mutant embryos is the first step toward analyzing the cause of death and can be accomplished with very few litters, provided the background level of unrelated embryonic death is low (Protocol 8.1: Establishing Control Frequency of Embryonic and Fetal Loss in a Mutant Mouse Colony [Papaioannou and Behringer 2023c]). If the mutant causes a phenotype displaying a variable time of death, it may take a while longer, a few more litters, and a lot of patience to determine the range of time of death. In either case, the idea is to pick several time points in gestation (Fig. 1) at which embryos can be easily genotyped and a morphological assessment of the embryonic and extraembryonic structures can be done to provide a maximum amount of information about phenotype. Accurate genotyping of the parents is of the utmost importance. Females genotyped as heterozygous should always be double checked at the time of their use in case of an error in the original genotyping. Once a stud male genotyped as heterozygous has been proven heterozygous by

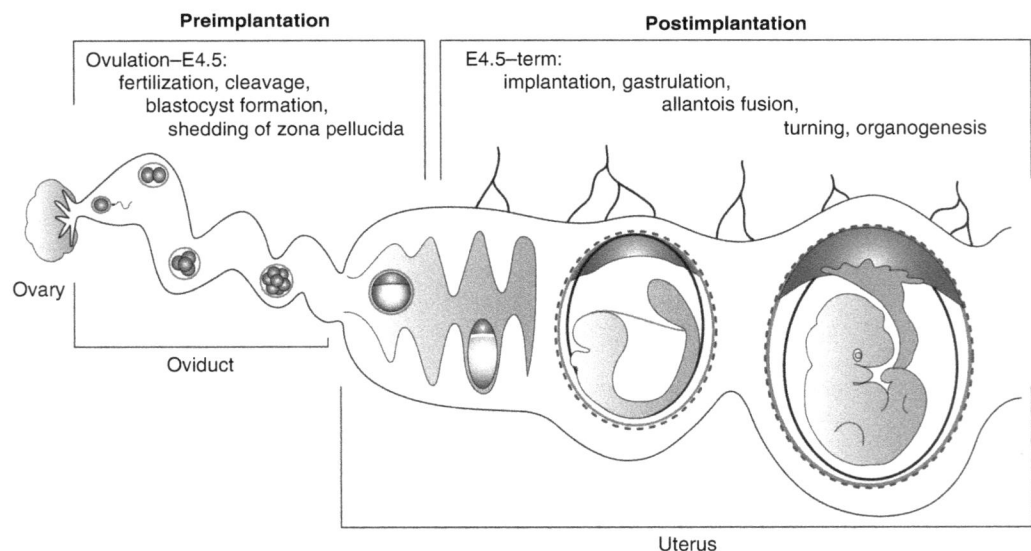

FIGURE 1. Schematic representation of embryonic and fetal development in the mouse through E12.5. The period of development before implantation in the uterus (preimplantation development) is characterized by fertilization, cleavage, compaction, blastocyst formation, and hatching from the zona pellucida. Embryos move through the oviduct and traverse the uterotubal junction (between the uterus and the oviduct) at about E2.5. Attachment and implantation into the uterus follow at E4.5 and, during postimplantation development, the embryos undergo gastrulation and organogenesis. Attachment of the allantois to the chorion to form the chorio-allantoic placenta is an important developmental landmark at about E8.5. Between E8.5 and E9.5, embryos turn from a dorsal flexure to a ventral flexure or fetal position. The period between E12.5 and term is characterized by organogenesis and growth. (Modified, with permission, from Papaioannou and Hadjantonakis 2003, © Humana Press.)

 Cite this chapter as *Cold Spring Harb Protoc*; doi:10.1101/pdb.over107970

TABLE 1. Establishing time of death of homozygous mutants by dissection of embryos at E12.5

Time of death of homozygous mutants		Representation of homozygous mutants at E12.5	Analysis
Embryonic day	Stage		
0–4.5	Between fertilization and implantation	"missing" ~25% of implantation sites compared with number of corpora lutea	Chapter 9: Phenotypic Analysis of Preimplantation Lethality in Mice (Papaioannou and Behringer 2023d)
4.5–9.5	Between implantation and establishment of allantoic circulation	Empty decidua or degenerating remains ~25% over control losses	Chapter 10: Phenotypic Analysis of Periimplantation to Mid-Gestation Lethality in Mice (Papaioannou and Behringer 2023e)
9.5–12.5	After establishment of allantoic circulation	~25% abnormal or developmentally delayed dead embryos[a]	Chapter 11: Analysis of Mid- to Late-Gestation Phenotypes in Mice (Papaioannou and Behringer 2023f)
12.5–term	During organogenesis	Viable, normal or viable but abnormal embryos[b]	Chapter 11: Analysis of Mid- to Late-Gestation Phenotypes in Mice (Papaioannou and Behringer 2023f)

After it is established that homozygous mutants are not born, dissection at a single time point can pinpoint a range of time of death by the appearance of what remains in the uterus.

[a]The developmental stage of dead embryos can help determine the exact time of death before E12.5.

[b]Abnormal or developmentally delayed homozygotes may be found alive with beating hearts at E12.5 indicating that the onset of the mutant effect is before this stage but the time of death is later.

the production of a litter containing homozygous mutants, he can be marked as a proven heterozygote and used with confidence.

Do not make the mistake of trying to cover every developmental stage in this initial screen. Concentrate your efforts on a few stages, starting with embryonic day (E)12.5 (also called 12.5 days post coitus [dpc]), to be confident of your observations and get enough data to do statistics. These dissections serve a dual function: First, dead or abnormal embryos may be discovered that, upon genotyping, are found to be the homozygous mutants, and thus the dissections provide material for analyzing the mutant phenotype. Second, the time of death and/or the time of onset of the abnormalities can be inferred from the developmental stage and condition of the mutant embryos, or alternatively, the absence of mutant embryos among those recovered can indicate very early lethality.

Start by sacrificing one to three pregnant heterozygous females, mated to heterozygous males, at E12.5. This stage is chosen because the embryos are easy to dissect and because it provides a lot of information about what has been going on during the first half of gestation (Table 1). Take a tissue sample from each mother to reconfirm the genotype (the kidney yields very nice DNA for PCR). Count any swellings of the uterus, which represent the sites of embryo implantation (Fig. 2), no matter what their size.

Next, examine each fetus and implantation site by careful dissection (for an excellent reference with good pictures and instructions, use the mouser's bible, *Manipulating the Mouse Embryo* [Behringer et al. 2014]). You will want to collect some fetal tissue for genotyping and also to assess the embryo, placenta, and fetal membranes for normality, so a systematic dissection is called for—first removing maternal tissue (uterus and decidua) and then progressively working through the fetal membranes (trophoblast, Reichert's membrane, placenta, yolk sac, and amnion) to get to the embryo itself (Fig. 3). Then examine the umbilical and vitelline vessels, and finally separate the embryo from the placenta. This will give you a chance to evaluate each conceptus thoroughly. Separate the embryos into individual dishes or drops of media, give them numbers right away, and take notes as you go. Make use of the various resources (Kaufmann 1992; Rossant and Tam 2002; Behringer et al. 2014) to learn embryo dissection and anatomy, but even without a lot of experience, careful observation and comparisons could turn up differences among embryos. Acquire images of anything that looks out of the ordinary. The subsequent genotyping of embryos will then reveal the significance of these differences.

At this point, collect embryonic tissue for further analysis. You will certainly want to do this if you see any potential abnormal phenotypes. It is very little work to preserve each numbered embryo and its fetal membranes and placenta separately in small vials or tubes; they can be discarded later if they are not needed. The fixative of choice will depend on the most likely analysis to be done. For genotyping, it

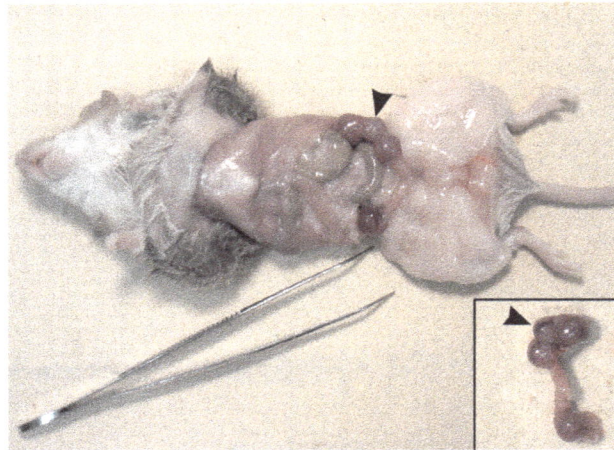

FIGURE 2. Implantation sites at E12.5. The abdomen has been opened and the fat pad reflected caudally to reveal four implantation sites, one in the right uterine horn and three in the left. The middle site (arrowhead) contains an abnormal embryo and is smaller than the others. (*Inset*) The uterus after dissection from the female.

is simplest to use a sample of yolk sac, but be careful to separate it completely from the placenta to avoid maternal tissue contamination. An alternative source of tissue for genotyping, especially if you think the yolk sac might have a phenotype that you want to investigate, is a leg of the embryo, leaving the other three legs as well as the yolk sac for future analysis. Whatever embryonic sample you take for genotyping, be sure to wash it extensively to avoid maternal contamination because the mother is, or should be, a heterozygote.

Once you have genotyped the fetal (and maternal) samples, you are very close to determining the time of embryonic lethality. If you have not recovered any homozygous mutant embryos among the implanted E12.5 embryos or degenerating remains, the mutant embryos probably died between

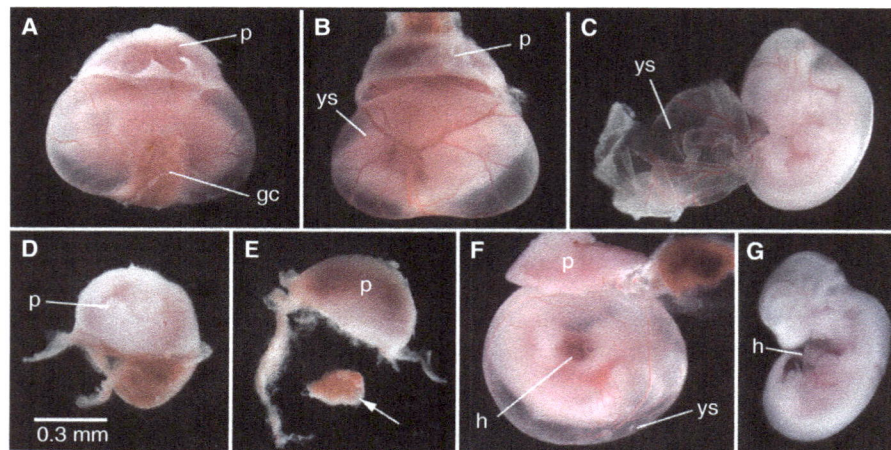

FIGURE 3. Successive stages of dissection of embryonic day (E) 12.5 normal and abnormal embryos. (*A*) A normal conceptus with Reichert's membrane intact and residual trophoblast giant cells external to Reichert's membrane forming a band around the yolk sac. (*B*) Normal conceptus with giant cells and Reichert's membrane reflected to the edge of the placenta clearly revealing the vascular yolk sac. (*C*) Normal conceptus with the placenta removed and the yolk sac reflected. The amnion is still intact, tightly surrounding the embryo but visible as a thin membrane anterior to the head. (*D,E*) Successive stages of dissection of an early resorption with placental structures, membranes and embryonic debris (arrow). (*F,G*) Successive stages of dissection of a *Tbx2* homozygous mutant embryo suffering from cardiac insufficiency. The mutant was still vital, as evidenced by a heartbeat, but there is a reduced amount of blood in the yolk sac vasculature, pooling of blood in the heart (*F*), and pericardial edema (*G*). Development is also somewhat delayed compared with the wild-type control (*C*), judging by developmental landmarks. (gc) giant cells, (h) heart, (p) placenta, (ys) yolk sac. (Modified from Papaioannou and Behringer 2012.)

fertilization and implantation at E4.5, leading you to Chapter 9: Phenotypic Analysis of Preimplantation Lethality in Mice (Papaioannou and Behringer 2023d). Do not be fooled by small numbers, however. Before you jump to this conclusion, double check the mothers' genotypes and gather enough data to get a respectable p value in a χ^2 test (see Box 1). Once you are sure that the homozygous mutants have been lost before implantation, go to Chapter 9: Phenotypic Analysis of Preimplantation Lethality in Mice (Papaioannou and Behringer 2023d). Otherwise, carry on with this assessment of the time of death.

If the embryos died between implantation at E4.5 and about E9.5, before the establishment of the allantoic circulation, they will be represented at E12.5 by implantation sites that contain only degenerating remains. These sites are variously referred to as empty decidua, moles, or resorption sites. They may actually be empty or they may contain remnants of extraembryonic structures: trophoblast giant cells, yolk sac, Reichert's membrane, and/or the ectoplacental cone. In the absence of an allantoic circulation, however, very little in the way of embryonic (as opposed to extraembryonic) remains will be recognizable by E12.5, because the embryo will be in an advanced state of degeneration. The number of these resorption sites should match the number of "missing" homozygous mutants if that is what they represent, but sufficient remains may allow you to genotype them for a positive identification. Go to Chapter 10: Phenotypic Analysis of Periimplantation to Mid-Gestation Lethality in Mice (Papaioannou and Behringer 2023e) for analysis of early postimplantation phenotypes.

If embryonic death occurred after the establishment of the allantoic circulation, the mutant embryos will either be alive at E12.5, as defined by the presence of a beating heart, or present in varying stages of degeneration depending on when they died (Fig. 3). If embryos have no heartbeat, their developmental stage can be used to estimate the time of death (see next section). Degenerating embryos can sometimes be genotyped if the DNA has not degraded too much, but otherwise, their number should match the number of "missing" homozygous mutants. Go to Chapter 11: Analysis of Mid- to Late-Gestation Phenotypes in Mice (Papaioannou and Behringer 2023f).

If all of the embryos including homozygous mutants are viable at E12.5, embryonic lethality must be occurring between E12.5 and term. Detailed examination of the homozygous mutants, numbered and preserved along with the littermate controls, should be done at this point, because the onset of the mutant effect may already be evident. Dissection of embryos at a later time point near term, say, E16.5 or earlier if a mutant phenotype is evident, will further delineate the time of death.

Finally, keep in mind that there may be variable penetrance of the phenotype and variable expressivity, especially if the embryos have a mixed genetic background. Be sure to look at enough embryos to determine if the homozygous mutant phenotype is uniform and the mutants are all dying at the same time. Again, concentrate on one stage at a time to get sufficient numbers to be confident of your conclusion before moving on.

USEFUL DEVELOPMENTAL LANDMARKS FOR EMBRYOS BETWEEN E4.5 AND TERM

You may want to determine the developmental stage of an embryo if you need to estimate gestational age or the time of death of an embryo, or determine if a mutation causes a developmental delay compared with wild-type littermates. Developmental rates and the length of gestation vary somewhat between different strains of mice, and some developmental asynchrony is often found within litters. However, certain easily scored structures provide landmarks for comparing developmental stages within or between litters and/or estimating chronological age of a litter if this is unknown. For an in-depth analysis, there are a number of modern and classic reference works available (e.g., Theiler 1989; Wanek et al. 1989; Rugh 1990; Fujinaga et al. 1992; Downs and Davies 1993; Kaufmann and Bard 1999; Le Garrec et al. 2017).

The day of the plug is called E0.5, making the assumption that fertilization took place around midnight, although this may not be a justified assumption if the mice were together all night. If a female looks bulgy and you think that you missed detecting a vaginal plug, you can look for the appearance of blood in the vagina of the mother, which occurs at E10.5 when the uterine lumen reopens around the embryos. Examine the vagina the same way you would check for a plug. Otherwise, the following pointers can help in estimating embryonic age, estimating the time of death of degenerating embryo and/or staging embryos of known gestational age:

- Implantation takes place at E4.5.

- E5.5–E8.5. Developmental stage at these ages is quite variable even within litters because of the rapid rate of growth and development, but the presence or absence of a circumferential embryonic-extraembryonic constriction, as well as the overall size of embryos, is easily scored. For detailed staging of embryos during the latter part of this period, see Staging Embryos in Chapter 10: Phenotypic Analysis of Periimplantation to Mid-Gestation Lethality in Mice (Papaioannou and Behringer 2023e) and Downs and Davies (1993).

- Fusion of the allantois to the chorion takes place around E8.5 when the embryo has five to eight somite pairs and allantoic circulation is established shortly thereafter.

- Embryos turn from a dorsal curvature (lordosis) to a ventral curvature (fetal position) by E9.5 (between six and 16 somite pairs) (Fig. 4).

- E8.5–E12.5. The number of somites can be easily counted for finely graded developmental staging (see Chapter 10: Phenotypic Analysis of Periimplantation to Mid-Gestation Lethality in Mice [Papaioannou and Behringer 2023e]).

- E8.5. Heart looping progression can be used to subdivide developmental stages at this age (Fig. 5).

- Neural tube closure begins at around E8.5 at the six- to seven-somite stage and is completed by E10.5 at the 29- to 30-somite stage (see Fig. 5 in Chapter 10: Phenotypic Analysis of Periimplantation to Mid-Gestation Lethality in Mice [Papaioannou and Behringer 2023e]).

- E9–E18.5. The shape of the limb buds reflects developmental stage: Forelimb buds first appear as lateral ridges at the level of somites seven through 12 at E9 and hindlimb buds appear a day later at the level of somites 23–28. This asynchrony in first appearance of the forelimbs and hindlimbs is maintained in their later development.

- E10.5. Eye pigment appears in a circular pattern, unless, of course, the embryo is albino. Pigment remains visible later even in severely degenerated embryos provided they reached E10.5.

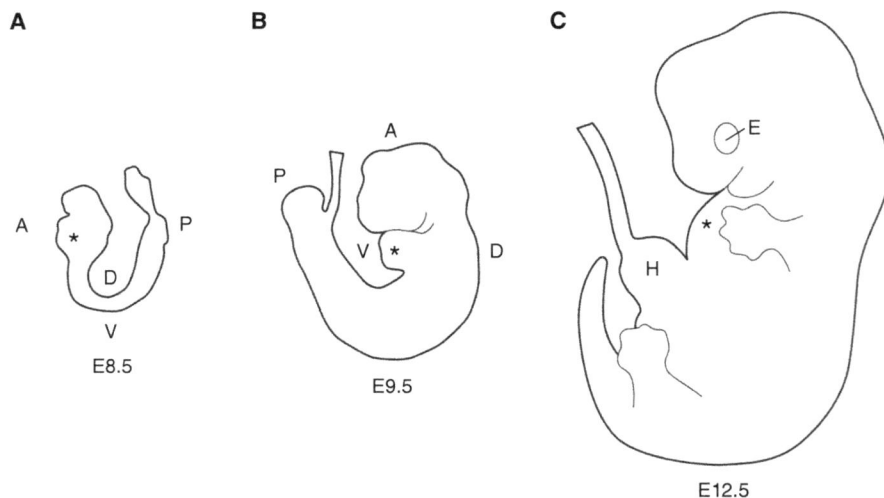

FIGURE 4. (A) Lordosis of an E8.5 embryo before turning. The embryo has a dorsal curvature. (B) Fetal position of a E9.5 embryo after turning. Embryos in the fetal position have a ventral curvature. (C) Umbilical hernia (H) at E12.5. Between E10.5 and E15.5, a loop of midgut is located outside the abdominal cavity in the base of the umbilical cord. Eye pigment is evident in a circular pattern from E10.5. The placenta and yolk sac are not shown. (A) Anterior, (P) posterior, (V) ventral, (D) dorsal, (E) eye, (*) location of the heart.

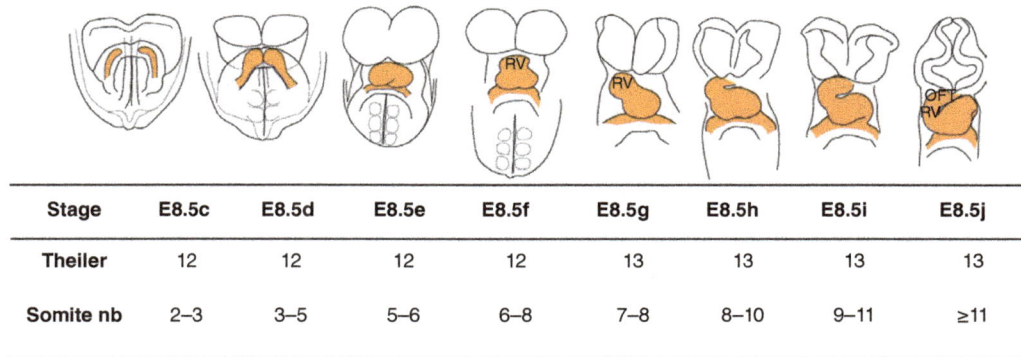

Stage	E8.5c	E8.5d	E8.5e	E8.5f	E8.5g	E8.5h	E8.5i	E8.5j
Theiler	12	12	12	12	13	13	13	13
Somite nb	2–3	3–5	5–6	6–8	7–8	8–10	9–11	≥11

FIGURE 5. Staging of the shape changes during the formation and looping of the heart tube (orange) in the E8.5 mouse embryo. The heart tube is initially bilaterally symmetrical and straight until stage E8.5f. This staging scale, focused on the heart, is finer than Theiler stages and not fully synchronous with the addition of somites. (RV) Right ventricle, (OFT) outflow tract, (nb) number. (Reprinted from Le Garrec et al. 2017.)

- E10.5. A loop of the midgut herniates into the umbilical cord (Fig. 4). This is known as a physiological umbilical hernia and it persists until the gut loop is withdrawn into the peritoneal cavity at E15.5.

- Features of the head including the branchial arches, the mouth and nose region, the eyes, and the external ears all follow a strict developmental progression and can be used to stage and compare later embryos.

Knowing the morphology of wild-type mouse embryos at landmark stages will serve you well for recognizing when an embryo is abnormal.

REFERENCES

Behringer RR, Gertsenstein M, Nagy KV, Nagy A. 2014. *Manipulating the mouse embryo: a laboratory manual.* Cold Spring Harbor Laboratory Press, Cold Spring Harbor, NY.

Downs KM, Davies T. 1993. Staging of gastrulating mouse embryos by morphological landmarks in the dissecting microscope. *Development* 118: 1255–1266. doi:10.1242/dev.118.4.1255

Fujinaga M, Brown NA, Baden JM. 1992. Comparison of staging systems for the gastrulation and early neurulation period in rodents: a proposed new system. *Teratology* 46: 183–190. doi:10.1002/tera.1420460211

Kaufmann MH. 1992. *The atlas of mouse development.* Academic, London.

Kaufmann MH, Bard JBL. 1999. *The anatomical basis of mouse development.* Academic, San Diego.

Le Garrec JF, Domínguez JN, Desgrange A, Ivanovitch KD, Raphaël E, Bangham JA, Torres M, Coen E, Mohun TJ, Meilhac SM. 2017. A predictive model of asymmetric morphogenesis from 3D reconstructions of mouse heart looping dynamics. *eLife* 6: e28951. doi:10.7554/eLife.28951

Papaioannou VE, Behringer RR. 2012. Early embryonic lethality in genetically engineered mice: diagnosis and phenotypic analysis. *Vet Pathol* 49: 64–70. doi:10.1177/0300985810395725

Papaioannou VE, Behringer RR. 2023a. Phenotypic analysis of dominant mutant effects in mice. *Cold Spring Harb Protoc* doi:10.1011/pdb.over107978

Papaioannou VE, Behringer RR. 2023b. Phenotypic analysis of perinatal lethality in mice. *Cold Spring Harb Protoc* doi:10.1011/pdb.over107975

Papaioannou VE, Behringer RR. 2023c. Establishing control frequency of embryonic and fetal loss in a mutant mouse colony. *Cold Spring Harb Protoc* doi:10.1011/pdb.over108083

Papaioannou VE, Behringer RR. 2023d. Phenotypic analysis of preimplantation lethality in mice. *Cold Spring Harb Protoc* doi:10.1011/pdb.over107971

Papaioannou VE, Behringer RR. 2023e. Phenotypic analysis of periimplantation to mid-gestation lethality in mice. *Cold Spring Harb Protoc* doi:10.1011/pdb.over107972

Papaioannou VE, Behringer RR. 2023f. Analysis of mid- to late-gestation phenotypes in mice. *Cold Spring Harb Protoc* doi:10.1011/pdb.over107973

Papaioannou VE, Hadjantonakis AK. 2003. *Stem cell handbook* (ed. S. Sell), pp. 19–31. Humana, Totowa, NJ.

Rossant J, Tam PP. 2002. *Mouse development: patterning, morphogenesis, and organogenesis.* Academic, San Diego.

Rugh R. 1990. *The mouse: its reproduction and development.* Oxford Science Publications, Oxford

Theiler K. 1989. *The house mouse: atlas of embryonic development.* Springer-Verlag, New York.

Wanek N, Nuneoka K, Holler-Dinsmore G, Burton R, Bryant SV. 1989. A staging system for mouse limb development. *J Exp Zool* 249: 41–49. doi:10.1002/jez.1402490109

Establishing Control Frequency of Embryonic and Fetal Loss in a Mutant Mouse Colony

In the analysis of prenatal lethal recessive mutations, one must account for embryonic losses that are not related to the mutant phenotype. This protocol details the way to determine what the background level of unrelated embryonic loss is by a simple backcrossing strategy in the particular mouse strain that carries the lethal recessive mutation.

INTRODUCTION

During the assessment of the time of death of prenatal lethal mutants, it is essential to have an estimate of the frequency of embryonic and fetal losses that occur for reasons unrelated to a specific mutant genotype—that is, the control frequency of embryonic loss in your mutant colony unrelated to the mutation. This can vary with different strains of mice and must be determined empirically from backcrosses, as detailed here. First, you will need to determine the gestational age of interest, which will usually correspond to the time of death of the mutant embryos as determined by the strategies in Chapter 8: Phenotypic Analysis: Assessing Timing of Recessive Prenatal Lethality in Mice (Papaioannou and Behringer 2023a). For preimplantation stages, counts of corpora lutea (CL) are recommended to determine the expected number of embryos (Protocol 4.1: Counting Murine Corpora Lutea to Determine the Number of Oocytes Ovulated [Papaioannou and Behringer 2023b]); for postimplantation stages the number of implantation sites provides a measure of embryos that implanted. The use of reciprocal backcrosses will control for the possibility of heterozygous effects of the mutation on reproductive capacity of either males or females.

MATERIALS

It is essential that you consult the appropriate Material Data Safety sheets and your institution's Environmental Health and Safety Office for proper handling of equipment and hazardous materials used in this protocol.

RECIPES: Please see the end of this protocol for recipes indicated by <R>. Additional recipes can be found online at http://cshprotocols.cshlp.org/site/recipes.

Reagents

Heterozygous and wild-type male and female mice from the mutant colony
Phosphate-buffered saline (PBS) <R>

Equipment

Dissecting microscope for embryo recovery and examination
Dissection instruments
Petri dishes

METHOD

1. Mate heterozygous males with wild-type females (backcross) and vice versa (heterozygous females with wild-type males).

2. Dissect embryos at the gestational stage you want to compare and count the dead, dying, or missing embryos from these control backcrosses.

3. Calculate a modified Mendelian ratio expected in the intercross (heterozygote × heterozygote).

   In the control backcross, the genotypes of the embryos will match the genotypes of embryos from an intercross, except for the absence of the homozygous mutants, and thus the control frequency of embryonic death can be used to calculate a modified, expected Mendelian ratio for the intercross. For example, if the control frequency of normal embryo recovery from a backcross is 92 normal embryos/100 CL or implantation sites detected (8% loss), then the expected recovery from the intercross, is 25% less than this value (i.e., 69 embryos/100 CL or implantation sites). This takes into account the 8% random loss (eight embryos) found in the backcross and a 25% loss of the remainder due to the homozygous mutant genotype assuming the mutation is recessive.

DISCUSSION

When determining the control frequency of embryonic loss, be on the lookout for dominant effects that could affect the heterozygous embryos or the reproductive capability of the heterozygous mothers. Major dominant effects on embryonic development should have been detected during earlier analyses (see Chapter 5: Recovering a Targeted Mutation in Mice from Embryonic Stem Cell Chimeras or CRISPR–Cas Founders [Papaioannou and Behringer 2023c]), but heterozygous effects on reproductive capacity might have been missed if they are subtle. To control for this possibility, use reciprocal matings of either heterozygous males or heterozygous females × wild type when determining the control frequency of embryonic loss.

RECIPE

Phosphate-Buffered Saline (PBS)

Reagent	Amount to add (for 1× solution)	Final concentration (1×)	Amount to add (for 10× stock)	Final concentration (10×)
NaCl	8 g	137 mM	80 g	1.37 M
KCl	0.2 g	2.7 mM	2 g	27 mM
Na_2HPO_4	1.44 g	10 mM	14.4 g	100 mM
KH_2PO_4	0.24 g	1.8 mM	2.4 g	18 mM

If necessary, PBS may be supplemented with the following:

$CaCl_2 \cdot 2H_2O$	0.133 g	1 mM	1.33 g	10 mM
$MgCl_2 \cdot 6H_2O$	0.10 g	0.5 mM	1.0 g	5 mM

PBS can be made as a 1× solution or as a 10× stock. To prepare 1 L of either 1× or 10× PBS, dissolve the reagents listed above in 800 mL of H_2O. Adjust the pH to 7.4 (or 7.2, if required) with HCl, and then add H_2O to 1 L. Dispense the solution into aliquots and sterilize them by autoclaving for 20 min at 15 psi (1.05 kg/cm^2) on liquid cycle or by filter sterilization. Store PBS at room temperature.

REFERENCES

Papaioannou VE, Behringer RR. 2023a. Phenotypic analysis: assessing timing of recessive prenatal lethality in mice. *Cold Spring Harb Protoc* doi:10.1011/pdb.over107970

Papaioannou VE, Behringer RR. 2023b. Counting murine corpora lutea to determine the number of oocytes ovulated. *Cold Spring Harb Protoc* doi:10.1011/pdb.prot107986

Papaioannou VE, Behringer RR. 2023c. Recovering a targeted mutation in mice from embryonic stem cell chimeras or CRISPR–Cas founders. *Cold Spring Harb Protoc* doi:10.1011/pdb.over107959

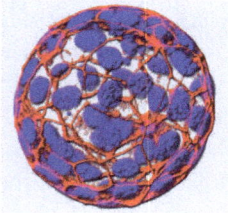

CHAPTER 9

Phenotypic Analysis of Preimplantation Lethality in Mice

Preimplantation development covers a period of ~4.5 d from fertilization to implantation in the uterus. If a homozygous mutant phenotype causes the death of embryos during this period, simple culture methods are available that support preimplantation development to allow a thorough morphological assessment. Embryos are recovered from the oviducts or uterus and examined for gross morphology, cell number, and progression through cleavage stages. Blastocysts can also be cultured over the implantation period and undergo a process analogous to implantation in vitro. Different categories of phenotypes such as failure of compaction, abnormal blastocyst formation, or failure to hatch from the zona pellucida and failure to attach and outgrow in vitro are discussed in relation to what each phenotype might portend. Further experimental procedures such as isolation and assessment of blastocyst inner cell mass and analysis of induced implantation delay in vivo may also be appropriate. Additional assessment of preimplantation embryos can involve histology and localization of mRNA or proteins either in sections or whole embryos.

INTRODUCTION

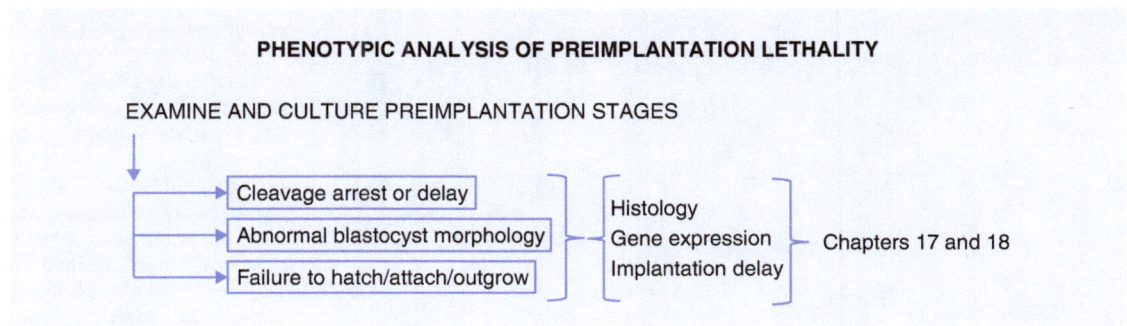

A preimplantation lethal phenotype is indicated if no homozygous mutant offspring are born, no homozygous mutants are detected among implanted embryos at postimplantation stages, and there is not an excess of empty or abnormal implantation sites (see Chapter 8: Phenotypic Analysis: Assessing Timing of Recessive Prenatal Lethality in Mice [Papaioannou and Behringer 2023a]). There should also be a 25% excess of corpora lutea (CL) over the number of implantation sites, by comparison with the number in a control backcross (heterozygous × wild type) to take normal variation into account

© 2024 Cold Spring Harbor Laboratory Press
Cite this chapter as *Cold Spring Harb Protoc*; doi:10.1101/pdb.over107971

(Protocol 8.1: Establishing Control Levels of Embryonic and Fetal Loss in a Mutant Mouse Colony [Papaioannou and Behringer 2023b]). For all these determinations, collect sufficient data to provide statistical support: chi square analysis (see Box 1. Worked Examples of a Chi Square Test of Mendelian Segregation in Chapter 8: Phenotypic Analysis: Assessing Timing of Recessive Prenatal Lethality in Mice [Papaioannou and Behringer 2023a]) using the backcross frequencies to estimate a corrected expected Mendelian frequency will be most accurate. Once you are sure that the homozygous mutant embryos are dying before they implant, you have narrowed the time window to ∼4.5 d—the time between fertilization and implantation. But it is still important to find the earliest time point at which the mutant departs from normal to determine the kind of effect the mutation is having.

During the preimplantation period (Fig. 1), the zygote should go through six or seven special mitotic divisions called cleavage (so called because no overall growth of the embryo occurs during this process, but simply a cleaving of the cytoplasm accompanied by nuclear division), resulting in smaller cells called blastomeres. These divisions are nearly synchronous resulting in the majority of embryos having an even number of blastomeres at any given time (Fig. 1C–E). Toward the end of cleavage, the blastomeres compact into a tight spherical ball of cells known as a compact morula (Fig. 1F) and cellular differentiation begins with the formation of an outer layer of trophectoderm and an inner cell mass (ICM) (Fig. 1G). The trophectoderm begins to pump fluid, and the process of cavitation results in a cystic ball of cells, the blastocyst, with the ICM located asymmetrically at one pole. The trophectoderm overlying the ICM is known as the polar trophectoderm, and this is the embryonic pole of the embryo; the trophectoderm on the opposite or abembryonic pole is called the mural trophectoderm. The cavity is called the blastocyst cavity (or blastocoel). Next, a third cell type, the primitive endoderm, forms on the blastocoelic surface of the ICM. All of these events take place on a fairly precise time

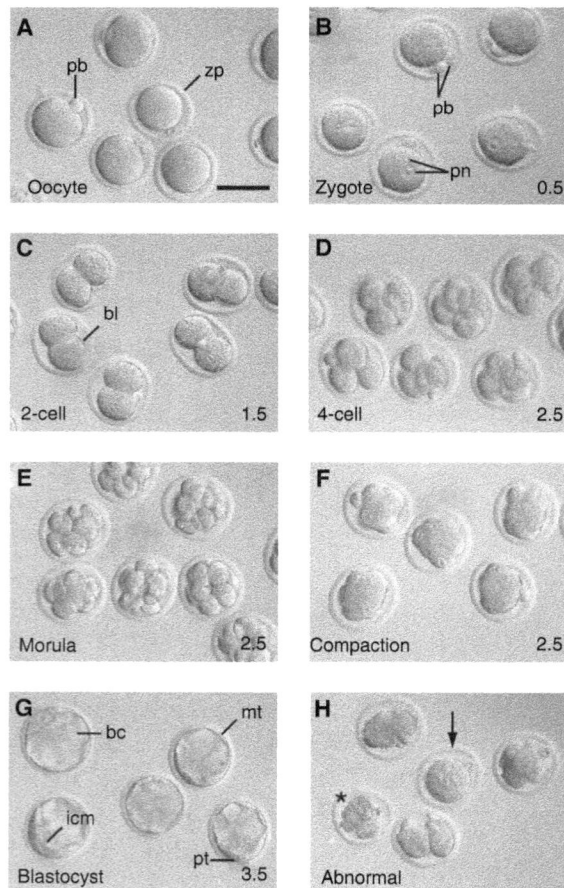

FIGURE 1. Preimplantation development. Differential interference contrast (DIC) images with embryonic days (E) indicated. Ovulated oocyte to morula stages are found in the oviduct, whereas late morulae to blastocysts are found in the uterus. Ovulated oocytes to blastocysts are surrounded by a zona pellucida. (A) Ovulated oocytes have a single polar body and a single haploid nucleus. (B) Zygotes are easy to distinguish from unfertilized oocytes by the presence of two polar bodies and two pronuclei and also by the shrinkage of the zygote, resulting in a larger subzona space. (C–E) Two- to eight-cell cleavage stages. (F) Compact morulae. The individual blastomeres are no longer distinguishable due to tighter cell–cell adhesion that occurs during compaction. Polar bodies are still visible. (G) Blastocyst-stage embryos consisting of an outer layer of trophectoderm cells surrounding an inner cavity, the blastocyst cavity, with an inner cell mass (ICM) located at one pole. The trophectoderm overlying the ICM is the polar trophectoderm, whereas the rest of the trophectoderm is the mural trophectoderm. Blastocysts expand and subsequently "hatch" out of the zona pellucida before implanting into the uterus. (H) Abnormal embryos that are fragmenting (*), shrunken (arrow), and degenerate. Scale bar, 100 μm. (bc) Blastocyst cavity, (bl) blastomere, (icm) inner cell mass, (mt) mural trophectoderm, (pb) polar body, (pn) pronucleus, (pt) polar trophectoderm, (zp) zona pellucida.

Cite this chapter as Cold Spring Harb Protoc; doi:10.1101/pdb.over107971

schedule, although there will be some minor asynchrony, and a morphological assessment is straight-forward (see Chapter 2 in Behringer et al. 2014).

Examination of embryos at intervals after fertilization will quickly tell you whether some embryos either lag behind or arrest their development. The object is to identify the point at which embryos first begin to stray from the straight and narrow path and then to genotype them to confirm that this phenotype correlates with the mutant genotype. Theoretically, even a single cell from a preimplantation embryo can be genotyped with a sensitive polymerase chain reaction (PCR) assay, but, depending on the time of death or arrest, you might have as many as 100 cells with which to work. Note that when using reverse transcriptase–PCR (RT-PCR) for genotyping preimplantation embryos, it may be necessary to remove the zona pellucida (ZP), a transparent, extracellular spherical envelope that is deposited around oocytes during oogenesis) to avoid possible contamination from cytoplasmic remnants of maternal cumulus cells trapped within it. Keep in mind that the zygotic genome in the mouse begins to function as early as the two-cell stage, so a mutant defect might become evident very early if the gene in question has a role in some basic cellular functions such as cell division or energy metabolism.

Start the examination at E3.5, when normal embryos can be flushed from the uterus as expanded blastocysts. The ZP will still be intact around the embryo and any embryos that arrested or died should be recoverable as ZP-contained remains. It is a good idea to count CL (see Protocol 4.1: Counting Murine Corpora Lutea to Determine the Number of Oocytes Ovulated [Papaioannou and Behringer 2023c]) to make sure that all of the ovulation products are recovered (this is especially important if for some technical reason there is no means of genotyping the preimplantation embryos, in which case conclusions will be based on expected Mendelian frequencies). Mice usually have a very low rate of preimplantation failure, although this may vary with the strain, and it might be worthwhile to analyze a backcross to establish the control level of preimplantation failure (Protocol 8.1: Establishing Control Levels of Embryonic and Fetal Loss in a Mutant Mouse Colony [Papaioannou and Behringer 2023b]). In fact, it is essential to know the control level of preimplantation failure if the embryos cannot be genotyped. To make an assessment of preimplantation development, sort the embryos into the following morphological categories (Fig. 1):

- degenerated or fragmented;
- unfertilized oocytes;
- zygotes;
- two cell, etc., up to eight cell;
- compact morulae;
- cavitating blastocysts;
- expanded blastocysts.

Then, genotype the embryos and correlate their genotype with their phenotype. If the homozygous mutants are among the normal-looking blastocysts, go to Analysis of Blastocysts and the Process of Implantation, below. If they are either delayed, abnormal, or among the degenerating embryos, continue to the next section.

ANALYSIS OF CLEAVAGE STAGES AND EARLY BLASTOCYSTS

Embryo Culture

Whether the mutant embryos are degenerating, arrested at a particular stage, or have abnormal morphology at the blastocyst stage, the object is to gain a dynamic picture of how they got that way. Fortunately, preimplantation embryos are very easy to culture from fertilization right through cavitation to the blastocyst stage and hatching from the zona pellucida (Protocol 9.1: Phenotypic Analysis of Preimplantation Mouse Embryos in Culture [Papaioannou and Behringer 2023d]) and watching a clutch of embryos progress through cleavage is a satisfying way of simply waiting for the mutants to reveal themselves.

Cleavage and the formation of a normal blastocyst involve several related, but not necessarily dependent, processes. First, cell number increases and cell volume decreases with time. A genetic program for development begins to unfold from activation of the zygotic genome starting at the two-cell stage through the onset of expression of specific genes. Concomitantly, cellular interactions take place among the blastomeres that influence the spatial distribution of cytoplasmic organelles and membrane components. Feedback from the position of a cell with respect to other cells influences gene expression. The result of all of this is the reproducible differentiation of different cell types in appropriate places at appropriate times within the embryo. Certain features, such as compaction of the blastomeres and the pumping of fluid by the trophectoderm, occur at set times after fertilization and are independent of cell number. Other features, such as the differentiation of the ICM, depend on cell position and microenvironment and thus require a sufficient number of cells to produce the necessary microenvironment for the formation of the ICM. Once you locate the defective process of your mutant embryos, you will be able to refer to the associated literature to determine what is known about the biochemistry and genetic control of each process to get at the mechanism of action of your mutation. Some general methods for additional analysis of preimplantation embryos may be found in Further Analysis of Preimplantation Embryos, below.

Some Possible Phenotypes

No Mutant Embryos

By this stage of the game, you should have ruled out a haploid effect on sperm or oocyte (see Chapter 5: Recovering a Targeted Mutation in Mice from Embryonic Stem Cell Chimeras or CRISPR–Cas Founders [Papaioannou and Behringer 2023e]), so a failure of fertilization should not be the problem. Some rare examples of sperm/oocyte incompatibility exist (e.g., DDK syndrome; Babinet et al. 1990), but this is an unlikely possibility. Double check the accuracy of the genotyping procedure and collect more embryos to ensure that the assessment of "no mutant embryos present" is correct. If it is, perhaps you did not follow all of the diagnostic steps outlined in Chapter 5: Recovering a Targeted Mutation in Mice from Embryonic Stem Cell Chimeras or CRISPR–Cas Founders (Papaioannou and Behringer 2023e). If you determined that the male (or female) offspring of the chimera or CRISPR–Cas founder were fertile, but did not check the other sex, it is just possible that you are only now turning up a dominant effect on oogenesis (or spermatogenesis). Go back to the section Are Heterozygous Offspring of the Chimeras or CRISPR–Cas Founders Normal, Fertile, and of Both Sexes? in Chapter 5: Recovering a Targeted Mutation in Mice from Embryonic Stem Cell Chimeras or CRISPR–Cas Founders (Papaioannou and Behringer 2023e).

Arrest or Delay during Cleavage

Arrest or developmental delay during cleavage (e.g., *Pes1*; Lerch-Gaggl et al. 2002) will be evident as (1) a failure of the mutants to progress beyond a certain number of cells or (2) significant lagging of the mutants behind the wild-type and heterozygous embryos. Although it seems very easy to count cells (blastomeres) of precompaction embryos under stereo dissecting microscope optics, the morphology can sometimes be deceptive if the embryo is fragmenting, and once compaction starts (Fig. 1F), it is very challenging to count cells this way. However, simple nuclear counting techniques with fluorescent dyes can be used that are much more accurate (Protocol 9.2: Cell Counting Techniques for Preimplantation Mouse Embryos Using Fluorescent DNA Dyes [Papaioannou and Behringer 2023f]). But remember that you may no longer be able to genotype an embryo that has been used for cell counting, so try to determine the suspected phenotype first. The nuclear counting technique provides a good indication of cell number and can be used to recognize cell division and cell death by nuclear morphology so that you can also assess proliferation and cell death.

Cite this chapter as *Cold Spring Harb Protoc*; doi:10.1101/pdb.over107971

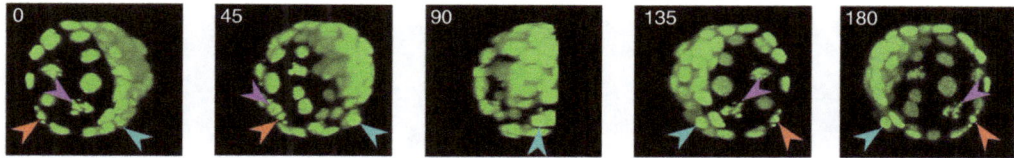

FIGURE 2. Nuclear-localized green fluorescent protein (GFP) as a cell counting tool. Three-dimensional reconstruction of half of a blastocyst transgenic for a histone-enhanced GFP fusion gene. Each panel shows a 45° rotation from the previous panel to show a 180° rotation of the half blastocyst. The inner cell mass is on the right of the *first* panel. With this chromatin-specific label, all nuclei can be counted in confocal images and, in addition, pycnotic cells (orange arrowhead), mitotic cells (purple arrowhead), and interphase nuclei can be distinguished. The polar body (blue arrowhead) is also visible. (Images courtesy of Kat Hadjantonakis and Virginia Papaioannou.)

An alternative to counting nuclei by using added nuclear dyes is to use certain nuclear-localized transgenic markers, such as histone-H2b-eGFP (Fig. 2). H2b-eGFP localizes to interphase nuclei, mitotic chromosomes, and fragmenting nuclei, to accurately assess cell number, mitosis, and cell death, particularly when combined with confocal microscopy. The advantage of this type of marker is that the embryos can be observed vitally and recovered for further analysis or genotyping. However, the transgenic marker will have to be crossed onto the strain of mice carrying the mutant allele by breeding before it can be used and this cross may alter the genetic background and, potentially, the mutant phenotype.

No Compaction

If no or abnormal compaction is observed, check cell numbers (Protocol 9.2: Cell Counting Techniques for Preimplantation Mouse Embryos Using Fluorescent DNA Dyes [Papaioannou and Behringer 2023f]) to determine if this is a case of developmental arrest before compaction, which should occur between the eight- to 32-cell stages, or whether cell number continues to increase in the absence of compaction. The compaction process relies on changes in cell adhesion, and if the blastomeres continue to divide but do not compact, alterations in cell adhesion could be the problem (e.g., *Cdh1*; Ohsugi et al. 1997).

Abnormal Blastocyst Morphology

There are several types of abnormal blastocyst morphology (e.g., *Nup214*; van Deursen et al. 1996). If the embryos compact but do not form a cavity, it could signal a developmental arrest or a defect in the pumping mechanism of the trophectoderm cells. Trophectoderm cells first form intracellular vesicles, which later coalesce to become the blastocyst cavity. Integrity of the blastocyst cavity depends on the tight junctions that form between adjacent trophectoderm cells. If a blastocyst cavity forms, but there is no ICM, this could be due to fewer-than-normal cells present at the time of compaction, in which case not enough cells are available to enclose cells in an inside microenvironment and all of the cells become trophectoderm. Alternatively, the ICM might have formed but undergone cell death. The cell counting technique should sort this out. For further analysis, see Further Analysis of Preimplantation Embryos, below.

ANALYSIS OF BLASTOCYSTS AND THE PROCESS OF IMPLANTATION

If your mutant embryos form normal-looking blastocysts but fail to implant, several in vitro assays can be used to determine what is wrong. From the E3.5 starting point (Fig. 3A), culture blastocysts under conditions that will allow them to "hatch" from the ZP and attach to and outgrow on a tissue culture dish (Protocol 9.1: Phenotypic Analysis of Preimplantation Mouse Embryos in Culture [Papaioannou

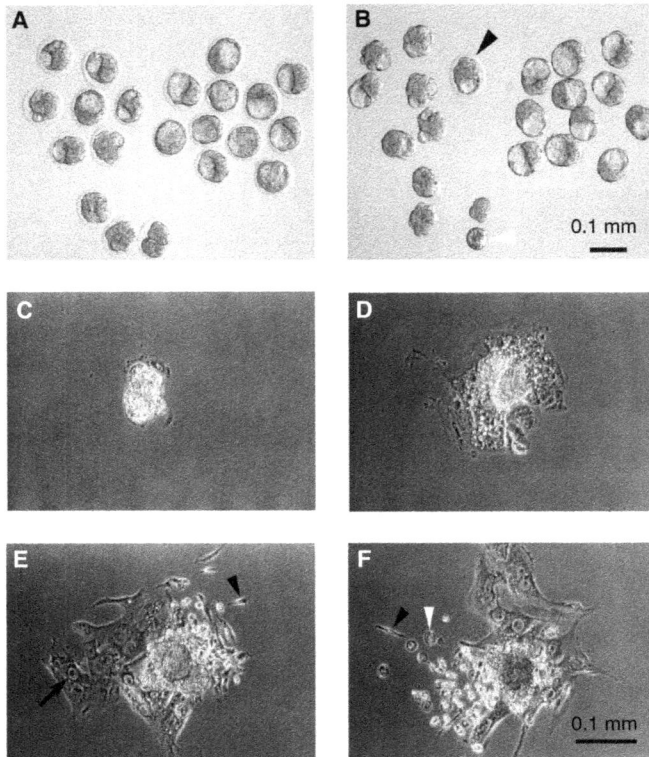

FIGURE 3. Zona pellucida removal and in vitro attachment and outgrowth of blastocysts. (A,B) Brightfield microscopy on a dissecting scope, (C–F) phase-contrast microscopy on an inverted microscope. (A) E3.5 embryos freshly flushed from the uterus. The embryos have been sorted by gross morphology: The 11 embryos on the *right* are expanded blastocysts; the seven on the *left* are beginning to cavitate or are partially expanded; the three on the *bottom* have not cavitated. (B) The same embryos after removal of the zona pellucida from all but one of the partially expanded blastocysts (black arrowhead). The embryo at *bottom center* has separated into a compact group of cells and a single arrested blastomere (white arrowhead). (C,D) Embryos explanted at E3.5 and grown in vitro for 2 d. The blastocyst in C has collapsed and attached to the culture dish and the trophoblast cells are beginning to grow out from the central clump. In D, the trophoblast cells form a monolayer on the culture dish surrounding the central clump of cells composed of ICM derivatives. (E,F) Embryos explanted at E3.5 and grown in vitro for 7–8 d. The large, clear nuclei with prominent nucleoli are the giant trophoblast cell nuclei (arrow) in the trophoblast monolayer surrounding the ICM. ICM derivatives form a compact central mass and some cells have begun to migrate away from the outgrowth (black arrowheads). In F, some rounded dead cells are evident (white arrowhead), a normal occurrence after this period of culture. Scale bar in B for A,B; scale bar in F for C–F.

and Behringer 2023d]). Then, determine which aspect of blastocyst function is affected by the mutation.

Some Possible Phenotypes during Blastocyst Culture

Failure to Hatch

In the uterus, embryos eventually escape from the ZP, probably through a combination of blastocyst expansion and zonalytic activity in the uterus. In the artificial culture situation in vitro, escape from the ZP depends on the continued expansion of the embryo until it cracks the ZP and escapes through the crack. This "hatching" process is taken as a measure of a healthy, functional trophectoderm, and thus failure to hatch in vitro could reflect a trophectoderm defect, either through decreased cell number or decreased pumping activity (e.g., *Smarc1b*; Guidi et al. 2001). Unfortunately, the rate of spontaneous failure to hatch in vitro is fairly high, but if you can genotype your embryos, you can still correlate this phenotype with genotypes. If, for example, 23% of wild-type embryos fail to hatch but 99% of homozygous mutants fail, this indicates a clear, genotype-driven difference. If you cannot genotype the embryos, it is best to do a control backcross and compare the incidence of nonhatching with the rate among intercross embryos under the same culture conditions (see Protocol 8.1: Establishing Control Levels of Embryonic and Fetal Loss in a Mutant Mouse Colony [Papaioannou and Behringer 2023b]). Failure to hatch by itself is not a very satisfactory end point. If this is your phenotype, continue on to the next section and eventually Analysis of Isolated ICM, below.

Failure to Attach

The next step is to determine if the mutant embryos attach to the culture dish (e.g., *Mcl1*; Rinkenberger et al. 2000). This process is mediated by the trophectoderm and is presumably the in vitro counterpart of trophectoderm attachment to the uterine epithelium, the first step in the implantation process. To give the embryos the best chance to show their capabilities, you may wish to eliminate the hatching hurdle if you have already established that your mutants hatch at a normal rate. In fact, even if your mutants show a hatching deficiency, it is worth removing the ZP to further test the functioning of the trophectoderm. To do this, simply remove the ZP with acidic Tyrode's solution (Nagy et al. 2006c) and see Box 1: Hints on Removal of the Zona Pellucida before you place the blastocysts in culture. Then sit back and wait for the mutants to appear. Culture the embryos individually, but identified by litter so that you can confirm the mothers' genotypes. Score the embryos daily for attachment to the culture dish. If one-quarter of the embryos continue to float while their sibs are attaching and outgrowing, you have probably found the phenotype. Genotype the embryos to confirm this and then go to Analysis of Isolated ICM, below.

Failure to Outgrow

The last phase of implantation in the uterus is the invasion of the maternal uterine epithelium by the trophectoderm and the concomitant transformation of the trophectoderm into trophoblast giant cells (cells that endoreduplicate their DNA in the absence of nuclear division, thus becoming very large in the process). The in vitro counterpart of this activity is the spreading of trophectoderm as a monolayer

---

BOX 1. HINTS ON REMOVAL OF THE ZONA PELLUCIDA

The quickest and easiest method of removing the zona pellucida (ZP), the glycoprotein coat surrounding the oocyte and early embryo, is to dissolve it in an acidic solution (see Nagy et al. 2006c). The only trick is to do so without exposing the oocytes or embryos to the acid long enough to cause them damage. To accomplish this, use buffered embryo culture medium to control the microenvironment around the embryo. Perform the whole procedure under a dissecting microscope, and your watchful eye, in two watch glasses or Petri dishes, one with a large volume of buffered medium and the other with a large volume of the acidic solution, keeping the following advice in mind:

- Use acidic Tyrode's salt solution (pH 2.5) at room temperature. This solution is very stable and can be made in advance and either frozen as aliquots or stored at 4°C indefinitely.

- Draw up a small amount of acidic Tyrode's solution into a mouth pipette, pick up the embryos a few at a time from a dish of buffered medium, and place them in the dish of acidic Tyrode's solution.

- Keep a close watch on them and, at the same time, empty the mouth pipette and draw up a small volume of buffered medium.

- As soon as you see the ZP start to disappear, expel a little of the buffered medium over the embryos and then collect them in the pipette and transfer them back to the dish of buffered medium. The ZP should be gone (see Fig. 3A,B).

If the ZP does not dissolve within 10–30 sec in acidic Tyrode's solution, you may have transferred the embryos into the solution along with too large a volume of buffered medium. Move them around in the dish and next time use a smaller bore pipette or less medium. Do not be tempted to work with more than a few embryos at a time, because it can be hard to keep track of them in the dish. If you lose an embryo in the acidic solution, do not waste time searching for it—it will be beyond rescue.

---

onto the culture dish and the loss of integrity of the blastocyst cavity. The ICM comes to lie on top of the trophectoderm monolayer in a small compact mound of cells, whereas the trophoblast giant cells, with their giant nuclei and refractile perinuclear granules, are easily visible in the surrounding monolayer. Failure of the trophoblast to outgrow (e.g., *Trrap*; Herceg et al. 2001) or failure of the ICM to be maintained in the outgrowth (e.g., *Fgf4*: Feldman et al. 1995) are some possible outcomes. Monitor explanted, individually cultured embryos daily and classify them into the following categories (Fig. 3):

- expanded and attached
- attached but not expanded
- trophoblast cells beginning to form a monolayer of giant cells with ICM clump on top
- monolayer of trophoblast giant cells with no ICM.

These classifications will help identify a developmental delay, problems with the attachment or invasive phase of trophectoderm function, giant cell transformation, and development of the ICM. You could take closer note of the morphology, but outgrowing blastocysts are extremely variable, and, unless you are very experienced or have a control group growing in parallel, finer classifications that take into account the size or morphology of the ICM, etc., may not be informative.

Starting at around E3.5 and continuing through the implantation process, trophectoderm cells are highly phagocytic, and this forms the basis for an in vitro test of trophectoderm differentiation and function. Simply add 1- to 3-μm fluorescent or colored-latex microspheres to blastocyst cultures for 8 h or more, wash the cultures, and determine whether the trophectoderm cells have phagocytosed any of the spheres. This test can be applied to zona-free blastocysts or to embryos that have attached to the culture dish.

Where to Go from Here

If you have reached this section with a diagnosis of a preimplantation lethality, we fully expect that you will have found your phenotype by this point, because it is unlikely that the homozygous mutant embryos even got as far as attaching to the culture dish. We say this because attachment usually stimulates the uterus to undergo a semiautonomous process called decidualization. Once the uterus has received the stimulus of an attaching embryo, the process does not depend on the continued presence or normality of an embryo; in fact, an oil drop, mechanical injury, and even certain bacteria can stimulate a decidual reaction. Thus, even a defective embryo could induce decidualization as long as it can hatch and/or the trophectoderm can start to attach. It is the decidua that you were looking for as a sign of the beginning of implantation in Chapter 8: Phenotypic Analysis: Assessing Timing of Recessive Prenatal Lethality in Mice [Papaioannou and Behringer 2023a]) and you have reached this section because an insufficient number of empty deciduae were present to account for your mutant embryos. Therefore, it stands to reason that the mutants will probably not attach or outgrow in this in vitro assay. However, you can still do several more tests, detailed in the next two sections, if blastocysts failed to attach. One is to isolate the ICM, the other is to induce implantation delay.

Analysis of Isolated ICM

Because there are two distinct tissues present in the blastocyst, it is possible that the gene you are investigating has its role in only one or the other cell type. If you have expression data, you might even have a good idea of which cell type is most likely to be affected. Blastocyst culture described above mostly tests the trophectoderm functions of attachment, invasion, and differentiation. But even if you have already identified a trophectoderm defect that resulted in failure of implantation, there may also be independent effects on the ICM (e.g., *Mcl1*; Rinkenberger et al. 2000). So now let us see what the ICM can do on its own. First remove the ZP (Nagy et al. 2006c and Box 1: Hints on Removal of the Zona Pellucida). Then isolate the ICM from the trophectoderm using a method called "immunosurgery" (Fig. 4) (Nagy et al. 2006a). The expanded blastocyst is exposed to anti-mouse antibody, washed, and then exposed to complement, which will lyse all the trophectoderm cells. Gentle pipetting with a

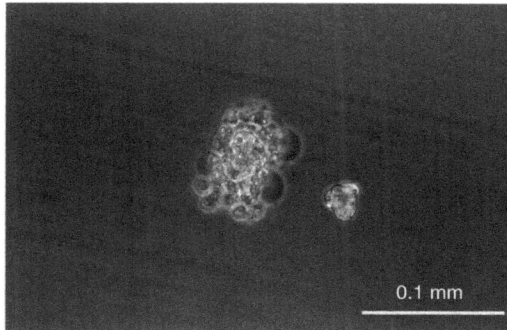

FIGURE 4. Isolation of the ICM from E3.5 blastocysts. Blastocysts were treated with rabbit anti-mouse antiserum, washed, and exposed to guinea pig complement. The outer trophectoderm cells in the embryo on the *left* are undergoing lysis. The isolated ICM on the *right* was separated from the layer of dead trophectoderm by gentle pipetting. Brightfield microscopy on a dissecting microscope.

finely drawn, mouth-controlled pipette will remove the trophectoderm debris, leaving a pure ICM that you can culture in vitro (Protocol 9.1: Phenotypic Analysis of Preimplantation Mouse Embryos in Culture [Papaioannou and Behringer 2023d]).

Around the time of implantation, the ICM forms a layer of primitive endoderm on its blastocoelic surface in the second cellular differentiation step of embryogenesis. This also occurs in isolated ICMs in culture, so that after ~2–3 d (E3.5 plus 2–3 d of culture), the primitive endoderm forms a complete rind around the spherical ICM. A basement membrane will be evident at the basal surface of the epithelial endoderm cells. A possible phenotype to look for is the absence of a primitive endoderm layer, which should be detectable either by morphology or by primitive endoderm marker gene expression. Its absence could indicate a developmental arrest or a failure of primitive endoderm differentiation. Another possibility is that the mutant ICMs fail to thrive in culture and either die or arrest cell division. The cell counting technique for preimplantation embryos (Protocol 9.2: Cell Counting Techniques for Preimplantation Mouse Embryos Using Fluorescent DNA Dyes [Papaioannou and Behringer 2023f]) can be applied to isolated ICMs as well. Keep in mind that in the intact embryo, failure of the ICM will have secondary effects on the trophectoderm because a signal from the ICM is required to keep the polar trophectoderm in a proliferative state to form the ectoplacental cone and eventually the placenta. In the absence of contact with a functional ICM, the trophectoderm forms only nondividing, terminally differentiated giant cells, which are able to elicit a decidual response but will not support further development.

Analysis of Implantation Delay

Many mammals, including mice, have a reproductive adaptation known as implantation delay, or diapause. Basically, preimplantation embryos pause in their developmental program, often because of some adversity, and await a more propitious time to resume development. This is actually a state imposed by the reproductive tract of the mother when nutritional or other conditions do not favor the embryo. In the case of mice, implantation delay can occur while the mother is nursing a previous litter. Other mammals such as some marsupials, bears, and seals make implantation delay a regular part of their reproductive cycles to coordinate reproduction with seasonal and environmental changes.

Because implantation delay is within the normal repertoire of behaviors for a mouse embryo, one can exploit this feature to help uncover cryptic phenotypes caused by mutant genes. Diapause can be induced experimentally by ovariectomy of the mother during the preimplantation phase of development. The lack of ovarian hormones causes the embryos to enter the physiological state of diapause, during which they are somewhat quiescent and do not initiate implantation. They do, however, hatch from the ZP and continue to increase their cell number, albeit slowly. Maintenance of viability and developmental potential during diapause has a genetic basis (like everything else) and mutation in at least one gene (*Lif*) reveals a role in maintenance of the ICM during diapause when all other aspects of the mutant embryo appear normal (Nichols et al. 2001). Thus, you may wish to try implantation delay in your investigation under two conditions: (1) if your gene is expressed during preimplantation or

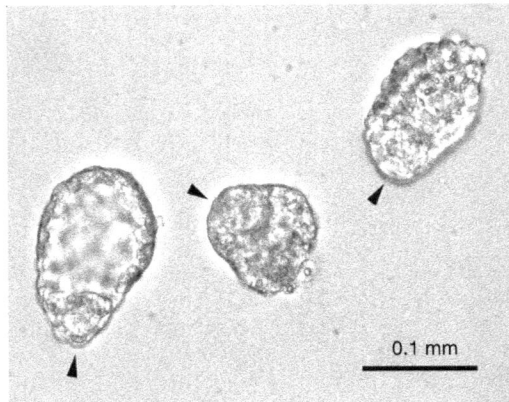

FIGURE 5. Blastocysts in implantation delay. The three blastocysts are in diapause or implantation delay after recovery from the uterus at E6.5 following ovariectomy at E2.5. The embryos have escaped from their zonae pellucidae and visible giant cell transformation has begun at the abembryonic pole in the two embryos on the *right*. Arrowheads indicate the embryonic pole with the inner cell mass.

periimplantation development but you do not see a mutant phenotype at that time or (2) as an additional test of function and viability in embryos that have a blastocyst or periimplantation phenotype.

Ovariectomize pregnant females bilaterally at E2.5 (Chapter 6 in Behringer et al. 2014) and recover the ovaries to count CL (see Protocol 4.1: Counting Murine Corpora Lutea to Determine the Number of Oocytes Ovulated [Papaioannou and Behringer 2023c]) so that you know the maximum number of embryos that you can expect to recover later. Ovariectomy will cause the embryos to arrest in diapause or implantation delay once they reach the blastocyst stage, although they will shed the ZP and increase cell number slowly. After 4–8 d, recover embryos by flushing the uterus as you would for normal blastocysts. Delayed embryos can survive longer in the uterus, but they will begin to decline after about the first week so the analysis should be done sooner rather than later. The morphology of delayed embryos can be quite variable. They might not be expanded and frequently giant cells will be evident in the mural trophectoderm (Fig. 5). Assessment of these embryos for possible mutant effects could include determining embryo loss (by comparing numbers recovered with the CL counts), simple or differential cell counts (see Protocol 9.2: Cell Counting Techniques for Mouse Preimplantation Embryos Using Fluorescent DNA Dyes [Papaioannou and Behringer 2023f]), and assessment of cell death; determining further developmental potential in vitro (see Protocol 9.1: Phenotypic Analysis of Preimplantation Mouse Embryos in Culture [Papaioannou and Behringer 2023d]); and assessing differentiation with molecular markers.

FURTHER ANALYSIS OF PREIMPLANTATION EMBRYOS

The previous sections rely heavily on morphological assessment of living embryos isolated at different preimplantation stages or cultured in vitro. Combined with staging, cell counting, and measurement of cell death, this analysis should provide a clear assessment of the nature of a preimplantation or implantation defect. Further analysis will depend on what you have learned and the nature of the gene mutated. The next level of analysis combines morphological analysis at the histological or ultrastructural level with various ways of determining the localization of mRNA and proteins, as well as the more global assessment of gene expression patterns. Whenever possible, ascertain the genotype of each embryo, and, if this is not possible, compare with a control group of embryos from heterozygous × wild-type mice (see Protocol 8.1: Establishing Control Levels of Embryonic and Fetal Loss in a Mutant Mouse Colony [Papaioannou and Behringer 2023b]) and test a large enough group of embryos to attain significance in a χ^2 analysis (see Box 1 in Chapter 8: Phenotypic Analysis: Assessing Timing of Recessive Prenatal Lethality in Mice [Papaioannou and Behringer 2023a]). Superovulation of females (Behringer et al. 2018) is a useful way of increasing embryo number for preimplantation analysis. If there is any reason to suspect that the sex of the embryo affects the phenotype or the genotype affects the sex, the sex of embryos can be determined (Protocol 5.1: Sex Genotyping Mice by Polymerase Chain Reaction [Papaioannou and Behringer 2023g]).

Histology of Preimplantation Embryos for Histochemistry and In Situ Localization of mRNA or Proteins

Cleavage-stage embryos can be fixed and processed within oviducts removed intact from pregnant females. Once embryos have moved through the uterotubal junction and dispersed throughout the uterus, however, it becomes impractical to section the entire uterus to find them. A good alternative is to collect blastocysts by flushing them from the uterus as usual and then transferring them into an isolated oviduct as if you were doing embryo transfer (Nagy et al. 2006b). For this purpose, you can use an oviduct from the same female, leaving a bit of uterus attached to prevent them from flushing out. The oviduct acts as a container for fixing and processing a large number of embryos in a compact package that is easily handled and sectioned. Either paraffin or plastic embedding can be used, depending on the intended use and level of detail required; the latter provides higher resolution of cellular detail. Sections can then be subjected to specific histochemical staining or immunocytochemistry to highlight cellular components and obtain a high-resolution morphological picture, perhaps by taking advantage of confocal microscopy. The expression pattern of specific genes, the localization of specific gene products, or the presence of apoptotic cells can be determined using in situ hybridization or immunolocalization (e.g., *Mdm2*, *Mcl1*; Rinkenberger et al. 2000; Chavez-Reyes et al. 2003). A number of molecular markers of different cell types or different states of differentiation will aid in characterizing the mutant embryos, depending on the phenotype.

Gene Expression in Whole Embryos

Gene expression analysis by in situ hybridization of mRNAs (e.g., *Eomes*; Hancock et al. 1999) or immunolocalization of proteins (e.g., POU5F1; Liu et al. 2004; NANOG, CDX2, GATA4; Kurowski et al. 2019) can also be done on whole preimplantation embryos (Fig. 6C,D), although the embryos can be tricky to handle through all steps of the procedures. An alternative approach is the hairpin

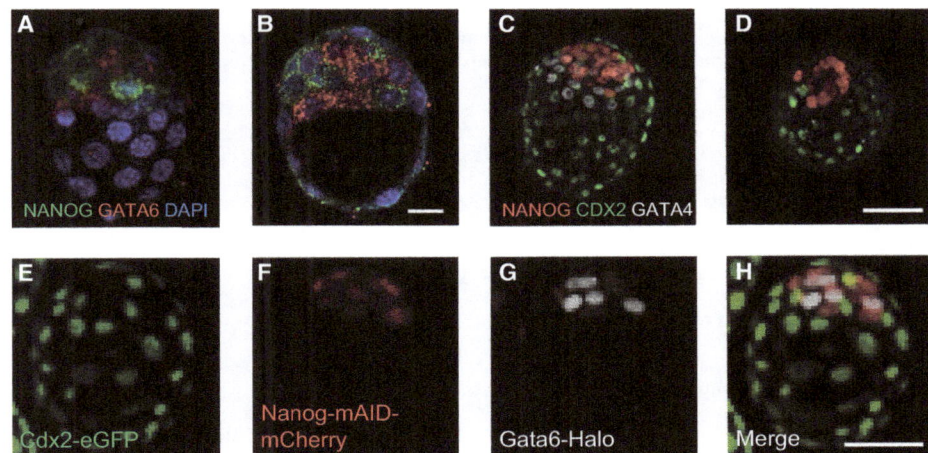

FIGURE 6. (*A,B*) Hairpin chain reaction (HCR) fluorescent in situ hybridization of a blastocyst. Multiplex hybridization and visualization by confocal microscopy for *Nanog* (green), *Gata6* (red), and DAPI counterstain (blue). (*A*) 3D projection, (*B*) optical section. Scale bar,15 μm. (*C,D*) Immunofluorescent staining of blastocysts for NANOG (red), CDX2 (green), and GATA4 (white). (*C*) *Fgfr1$^{+/+}$*; *Fgfr2$^{+/+}$*; (*D*) *Fgfr1$^{-/-}$*; *Fgfr2$^{-/-}$*. Scale bar, 50 μm. (*E–H*) Live extended focus images of triple fluorescent reporter alleles in a blastocyst. Cdx2-eGFP (green), Nanog-mAID-mCherry (red), Gata6-Halo (white). Scale bar, 40 μm. (*A,B*, Courtesy of Marta Portela Martinez, Mohamed Gatie, and Kat Hadjantonakis; *C,D*, modified, with permission, from Kurowski et al. 2019; *E–H*, modified, with permission, from Gu et al. 2018.)

chain reaction fluorescent in situ hybridization method that is imaged using confocal microscopy and can be multiplexed to examine combinations of gene expression at cellular resolution (Fig. 6A,B; Huss et al. 2015). Embryos processed as whole mounts are available for genotyping afterward, making the number requirement much lower for a definitive assessment of phenotype. Alternative ways of assessing mRNA or proteins in individual embryos include quantitative RT-PCR to detect specific transcripts of interest (e.g., *Fgf4*; Goldin and Papaioannou 2003), or two-dimensional gel electrophoresis of radiolabeled proteins (Shi et al. 1994). Because the analysis must be done on a single-embryo basis, the sensitivity of the different assays can be limiting. Every attempt should be made to develop a way of genotyping embryos at the same time to attribute differences definitively to the homozygous mutant phenotype.

New mouse lines have been generated using CRISPR–Cas to add fluorescent tags to proteins expressed in the various cell types of the blastocyst, including epiblast, primitive endoderm, and trophectoderm (Fig. 6 E–H; Gu et al. 2018). The fluorescent fusion proteins retain wild-type function and provide cell type–specific markers of the different cell lineages. These knock-in alleles can be bred into mice carrying your mutation to generate double heterozygotes, which can then be crossed to heterozygotes for your mutation to generate homozygous mutants with the cell type-specific fluorescent allele(s). Live imaging of these embryos will reveal the phenotypic state of different cell types in the mutant preimplantation embryos as well as the gene expression of the particular fluorescent markers used. In contrast to in situ hybridization and immunolocalization of a marker, this fluorescent reporter approach enables time-lapse imaging for dynamic analyses. One caveat using this approach is that these fluorescent tag alleles may be on genetic backgrounds that differ from that of your mutant allele, potentially introducing genetic background effects.

REFERENCES

Babinet C, Richoux V, Guenet JL, Renard JP. 1990. The DDK inbred strain as a model for the study of interactions between parental genomes and egg cytoplasm in mouse preimplantation development. *Dev Suppl* 1990: 81–87. doi:10.1242/dev.108.Supplement.81

Behringer RR, Gertsenstein M, Nagy KV, Nagy A. 2014. *Manipulating the mouse embryo: a laboratory manual*. Cold Spring Harbor Laboratory Press, Cold Spring Harbor, NY.

Behringer R, Gertsenstein M, Nagy KV, Nagy A. 2018. Administration of gonadotropins for superovulation in mice. *Cold Spring Harb Protoc* doi:10.1011/pdb.prot092403

Chavez-Reyes A, Parant JM, Amelse LL, de Oca Luna RM, Korsmeyer SJ, Lozano G. 2003. Switching mechanisms of cell death in *mdm2-* and *mdm4*-null mice by deletion of p53 downstream targets. *Cancer Res* 63: 8664–8669.

Feldman B, Poueymirou W, Papaioannou VE, DeChiara TM, Goldfarb M. 1995. Requirement of FGF-4 for postimplantation mouse development. *Science* 267: 246–249. doi:10.1126/science.7809630

Goldin SN, Papaioannou VE. 2003. Paracrine action of FGF4 during peri-implantation development maintains trophectoderm and primitive endoderm. *Genesis* 36: 40–47. doi:10.1002/gene.10192

Gu B, Posfai E, Rossant J. 2018. Efficient generation of targeted large insertions by microinjection into two-cell-stage mouse embryos. *Nat Biotechnol* 36: 632–637. doi:10.1038/nbt.4166

Guidi CJ, Sands AT, Zambrowicz BP, Turner TK, Demers DA, Webster W, Smith TW, Imbalzano AN, Jones SN. 2001. Disruption of Ini1 leads to peri-implantation lethality and tumorigenesis in mice. *Mol Cell Biol* 21: 3598–3603. doi:10.1128/MCB.21.10 .3598-3603.2001

Hancock SN, Agulnik SI, Silver LM, Papaioannou VE. 1999. Mapping and expression analysis the mouse ortholog of *Xenopus Eomesodermin*. *Mech Dev* 81: 205–208. doi:10.1016/S0925-4773(98)00244-5

Herceg Z, Hulla W, Gell D, Cuenin C, Lleonart M, Jackson S, Wang ZQ. 2001. Disruption of *Trrap* causes early embryonic lethality and defects in cell cycle progression. *Nat Genet* 29: 206–211. doi:10.1038/ ng725

Huss D, Choi HM, Readhead C, Fraser SE, Pierce NA, Lansford R. 2015. Combinatorial analysis of mRNA expression patterns in mouse embryos using hybridization chain reaction. *Cold Spring Harb Protoc* 2015: 259–268. doi:10.1101/pdb.prot083832

Kurowski A, Molotkov A, Soriano P. 2019. FGFR1 regulates trophectoderm development and facilitates blastocyst implantation. *Dev Biol* 446: 94–101. doi:10.1016/j.ydbio.2018.12.008

Lerch-Gaggl A, Haque J, Li J, Ning G, Traktman P, Duncan SA. 2002. Pescadillo is essential for nucleolar assembly, ribosome biogenesis, and mammalian cell proliferation. *J Biol Chem* 277: 45347–45355. doi:10.1074/jbc.M208338200

Liu L, Czerwiec E, Keefe DL. 2004. Effect of ploidy and parental genome composition on expression of Oct-4 protein in mouse embryos. *Gene Expr Patterns* 4: 433–441. doi:10.1016/j.modgep.2004 .01.004

Nagy A, Gertsenstein M, Vintersten K, Behringer R. 2006a. Immunosurgery: isolating the inner cell mass (ICM) of blastocysts. *Cold Spring Harb Protoc* doi:10.1011/pdb.prot4428

Nagy A, Gertsenstein M, Vintersten K, Behringer R. 2006b. Oviduct transfer. *Cold Spring Harb Protoc* doi:10.1011/pdb.prot4379

Nagy A, Gertsenstein M, Vintersten K, Behringer R. 2006c. Removal of zona pellucida. *Cold Spring Harb Protoc* doi:10.1011/pdb .prot4421

Nichols J, Chambers I, Taga T, Smith A. 2001. Physiological rationale for responsiveness of mouse embryonic stem cells to gp130 cytokines. *Development* 128: 2333–2339. doi:10.1242/dev.128 .12.2333

Ohsugi M, Larue L, Schwarz H, Kemler R. 1997. Cell-junctional and cytoskeletal organization in mouse blastocysts lacking E-cadherin. *Dev Biol* 185: 261–271. doi:10.1006/dbio.1997.8560

Papaioannou VE, Behringer RR. 2023a. Phenotypic analysis: assessing timing of recessive prenatal lethality in mice. *Cold Spring Harb Protoc* doi:10.1011/pdb.over107970

Papaioannou VE, Behringer RR. 2023b. Establishing control levels of embryonic and fetal loss in a mutant mouse colony. *Cold Spring Harb Protoc* doi:10.1011/pdb.prot108083

Papaioannou VE, Behringer RR. 2023c. Counting murine corpora lutea to determine the number of oocytes. *Cold Spring Harb Protoc* doi:10.1011/ pdb.prot107986

Papaioannou VE, Behringer RR. 2023d. Phenotypic analysis of preimplantation mouse embryos in culture. *Cold Spring Harb Protoc* doi:10.1011/pdb.prot108090

Papaioannou VE, Behringer RR. 2023e. Recovering a targeted mutation in mice from embryonic stem cell chimeras or CRISPR–Cas founders. *Cold Spring Harb Protoc* doi:10.1011/pdb.tover107959

Papaioannou VE, Behringer RR. 2023f. Cell counting techniques for preimplantation mouse embryos using fluorescent DNA dyes. *Cold Spring Harb Protoc* doi:10.1011/pdb.prot108091

Papaioannou VE, Behringer RR. 2023g. Sex genotyping mice by polymerase chain reaction. *Cold Spring Harb Protoc* doi:10.1011/pdb.prot108062

Rinkenberger JL, Horning S, Klocke B, Roth K, Korsmeyer SJ. 2000. Mcl-1 deficiency results in peri-implantation embryonic lethality. *Genes Dev* **14:** 23–27. doi:10.1101/gad.14.1.23

Shi CZ, Collins HW, Buettger CW, Garside WT, Matschinsky FM, Heyner S. 1994. Insulin family growth factors have specific effects on protein synthesis in preimplantation mouse embryos. *Mol Reprod Dev* **37:** 398–406. doi:10.1002/mrd.1080370406

van Deursen J, Boer J, Kasper L, Grosveld G. 1996. G_2 arrest and impaired nucleocytoplasmic transport in mouse embryos lacking the proto-oncogene CAN/Nup214. *EMBO J* **15:** 5574–5583. doi:10.1002/j.1460-2075.1996.tb00942.x

Phenotypic Analysis of Preimplantation Mouse Embryos in Culture

Preimplantation embryo culture is a valuable approach to investigate a preimplantation lethal phenotype. Standard culture methods have been perfected such that the entire preimplantation period and the process of implantation can be followed in vitro. This protocol provides modifications for the analysis of clutches of embryos from heterozygous matings specifically for the purpose of distinguishing a preimplantation phenotype in homozygous mutants.

INTRODUCTION

In vitro culture of preimplantation mouse embryos is an excellent way of observing events that normally take place in the mother's reproductive tract. Superovulation of female mice can be used to increase the number of embryos obtained per female (Behringer et al. 2018). Protocols for mouse embryo collection, culture media, and culture are fairly standard (see Chapter 4 in Behringer et al. 2014). Preimplantation embryos are quite robust and will survive a range of culture conditions; however, their development is compromised in suboptimal conditions. To recapitulate in vivo development as closely as possible so that a mutant phenotype can be distinguished, rigorous attention must be paid to the details of culture, particularly the culture medium, pH, temperature, and humidity. Although several popular medium formulations have been in use for decades, some of these clearly do not support optimal development and should be avoided. As embryos develop, their culture requirements change. For attachment and further growth of blastocysts in vitro, it is necessary to have serum in the medium and to culture the embryos on tissue culture plastic (as opposed to bacteriological Petri dishes). This protocol provides hints for culturing embryos from the zygote stage to the blastocyst stage for the purpose of assessing mutant phenotypes during the preimplantation period.

MATERIALS

It is essential that you consult the appropriate Material Data Safety sheets and your institution's Environmental Health and Safety Office for proper handling of equipment and hazardous materials used in this protocol.

Reagents

Embryo culture medium

FHM (flushing and handling medium, which contains HEPES and reduced bicarbonate to maintain the correct pH in air) for collecting and handling embryos in air (e.g., MilliporeSigma; also see Chapter 4 in Behringer et al. 2014)

KSOM + AA for zygote through blastocyst culture (e.g., MilliporeSigma; also see Chapter 4 in Behringer et al. 2014)

Heterozygous male and female mice

Pregnant mice at embryonic day (E) 0.5

Equipment

Dissection instruments

Embryo handling pipette (see Troubleshooting for preparation)

Incubator, humidified, gassed 5% CO_2 in air or 5% CO_2/5% O_2/90% N_2, 37°C

Light mineral oil

Microscopes

Dissecting microscope for embryo handling and evaluation

Phase or DIC microscope for embryo evaluation

Plastic Petri dishes, 35-mm (or glass watch glasses) for embryo collection

Tissue culture plastic Petri dishes, 35- to 100-mm, for embryo culture

METHOD

1. Set up heterozygous matings and check plugs daily.

   Either select heterozygous females in estrus and mate with heterozygous males or, alternatively, super-ovulate immature heterozygous females (Behringer et al. 2018) and set them up with heterozygous males. Check for plugs the following day.

2. On the day of the plug (E0.5), prepare culture dishes. When comparing embryos of different genotypes, each embryo must be cultured separately for identification purposes. This is easily done by setting up a series of numbered drops of KSOM + AA medium (10–20 μL) in tissue culture Petri dishes and covering them with light mineral oil to keep them separate and minimize evaporation. Equilibrate culture medium drops in a gassed incubator for an hour or so before collecting the zygotes.

3. Collect zygotes from the oviducts using FHM and remove cumulus cells (see Chapter 4 in Behringer et al. 2014).

4. Take a sample of tissue from the mother to confirm her genotype.

5. Place the zygotes individually into equilibrated culture medium drops but identified by litter (this way you can check the genotype of the mother in the interim and discard any litters from mothers who were misgenotyped).

6. Score each embryo according to the presence or absence of two pronuclei (see Fig. 1B in Chapter 9: Phenotypic Analysis of Preimplantation Lethality in Mice [Papaioannou and Behringer 2023a]) and place in the incubator.

7. Score the embryos daily with a dissecting microscope or with phase or DIC microscopy, using the classifications listed in the Introduction of Chapter 9: Phenotypic Analysis of Preimplantation Lethality in Mice (Papaioannou and Behringer 2023a). Acquire photographic images for documentation.

   For embryos within the zona pellucida, focus on the zona pellucida until you see a sharp edge, which will be the equatorial plane. For general tips on embryo imaging, see Photodocumentation of Embryos in Chapter 10: Phenotypic Analysis of Periimplantation to Mid-Gestation Lethality (Papaioannou and Behringer 2023b).

8. At the end of 4–5 d of culture (or less, if an abnormal mutant phenotype is evident earlier), genotype each embryo by polymerase chain reaction (PCR) and correlate genotype with phenotype to confirm the mutant status of abnormal embryos.

As always, if something technical prevents you from genotyping the embryos, you will need to know the normal variation in a control backcross to make a valid statistical comparison (see Protocol 8.1: Establishing Control Frequency of Embryonic and Fetal Loss in a Mutant Mouse Colony [Papaioannou and Behringer 2023c]).

TROUBLESHOOTING

Handling embryos is best done with a mouth-controlled Pasteur pipette that has been pulled out over a flame and broken off at an internal diameter of ∼120–150 μm (or just large enough that an embryo fits inside the pipette easily without distortion). Before picking up the embryos, fill the pipette with medium up to the first widening of the pipette bore. You can pick up a large number of embryos at a time in a small volume, but keep them near the tip of the pipette so that they are not lost on the air/medium interface. Monitor all embryo manipulations under a dissecting microscope. Once the embryos are in their individual culture drops, they can be monitored and scored over a period of days. During monitoring, minimize the time out of the incubator to avoid extreme temperature and pH fluctuations.

Make sure the culture drops are completely covered by the mineral oil. If the drop is at the air/oil interface, it will quickly evaporate. Keep the volume of the culture drops low to prevent bubbles of medium from lifting off into the oil when the dish is moved.

DISCUSSION

The starting point for embryo culture can be any time during preimplantation development and can be extended for longer periods to observe the attachment and outgrowth of blastocysts. Isolated inner cell masses (ICMs) can also be cultured to observe the formation of the primitive endoderm. For extended periods of culture and for ICM culture, embryonic stem (ES) cell medium with 10% fetal calf serum should be used as the embryos nutritional requirements change around the time of implantation. If you are culturing from the zygote through blastocyst attachment and outgrowth on the culture dish, the medium should be changed after attachment.

REFERENCES

Behringer RR, Gertsenstein M, Nagy KV, Nagy A. 2014. *Manipulating the mouse embryo: a laboratory manual.* Cold Spring Harbor Laboratory Press, Cold Spring Harbor, NY.

Behringer R, Gertsenstein M, Nagy KV, Nagy A. 2018. Administration of gonadotropins for superovulation in mice. *Cold Spring Harb Protoc* doi:10.1101/pdb.prot092403

Papaioannou VE, Behringer RR. 2023a. Phenotypic analysis of preimplantation lethality in mice. *Cold Spring Harb Protoc* doi:10.1011/pdb.over107971

Papaioannou VE, Behringer RR. 2023b. Phenotypic analysis of periimplantation to mid-gestation lethality. *Cold Spring Harb Protoc* doi:10.1011/pdb.over107972

Papaioannou VE, Behringer RR. 2023c. Establishing control frequency of embryonic and fetal loss in a mutant mouse colony. *Cold Spring Harb Protoc* doi:10.1011/pdb.prot108083

Cell Counting Techniques for Preimplantation Mouse Embryos Using Fluorescent DNA Dyes

Counting cells in preimplantation embryos by light microscopy is straightforward until morula compaction, when cell boundaries in living embryos become indistinct. An alternative to morphological assessment of cell number is to use fluorescent DNA dyes. This protocol details simple nuclear counting with a single DNA dye (Hoechst) or a more complicated procedure in which differential nuclear counts of the trophectoderm and inner cell mass (ICM) can be made using immunosurgery of the blastocyst and two DNA dyes (Hoechst and propidium iodide).

INTRODUCTION

A fluorescent nuclear marker, such as green fluorescent protein incorporated into the genome of mice provides a simple way of counting cells in preimplantation embryos (see Chapter 9: Phenotypic Analysis of Preimplantation Lethality in Mice [Papaioannou and Behringer 2023a]). However, if the mice you are interested in do not have such a marker, fluorescent DNA dyes provide a way of determining cell number. Simple nuclear counts can be done on preimplantation embryos for an accurate assessment of total cell number. Alternatively, a more complicated technique for differential counts of trophectoderm and inner cell mass (ICM) cell numbers can be done on blastocysts or even late compact morulae (e.g., Goldin and Papaioannou 2003). The cell counting techniques described in this protocol have the advantage that you can distinguish interphase nuclei, mitotic cells, and apoptotic cells by their characteristic chromatin patterns and thus quantify mitosis and cell death. For simple total cell counts, the fluorochrome Hoechst is used. For differential cell counts of trophectoderm and ICM, two DNA dyes are used: one that enters vital cells (Hoechst) and one that is excluded from vital cells (propidium iodide [PI]). The technique uses a procedure called immunosurgery (Nagy et al. 2006a), which lyses only the outer cells of a compact morula or blastocyst. Protection of the inner cells depends on the existence of tight junctions between the outer cells of trophectoderm. These tight junctions form during the compaction phase of preimplantation development, and thus differential cell counts can only be made on morulae that are compacted or on blastocysts. The zona pellucida can be left intact during these procedures to facilitate embryo handling.

MATERIALS

It is essential that you consult the appropriate Material Data Safety sheets and your institution's Environmental Health and Safety Office for proper handling of equipment and hazardous materials used in this protocol.

RECIPES: Please see the end of this protocol for recipes indicated by <R>. Additional recipes can be found online at http://cshprotocols.cshlp.org/site/recipes.

Simple Total Cell Counts (Steps 1–5)

Reagents

Acid Tyrode's solution (optional) (e.g., MilliporeSigma)
Embryo culture medium (e.g., FHM; see Chapter 4 in Behringer et al. 2014)
Hoechst 33258, 10 µg/mL (final concentration in culture medium)
Petroleum jelly
Preimplantation embryos
Triton X-100, 1% (optional)

Equipment

Embryo handling pipette (see Troubleshooting in Protocol 9.1: Phenotypic Analysis of Preimplantation Mouse Embryos in Culture [Papaioannou and Behringer 2023b])
Forceps
Glass microscope slides (precleaned)
Small square coverslips (22-mm × 22-mm)
UV fluorescence microscope (epifluorescence)

Differential ICM/Trophectoderm Cell Counts (Steps 6 and 7)

Reagents and Equipment

All of the above plus:
Anti-mouse antiserum (e.g., rabbit anti-mouse RBC; rabbit anti-mouse embryo; see Troubleshooting, below)
Complement, Guinea Pig Serum <R>
PI, 10 µg/mL (final concentration in culture medium)

METHOD

Simple Total Cell Counts

1. Expose the embryos to the DNA dye Hoechst 33258 in culture medium for 30 min at 37°C.

2. Mount the embryos as follows to allow for the possible need to squash them if all of the nuclei cannot be easily visualized.

   i. Have precleaned microscope slides, square coverslips, and petroleum jelly handy. Rub a little petroleum jelly into the heel of your hand and scrape the edge of the coverslip gently across your hand to collect a thin ridge of jelly along one side; repeat on the opposite side and set the coverslips aside (jelly-side up).

   ii. Pick up an embryo with an embryo handling pipette and place it on the microscope slide with ~10 µL of medium (with Hoechst dye). Carefully place the coverslip over the embryo, making contact with the drop of medium but letting the coverslip rest on the petroleum jelly ridges, creating a column of medium containing the embryo between the coverslip and slide.

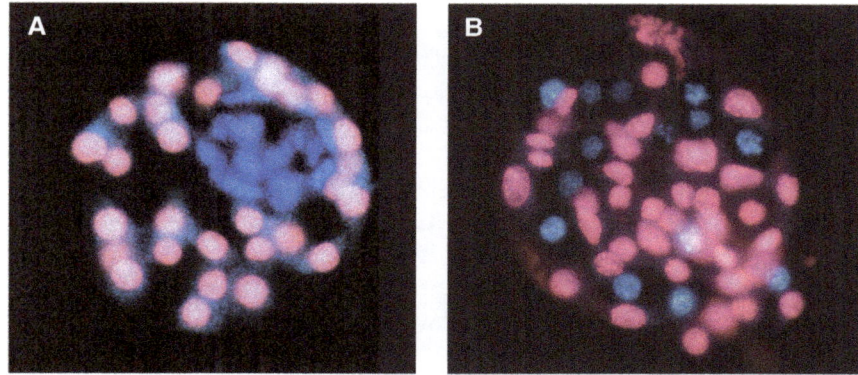

FIGURE 1. Differential cell counts of the inner cell mass (ICM) and trophectoderm of a double-dyed blastocyst before (*A*) and after (*B*) squashing. The trophectoderm nuclei appear reddish pink due to the propidium iodide (PI) and Hoechst dyes, whereas the vital ICM nuclei appear blue. (Reprinted, with permission, from Goldin and Papaioannou 2003.)

3. Observe the embryo under the fluorescence microscope and count nuclei. You may have to squash the more advanced embryos to see all of the nuclei (see Step 5), in which case you will be unable to easily recover material for subsequent genotyping.

4. To recover the embryos for genotyping, gently lift the coverslip with forceps and collect the embryo with an embryo handling pipette.

5. If you need to squash the embryo for better observation of nuclei, carefully press on the coverslip with a pair of forceps, keeping your eyes on the embryo under the microscope until it is sufficiently squashed.

    A few nuclei may scatter if cells lyse, but you can scan the entire field under the coverslip for errant nuclei. If the drop is too big, it will be difficult to squash the embryo sufficiently because of capillary forces once the fluid fills the space under the coverslip. If the drop is too small, the capillary force might overcome the resistance of the petroleum jelly and the embryo may be squashed with too much force.

Differential ICM/Trophectoderm Cell Counts

6. Perform immunosurgery on the embryos by exposing them to an anti-mouse antibody as detailed in Nagy et al. (2006a).

7. Wash the embryos thoroughly in culture medium to remove free antibodies and then expose them to a source of complement (guinea pig serum, as per Nagy et al. 2006a) but with the addition of the two fluorochromes, Hoechst and PI, for 30 min at 37°C). The outer cells will lyse and become Hoechst- and PI-positive, but the inner cells, protected from exposure to the antibody by the tight junctions in the trophectoderm, will remain intact and stain with Hoechst and exclude PI. The nuclei are detected as for simple cell counts above, but this time the trophectoderm cell nuclei are pink (red plus blue) and the ICM cell nuclei are blue (Fig. 1).

TROUBLESHOOTING

Simple Total Cell Counts

A brief pretreatment (20–30 sec) with 1% Triton X-100 will make the embryos easier to squash and prevent the distortion of nuclei associated with cell lysis. Similarly, to facilitate squashing, the zona pellucida can be removed by treatment with acidic Tyrode's solution (Nagy et al. 2006b) before squashing and counting.

Differential ICM/Trophectoderm Cell Counts

Because the immunosurgically treated embryos are delicate and tend to fall apart, it is best not to put them through a wash step following complement treatment. However, this means that the dyes will still be present in the medium, and, if you squash the embryos, the PI will rapidly enter the ICM nuclei because squashing damages the cell membranes. Count all of the pink trophectoderm cell nuclei before squashing and then be prepared to count ICM nuclei quickly, or capture an image as soon as you squash before the blue ICM cell nuclei change color. Alternatively, confocal microscopy, if available, can image through the entire volume of the embryo and alleviate this problem.

We have yet to find a commercial anti-mouse antibody source that works well for immunosurgery, so you may have to beg, borrow, or make it yourself (Nagy et al. 2006a). Be sure to test it well before use on valuable embryos.

DISCUSSION

Treatment of preimplantation embryos with Hoechst and brief exposure to UV irradiation during fluorescent imaging is compatible with viability (Ebert et al. 1985); thus, following simple total nuclear counting, whole embryos can be returned to the maternal reproductive tract for continued development if required.

RECIPE

Complement, Guinea Pig Serum

Reconstitute 5 mL of Complement, Guinea Pig serum (lyophilized; Calbiochem, Merck 234395) in 5 mL of sterilized MilliQ H_2O and filter. Make 100-µL aliquots on ice and store at $-80°C$. Thaw immediately before use.

REFERENCES

Behringer RR, Gertsenstein M, Nagy KV, Nagy A. 2014. *Manipulating the mouse embryo: a laboratory manual*. Cold Spring Harbor Laboratory Press, Cold Spring Harbor, NY.

Ebert KM, Hammer RE, Papaioannou VE. 1985. A simple method for counting nuclei in the preimplantation mouse embryo. *Experientia* 41: 1207–1209. doi:10.1007/BF01951732

Goldin SN, Papaioannou VE. 2003. Paracrine action of FGF4 during peri-implantation development maintains trophectoderm and primitive endoderm. *Genesis* 36: 40–47. doi:10.1002/gene.10192

Nagy A, Gertsenstein M, Vintersten K, Behringer R. 2006a. Immunosurgery: isolating the inner cell mass (ICM) of blastocysts. *Cold Spring Harb Protoc* doi:10.1011/pdb.prot4428

Nagy A, Gertsenstein M, Vintersten K, Behringer R. 2006b. Removal of zona pellucida. *Cold Spring Harb Protoc* doi:10.1011/pdb.prot4421

Papaioannou VE, Behringer RR. 2023a. Phenotypic analysis of preimplantation lethality in mice. *Cold Spring Harb Protoc* doi:10.1011/pdb.over107971

Papaioannou VE, Behringer RR. 2023b. Phenotypic analysis of preimplantation embryos in culture. *Cold Spring Harb Protoc* doi:10.1011/pdb.prot108090

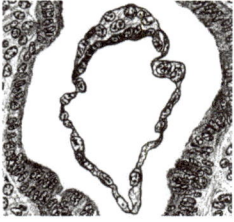

CHAPTER 10

Phenotypic Analysis of Periimplantation to Mid-Gestation Lethality in Mice

Periimplantation to mid-gestation lethality is indicated if no living homozygous mutants are recovered at E12.5 and the number of empty implantation sites or degenerating/abnormal embryos fits the expected number of homozygous mutants. To determine the time of death, this chapter details the characteristic features of lethality shortly after implantation (E4.5–E5.5) or lethality between gastrulation and allantoic fusion (E6.5–E9.5). Determining the phenotype of the mutants involves making a gross morphological assessment, staging the embryos, and photodocumenting any abnormalities. Further levels of analysis discussed are histological assessment, molecular characterization of gene expression in the mutant embryos, and measurements of cell proliferation and cell death.

INTRODUCTION

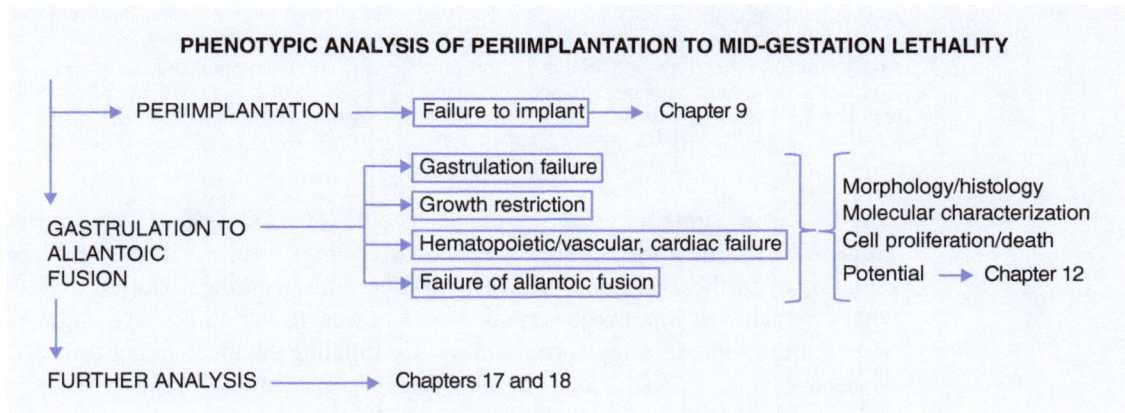

PHENOTYPIC ANALYSIS OF PERIIMPLANTATION TO MID-GESTATION LETHALITY

Indications for embryonic lethality during early gestation, between embryonic day (E) 4.5 and E9.5, are that (1) no normal, living homozygous mutant embryos are recovered from dissections at E12.5 and (2) the number of implantation sites that are either empty of embryonic remains or have only extraembryonic membranes and degenerated embryos fits the expected number of homozygous mutants. If you were successful in genotyping any of these abnormal remains, they should be the

© 2024 Cold Spring Harbor Laboratory Press
Cite this chapter as *Cold Spring Harb Protoc*; doi:10.1101/pdb.over107972

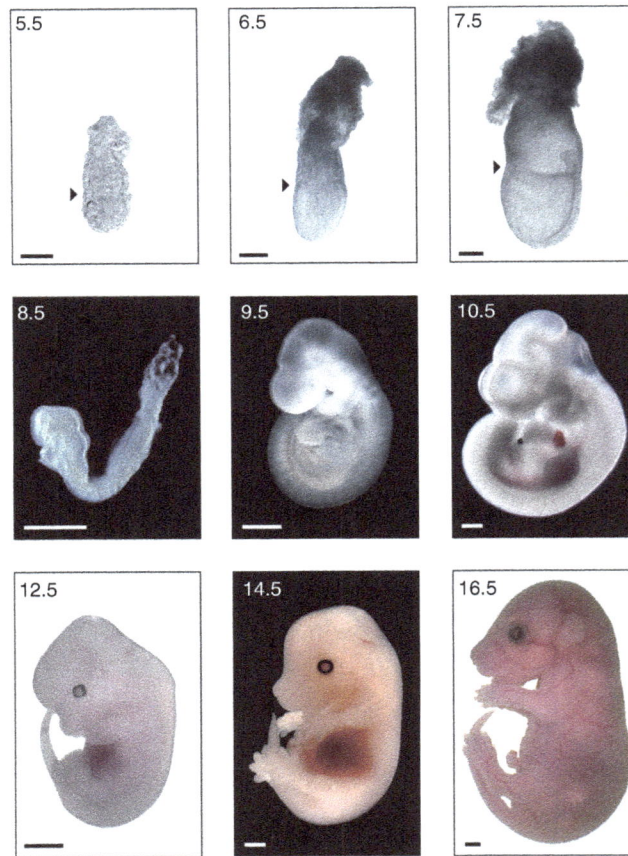

FIGURE 1. Postimplantation stages of mouse development. Dissecting microscope with top (reflective) and/or bottom (transmitted) lighting. Lateral views of freshly dissected whole-mount embryos at various embryonic days (E), indicated in *upper left* corner of each panel. Embryos (E5.5–E8.5) are oriented with anterior to the *left*. Arrowheads in E5.5–E7.5 images indicate junctions between the embryonic (*below*) and extraembryonic (*above*) regions. The trophoblast and Reichert's membrane have been dissected away from the embryonic region of the E5.5–E7.5 embryos. The placenta, Reichert's membrane, yolk sac, and amnion have been removed from all of the other embryos. Bars, 0.1 mm, E5.5–E7.5; 0.5 mm, E8.5–E10.5; 1 mm, E12.5–E16.5. (Image of E5.5 embryo provided by Aya Wada and Richard Behringer.)

homozygous mutants, although there is likely to be some low level of embryonic wastage due to causes unrelated to the mutation. The next step is to determine the time of onset of the phenotypic effects of the mutation by further dissections and embryo collections, working backward to find the time at which all embryos, including homozygous mutants, are normal in size and morphology (Fig. 1). Even if you do not reach the point of seeing normal embryos, estimating the approximate time of death from the nature of the embryonic remains (see Chapter 8: Phenotypic Analysis: Assessing Timing of Recessive Prenatal Lethality in Mice [Papaioannou and Behringer 2023a]) will indicate how you should proceed: With no embryonic remains or only a few giant trophoblast cells and membranes in a small decidual swelling, periimplantation lethality is indicated. Later times of death are indicated by the presence of recognizable dead embryos or degenerating remains. The following sections provide guidance for analyzing different times of embryonic death and where to go from there.

PERIIMPLANTATION DEATH, E4.5–E5.5

If small decidua with no embryonic remains or only a few giant trophoblast cells and some membranes are found at E12.5, the indication is for death shortly after implantation (e.g., *Mdm2*; Montes de Oca Luna et al. 1995; *Fgf4*; Feldman et al. 1995). The most common cause of lethality at this stage is failure of the embryo to implant properly in the uterus. Because trophectoderm growth and continuing function depend on the presence of a viable inner cell mass (ICM), lethality at this stage could be due to a mutant effect in either trophectoderm or ICM.

To investigate this phenotype, gross dissections will be impractical for all but the experts. You can, however, flush E4.5 embryos from the uterus in the same way that E3.5 blastocysts are flushed (see Chapter 4 in Behringer et al. 2014). Depending on the developmental stage, some or all of the embryos may be in the process of attaching to the uterine wall, so a more forceful flow of medium through the uterus may be necessary to dislodge them. Typically, the uterus will balloon with the pressure of the flushing medium and then the fluid will squirt out as the implanting embryos are dislodged. The rate of recovery of embryos using this method is not as high as that for preimplantation stages and may be lower than 50%, but corpora lutea (CL) counts (Protocol 4.1: Counting Murine Corpora Lutea to Determine the Number of Oocytes Ovulated [Papaioannou and Behringer 2023b]) can still provide an estimate of the maximum number of embryos to expect, thus providing a good estimate of the recovery rate.

Keep in mind that the mutant embryos may be lagging in development and may not have begun to attach, and thus could be overrepresented among the recovered embryos. Accordingly, genotyping is as important as ever. At this stage, wild-type embryos are no longer contained in the zona pellucida (ZP) and will have giant trophectoderm cells at the abembryonic pole (Fig. 2). They can be morphologically quite variable, which makes identifying a defect difficult unless it is striking (e.g., no ICM or severe growth restriction). Differential cell counts (Protocol 9.2: Cell Counting Techniques for Preimplantation Mouse Embryos Using Fluorescent DNA Dyes [Papaioannou and Behringer 2023c]) might provide a clue, but the technical difficulties of counting cells and genotyping the same embryo means that a large number of embryos will need to be counted and possibly compared with a control population.

If everything looks relatively normal at E4.5, but you know that the embryos die periimplantation, either the mutant effect appears later or it is not evident by gross morphology (or cell number). Rather than trying to dissect E5.5 embryos, which is difficult but not impossible, skip directly to histological evaluation of the embryos at periimplantation stages in utero (Protocol 10.1: Orientation of Implanted or Dissected Mouse Embryos for Histological Sectioning [Papaioannou and Behringer 2023d]) and carry out a functional assessment of the preimplantation and periimplantation embryos to uncover hidden defects and determine the time of onset of the mutant effect as outlined in Chapter 9: Phenotypic Analysis of Preimplantation Lethality in Mice (Papaioannou and Behringer 2023e). Start with an assessment of blastocysts at E3.5. If they appear normal, continue by testing in vitro outgrowth, ICM development, and possibly implantation delay. If, on the other hand, the blastocysts already appear abnormal by E3.5, they have preimplantation problems even though they are capable

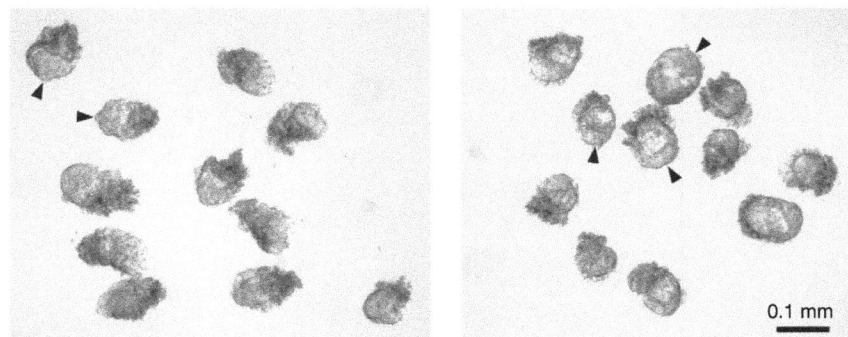

FIGURE 2. Blastocysts flushed from the uterus at E4.5. (Left) Implanting embryos that were forcibly flushed from the uterus. (Right) The same embryos after 1 h of culture. At initial recovery, the embryos are collapsed, but many pump up the blastocoel to a more recognizable blastocyst morphology following culture. The giant cells are evident at the abembryonic pole, opposite the smooth embryonic pole containing the ICM (marked with an arrowhead in several embryos). Some maternal uterine epithelial cells are still attached to the trophectoderm.

of eliciting a decidual response. In this case, go to Analysis of Cleavage Stages and Early Blastocysts in Chapter 9: Phenotypic Analysis of Preimplantation Lethality in Mice (Papaioannou and Behringer 2023e).

GASTRULATION TO ALLANTOIC FUSION LETHALITY, E6.5–E9.5

If embryonic lethality after the periimplantation period is indicated by the presence of degenerating embryos or extraembryonic tissue at E12.5, select earlier alternate days (e.g., E7.5 and E9.5) for the next dissections. Dissect and examine each implantation site, using techniques appropriate for each stage (Behringer et al. 2014), and genotype from whatever embryonic or extraembyronic tissue is available, except trophoblast, which is likely to be contaminated with maternal tissue. When you find a stage at which all homozygous mutants are normal in size and morphology, you have identified the starting point for further analysis and can move later in development until you encounter the onset of the abnormality. Start with a detailed gross morphological examination. Subsequently, the embryos can be used for other purposes including histological examination and molecular characterization such as marker gene expression using whole-mount RNA in situ hybridization, immunohistochemistry/fluorescence, or RNA-seq, provided that the embryos have been fixed and treated properly (see Molecular Characterization of Mutant Phenotypes, below). Cell proliferation and cell death can also be assessed.

During the time frame of E6.5–E9.5, some of the common causes of lethality are failure of gastrulation (e.g., *Wnt3*; Liu et al. 1999), growth restriction (e.g., *Myc*; Davis et al. 1993), failure of extraembryonic membranes to function correctly (e.g., *Dlx3*; Morasso et al. 1999), including failure of vasculogenesis or hematopoiesis (e.g., *Jag1*; Xue et al. 1999), failure of the allantois to fuse with the chorion (e.g., *Tbx4*; Naiche and Papaioannou 2003), and lack of proper cardiac specification or cardiovascular failure (e.g., *Mef2c*; Lin et al. 1997), so be on the lookout particularly for these defects. The following techniques are some of the most common and useful ways of analyzing these early-gestation stages. As you progress in the analysis of a phenotype, you may find that you wish to examine an earlier or later stage, so use these methods selectively but freely to investigate any features, or follow any intuitions you might have.

Gross Morphological Assessment

The dissections done to determine the time of death provide a gross morphological assessment of the phenotype over the time course of the embryonic lethality. This phenotypic and genotypic classification results in numbers that can be used to determine whether the onset of mutant effect occurs at a specific stage or is variable. The value of a good gross morphological assessment of freshly dissected living embryos should not be underestimated. The more knowledgeable you are in embryo development, the better prepared you will be to identify deviations.

Careful observation of embryos during dissection can reveal features that are not evident once they are fixed, including, for example, the presence, strength, and regularity of a heart beat; blood circulation and the vascularity of the placenta and yolk sac as seen by their color and the presence of blood in vessels; the presence of extravascular blood, which could indicate hemorrhage but might be washed away during tissue processing; or the condition of advanced necrotic embryos that cannot be further processed. It should be noted that dissection in cold phosphate-buffered saline (PBS) can result in a temporary slowing or cessation of the heartbeat, which can be restored as the embryos warm to room temperature. The normal rate is 100–110 beats/minute. Note that because of the inherent contractile nature of heart tissue, the presence of a heartbeat is not by itself proof of a healthy embryo. Take detailed notes during dissections and acquire images of the embryos to document any morphological defects.

The developmental stage of each embryo should also be determined (see below and Useful Developmental Landmarks for Embryos between E4.5 and Term in Chapter 8: Phenotypic Analysis: Assessing Timing of Recessive Prenatal Lethality in Mice [Papaioannou and Behringer 2023a]). At the beginning of this time range (E6.5–E7.5), the shape and size of the embryo can be used for staging as

Cite this chapter as *Cold Spring Harb Protoc*; doi:10.1101/pdb.over107972

well as the extent of the primitive streak, the presence of the node, and the presence and size of the allantois (Fig. 3). Toward the later end of the range (E8.0–E9.5), somite number is an easily assessed and accurate indicator of developmental stage (see Staging Embryos, below).

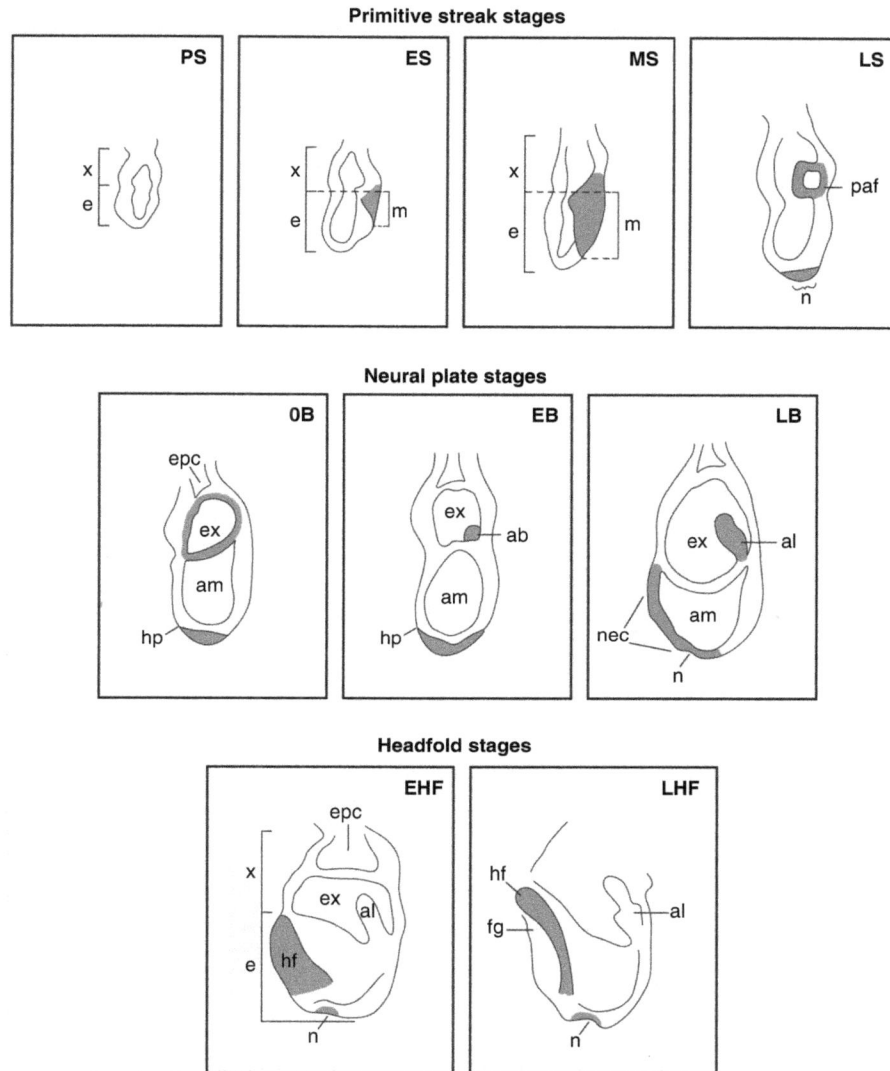

FIGURE 3. Downs and Davies staging system for E6.5–E8.0 embryos. All embryos are shown in side view with anterior to the left and ectoplacental cone removed. Preprimitive-streak (PS) embryos consist of two layers of cells with a circumferential constriction between the embryonic (e) and extraembryonic (x) regions, but no primitive streak. Early-streak embryos (ES) have a primitive streak at the posterior pole and a small wedge of mesoderm (m, *gray*) at the posterior pole. This is evident as an area in which the two layers, primitive endoderm and epiblast, are no longer distinct. Midstreak (MS) embryos are characterized by a primitive streak that has extended 50%–100% of the length of the embryonic portion at the posterior pole. Late-streak (LS) embryos have a node (n, gray) at the anterior end of the primitive streak, and also a posterior amniotic fold (paf, gray) at the posterior end of the primitive streak, just at the boundary of the embryonic and extraembryonic regions. Neural-plate-stage embryos are staged on the basis of the allantoic bud: No bud (OB) embryos have no allantois but are characterized by the fusion of the amnionic folds to form three separate cavities: the amniotic cavity (am), the extraembryonic coelom (ex), and the ectoplacental cavity (epc), and an anterior extension of the node to form the head process (hp, gray). Early allantoic bud (EB) embryos have a small allantoic bud (ab, gray) pushing into the extraembryonic coelom that is best seen in a posterior view (not shown). Late bud (LB) embryos have a long allantois (al, gray) extending through the extraembryonic coelom and a thickened neural ectoderm (nec, gray). The headfold (hf, gray) stage can be subdivided into early headfold (EHF), which has bilaterally thickened headfolds anterior to the crescent-shaped node, and late headfold (LHF), which is characterized by larger, sigmoid-shaped headfolds. (fg) Foregut. (Adapted, with permission, from Downs and Davies 1993, © Company of Biologists.)

A range of developmental stages is commonly observed between and even within litters of a given gestational age, but a large disparity could indicate that the mutation is causing a developmental delay. To the extent possible, you may also wish to determine that the rate of development of different organs within embryos is appropriate, particularly if the expression of your gene is localized there. For example, if the somite number is appropriate for a given age, is eye development also appropriate for that age or does eye development lag behind? Be on the lookout for size differences within a given developmental stage. A mutation might cause a reduction in overall cell number, resulting in an embryo of small size but otherwise normal morphology.

Growth restriction can also be the result of cardiovascular insufficiency. Examine the heart to determine whether the bilateral heart tubes have coalesced at the midline and fused to form a single heart tube by E8.5–E9.0 (see Fig. 5 in Chapter 8: Phenotypic Analysis: Assessing Timing of Recessive Prenatal Lethality in Mice [Papaioannou and Behringer 2023a] and Figure 3 in Chapter 11: Analysis of Mid- to Late-Gestation Phenotypes in Mice [Papaioannou and Behringer 2023f]). Irregular beats begin in the endothelial strands at about E8.0 and the heartbeat should be strong and regular by E9.5. Looping of the embryonic heart tube takes place starting at around E8.5, although defects in looping per se are not life-threatening at this stage.

Examine embryos for laterality defects, which is best done at E9.5 or later, after axial rotation from dorsal to ventral flexure (e.g., *Invs*; Morgan et al. 1998; *Tbx6*; Hadjantonakis et al. 2008). The most obvious asymmetry is cardiac looping but several other laterality characteristics can be scored (Fig. 4, and also see Fig. 3 in Chapter 11: Analysis of Mid- to Late-Gestation Phenotypes in Mice [Papaioannou and Behringer 2023f]):

- the heart makes a C-shaped loop to the right side of the embryo
- the tail and the placenta are normally on the right side of the embryo
- the vitelline vessels emerge from the left side of the embryo and are on the opposite side from the placenta
- the spleen and stomach are located within the left side of the abdomen.

Finally, determine whether neural tube closure is taking place on schedule (e.g., *Twist1* [Chen and Behringer 1995]; *Cart1* [Zhao et al. 1996]; *Shroom3* [Hildebrand and Soriano 1999]). Closure of the neural tube is a continuous process starting at the six- to seven-somite stage (~E8.5) at the level of the hindbrain (closure 1). Two other points of closure appear at the 12- to 15-somite stage (E8.5–E9.5) at the forebrain/midbrain boundary (closure 2) and at the rostral extremity of the neural tube (closure 3)

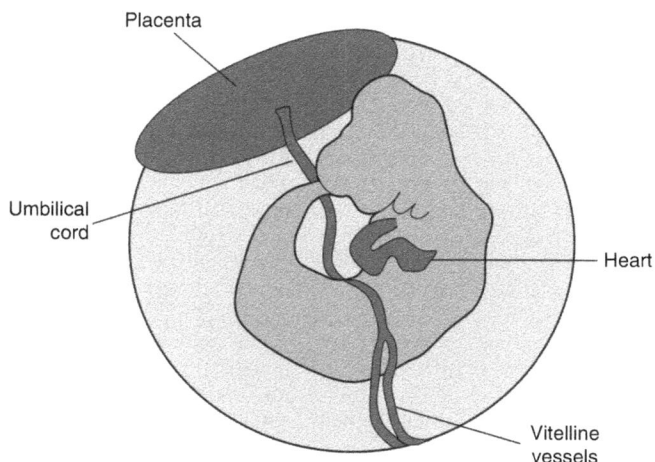

FIGURE 4. Laterality features of an E9.5 embryo. The placenta and tail are on the embryo's *right* side and the vitelline vessels emerge on the *left* side. The heart has a C-shaped loop toward the *right*. The outflow tract of the heart is on the *right* at the anterior pole of the heart tube. The inflow tract is at the posterior pole of the heart on the embryo's *left*.

Cite this chapter as *Cold Spring Harb Protoc*; doi:10.1101/pdb.over107972

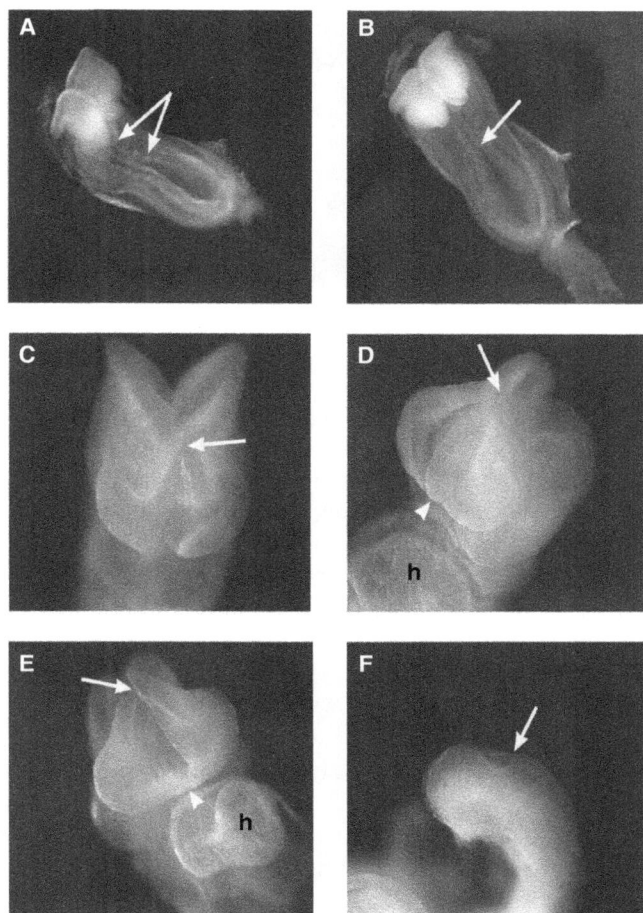

FIGURE 5. Neural tube closure. (*A,B*) Oblique and dorsal images of a six- to eight-somite-pair embryo showing the first neural tube closure at the level of the future hindbrain–cervical junction (arrows). (*C–E*) Three different views of the head of a 12- to 16-somite-pair embryo showing the second neural tube closure at the level of the forebrain–midbrain junction (arrow) and a third point of closure at the rostral end of the neural tube (arrowhead). (*F*) The posterior neuropore at the caudal extremity of the neural tube. Closure of the spinal neural tube proceeds from the hind-brain–cervical junction caudally and eventually closes the posterior neuropore at about the 24- to 26-somite-pair stage. Arrow indicates the extent of closure in this embryo. (h) Heart. (Images courtesy of Saadi Ghatan.)

(Fig. 5). The tube then zips up caudally from closure 3 and in both directions from closures 1 and 2, eventually reaching the posterior extremity of the neural tube at the posterior neuropore. Failure or disruption of neural tube closure can have many causes, both intrinsic and extrinsic to the neural tube. The process is also a delicate one and the closure points can reopen if the embryo is handled roughly or sits in the dissection dish for too long, so be aware of these artifacts.

Staging Embryos

Between the preprimitive-streak stage, when the embryo consists of two cell layers in a simple cup shape, and the headfold stage, when the embryo has three germ layers and all of the extraembryonic membranes, growth and morphogenesis are rapid and embryos within a litter or age range can vary considerably. Because so many new tissues are appearing, it may be important to stage the embryos before making comparisons between mutant and wild type to detect a phenotype. Downs and Davies (1993) have described a widely used staging system for preprimitive-streak to headfold-stage embryos (E6.5–E8.0) that divides this period into nine stages based on major morphological characteristics that are easily identified in intact embryos under a dissecting microscope, once Reichert's membrane has been removed. The primitive-streak stage is divided into four substages based on the presence and

TABLE 1. Somite pairs visible morphologically at different embryonic days (E), numbered from the anterior-most somite pairs

Embryonic day	E8.5	E9.5	E10.5	E11.5	E12.5
Theiler stage[a]	13	15	17	19	20–21
Somite pairs visible	1–13	5–30[b]	5–39[c]	26–48[c]	40–55

Data compiled from Kaufmann and Bard (1999).
Some variation is to be expected.
[a]See https://www.emouseatlas.org/emap/home.html
[b]Anterior-most somites differentiate and are no longer visible as discrete blocks of tissue.
[c]Somite pairs 31–48 are in the tail.

extent of the primitive streak; the neural-plate stage is subdivided into three substages based on the presence and growth of the allantois; and the headfold stage is divided into early and late stages depending on the morphological appearance of the headfolds (Fig. 3).

Once somites begin to form, they can be used as a measure of developmental stage. Somites are the transitory, paired, segmental blocks of paraxial mesoderm that differentiate into the dermatome, sclerotome, syndetome, and myotome, and eventually give rise to the dermis, vertebral column, ribs, associated musculature, tendons, and other structures. They form and then differentiate in an anterior-to-posterior progression. Between the ages of E7.5 and E11.5, their number provides a useful and fine-grained indicator for comparing developmental stages and is especially useful when combined with measurement of crown-rump length to access body size (Table 1). The first 10 somite pairs form at about one per hour, the trunk somites take between 1.5 and 2 h each, and the tail somites form at a rate of about one every 2–3 h. The first somite pair is located just caudal to the otic placodes (later, the otic vesicles) and counting somites in embryos of up to about 10 somite pairs (approximately E8.5) is straightforward. At later stages, the differentiation of the most anterior somites makes them difficult to discern; however, the position of somites relative to the emerging limb buds is constant, so the numbers can be accurately estimated. Somite pairs 7–12 are in line with the forelimb bud, and somite pairs 23–28 are in line with the hindlimb bud (Fig. 6). Thus, in later embryos, you can start counting at one of these landmarks. These relative positions shift later in development but the method is valid for embryos up to about 40–45 somite pairs (E11.5). If in doubt, you can always count somites after whole mount in situ hybridization with a marker of somites such as *Meox1* (*Mox1*).

Photodocumentation of Embryos

High-quality images enhance publications and are useful for documentation and analysis. When you are dissecting fresh embryos, they are usually destined for additional analyses and will never look the

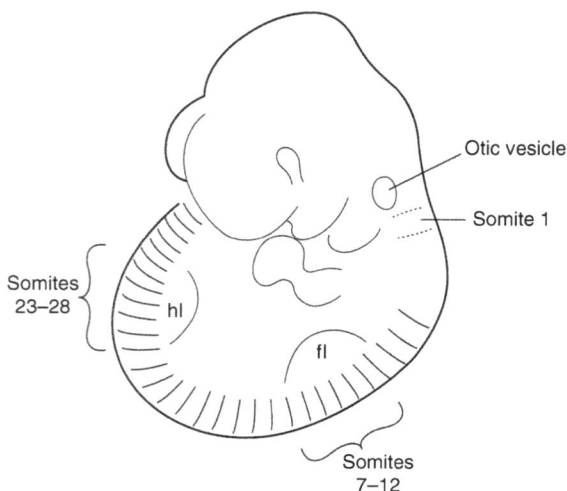

FIGURE 6. Counting somites to stage embryos up to E11.5. The first somite (indicated by dotted lines) is located just posterior to the otic vesicle but has differentiated by the stage shown and is no longer visible. Somites 7–12 are located at the level of the forelimb (fl) bud and somites 23–28 are located at the level of the hindlimb (hl) bud.

Cite this chapter as *Cold Spring Harb Protoc*; doi:10.1101/pdb.over107972

same again. It is not necessary to acquire images of every embryo, but it is worthwhile to acquire an image of a representative wild-type phenotype as well as any abnormalities you detect. Even if you do not yet know what you are looking for, you can always refer back to images that you have taken after the embryos are genotyped. Documentation of in situ hybridization and other types of analysis is also worthwhile, even if the preparations are more or less permanent. Take the best-quality images you can each time (so that you will have publication-quality images when the time comes) and maintain a detailed and rigorous bookkeeping system to keep track of them. The most convenient way to do this is with a computer with a high-quality digital camera that can be attached to both a dissecting and a compound microscope. As you acquire images, store each image with a unique file name and record the details on a spreadsheet linked to detailed laboratory notes including date and time of dissection.

Dissected postimplantation embryos are best photographed on a dissecting microscope where lower magnification will provide greater depth of field so that more of the three-dimensional embryo is in focus. Embryos should be completely immersed in clean medium or PBS to avoid reflections from the surface of the liquid. Experiment with backgrounds and lighting to get the best out of your particular sample and microscope. A useful setup for whole embryos is a transmitted light base with a tilted mirror and frosted glass stage, with additional light provided by a fiber optic illuminator with adjustable "gooseneck" arms. Adjust the mirror to provide uniform background lighting and the fiber optic lights to provide appropriate top or side lighting. Then, adjust the light levels until the image shows the features that you wish to illustrate: Internal features will benefit from more transmitted light; surface features will benefit from more top (reflective) lighting.

A neat trick that results in a nice uniform background and allows for flexible positioning of embryos is the use of a petri dish containing semisolid agarose. Prepare 60-mm plastic culture dishes with a layer of 0.5%–1% agarose (~5-mm-deep) ahead of time and store them at 4°C. When you are ready to acquire images of embryos, simply flood an agarose dish with PBS or medium and introduce the embryos. To position the embryos, scoop out a shallow hole in the agarose and wedge the embryo into it in the desired position. This is especially useful for photographing at odd angles or for producing a lineup of embryos all in the same position, because it prevents one embryo from floating off as you position another.

Histological Assessment

Following this detailed gross morphological characterization, but before moving on to sophisticated techniques such as marker gene analysis, a higher-resolution morphological assessment using histological sections and hematoxylin and eosin (H&E) staining is in order. If the time of death is between E4.5 and, say, E8.5, it is easiest to section the embryos while they are still in the uterus (Protocol 10.1: Orientation of Implanted or Dissected Mouse Embryos for Histological Sectioning [Papaioannou and Behringer 2023d]). Furthermore, keeping the embryos and membranes intact is especially important if you suspect or know of a problem with the extraembryonic membranes or placental development. At later stages, the embryos can be fixed, embedded, and sectioned after they are dissected out of the uterus (Protocol 10.1: Orientation of Implanted or Dissected Mouse Embryos for Histological Sectioning [Papaioannou and Behringer 2023d]), but extraembryonic structures should also be processed so as not to lose information about the membranes and placenta.

Fixing and embedding later-stage embryos within the intact uterus is also an option, even though assigning a genotype will be more difficult. The importance of collecting serial sections with a known orientation cannot be overemphasized. In our experience, it is rare to find a histological service facility that can adequately prepare histological sections from embryos, so it is probably best to do it yourself, particularly if you are concerned about the plane of sectioning. The main source for comparison and analysis of histological morphology is the excellent book *The Atlas of Mouse Development* (Kaufmann 1992) and its supplement (Baldock et al. 2015), as well as the online resource EHistology (https://www.emouseatlas.org/emap/home.html); use these resources to see what you are aiming for.

A good morphological assessment of a serially sectioned embryo will provide an immense amount of information about the mutant phenotype. Presence and normality of tissue types, organ developmental stages, cell death, and proliferation can all be determined. As a supplement to H&E staining, specialized histochemical stains can be used to highlight specific cell types. If indicated by the mutant phenotype, scanning or transmission electron microscopy could be used to provide a more detailed ultrastructural analysis. But the embryo is complicated, and interpretation of early embryonic stages is a skill that must be learned. Consultation with an expert or at least a few good books (Kaufmann 1992; Kaufmann and Bard 1999; Rossant and Tam 2002; Baldock et al. 2015) or other resources (https://www.emouseatlas.org/emap/home.html) may be a good idea at this time.

When abnormalities are detected, care must be taken in attributing them to the mutant genotype, because abnormalities are frequently observed in nonmutant embryos. If you collected the yolk sac or a limb for genotyping, you will be able to make a positive identification of the mutants. But if you sectioned embryos in utero, you will not know the genotypes. One option is to examine enough embryos to trust that a Mendelian ratio of phenotypes reflects the Mendelian ratio of genotypes, but you will always have to describe the abnormal embryos as "putative mutants" and accept that the occasional abnormal but nonmutant embryo may confound your determination of the phenotype. If the phenotype of the homozygous mutants was well characterized on the basis of gross morphology of genotyped embryos, it may not matter too much if you do not genotype embryos from histological sections.

Alternatively, you can genotype the embryos from sections after you have finished the morphological analysis (e.g., *Hdh*; Zeitlin et al. 1995). Basically, this involves removing the coverslips, or using a slide containing adjacent sections that has not been stained and coverslipped, and collecting a sample of the embryo section for polymerase chain reaction (PCR) analysis, carefully avoiding contamination with maternal tissue. It is possible to scrape away the maternal tissue with a razor blade or scalpel under a dissecting microscope and extract DNA for PCR analysis from the isolated embryo section using cloning rings placed directly on the slide over the section. A more sophisticated, and probably more reliable, method is laser-assisted microdissection, or laser capture, in which a defined area of the section is laser dissected and "captured" in a tube for PCR. The limitation might be the availability of laser microdissection equipment, although the technology is increasingly used in cancer research and may be available in pathology departments.

Molecular Characterization of Mutant Phenotypes

Strategy

Analysis of the morphology of an embryo, placenta, or extraembryonic membranes will provide a good indication of the affected tissues. When this information is combined with information on the normal gene expression pattern of the mutated gene during the period at which the abnormality first appears, it may suggest a hypothesis regarding the developmental role of the gene. One of the most popular and effective ways of testing such hypotheses is the molecular characterization of the mutant embryos to determine patterns of gene and protein expression (e.g., *Tbx6*; Chapman and Papaioannou 1998). For this analysis, a variety of molecular markers of differentiation is used to determine precisely which tissues have differentiated molecularly, as well as the nature of any abnormal structures. Many gene products provide markers of specific tissues, particular stages of development, or particular cell states at different times in development. Judicious selection of markers appropriate to the abnormality observed will give your gross morphological and histological assessment the added strength of molecular backup. Several markers per affected tissue are usually adequate, provided that you have chosen them well to address the specific abnormality. Be aware that the absence of a marker could be the result of a lack of differentiation within the tissue or a lack of the tissue itself. In addition, you may be able to identify specific signaling pathways that are disrupted in the mutant embryo by the pattern of genes expressed.

Most commonly, probes for mRNA are used to detect molecular markers, although antibodies to specific proteins are also very useful and provide slightly different information (i.e., the location of the protein, rather than the message). Both can be used to provide temporal and spatial gene expression pattern data that will be informative when mutants are compared with controls.

Analysis of Cell Type–Specific Gene Expression

There are many options for detecting gene expression in specific cell types or tissues at the level of mRNA, protein, or protein activity. Whole embryos or partially dissected tissue (i.e., organs) can be processed intact for whole-mount nonradioactive mRNA in situ hybridization, immunohistochemistry, immunofluorescence, or histochemistry (e.g., β-galactosidase or alkaline phosphatase activity) to see the overall pattern of expression. The size of the embryo or organ can present a limitation due to penetration of the reagents, causing high background or trapping of reagents in cavities of whole mounts, but whole mounts work well for embryos up to about E12.5. Processed whole mounts can subsequently be sectioned (cryostat or vibratome) and examined with or without counterstaining for a more detailed histological picture. Differential interference contrast (DIC) microscopy can provide tissue structure information, obviating the need for a counterstain. If you do use a counterstain, choose one that does not obscure the gene expression signal.

Alternatively, fixed embryos or tissues can be sectioned first and then processed for mRNA in situ hybridization, chromogenic or fluorescent immunohistochemistry (standard or confocal microscopy), or histochemistry. Sections are especially advantageous when embryos or tissues are too large for reagents to penetrate efficiently. Sections of varying thicknesses can be cut for different purposes from tissues that have been embedded in different media: Plastic media are used for very thin sections (0.2–2 μm), paraffin wax is used for sections typically between 4 and 8 μm, OCT compound (Sakura Finetechnical Co.) is used for cryosectioning at 6–20 μm, and agarose can be used to embed organs such as the brain to cut 10- to 200-μm-thick sections on a vibratome. Also, metal or acrylic blocks, called mouse brain matrices, are specifically designed to hold a mouse brain, allowing you to make 1-mm sagittal or coronal brain slices using razor blades. These matrices could be used with other tissues as well.

You are not limited to detecting a single mRNA or protein per sample. mRNA and protein can be colocalized by sequential colorimetric in situ hybridization and immunohistochemistry (or histochemistry in the case of β-galactosidase) with different substrates to produce different color precipitates. Similarly, two different mRNAs can be detected using probes that are differentially labeled. Finally, the hairpin chain reaction (HCR) fluorescent in situ hybridization method can be used for multiple probes at once and, combined with fluorescent microscopy, provides cellular resolution (Fig. 7; Huss et al. 2015; Anderson et al. 2020).

The following are a few helpful hints.

- When sectioning, mount a fairly small number of sections on each slide so that you have plenty of slides for controls.
- Use alternate slides for controls to have comparable sections.

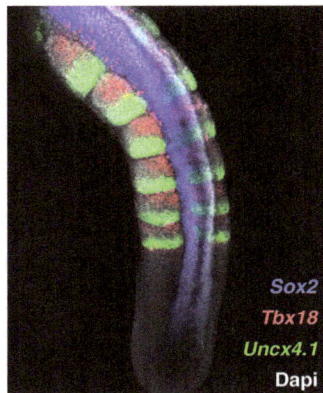

FIGURE 7. Hairpin chain reaction (HCR) fluorescent in situ hybridization of E10.5 tailbud region, showing neural tube (blue) and forming somites (red and green). Multiplex hybridization and visualization by confocal microscopy for *Sox2* (blue), *Tbx18* (red), *Uncx4.1* (green), and DAPI counterstain (gray). (Images courtesy of Matthew Anderson and Mark Lewandoski.)

- For mRNA detection, be sure to work under RNase-free conditions at all times, even during initial dissection.

- In whole-mount embryos, make holes with a syringe needle in large cavities, such as the neural tube and peritoneum, with small punctures to prevent trapping and assist penetration of reagents.

For standard methods for mRNA in situ hybridization or immunohistochemistry, see Wilkinson (1992), Wassarman and DePamphilis (1993), Tuan and Lo (2000), Nagy et al. (2003), Behringer et al. (2014), and Huss et al. (2015).

Cell Proliferation and Cell Death

Embryonic growth and development are not only the result of rapid cell division but also cell migration, regional differences in mitotic rate, as well as specific sites of programmed cell death (apoptosis). Mutant phenotypes frequently include alterations in proliferation or apoptosis, either as a primary result of the mutation or as the result of secondary effects on embryo patterning or viability (e.g., *Hdh*; Zeitlin et al. 1995). Provided that you have stage- and age-matched controls for comparison, the assessment of proliferation and cell death in mutant embryos can contribute to an explanation of the phenotype. You may wish to analyze the entire embryo or concentrate on a specific area of expression of your gene or a structure that shows a phenotype. These two measures are generally complementary in the analysis of a mutant phenotype. For example, a mutant phenotype characterized by the underdevelopment of some structure could result from hypoproliferation, failure of cell migration, and/or an increase in cell death. Likewise, an apparent overgrowth of a structure might be caused by hyperproliferation, or more rarely, hypertrophy, or by a lack of apoptosis (assuming that apoptosis is normally part of the development of the structure). Because apoptosis is usually restricted to very specific areas in the normal embryo, elevated levels can indicate a generalized detrimental effect (e.g., vascular insufficiency leading to cell death) or specific effects (e.g., aberrant signaling triggering the apoptotic response in a specific tissue). Many methods can detect proliferation or cell death, and they can be applied to whole embryos or sections, depending on embryonic age and level of detail desired.

Measuring Cell Proliferation

Because embryonic cells divide rapidly, the problems associated with obtaining measurable mitotic indices from a slow-growing tissue do not apply. Thus, although it is possible to administer an extrinsic mitotic label such as bromodeoxyuridine (BrdU) or 5-ethynyl-2′-deoxyuridine (EdU) to an embryo over a period of time by administering it to the pregnant mother, or to inject the mother with colcemid to arrest and accumulate mitoses in the embryo, this is usually not necessary. Intrinsic markers of mitosis are adequate, and it is nice to be able to avoid problems associated with hazardous chemicals. The simplest method to quantitate proliferation is to assess cell division in hematoxylin and eosin (H&E)-stained histological sections. Different tissue types can be easily distinguished and mitoses are clearly evident by the densely stained chromosomes in mitotic figures. The high mitotic index of the ependymal cells lining the lumen of the brain and spinal cord provides a clear example (Fig. 8A). Because examination of histological sections is frequently part of the analysis of a mutant phenotype anyway, this is often the easiest and most convenient way to assess whether additional analysis is necessary. Incidentally, cell death can also be recognized by nuclear morphology in histological sections by the presence of darkly staining pyknotic (degenerating) nuclei, so this also provides an assessment of possible alterations in apoptosis; see the next section, Measuring Cell Death.

Antibodies to proteins involved in the mitotic cycle offer another convenient way of visualizing and quantitating cell proliferation, both in sections and in whole embryos. When you are first working out the protocols, sections of adult intestinal epithelium provide a convenient, rapidly dividing control tissue. For the analysis, always compare homozygous mutants with wild-type embryos of the same developmental stage. Antibodies to proliferating nuclear cell antigen (PCNA) are commer-

Cite this chapter as *Cold Spring Harb Protoc*; doi:10.1101/pdb.over107972

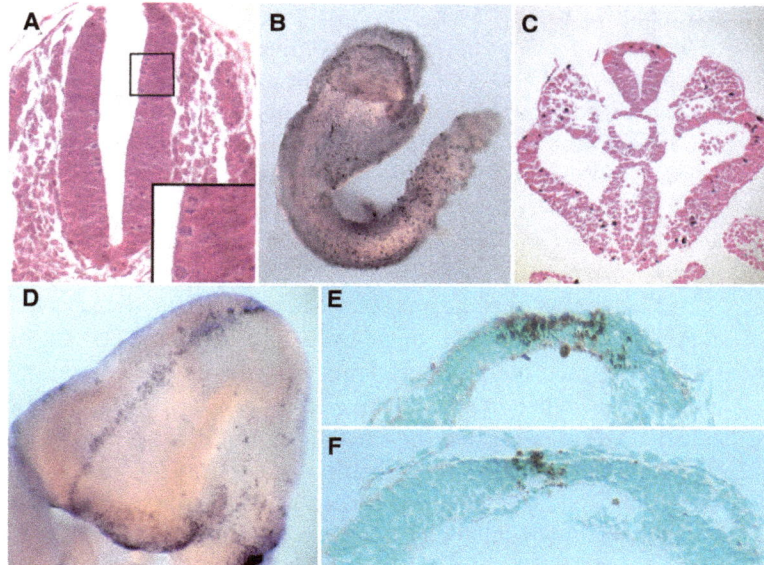

FIGURE 8. Detection of cell proliferation and cell death. (*A*) Hematoxylin and eosin (H&E) staining of a section of the neural tube of an E10.5 embryo. Numerous, darkly staining mitotic nuclei are visible at the luminal surface of the neuroepithelium. The boxed area shown at higher magnification in the inset shows a nucleus in anaphase and two nuclei in metaphase. (*B*) Whole mount of an E8.5 embryo with mitotic cells shown by immunolocalization of phosphohistone H3. (*C*) Transverse section through the midgut region of an E9.5 embryo with mitotic cells visualized by immunolocalization of phosphohistone H3. Immunolocalization was done on the whole embryo, which was then embedded, sectioned, and counterstained with nuclear fast red. (*D*) Detection of cell death in the head region of an E9.5 embryo using the terminal dUTP nick end labeling (TUNEL) assay on a whole mount. The midline streak of apoptotic cells marks the point of closure of the neural tube; cell death is a normal part of this process. (*E,F*) Detection of apoptotic cells at the site of neural tube closure by immunolocalization of activated caspase in sections of the rhombencephalon and prosencephalon, respectively, counterstained with methyl green. (*B,C,* Courtesy of Amalene Cooper-Morgan and Virginia Papaioannou; *D–F,* courtesy of Saadi Ghatan.)

cially available and often used in histochemical staining procedures on sectioned tissue to classify cells into phases of the mitotic cycle. PCNA is present in all mitotic stages, but is differentially expressed during different phases of the cell cycle such that G_1, S, G_2, and M can be recognized. However, this requires interpretation of staining density and may be a bit more elaborate than necessary for an initial survey of cell proliferation in mutant embryos, unless, of course, you have reason to suspect that an effect on cell cycle progression could be part of the phenotype.

Immunohistochemical staining with antiphosphohistone H3 (PH3) antibody, which characterizes mitosis (and meiosis) and is also commercially available, offers a suitable marker for embryonic cell proliferation and can be applied to early whole embryos or to sections (Fig. 8B,C). The advantage of doing whole mounts, of course, is that overall patterns can be seen in the intact embryo. However, the limitation is the depth of penetration of the reagents, so this procedure works best on young embryos ($\leq$E10.5) or on dissected tissues from older embryos. Whole embryos can later be embedded and sectioned for detailed cell counts and determination of the mitotic index (MI) (MI = number of mitotic cells/total number of cells examined) once the overall patterns have been assessed. A counterstain such as nuclear fast red will be necessary so that the MI can be calculated and compared. PH3 can also be visualized by immunofluorescence and nuclei counterstained with DAPI. The MI can be calculated for specific tissues or for the embryo as a whole, provided that the total number of cells examined is reasonably large (several thousand cells).

Measuring Cell Death

Alterations in tissue dynamics may involve changes in both proliferation and cell death and so assays for both are often considered together. Apoptosis, also called programmed cell death, is a normal, reproducible part of embryogenesis and may be disturbed in a mutant phenotype. Necrotic cell death may also be part of a mutant phenotype if placental or vascular insufficiency leads to tissue necrosis. Not all assays of cell death allow you to distinguish between apoptosis and necrosis. Apoptosis can be recognized morphologically by characteristic features in H&E sections that include condensation of chromatin (sometimes in association with the nuclear membrane), nuclear fragmentation, blebbing, and fragmentation of cytoplasm into apoptotic bodies, often surrounded by a clear halo. Apoptotic indices can be calculated in the same way as mitotic indices, and histological sections will sometimes provide adequate information to rule out (or rule in) apoptosis as a factor in a mutant phenotype.

Alternatively, specific markers of apoptotic or dying cells can be used for quantitation. The commonly used terminal dUTP nick end labeling (TUNEL) assay is commercially available as a kit, but the method has limitations involving tissue penetration in embryo whole mounts and does not distinguish between apoptotic and necrotic cell death. Several alternative apoptotic markers, such as activated caspase, can be detected immunohistochemically with good reproducibility. A convenient test tissue for working out the method and to use as a control is a E12.5 limb bud (either whole mount or sections) because apoptosis takes place in the interdigital web tissue. In addition, several places in the embryo in which apoptosis occurs normally (e.g., along the ridges of the closing neural tubes at E9.5 [Fig. 8D–F], in the dorsal root ganglia at E12.5, and in the central region of the presumptive retina at E11.5) can be used as internal controls. As with measurements of cell proliferation, both the pattern of apoptosis and the frequency as measured by the apoptotic index can provide information on mutant phenotypes.

The histological features of necrotic cell death closely resemble apoptosis. However, necrosis generally begins in the embryo in foci that expand to encompass large areas. If this pattern of cell death occurs in the homozygous mutant embryos, you might have recovered mushy, degenerating remains, which could not be processed for histology. A necrotic embryo would immediately suggest placental or circulatory insufficiency, and thus, generalized embryonic cell death may be secondary to a developmental or functional problem with the extraembryonic structures or the circulatory system. In this case, the development of the placenta and all of the extraembryonic structures should be examined.

REFERENCES

Anderson MJ, Magidson V, Kageyama R, Lewandoski M. 2020. *Fgf4* maintains *Hes7* levels critical for normal somite segmentation clock function. *eLife* **9**: e55608. doi:10.7554/eLife.55608

Baldock R, Bard J, Davidson DR, Morriss-Kay G. 2015. *Kaufman's atlas of mouse development supplement: with coronal sections.* Elsevier, Amsterdam.

Behringer RR, Gertsenstein M, Nagy KV, Nagy A. 2014. *Manipulating the mouse embryo: a laboratory manual*, 4th ed. Cold Spring Harbor Laboratory Press, Cold Spring Harbor, NY.

Chapman DL, Papaioannou VE. 1998. Three neural tubes in mouse embryos with mutations in the T-box gene, *Tbx6*. *Nature* **391**: 695–697. doi:10.1038/35624

Chen ZF, Behringer RR. 1995. *twist* is required in head mesenchyme for cranial neural tube morphogenesis. *Genes Dev* **9**: 686–699. doi:10.1101/gad.9.6.686

Davis AC, Wims M, Spotts GD, Hann SR, Bradley A. 1993. A null *c-myc* mutation causes lethality before 10.5 days of gestation in homozygotes and reduced fertility in heterozygous female mice. *Genes Dev* **7**: 671–682. doi:10.1101/gad.7.4.671

Downs KM, Davies T. 1993. Staging of gastrulating mouse embryos by morphological landmarks in the dissecting microscope. *Development* **118**: 1255–1266. doi:10.1242/dev.118.4.1255

Feldman B, Poueymirou W, Papaioannou VE, DeChiara TM, Goldfarb M. 1995. Requirement of FGF-4 for postimplantation mouse development. *Science* **267**: 246–249. doi:10.1126/science.7809630

Hadjantonakis A-K, Pisano E, Papaioannou VE. 2008. *Tbx6* regulates left/right patterning in mouse embryos through effects on nodal cilia and perinodal signaling. *PLoS ONE* **3**: e2511. doi:10.1371/journal.pone.0002511

Hildebrand JD, Soriano P. 1999. Shroom, a PDZ domain-containing actin-binding protein, is required for neural tube morphogenesis in mice. *Cell* **99**: 485–497. doi:10.1016/S0092-8674(00)81537-8

Huss D, Choi HM, Readhead C, Fraser SE, Pierce NA, Lansford R. 2015. Combinatorial analysis of mRNA expression patterns in mouse embryos using hybridization chain reaction. *Cold Spring Harb Protoc* **2015**: 259–268. doi:10.1101/pdb.prot083832

Kaufmann MH. 1992. *The atlas of mouse development.* Academic, London.

Kaufmann MH, Bard JBL. 1999. *The anatomical basis of mouse development.* Academic, San Diego.

Lin Q, Schwarz J, Bucana C, Olson EN. 1997. Control of mouse cardiac morphogenesis and myogenesis by transcription factor MEF2C. *Science* **276**: 1404–1407. doi:10.1126/science.276.5317.1404

Liu P, Wakamiya M, Shea MJ, Albrecht U, Behringer RR, Bradley A. 1999. Requirement for *Wnt3* in vertebrate axis formation. *Nat Genet* **22**: 361–365. doi:10.1038/11932

Montes de Oca Luna R, Wagner DS, Lozano G. 1995. Rescue of early embryonic lethality in *mdm2*-deficient mice by deletion of *p53*. *Nature* **378**: 203–206. doi:10.1038/378203a0

Morasso MI, Grinberg A, Robinson G, Sargent TD, Mahon KA. 1999. Placental failure in mice lacking the homeobox gene *Dlx3*. *Proc Natl Acad Sci* **96**: 162–167. doi:10.1073/pnas.96.1.162

Morgan D, Trunpenny L, Goodship J, Dai W, Majumder K, Matthews L, Gardner A, Schuster G, Vien L, Harrison W, et al. 1998. Inversin, a

Cite this chapter as *Cold Spring Harb Protoc*; doi:10.1101/pdb.over107972

novel gene in the vertebrate left-right axis pathway, is partially deleted in the *inv* mouse. *Nat Genet* **20:** 149–156. doi:10.1038/2450

Nagy A, Gertsenstein M, Vintersten K, Behringer R. 2003. *Manipulating the mouse embryo: a laboratory manual*, 3rd ed. Cold Spring Harbor Laboratory Press, Cold Spring Harbor, NY.

Naiche LA, Papaioannou VE. 2003. Loss of *Tbx4* blocks hindlimb development and affects vascularization and fusion of the allantois. *Development* **130:** 2681–2693. doi:10.1242/dev.00504

Papaioannou VE, Behringer RR. 2023a. Phenotypic analysis: assessing timing of recessive prenatal lethality in mice. *Cold Spring Harb Protoc* doi: 10.1011/pdb.over107970

Papaioannou VE, Behringer RR. 2023b. Counting murine corpora lutea to determine the number of oocytes ovulated. *Cold Spring Harb Protoc* doi:10.1011/pdb.prot107986

Papaioannou VE, Behringer RR. 2023c. Cell counting techniques for pre-implantation mouse embryos embryos in culture. *Cold Spring Harb Protoc* doi:10.1011/pdb.prot108091

Papaioannou VE, Behringer RR. 2023d. Orientation of implanted or dissected mouse embryos for histological sectioning. *Cold Spring Harb Protoc* doi:10.1011/pdb.prot108094

Papaioannou VE, Behringer RR. 2023e. Phenotypic analysis of preimplantation lethality in mice. *Cold Spring Harb Protoc* doi:10.1011/pdb.over107971

Papaioannou VE, Behringer RR. 2023f. Analysis of mid- to late-gestation phenotypes in mice. *Cold Spring Harb Protoc* doi:10.1011/pdb.over107973

Rossant J, Tam PP. 2002. *Mouse development: patterning, morphogenesis, and organogenesis.* Academic, San Diego.

Tuan RS, Lo CW. 2000. *Developmental biology protocols.* Humana, Totowa, NJ.

Wassarman PM, DePamphilis ML. 1993. *Guide to techniques in mouse development.* Academic, San Diego

Wilkinson DG. 1992. *Whole mount in situ hybridization of vertebrate embryos.* IRL, Oxford.

Xue Y, Gao X, Lindsell CE, Norton CR, Chang B, Hicks C, Grendron-Maguire M, Rand EB, Weinmaster G, Gridley T. 1999. Embryonic lethality and vascular defects in mice lacking the Notch ligand Jagged1. *Hum Mol Genet* **8:** 723–730. doi:10.1093/hmg/8.5.723

Zeitlin S, Liu J-P, Chapman DL, Papaioannou VE, Efstratiadis A. 1995. Increased apoptosis and early embryonic lethality in mice nullizygous for the Huntington's disease gene homologue. *Nat Genet* **11:** 155–163. doi:10.1038/ng1095-155

Zhao Q, Behringer RR, de Crombrugghe B. 1996. Prenatal folic acid treatment suppresses acrania and meroanencephaly in mice mutant for the *Cart1* homeobox gene. *Nat Genet* **13:** 275–283. doi:10.1038/ng0796-275

Protocol 10.1

Orientation of Implanted or Dissected Mouse Embryos for Histological Sectioning

Histological analysis is an informative part of any mutational analysis. Interpretation of histological sections is much easier, and comparisons of mutant and wild-type embryos are more reliable if the plane of sectioning is precisely controlled. This protocol provides methods for obtaining sections with reproducible orientation for embryos between E4.5 and E9.5 still in the uterus and for dissected embryos E9.5 and older.

INTRODUCTION

Serial histological paraffin sections of embryos are extremely useful in determining morphological abnormalities in mutants. For the interpretation of histological sections, it is critical to know the orientation of the embryo and the plane of section, and nowhere is this more important than when comparing mutant and wild-type embryos in the phenotypic analysis of induced or naturally occurring mutations. Standard orientation planes for sectioning are sagittal, transverse, and frontal, which all depend on the position of the embryo in the block of paraffin during embedding and sectioning (see Chapter 17 in Behringer et al. 2014 and Nagy et al. 2007c). This protocol provides guidance for achieving these orientations in sections of embryos that have either been fixed while still intact in the uterus (Steps 1–6) or have been dissected out of the uterus and away from the placenta and extraembryonic membranes (Steps 7–11).

Because implanted embryos have a predictable orientation of the placenta with respect to the uterus (see Chapter 2 in Behringer et al. 2014; see also Fig. 1 in Chapter 8: Phenotypic Analysis: Assessing Timing of Recessive Prenatal Lethality in Mice [Papaioannou and Behringer 2023]), the key to preparing serial sections of embryos during early gestation is to leave the embryos in the uterus and use uterine landmarks to orient the tissue in the paraffin block during embedding. An added advantage is that the placenta and extraembryonic membranes will remain intact and in a consistent orientation for comparison and analysis. At later stages, dissection of embryos out of the uterus provides the advantages of gross morphological assessment and genotyping to identify the mutants before histological processing. This can result in a considerable savings in effort as only a sample of the control embryos need be processed for comparison with the mutants. The placenta and yolk sac can also be sectioned following dissection.

© 2024 Cold Spring Harbor Laboratory Press
Cite this protocol as Cold Spring Harb Protoc; doi:10.1101/pdb.prot108094

MATERIALS

It is essential that you consult the appropriate Material Data Safety sheets and your institution's Environmental Health and Safety Office for proper handling of equipment and hazardous materials used in this protocol.

RECIPES: Please see the end of this protocol for recipes indicated by <R>. Additional recipes can be found online at http://cshprotocols.cshlp.org/site/recipes.

Reagents

Fixative (e.g., Bouin's fixative <R> or Paraformaldehyde (PFA; 4%) <R>)
Paraffin for embedding

Equipment

Dissection instruments
Dissecting pins
Dissecting stereomicroscope
Embedding molds
Paper index cards
Pregnant female mice

METHODS

Implanted Embryos within the Uterus (∼E4.5–E9.5)

1. Dissect out the uterine horns along with a bit of the cervix and the oviducts, leaving some of the mesometrial mesentery attached for orientation.

2. Spread the uterus onto a small piece of dry index card and, with dissecting pins, pin it in place through the cervix and oviducts (Fig. 1A).

   The tendency of the uterine muscle is to contract, so stretch each uterine horn out straight (but not under too much tension) and spread the mesentery out on the dry card to anchor the uterus on the paper. The placenta is located on the mesometrial side of the uterus, and thus this will result in all of the embryos being in more or less the same orientation.

3. Immerse the uterus, card and all, into fixative.

   *Bouin's is our favorite for general hematoxylin & eosin (H&E) histology, but other fixatives such as 4% PFA may be required for other procedures such as in situ hybridization or immunostaining (Protocol: **Fixation of Mouse Embryos and Tissues** [Nagy et al. 2007b]).*

4. After fixation and before clearing and embedding (Nagy et al. 2007a), cut the uterine horns with a scalpel into convenient lengths for sectioning (Fig. 1A,B).

   At E4.5, the decidual swellings may not be visible, so each horn can be cut into segments ∼1.5 cm in length. At later stages, two to five decidual swellings can be included in each segment, depending on their size, thus including two to five embryos in each paraffin block. This greatly decreases the amount of sectioning that needs to be done.

5. Label one side of the embedding mold and position the uterine segments. We recommend sectioning parallel to the long axis of the uterus because sectioning perpendicular to the long axis will require many more sections to be cut. To achieve this orientation, position the uterine segments in the embedding mold as shown in Figure 1B using the mesometrium for orientation, with the label on the left-hand side.

6. When the uterine segments are embedded in paraffin and removed from the embedding molds, mount the blocks in a microtome with the label on the left and make a diagonal cut in the top right-hand corner of the paraffin block (Fig. 1C) to indicate the orientation of the ribbon after

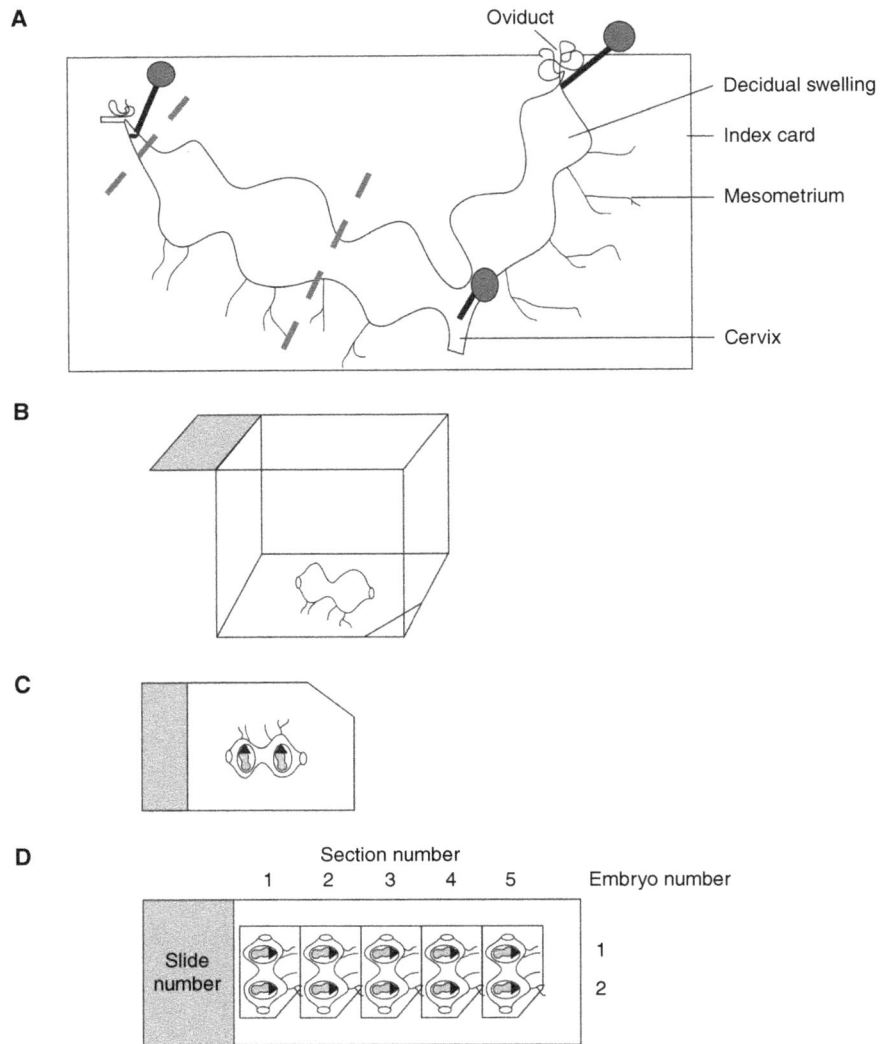

FIGURE 1. Orientation of embryos in the uterus for histology. (A) Pin down the dissected uterus on a piece of index card with dissecting pins, spreading the mesentery on the paper. After fixation, cut the uterus into conveniently sized segments (dashed lines); two decidual swellings are shown here. (B) Label embedding molds on the left-hand side and position the uterine segments with the mesentery at the front. (C) For sectioning, mount the block with the label on the left and trim the block, putting a notch in the top right-hand corner for orientation. (D) Mount the ribbon on labeled slides as shown. The first section is on the left and each of the embryos is in a horizontal row with the mesometrial pole, or placenta, on the right.

cutting. Mounting the ribbon on the slide as shown in Figure 1D results in sections of two (or more) embryos in rows on the same slide.

The long axis of the mouse embryo, before turning, is more or less at right angles to the long axis of the uterus, with the dorsal side of the embryo toward the mesometrial pole. Thus, sections oriented as shown will be transverse or sometimes sagittal. Once embryos have turned, their orientation will be more variable, but the placenta is always at the mesometrial pole and will be nicely sectioned for analysis.

Dissected Embryos (~E9.5 and Later)

7. Dissect embryos out of the uterus and remove the placenta and extraembryonic membranes (take a sample for genotyping).

8. Start by labeling the left side of the embedding mold.

Cite this protocol as *Cold Spring Harb Protoc*; doi:10.1101/pdb.prot108094

9. Position the embryos, one per block, for the plane of section you want (Fig. 2A).

   i. For sagittal sections, lay the embryo on its right side in the mold with its head to the left, toward the label. Sections will start on the right side of the embryo and go toward the left, because the deepest aspect of the mold will be the front of the paraffin block (Fig. 2B).

   ii. For transverse sections, position the embryo head down with its back to the label on the left side of the mold. Sections will start at the top of the head. On the slides, the right side of the embryo will be on the left (Fig. 2C).

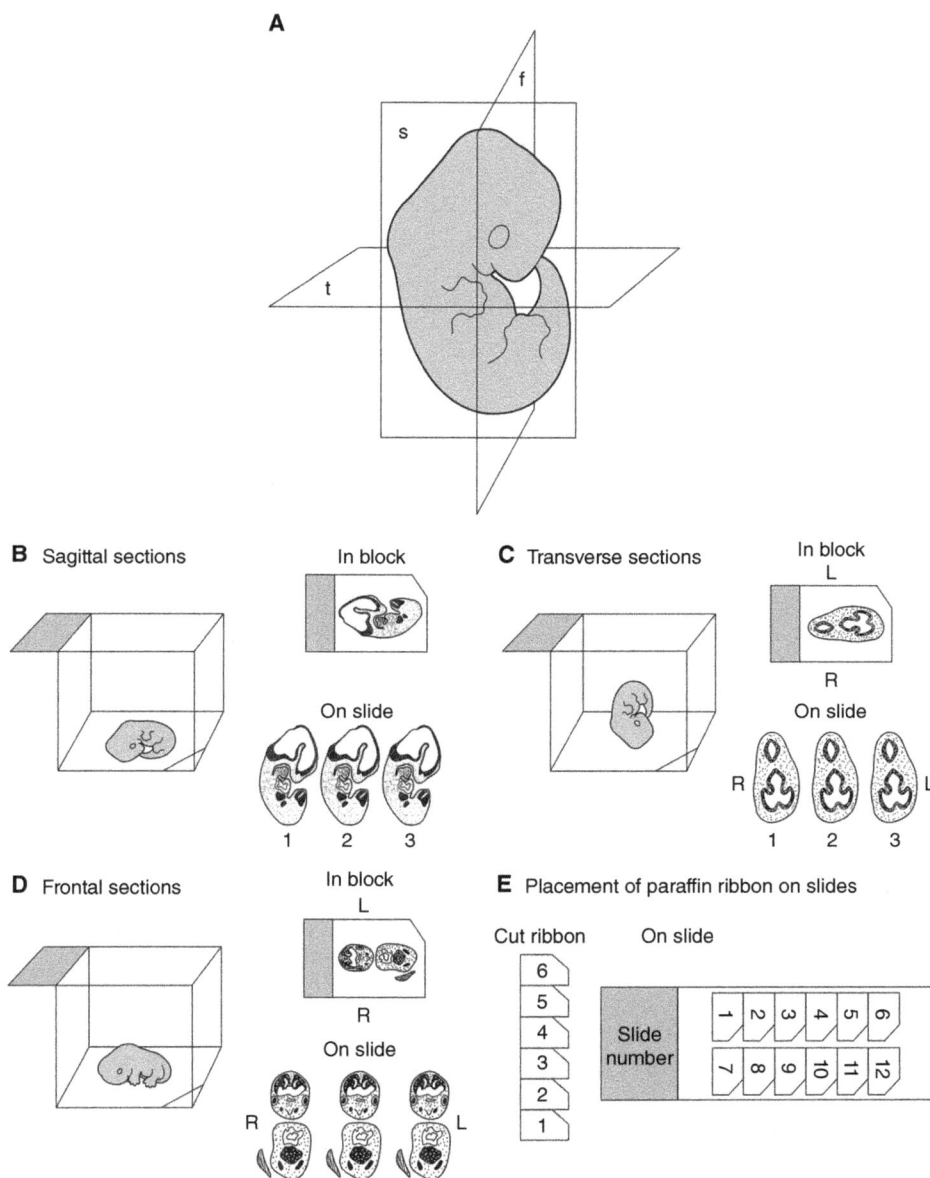

FIGURE 2. Orientation of dissected mid-gestation to late-gestation embryos for histology. (A) Standard sagittal (s), transverse (t), and frontal (f) planes of sectioning. After fixation and clearing the embryo, label an embedding mold on the left-hand side and position the embryos as shown for sagittal sections (B), transverse sections (C), or frontal sections (D). For sectioning, mount the paraffin block on the microtome with the label on the left and trim the block, putting a notch in the top right-hand corner for orientation. For transverse and frontal sections, the left side of the embryo is closest to the notch in the paraffin block. Mounting the ribbons on labeled slides as shown (E) will result in the embryo orientations indicated in B–D. If they are narrow enough, several rows of ribbon can be mounted on each slide. The numbers indicate the order of serial embryo sections. (R) Right side of the embryo; (L) left side of the embryo; (1–12) the order of sections.

iii. For frontal orientation, place the embryo face down with the head toward the left of the mold. The front of the embryo will be sectioned first. On the slides, the right side of the embryo will be on the left (Fig. 2D).

10. When sectioning the embryos (Nagy et al. 2007c), mount the block on the microtome with the label on the left. Trim the block, placing a notch on the upper right-hand corner to maintain the orientation of the sections (and in order to recognize inadvertent inversions of the paraffin ribbons).

11. Mount on labeled slides as shown (Fig. 2E), keeping the shiny side of the ribbon down.

RECIPES

Bouin's Fixative

2 g picric acid powder *or* saturated aqueous picric acid (see below)

Picric acid is explosive in crystal form!

20 g paraformaldehyde
Formalin (40% w/v formaldehyde), as an alternative to paraformaldehyde (see below)
NaOH (1 N)
2× Phosphate-buffered saline (PBS)
Glacial acetic acid (to be used with saturated picric acid and formalin; see below)

To prepare 1 L of Bouin's fixative, dissolve 2 g of picric acid in 500 mL of H_2O. Filter through a Whatman No. 1, or equivalent. Add 20 g of paraformaldehyde, and heat to 60°C in a fume hood. Add a few drops of 1 N NaOH to dissolve. Cool and add 500 mL of 2× PBS.

Alternatively, combine 75 mL of saturated picric acid, 25 mL of formalin, and 5 mL of glacial acetic acid. Store at 4°C.

Paraformaldehyde (PFA; 4%)

Reagent	Quantity (for 100 mL)	Final concentration
Paraformaldehyde (PFA)	4 g	4%
Phosphate-buffered saline for immunohistochemistry (10×), diluted to 1× <R>	100 mL	

Slowly dissolve PFA in 1× phosphate-buffered saline over low heat until solution clears. Cool. Use immediately or aliquot and store for up to 1 yr at −20°C.

REFERENCES

Behringer RR, Gertsenstein M, Nagy KV, Nagy A. 2014. *Manipulating the mouse embryo: a laboratory manual.* Cold Spring Harbor Laboratory Press, Cold Spring Harbor, NY.

Nagy A, Gertsenstein M, Vintersten K, Behringer R. 2007a. Embedding mouse embryos and tissues in wax. *Cold Spring Harb Protoc* doi:10.1011/pdb.prot4815

Nagy A, Gertsenstein M, Vintersten K, Behringer R. 2007b. Fixation of mouse embryos and tissues. *Cold Spring Harb Protoc* doi: 10.1011/pdb.prot4702

Nagy A, Gertsenstein M, Vintersten K, Behringer R. 2007c. Sectioning mouse embryos. *Cold Spring Harb Protoc* doi:10.1011/pdb.prot4703

Papaioannou VE, Behringer RR. 2023. Phenotypic analysis: assessing timing of recessive prenatal lethality in mice. *Cold Spring Harb Protoc* doi:10.1011/pdb.over107970

Cite this protocol as *Cold Spring Harb Protoc*; doi:10.1101/pdb.prot108094

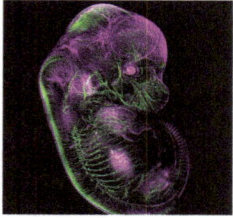

CHAPTER 11

Analysis of Mid- to Late-Gestation Phenotypes in Mice

Mid- to late gestation is characterized by tissue differentiation, maturation, organogenesis, and growth, and many mutant genes have detrimental effects during this phase of development. The outcome may be lethal before birth or may be compatible with life but result in birth defects. Some of the common causes of death during late gestation are hematopoietic defects, cardiovascular problems, and placental insufficiency. Many morphological abnormalities, lethal or not, can be investigated with gross and histological analyses or by visualization of the developing skeleton. Molecular characterization of mutant phenotypes, guided by the expression pattern of the mutant gene, can reveal disruptions in gene expression patterns of known developmental genes. Cell proliferation and cell death assays will reveal disruptions in cellular dynamics. Various modalities of 3D imaging of intact embryos can provide volumetric information about mutant phenotypes.

INTRODUCTION

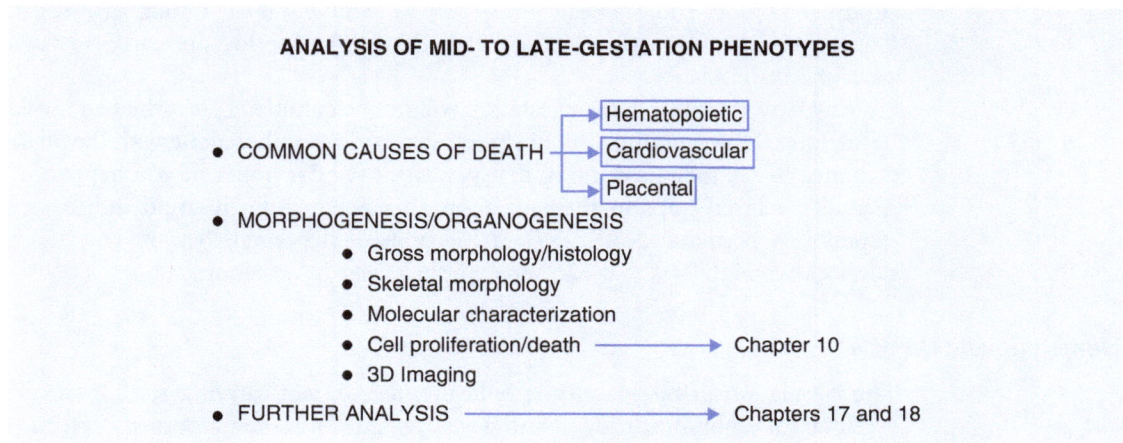

ANALYSIS OF MID- TO LATE-GESTATION PHENOTYPES

- COMMON CAUSES OF DEATH → Hematopoietic
    → Cardiovascular
    → Placental
- MORPHOGENESIS/ORGANOGENESIS
    - Gross morphology/histology
    - Skeletal morphology
    - Molecular characterization
    - Cell proliferation/death → Chapter 10
    - 3D Imaging
- FURTHER ANALYSIS → Chapters 17 and 18

Mid gestation to late gestation in the mouse (embryonic day [E] 9.5 until birth) is characterized by a period of tissue differentiation and maturation, organogenesis, and fetal growth. Thus, many mutant phenotypes that are first found during this period of development include those that affect viability and

Opening artwork: Courtesy of Evan Bardot and Kat Hadjantonakis.

organogenesis. If mutant fetuses survive to late fetal stages before succumbing, most of their development is available for study. However, there is also more time for early, nonlethal mutant phenotypes to become compounded through a developmental cascade of secondary and tertiary effects. Just because a fetus dies late during gestation does not mean that the mutant effects occur late, so determining the time of death is only a first step. In fact, the immediate cause of death may be accompanied by many developmental abnormalities that had their origin much earlier but were not life-threatening.

During late fetal stages, the essential organ systems are the heart, circulation, blood-forming tissues, and placenta. Abnormalities in virtually all other organ systems can be tolerated at least until birth. Thus, not all phenotypes that have their origin during the second half of gestation will be lethal, and many will be evident as birth defects in viable animals postnatally. If you have viable mutants with postnatal defects the techniques for early detection and analysis are similar whether the defects lead to lethality or not. The first section in this chapter deals with mutants that die between E9.5 and birth and the second section deals with the situation in which a phenotype detected postnatally indicates that the origin of an abnormality occurred during late gestation. Further analysis of mid- to late-gestation lethal phenotypes might include chimeras (Chapter 17: Getting around an Early Lethal Phenotype in Mice with Chimeras [Papaioannou and Berhringer 2023a]) and conditional alleles (Chapter 18: Tissue- and/or Temporal-Specific Mutations in Mice Using Conditional Alleles [Papaioannou and Behringer 2023b]).

ANALYZING EMBRYONIC PHENOTYPES LETHAL AFTER E9.5

If an embryo is present at the E12.5 dissection point, but is dead and possibly degenerating, death sometime after the establishment of the chorioallantoic placenta is indicated. With the fusion of the allantois to the chorion at around E8.5 and the formation of umbilical vessels within the allantois, the chorioallantoic placenta takes over the jobs of supplying the embryo with nutrients and oxygen, as well as waste removal, functions that up to this point were satisfied by diffusion or active transport across the extraembryonic membranes. At E12.5, the fetal liver takes over from the yolk sac blood islands in producing blood, and the heart pumps blood throughout the rapidly growing embryo and placenta. Consequently, common causes of death during this period are cardiovascular, hematopoietic, or placental insufficiency.

Any fetuses that die late in gestation will not be completely resorbed and will be born with the rest of the litter. However, they will likely be eaten by the mother along with the placentas before you have a chance to see them, assuming that you work regular hours. If you happen to find dead, developmentally delayed pups in the cage soon after parturition, it could indicate death before birth, as opposed to perinatal death. Even if the pups died perinatally, you may use this chapter to trace developmental defects that had their origin earlier in development.

Hematopoietic Defects

The circulation of blood cells, specifically, the oxygen-carrying red blood cells or erythrocytes, is essential for embryonic growth and development. Thus, hematopoietic defects that cause embryonic lethality after E9.5 are most likely those that compromise erythrocyte formation or integrity (e.g., *Tal1*; Mikkola et al. 2003). A deficiency of functional erythrocytes will create an oxygen deficit, leading to tissue growth restriction, developmental delay, and, if severe, death of the embryo. Embryos with hematopoietic defects will usually appear pale because of anemia and perhaps growth restricted in comparison to control littermates.

Hematopoiesis occurs at different locations within the mouse embryo at different stages of development (Fig. 1). Blood cell formation initially takes place in the blood islands of the yolk sac starting at E7.5 and continues there until E9.0. In contrast to the enucleated erythrocytes formed at other sites, the early yolk sac–derived circulating erythrocytes retain their nuclei. At E9.5, sites of hematopoiesis are found in the embryo—namely, in the aorta-gonad-mesonephros (AGM) region

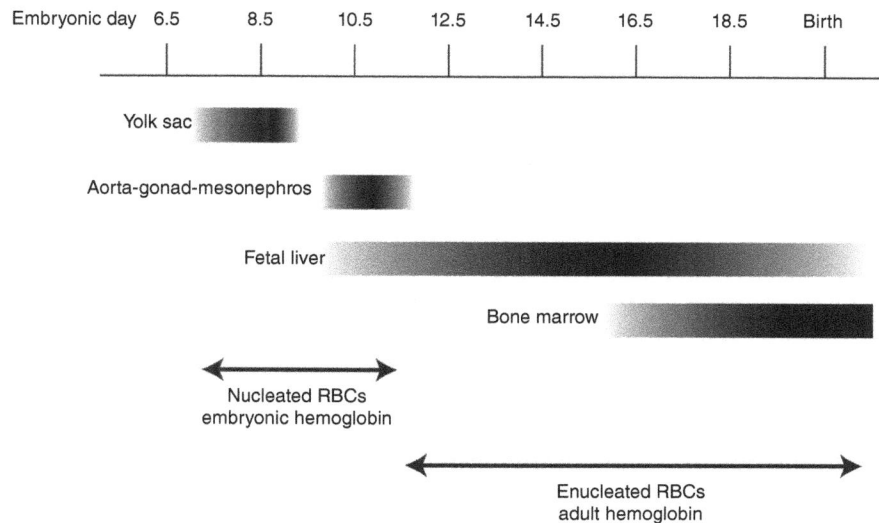

FIGURE 1. Developmental timing, tissue distribution, and erythrocyte characteristics during blood cell formation in the mouse embryo. (Shaded bars) The time period during which hematopoiesis occurs in the designated tissue; (intensity of shading) relative amounts of hematopoiesis in the tissue; (arrows) timing of formation of primitive nucleated erythrocytes (red blood cells [RBCs]) expressing embryonic hemoglobins, or definitive enucleated erythrocytes expressing adult-type hemoglobins (data from Baron 2003).

and fetal liver, which remain prominent sites of fetal hematopoiesis until just after birth. By E15.5, hematopoiesis initiates in the bone marrow, which remains the site of hematopoiesis throughout adult life. The erythrocytes generated by the fetal liver and bone marrow are enucleated. The type of hemoglobin expressed by erythrocytes also changes during development (Fig. 1), each with its own oxygen-carrying characteristics. Blood cell formation is a collaboration between the blood-forming progenitor cells (Fig. 2), the stromal microenvironment, or growth factors produced in other tissues (e.g., erythropoietin production from the kidneys) (Baron et al. 2012). Thus, defects in any of these tissues can result in red blood cell mutant phenotypes.

You may have been expecting a hematopoietic defect based on your knowledge of the protein encoded by your gene of interest and/or its expression pattern (e.g., *Smad5*; Chang et al. 1999). However, if the protein is novel or the expression pattern uninformative (e.g., ubiquitous expression), then you will likely be considering a hematopoietic defect if you noticed that your homozygous mutants, including their livers, appear pale and growth-restricted. A very simple first step in the analysis of a suspected hematopoietic defect is to generate a blood smear or cytospin preparation of the circulating blood to examine the morphology and staining characteristics of the blood cells and whether or not they are nucleated. Abnormal morphology may indicate a defect in formation that could cause the mutant erythrocyte to be more fragile or less hemodynamic for transit through the vasculature. With sufficient numbers of blood cells, you may also be able to determine the hemoglobin concentration per cell. Reduction in the hemoglobin concentration in erythrocyte will result in a lowered ability to carry oxygen to the tissues. You can also examine the type of hemoglobin (embryonic or adult) contained within the homozygous mutant erythrocyte by protein or mRNA expression analysis.

A blood smear will provide an initial piece of evidence to indicate a hematopoietic defect. Next, determine when the homozygous mutant phenotype first appears during development. Analysis of the gross morphology and histology of the homozygous mutant embryos in comparison to controls is the subsequent step in the analysis. Pay particular attention to the red blood cell–forming tissues, noting a presence or deficiency of blood cells. If there appears to be a reduction in the blood cell–forming tissues, then erythrocyte progenitors can be examined. Some of these progenitors can be grown in vitro in semisolid media, where they will give rise to colonies of erythrocytes (i.e., a clonogenic assay) and thus can be quantified (e.g., number of colony-forming units-erythroid or CFU-E, number of burst-forming units-erythroid or BFU-E) (Fig. 2). It is also possible that even more primitive multi-

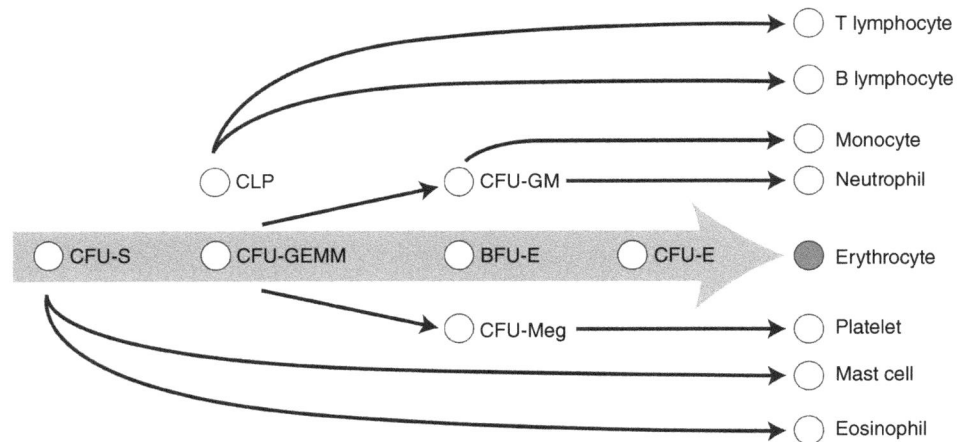

FIGURE 2. Blood cell development. Pluripotent hematopoietic stem cells differentiate into multipotent and unipotent progenitors with distinct differentiation potentials. Erythrocytes are derived from multipotent colony-forming unit-granulocyte, erythroid, macrophage, megakaryocyte (CFU-GEMM), which gives rise to unipotent erythroid progenitors: burst-forming unit-erythroid (BFU-E), and colony-forming unit-erythroid (CFU-E). Other blood cell progenitors are colony-forming unit-spleen (CFU-S), common lymphoid progenitor (CLP), colony-forming unit-granulocyte, macrophage (CFU-GM), and colony-forming unit–megakaryocyte (CFU-Meg). (Modified from Speck et al. 2002, with permission from Elsevier.)

potent hematopoietic stem cells may be defective, and these can also be examined in vitro using clonogenic assays. Subsequent molecular marker, cell proliferation, and cell death analyses (see Cell Proliferation and Cell Death in Chapter 10: Phenotypic Analysis of Periimplantation to Mid-Gestation Lethality in Mice [Papaioannou and Behringer 2023c]) can also be used to further define the tissue and stage at which the mutation acts.

Other avenues of analysis of the mutation include a chimera analysis (Chapter 17: Getting around an Early Lethal Phenotype in Mice with Chimeras [Papaioannou and Behringer 2023a]), in vitro differentiation of homozygous mutant embryonic stem (ES) cells into blood cells (Chapter 12: Uncovering Phenotypes in Mutant Mice by Determining Embryo, Organ, Tissue, and Cell Developmental Potential [Papaioannou and Behringer 2023d]), and, depending on the expression pattern, conditional knockouts in specific tissues (Chapter 18: Tissue- and/or Temporal-Specific Mutations in Mice Using Conditional Alleles [Papaioannou and Behringer 2023b]) to determine which defective tissue is causing the hematopoietic abnormality.

Cardiovascular Problems

Cardiovascular problems are a major cause of postimplantation lethality. Failure in heart specification or early differentiation can cause lethality before E9.5. After that time, insufficient cardiac function, structural defects, or conduction system defects can result in late-gestation death (e.g., *Tbx1* [Jerome and Papaioannou 2001]; *Tbx2* [Harrelson et al. 2004]). From an early stage, the cardiovascular system depends on all of the various components developing and working in synchrony. A problem in one area can rapidly lead to a cascade of secondary morphological and functional defects. Thus, when investigating cardiovascular defects, not only the heart but also the yolk sac, fetal, and umbilical vasculature, as well as the endoderm and cardiac neural crest, must all be considered (e.g., *Tbx3*; Davenport et al. 2003). Several telltale indicators of cardiac insufficiency will be evident on dissection of embryos during the second half of gestation: growth restriction, a blood-engorged liver, tissue edema, and pericardial effusion, evidenced by a swollen, fluid-filled pericardial sac (Fig. 3C). This pericardial edema can be the result of local problems in the heart or other defects, such as failed vascular connections, in more distal sites that lead to fluid buildup in the heart and pericardial cavity. In addition, alterations in the looping pattern of the heart (Fig. 3A) as well as the lack or irregularity of a heartbeat can be detected at dissection. Lack of a heartbeat will quickly lead to widespread cell death

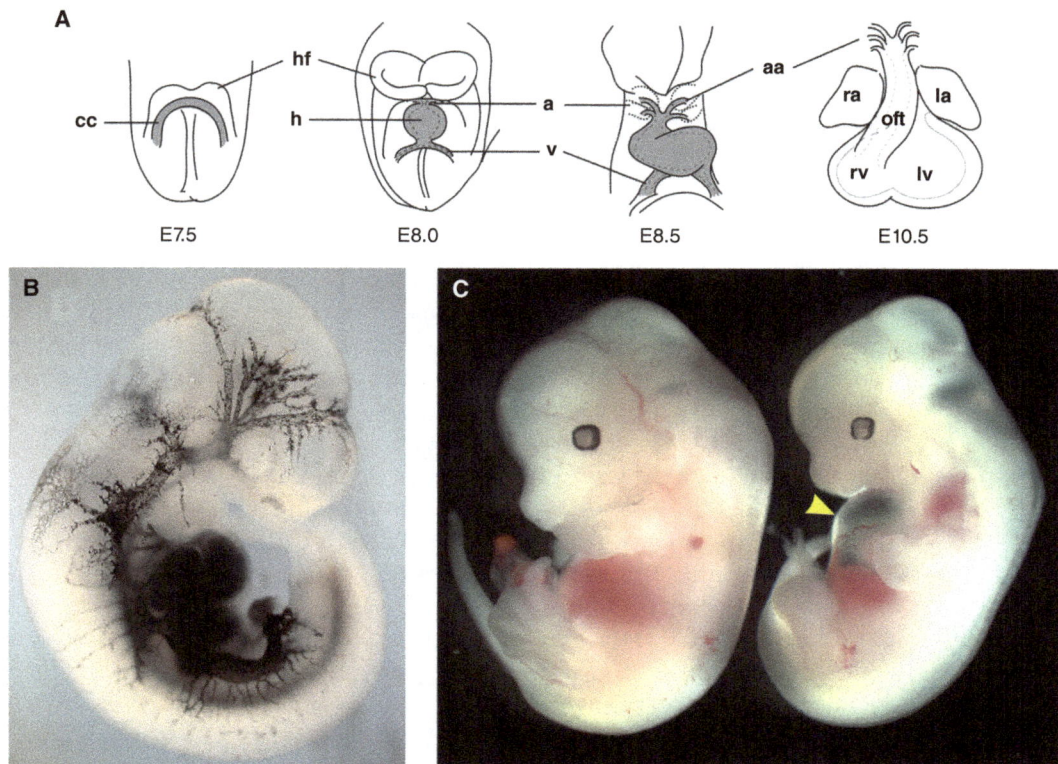

FIGURE 3. Cardiovascular development. (*A*) Ventral views of stages of heart development between E7.5 and E10.5. Cardiomyocytes first appear in the cardiac crescent (cc) at the headfold (hf) stage. The endocardial tubes fuse in the ventral midline to form the tubular heart (h) with an arterial and venous (a and v, respectively) pole. This tube loops into a C shape and then forms the four chambers of the heart: the right atrium (ra), left atrium (la), right ventricle (rv), and left ventricle (lv). At the arterial pole of the heart, the outflow tract (oft) connects with the aortic arch (aa) arteries in the pharyngeal region. (*B*) India ink injection into the heart of a E10.5 embryo, viewed from the right side. The ink particles fill the heart and outline vessels throughout the body. (*C*) A wild-type E12.5 embryo on the *left* and a *Tbx2* homozygous mutant embryo on the *right*. The mutant embryo displays cardiac effusion or a swollen pericardial sac (yellow arrowhead). (*A,B,* Courtesy of Robert Kelly; *C,* provided by Zach Harrelson and Virginia Papaioannou.)

in other parts of the embryo, including the vasculature. In these cases, knowing the wild-type gene expression pattern of the mutated gene will help identify the site of the primary defect.

Vasculogenesis takes place both within the extraembryonic tissues, starting as early as E7.0 in the yolk sac, and a little later within the embryo, in the heart, dorsal aorta, and branchial arches. The vessels gradually coalesce into a complete circulatory system connecting the embryo with the yolk sac through the vitelline circulation and with the placenta through the umbilical circulation. Nucleated primitive erythrocytes, formed in the yolk sac blood islands, will circulate throughout. The development of vasculature depends critically on the function of the heart, not only for the delivery of oxygen to the tissues but also through hemodynamic forces that affect vascular morphogenesis. Lethality that occurs after fusion of the allantois to the chorion could result from failure of blood formation or from vascular defects, possibly, but not necessarily, associated with cardiac insufficiency.

Once the time of death has been established and cardiovascular involvement is indicated, collect embryos at earlier stages and examine the gross morphology of the hearts for size, the presence of differentiated chambers, vascular connections, and laterality. Also, examine the extraembryonic membranes for the vascular pattern and presence of blood, noting any areas of bleeding or effusion of blood into tissues. To distinguish primary from secondary effects, it will be necessary to determine the earliest morphological abnormality. A simple way of visualizing the vasculature of mid-gestation to late-gestation embryos is to inject India ink or fluorescent beads or fluid either into the heart or a branch of the vitelline vein of the yolk sac, preferably while the heart is still beating (Nagy 2010) (Fig. 3B; also see Figure 4F–H in Chapter 16: Phenotypic Analysis of Dominant Mutant Effects in Mice [Papaioannou and Behringer 2023e]). Alternatively, the great vessels or the vasculature of the placenta or other organs can be detected by injecting with liquid plastic that solidifies to form a permanent cast of the vessels (Liu and Martin 2010).

Observations of the external morphology of the heart and vasculature should be followed up by serial section histology, of either the intact embryo or isolated hearts. The advantage of sectioning the intact embryo, including the yolk sac, is that vascular connections and other areas of the embryo can be examined at the same time (see Protocol 10.1: Orientation of Implanted or Dissected Mouse Embryos for Histological Sectioning [Papaioannou and Behringer 2023f]). Compare the homozygous mutant embryos with wild-type, stage-matched, or age-matched controls to determine if the four heart chambers are differentiating correctly and the endocardial cushions and septa are forming properly to separate the chambers. During later development, the cardiac valves that form from the endocardial cushions are necessary to ensure unidirectional blood flow and that the cardiac septa separate the right from left circulation (see Fig. 2 in Chapter 13: Phenotypic Analysis of Perinatal Lethality in Mice [Papaioannou and Behringer 2023g]). The ventricles are normally completely separated by the ventricular septum by E13.5, and ventricular septal defects are a common observation in cases of cardiac insufficiency. Nucleated erythrocytes should be evident in the heart and blood vessels from about E8.5.

Microcomputed tomography (microCT) is a very useful method to reconstruct the complex three-dimensional structures of the heart and great vessels in fixed specimens (Hsu et al. 2016). The entire embryo or fetus can be scanned without dissection. The heart can then be computationally segmented from the body (digital dissection) and viewed whole or as planar sections in any orientation (see Figure 3C,D in Chapter 13: Phenotypic Analysis of Perinatal Lethality in Mice [Papaioannou and Behringer 2023g]).

In addition to doing routine histology and/or volumetric imaging, a molecular analysis should be undertaken using molecular markers, in whole mounts and sections, that are specific for different developmental stages or indicators of the differentiation of the vasculature or specific chambers of the developing heart. Alterations in cell death or cell proliferation can play an important part in cardiac abnormalities. In addition, physiological measures such as calcium flux and electrocardiograms can be taken. Sophisticated applications of technologies such as ultrasound and magnetic resonance imaging (MRI) make it possible to visualize embryos in utero. These technologies can be used to analyze blood flow, heartbeat, brain development, and other morphometric parameters of live embryos in utero (e.g., NFATc1; Phoon et al. 2004; Golden et al. 2012; Zhang et al. 2018).

Placental Insufficiency

Another major cause of death after E9.5 is placental insufficiency (e.g., *Dlx3*; Morasso et al. 1999). You have reached this chapter because you have determined that your homozygous mutant embryos die after fusion of the allantois to the chorion. The placental abnormalities that lead to embryo lethality after this event are primarily those of the visceral yolk sac (vitelline) placenta or chorioallantoic placenta, for simplicity called the yolk sac and placenta, respectively (e.g., *Vhlh*; Gnarra et al. 1997). Defects in the parietal yolk sac, which consists of trophectoderm and parietal endoderm, could also cause lethality, but most likely earlier than this time point. Considering the essential nature of the yolk sac and placenta for the development of the mouse embryo to term, it is surprising how often these tissues are overlooked in expression studies. Thus, if you have identified a homozygous mutant lethal phenotype after E9.5, it is a good idea to determine if your gene of interest is expressed in the yolk sac and/or placenta, if you have not already done so. Yolk sac defects are generally easy to identify because this is one of the prominent extraembryonic membranes that you must dissect to reveal the embryo. Often, yolk sac abnormalities cause alterations in the yolk sac vasculature that are evident by simple visual inspection (e.g., a pale yolk sac). In contrast, the placenta,

Cite this chapter as *Cold Spring Harb Protoc*; doi:10.1101/pdb.over107973

which is also mutant, is sometimes discarded without examination and thus, at least initially, overlooked as a site of a primary defect when examining an embryonic lethal phenotype (e.g., *Rb*; Lee et al. 1992; Wu et al. 2003).

The visceral yolk sac is a bilaminar structure that envelops the developing embryo. The two tissue layers are of distinct origins: The outer endoderm layer is derived from the visceral endoderm and the inner mesoderm layer from the extraembryonic mesoderm generated by the primitive streak. Common gross mutant phenotypes associated with the yolk sac include a thin and transparent appearance or a pale appearance with an apparent absence of vasculature/blood, or an abnormal vascular pattern. The circulating blood cells in the yolk sac provide a visual marker of the yolk sac vasculature. However, if a blood flow problem causes a lack of blood cells in the yolk sac, then determining the presence of yolk sac vasculature defects requires more study. Histological analysis provides useful information with regard to the presence of the vasculature in the yolk sac. In addition, whole-mount immunostaining or in situ hybridization of the yolk sac using a vascular marker such as platelet endothelial cell adhesion molecule (PECAM) is useful for determining the presence, extent, and pattern of yolk sac vasculature. The two layers of the yolk sac can be physically separated using enzyme treatments (see Yoder et al. 1994; Chapter 5 in Behringer et al. 2014) providing a way to analyze each tissue in isolation (e.g., *Nodal*; Varlet et al. 1997). A note of caution: Yolk sac abnormalities can be intrinsic to the yolk sac tissues or secondary to cardiac defects.

ES cell–derived chimeras provide a very useful system to determine if a yolk sac defect is caused by primary abnormalities in the endoderm- or mesoderm-derived tissues. This is because one can generate polarized chimeras in which the yolk sac endoderm is either entirely mutant or entirely wild-type. The yolk sac phenotypes of such chimeras can be analyzed to define which tissue is altered by your mutation (see Chapter 17: Getting around an Early Lethal Phenotype in Mice with Chimeras [Papaioannou and Behringer 2023a]).

The placenta is composed of different tissue types of both maternal and fetal origin (Fig. 4). From the maternal side of the placenta moving to the fetal side are the maternally derived decidua, the embryo-derived trophoblast giant cells, spongiotrophoblast, labyrinth, and umbilical blood vessels. The trophoblast giant cells and spongiotrophoblast are derived from the trophoblast, whereas the labyrinth is a composite tissue derived from trophoblast and extraembryonic mesoderm. The labyrinth is the site of maternal–fetal exchange. The umbilical blood vessels are derived from the allantois, an extraembryonic mesoderm derivative. Defects in any of these placental tissues can compromise embryo growth and viability.

What types of placental phenotypes are possible? If the placenta is too small, it may not be able to fully support the growth of the fetus. Interestingly, a large placenta may also be insufficient to support fetal growth effectively (e.g., *Esx1*; Li and Behringer 1998). Enlarged placentas can be a secondary consequence of a primary tissue defect, and enlargement may represent a physiological response of the organ to compensate for these primary defects. Thus, alterations in placental size can change over time. Placental size differences and the corresponding functional abilities of abnormal placentas lead to alterations in fetal size and viability. Accordingly, a simple analysis would be to examine the gross morphology of the placenta and determine placental and fetal weights of the homozygous mutants in comparison to controls at different stages of development. Remember to blot excess liquid from the placentas and fetuses to obtain an accurate measure of their weights. Another simple gross analysis of the placenta is to examine it flat from the fetal side (Fig. 4A,B). A wild-type placenta will have concentric rings of healthy-appearing tissue. An abnormal placenta might show a disruption in these concentric rings of tissue, and, because the labyrinth is normally highly vascularized and red in color, you may find differences in color and morphology of this region.

Likely placental phenotypes include small placenta and small fetus, normal-sized placenta and small fetus, large placenta and small fetus, and large placenta and large fetus. A hypoplastic placenta could indicate a placental defect that cannot be compensated by increased growth. A placenta of normal size with a growth-restricted fetus may indicate a defect in the placenta that does not alter its gross morphology but compromises its function. An enlarged placenta but small fetus suggests a defect that triggers compensatory placental growth, but also that the enlarged placenta does not

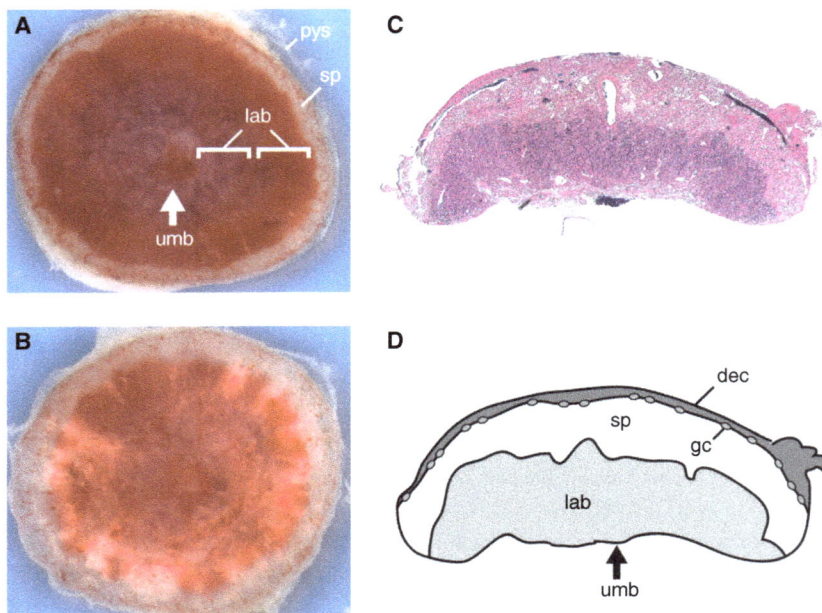

FIGURE 4. Placental morphology and histology. (*A*) An E18.5- wild-type placenta viewed from the fetal side, showing the concentric tissue organization from the central connection of the umbilical vessels. The brackets indicate the two areas of the labyrinth inside and outside the ring of attachment of the yolk sac to the placenta. (*B*) An E18.5 mutant placenta, showing disruption of tissue organization and heterogeneous morphology within the labyrinth. (*C*) Cross section of E18.5 wild-type placenta, showing various tissue layers of the organ; hematoxylin and eosin (H&E) staining. (*D*) Diagram of placenta shown in C. (dec) Maternal decidua, (gc) giant trophoblast cells, (lab) labyrinthine layer, (pys) parietal yolk sac, (sp) spongiotrophoblast, (umb) site of attachment of the umbilicus. (*A–C*, provided by Laurel Fohn and Richard Behringer.)

function sufficiently for normal fetal growth. An enlarged placenta and large fetus might suggest a general overgrowth phenotype rather than a defect limited to the placenta. Analyze timed matings at earlier and earlier stages of development to determine when you can first detect the gross morphological changes associated with the homozygous mutants.

Histological analysis of the placenta is very informative because of its stereotypic layered structure (Fig. 4C,D). Ultrastructural analysis (transmission electron microscopy) can also be valuable because standard light microscopy of paraffin-embedded histological sections may not provide the resolution required to see the specific details of the complex structures of the placenta, notably in the multilayered labyrinth. The three-dimensional structure of the maternal or fetal vasculature of the placenta can be examined by creating a plastic cast (Liu and Martin 2010) that provides information about the maternal–fetal interface. The trophoblast giant cells, spongiotrophoblast, and labyrinthine layers are histologically distinct, and each layer has specific molecular markers that can be used for their identification and determination of their differentiation status. Depending on the tissue defects that you uncover, you may also consider examining cell proliferation and cell death (Chapter 10: Phenotypic Analysis of Periimplantation to Mid-Gestation Lethality in Mice [Papaioannou and Behringer 2023c]).

A placental defect can lead to embryonic death, precluding your ability to study the effects of a mutation in the fetus at later stages. A useful method to bypass defects in placental structures derived from the trophoblast is to generate chimeras using wild-type tetraploid morulae or blastocysts in combination with mutant ES cells or diploid morulae (Chapter 17: Getting around an Early Lethal Phenotype in Mice with Chimeras [Papaioannou and Behringer 2023a]). Thus, wild-type function can be provided in the placenta by the tetraploid component of the chimera, allowing the mutation to be analyzed in isolation in the developing fetus.

 Cite this chapter as *Cold Spring Harb Protoc*; doi:10.1101/pdb.over107973

DEVELOPMENTAL DEFECTS IN MORPHOGENESIS AND ORGANOGENESIS

There are relatively few primary causes of death after E9.5 but many causes of morphological abnormalities, which can be associated with lethal or nonlethal mutations. You may have uncovered a morphological defect during the study of your homozygous mutants either prenatally or postnatally. You are now interested in determining the molecular, cellular, and embryological mechanisms that lead to the mutant defects. Your goal will be to determine when and in which cells or tissues the primary defect initiates the mutant phenotype that you have identified. Thus, it is essential to become familiarized with what is known about the development of the tissue or organ in question. Accordingly, as you begin your detailed analysis, it will be useful to answer some fundamental questions to provide a framework for a detailed analysis of the mutant phenotype in order to identify the primary defect.

- Which tissue(s) expresses the gene product in question?
- When does the tissue or organ form during embryogenesis?
- At what stage does expression of the gene in question initiate in the tissue?
- When does the mutant phenotype first become apparent in the forming organ?
- Is the mutant phenotype completely penetrant and fully expressed or is there variability?

Morphology: Gross and Histological

The initial step in the analysis is to document the development of the gross morphological defect in the homozygous mutants. Analyze sufficient numbers of mutant and control embryos at different embryonic stages to determine when the defect is first apparent (see Fig. 1 in Chapter 10: Phenotypic Analysis of Periimplantation to Mid-Gestation Lethality in Mice [Papaioannou and Behringer 2023c]). This will provide a starting point for the analysis because it indicates when the outcome of the mutation first has a detectable impact on normal development. The object is to correlate abnormal phenotypes with a mutant genotype to conclude that the mutation leads to the development of the abnormalities. At a minimum, one would want to analyze at least three mutants and three controls for each time point examined. Of course, analyzing greater numbers of mutants will make you more confident in your understanding of the mutant phenotype. If you have a mutant phenotype with variable penetrance and expressivity, you will potentially need to analyze many more mutants to correlate the phenotype with the genotype and also to determine the incidence of particular aspects of the phenotype among homozygous mutants.

Examination of fetuses should include both external and internal gross morphology. Using the expression pattern of the gene as a guide, dissection of different organ systems can be very informative. Do not be afraid to disassemble the embryos to isolate relevant internal organs. The skeleton can be readily visualized intact at later stages of gestation (see next section). Knowing the sex genotype and phenotype of embryos can be essential if your mutant gene is suspected of affecting sexual development or could in any way be sex limited (see Box 1. Sexing Mouse Embryos).

Histological analysis provides essential information on the microanatomical defects of the tissue or organ of interest. Depending on the situation, you may want to fix and section the embryos or fetuses within the uterus (Protocol 10.1: Orientation of Implanted or Dissected Mouse Embryos for Histological Sectioning [Papaioannou and Behringer 2023f]). However, with late-gestation fetuses, you will usually dissect the conceptus away from the uterus and perhaps dissect the embryo away from the extraembryonic tissues. Use irrelevant pieces of the conceptus for genotyping, taking care not to contaminate extraembryonic tissues with maternal tissues. The choice of fixative to use will depend on the analysis you intend to perform. For routine hematoxylin and eosin (H&E) histology, Bouin's fixative works very well. For RNA in situ hybridization studies, 4% paraformaldehyde (PFA) is best (see Chapter 17 in Behringer et al. 2014). For immunohistochemistry, 4% PFA may be adequate, but you must determine which fixative preserves the relevant epitopes recognized by each particular antibody. For frozen sections, the fixed specimen is generally equilibrated into sucrose solution

BOX 1. SEXING MOUSE EMBRYOS

Knowing the sex genotype and phenotype of embryos can be essential if your mutant gene is suspected of affecting sexual development or could in any way be sex limited. An XX genotype should be female (i.e., have ovaries) and an XY genotype should be male (i.e., have testes). Mice can have sex karyotype variants, including XO females and XXY males. There are also many instances of intersex phenotypes including true hermaphroditism (mice with ovotestes) and pseudohermaphroditism (mice with either testes or ovaries but both male and female reproductive tract organs). In most instances, a complete sex reversal (XX male or XY female) or an intersex phenotype will be infertile. Many methods exist for determining the sex of embryos, some that are appropriate for any age and some that are age-specific.

Genotype (XX vs. XY)

- Polymerase chain reaction (PCR) primers for genes located on the Y chromosome can be used on any embryonic tissue at any age (e.g., *Sry*) (see Protocol 5.1: Sex Genotyping Mice by Polymerase Chain Reaction [Papaioannou and Behringer 2023j]).

- The Y-chromosome and chromosome-19 autosomes are the smallest mouse chromosomes. Thus, a male karyotype will have three very small chromosomes and a female karyotype only two (Nagy et al. 2008).

- Y- or X-chromosome-specific probes can be used on chromosomal spreads from any age embryo using fluorescence in situ hybridization (FISH) (Otaka et al. 2015).

- Between E9.5 and about E12.5, sex chromatin can be detected in the amnion of female embryos (Fig. 5), leaving the rest of the embryo intact for additional analysis.

- XX and XY embryos can be distinguished after the blastocyst stage in crosses using males carrying an X-linked *eGFP* transgene crossed to wild-type females. All of the daughters from this cross will be eGFP-positive ($X^{eGFP}X$) and all of the sons eGFP-negative (XY). This method of genotyping is useful because you can easily and quickly sort living embryos under a fluorescent dissecting scope to pool the different sexes for further analysis (Hadjantonakis et al. 2001).

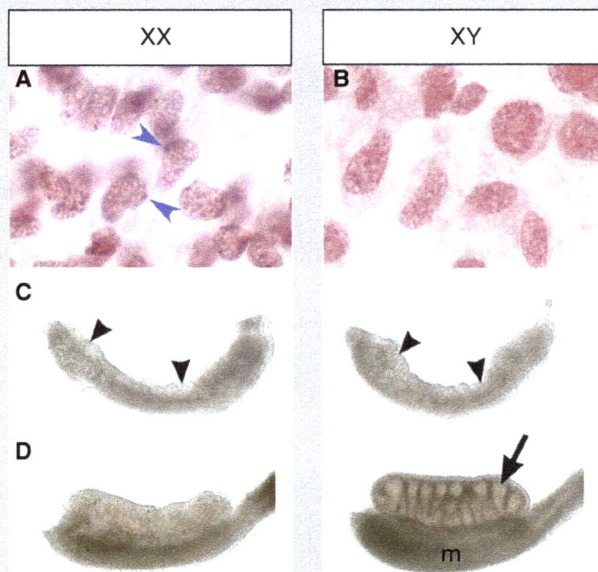

FIGURE 5. Sexing mouse embryos. (*A,B*) Sex chromatin in the amnion of E10 embryos. Isolated amnions are spread on a glass slide and stained with a drop of 2% orcein in 60% acetic acid and then coverslipped. Nuclei of the amnion cells of the female (*A*, XX) have a spot of darkly staining chromatin on the nuclear membrane (arrowheads), which represents the inactive X chromosome. In humans, this chromatin is known as the Barr body. No such spots are seen in the male nuclei (*B*, XY). (*C,D*) Dissected urogenital ridges from E11.5 (*C*) and E13.5 (*D*) female (XX) and male (XY) embryos. The gonads, which are located on the medial surface of the mesonephroi, have a more transparent appearance. (*C*) At E11.5, the male and female gonads are morphologically identical. (Arrowheads indicate the rostral–caudal extent of the gonads.) (*D*) E13.5 female and male gonads with attached mesonephroi (m). At this stage, the male and female gonads are morphologically distinct. The testes have cords (*arrow*), whereas the ovaries do not. (*C,D*, provided by Akio Kobayashi and Richard Behringer.) *Continued*

BOX 1. *Continued.*

Phenotype (Female vs. Male)

From about E12.5, the seminiferous cords become apparent as the testes differentiate. These are readily visualized using a dissecting microscope. At this stage, the fetal ovaries appear homogeneous in structure. The exact timing of the differentiation of the testis can be somewhat variable, depending on genetic background and the light schedule of your mouse colony. From E13.5 onward, the morphological differences between testes and ovaries are clear (Fig. 5C,D). At birth and at all subsequent stages, morphological features of the external genitalia distinguish male and female mice (see Fig. 3 in Chapter 5: Recovering a Targeted Mutation in Mice from Embryonic Stem Cell Chimeras or CRISPR–Cas Founders [Papaioannou and Behringer 2023k]).

before freezing in OCT embedding medium in plastic molds. Orienting the specimen for histological sectioning is very important. Your mutant and control specimens must be consistently cut in the same plane to make useful comparisons to convince yourself and others (see Figs. 1 and 2 in Protocol 10.1: Orientation of Implanted or Dissected Mouse Embryos for Histological Sectioning [Papaioannou and Behringer 2023f]).

Serial sections of the specimen are also very important to understand the three-dimensional morphology of the region in which the tissue is present. Cut sections of corresponding regions in mutants and controls. This may be difficult in the mutant because, by definition, the tissue or organ of interest is abnormal. Use unaffected regions of the mutant as reference points to compare with controls. When you generate images of the histological sections, it is a good idea to take images at low magnification to orient the viewer and also at high magnification to show specific details. Make sure that the histological staining or immunostaining is of consistent quality between mutants and controls so that you can focus on the genotype-specific, not technical, differences.

Visualization and Analysis of the Forming Skeleton

The mature skeleton is composed of bone and permanent cartilages. Bone is formed by two different processes, endochondral and membranous ossification. Endochondral ossification is bone formation that occurs from mesenchyme condensations through a cartilage template. Membranous ossification is bone formation directly from a mesenchymal condensation. It is possible to show the developing cartilaginous skeleton starting at ∼E12.0 by whole-mount alcian blue staining (e.g., *Tbx1* [Jerome and Papaioannou 2001]; *Tbx3* [Davenport et al. 2003]) (Nagy et al. 2009). Embryos are fixed whole without removal of the visceral organs, stained with alcian blue, and then cleared (see Chapter 17 in Behringer et al. 2014). Mineralization of the skeleton begins at later stages of development (∼E14.0). At these stages, one can stain for bone and mineralized tissues and also cartilage using a different method in which the fetus is skinned and visceral organs removed, fixed, and then processed for detection of bone and other mineralized tissues by alizarin red staining and cartilage by alcian blue staining (Fig. 6) (Ovchinnikov 2009). The nonskeletal tissues are digested in a base solution and cleared in glycerol (Behringer et al. 2014). The entire skeleton or specific regions of the skeleton can then be examined and documented.

For photodocumentation, place the skeleton preparations in a glass Petri dish in the clearing solution (for cartilage preparations) or plastic Petri dish in glycerol (for alizarin red/alcian blue preparations). Acquire images of the skeleton preparations on a dissecting microscope with transillumination lighting. Image the mutant and control skeletons in the same orientation. You can take an image of the entire skeleton, usually a lateral view (Fig. 6A,C), although it is also possible to take dorsal and ventral views. The limbs tend to obscure the neck and thorax regions. However, if you are particularly interested in these regions, you can use forceps to remove the limbs, which can be imaged separate from the body, showing either dorsal or ventral views (Fig. 6B,C). The rib cage can also be imaged in isolation. Cut the ribs close to the vertebrae to release the rib cage from the rest

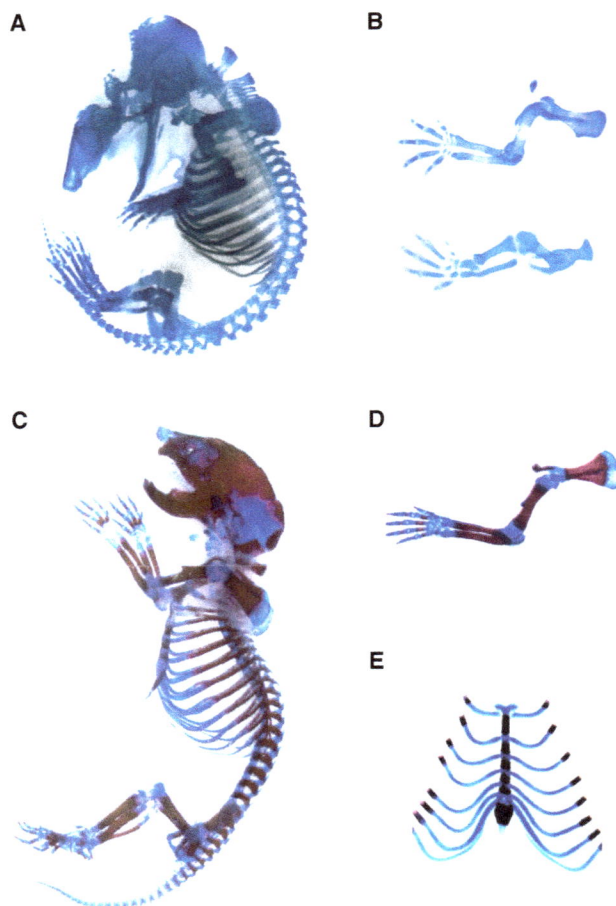

FIGURE 6. Presentation of skeleton preparations. (*A,B*) Cartilaginous skeleton preparation of E14.5 embryo stained with alcian blue. The embryo is imaged in a glass dish in BABB (one part benzyl alcohol [BA] and two parts benzyl benzoate [BB]); note that this medium dissolves plastic. Ammonium hydroxide treatment after fixation eliminates the yellow color caused by Bouin's fixative. (*A*) Intact embryo. If a dorsal or ventral image of the skeleton is desired, you can impale the lateral flank of the prepared embryo with a glass needle affixed to a support located to the left or right. (*B*) Forelimb (*top*) and hindlimb (*bottom*) were cut away from the skeleton preparation using a scalpel. (*C–E*) Skeleton preparation of newborn mouse stained with alizarin red and alcian blue. Skeleton preparation is imaged in a large plastic Petri dish in 1:1 glycerol:ethanol. (*C*) Intact neonate skeleton. (*D*) Forelimb was removed from the prepared skeleton using forceps. (*E*) Rib cage was isolated by cutting rib bones next to the vertebral column and a glass slide was placed on top of the rib cage to flatten it. Wild-type mice have seven ribs attached to the sternum (true ribs) and six ribs that are not attached to the sternum (false floating ribs). Not all of the false ribs are shown in this preparation. (*A–D*, provided by Dmitry Ovchinnikov and Richard Behringer; *E*, reprinted, from Rivera-Peréz et al. 1995, with permission from The Company of Biologists.)

of the skeleton. Then flat-mount the dissected rib cage by placing a glass slide on top of it for subsequent imaging (Fig. 6E). The skull can be imaged as described in Chapter 13: Phenotypic Analysis of Perinatal Lethality in Mice (Fig. 4) (Papaioannou and Behringer 2023g).

Molecular Characterization of Mutant Phenotypes

Molecular markers combined with knowledge of morphological defects and the expression pattern of the gene of interest provide essential information regarding the presence or absence of tissues or their state of differentiation. At later stages of embryogenesis, some methods of analysis become technically difficult. Whole-mount in situ hybridization methods must be modified to accommodate the larger size of the embryo and organs that hinders reagent access and results in probe trapping in body cavities. It is possible to bisect embryos, providing access of reagents to internal tissues. If you are studying a particular organ, such as the kidney (e.g., *Lhx1*; Shawlot and Behringer 1995), lung or limbs (e.g., *Fgf10*; Sekine et al. 1999), or heart (e.g., *Tbx2*; Harrelson et al. 2004), you can dissect it away from the embryo for whole-mount in situ hybridization. In situ hybridization of sectioned material circumvents these problems associated with whole-mount preparations, although three-dimensional information may be harder to obtain.

Remember that the lack of detectable expression of a molecular marker can be explained by two different mechanisms. First, the tissue may be present but not express the marker. Alternatively, the tissue may not be present and therefore there is no marker. To distinguish between these two possibilities, you need an independent assessment for the presence or absence of the tissue in question. This is facilitated if the null alleles have an associated expression reporter (e.g., *lacZ* or GFP) and the reporter can be used to identify the mutant tissue. Alternatively, either morphological features or the expression of an unrelated molecular marker may indicate the presence of a tissue. Molecular marker studies are, for the most part, qualitative assessments.

Often one would like a quantitative assessment of gene expression within a particular tissue of a developing embryo at the mRNA and/or protein level. However, such assessments must be made very carefully. If you can isolate the specific tissue or cells of interest, then you can use a variety of quantitative assays, including northern blot, real-time reverse transcription-polymerase chain reaction (RT-PCR), and western blot. However, the tissue of interest is often very small and not easily isolated. In these cases, expression is usually assayed by whole-mount or section in situ hybridization or immunostaining. Can these in situ methods be used for quantitative assessments of gene expression? The answer is generally no, unless stringent controls are performed, and, even then, the assessment can only be semiquantitative. All samples must be processed using the same reagents, at the same time, for the same amount of time. This is especially important for nonradioactive detection methods that rely on color reactions that become more intense and saturated with time. If your gene of interest is also expressed in an unaffected tissue, you could potentially use that as an internal control for differences noted in the tissue of interest. Expression in the unaffected tissue should be the same between the mutants and controls. For a more quantitative approach, hairpin chain reaction (HCR) RNA fluorescent in situ hybridization (FISH) on whole mounts or sections can be multiplexed for different probes and expression levels quantified (Anderson et al. 2020).

Cellular Characterization: Proliferation and Death

Alterations in cell behaviors, including proliferation and death, can result in tissues that are absent, hypoplastic, or hyperplastic. Your morphological and histological analyses have been used to determine when the first tissue defects were observed. This time point serves as the starting point to examine cell proliferation and death (e.g., *Fos* [Roffler-Tarlov et al. 1996]; *Foxc2* [Winnier et al. 1997]) (see Chapter 10: Phenotypic Analysis of Periimplantation to Mid-Gestation Lethality in Mice [Papaioannou and Behringer 2023c]). Alterations in cell proliferation or death may be transient rather than chronically present, and you may therefore have to perform your analyses at time intervals that are less than 24 h to feel confident that you have not missed a transient alteration in these processes.

3D Imaging of Embryos

Imaging of mouse embryos is a primary tool for visualizing morphological abnormalities and disturbances in gene expression caused by mutations. There are numerous imaging modalities available to visualize embryos, either as fixed or live specimens. Table 1 summarizes modalities that are useful in assessing the phenotype of mutant mice in three dimensions (3D) including time lapse imaging (the fourth dimension) (Sharpe and Wong 2011). The most suitable type of imaging modality for your purposes will depend on a number of considerations:

- stage and size of embryo, organ or region to be imaged
- imaging within or outside the reproductive tract

TABLE 1. Imaging modalities used in phenotyping mouse embryos that support digital three-dimensional (3D) reconstruction of embryos or organs

Imaging method	Detection	Level of resolution	Imaging depth	Primary uses	Notes	Protocol reference(s)
Epifluorescence microscopy	Fluorescence	Subcellular	500 μ	Detecting immunofluorescent staining, fluorescent cellular reporters, live and time-lapse imaging	Time-lapse imaging requires environmental chamber	Srinivas 2010
Confocal/multiphoton fluorescent microscopy	Fluorescence	Subcellular	500–1000 μ up to 2 mm	Detecting immunofluorescent staining, fluorescent cellular reporters, live and time-lapse imaging	Time-lapse imaging requires environmental chamber	Garcia et al. 2011; Noctor 2011; Williams et al. 2013
Light sheet fluorescent microscopy	Fluorescence	Single cell/tissue	6.5 mm (with clearing)	Detecting immunofluorescent staining, fluorescent cellular reporters, live and time-lapse imaging	Embedding in agarose may be required; time-lapse imaging requires environmental chamber	Udan et al. 2014; McDole et al. 2018
Optical coherence tomography (OCT)	Light reflectivity	Tissue	1–3 mm	Unstained specimens, live and time-lapse imaging	Time-lapse imaging requires environmental chamber; not suitable for fluorescent-based assays of gene expression	Larina et al. 2012
X-ray micro-computed tomography (microCT)	X-rays	Tissue	Whole embryo isolated or in utero	3D reconstructions of entire embryos	Generally, fixed specimens; various contrast agents required for soft tissue visualization; not for gene expression studies	Metscher 2011
High-frequency ultrasound imaging	High-frequency sound waves	Tissue	Whole embryo in utero	Imaging live embryos in utero, guided microinjections into cavities	Not for gene expression studies	Foster and Brown 2012
Microscopic magnetic resonance imaging (microMRI)	Water protons interacting with magnetic field	Tissue	Whole embryo or organ	Building 3D atlases of entire embryos or organs	Not for gene expression studies	Ruffins and Jacobs 2011; Bamforth et al. 2012
Optical projection tomography (OPT)	Light absorption or fluorescence	Tissue	10 mm	Imaging live embryos or organs, fluorescent and histochemical gene expression patterns		Colas and Sharpe 2009; Quintana and Sharpe 2011
Photoacoustic computed tomography (PACT)	Light absorption-generated acoustic waves	Tissue	10 mm	Imaging live embryos in utero, measuring blood oxygenation		Wang et al. 2020
Block face imaging: episcopic fluorescence image capturing (EFIC) and high-resolution episcopic microscopy (HREM)	Auto fluorescence or fluorescent dye	Subcellular/tissue	Whole embryo	3D reconstructions of entire embryos, histochemical stains detected	Nonliving tissues only	Mohun and Weninger 2012

Cite this chapter as *Cold Spring Harb Protoc*; doi:10.1101/pdb.over107973

- fixed or live specimen
- availability of equipment including culture chamber for time-lapse imaging
- resolution and depth of imaging required
- marker or marker-free imaging.

A common starting place is histological analysis of fixed specimens, which we have covered extensively (see Chapter 9: Phenotypic Analysis of Preimplantation Lethality in Mice [Papaioannou and Behringer 2023h]; Chapter 10: Phenotypic Analysis of Periimplantation to Mid-Gestation Lethality in Mice [Papaioannou and Behringer 2023c]; Chapter 11: Analysis of Mid- to Late-Gestation Phenotypes in Mice [Papaioannou and Behringer 2023i]). Histology provides resolution at the cellular and tissue level and can be combined with immunostaining and in situ hybridization for gene expression studies. A detailed database of mouse development is available for reference in the Edinburgh Mouse Atlas Project (eMAP; https://www.emouseatlas.org/emap/home.html), which includes histological sections of mouse embryos throughout gestation (eHistology) as well as an atlas of 3D volume and surface-rendered anatomy models with spatially mapped gene expression patterns derived from histological sections (eMAGE) (Armit et al. 2017). However, histology-based analysis is destructive of the embryo, cannot be used for live imaging, and 3D reconstructions based on sectioned material can be distorted.

Fluorescent imaging approaches are widely used in developmental studies and are particularly powerful analytical tools. Myriad fluorescent imaging modalities are used to investigate mutant phenotypes in the mouse. We have mentioned several optical imaging methods for embryos, including epifluorescent and confocal microscopy, that lend themselves to high-resolution imaging and accurate 3D depictions of whole, live or fixed embryos. In addition, Table 1 lists other methodologies, some of which use fluorescence, that allow the creation of 3D digital models that can be sectioned computationally and viewed in any plane. Specific embryonic regions or organs can be segmented (digitally dissected) from the entire data set using software to view these in isolation. Volumetric data sets can even be 3D-printed as physical objects. With the exception of block face imaging, all of these methods can be adapted for fixed or live embryos or tissue.

Figure 7 is a sampling of some of the images obtained by different methods. Epifluorescence microscopy (Fig. 7A) can be used in conjunction with embryo or organ culture to track cell movement with time-lapse image capture in living tissue. Confocal microscopy (Fig. 7B) provides excellent spatial resolution and 3D optical sectioning for detecting fluorescent markers of different colors in living or fixed tissue (e.g., *Wnt5a*; Arora et al. 2016). Light sheet microscopy (Fig. 7C), which depends on a plane of fluorescence excitation, results in minimal sample photodamage and minimal reporter photobleaching and is useful for whole, fixed specimens (Bunce et al. 2021) or extended time-lapse imaging of living mouse embryos (Udan et al. 2014; McDole et al. 2018).

Several imaging methods do not depend on fluorescence labeling or colorimetric gene expression assays. Optical coherence tomography (OCT) (Fig. 7D) can be used for static and time-lapse imaging of whole living embryos in culture (e.g., *Wdr19*; Wang et al. 2017). X-ray microcomputed tomography (μCT) (Fig. 7E) produces virtual sections of fixed or live embryos (e.g., *Plxnd1*; Degenhardt et al. 2010). Micro high-frequency ultrasound imaging (Fig. 7F) can be used for nondestructive, real-time imaging of embryos within the uterus (Foster et al. 2003). Microscopic magnetic resonance imaging (μMRI) (Fig. 7G), which shows the contrasts between tissues of untreated embryos but lacks cellular resolution, has also been used in phenotyping (e.g., *Mbnl1/Mbnl2*; Sta Maria et al. 2021). Optical projection tomography (OPT) (Fig. 7H) can be performed on live, cultured, or fixed embryos and used with immunofluorescence or histochemical stains to show the complete distribution of a signal throughout a relatively large specimen. OPT has been used to create developmental atlases (Wong et al. 2015). Photoacoustic tomography can be combined with ultrasound imaging and allows in utero embryonic imaging for morphology as well as blood oxygenation and other metabolic parameters (Wang et al. 2020). Finally, there are methods that

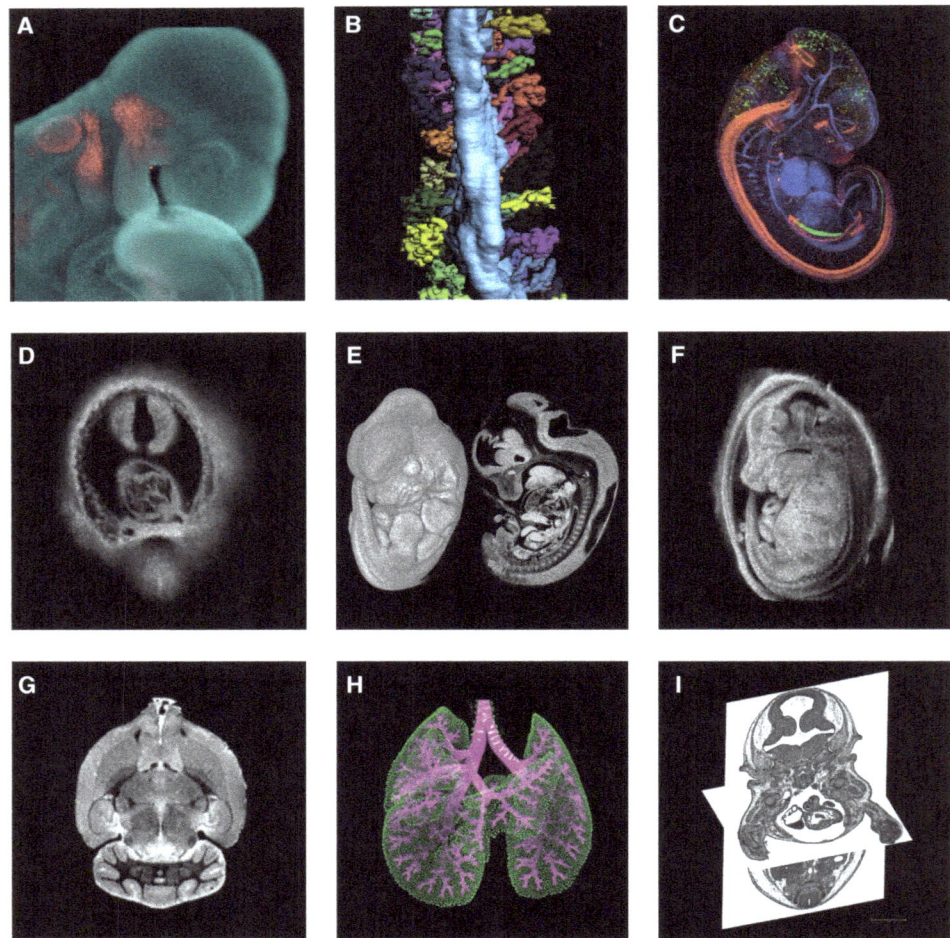

FIGURE 7. Imaging modalities for embryos. (A) Epifluorescent image of an E9.5 embryo immunostained for SOX10 (red) and DAPI (cyan). (B) Confocal microscopy and 3D renderings of E-CAD and FOXA2 immunostained uterine lumen (blue) and endometrial glands (randomly pseudocolored). (C) Light sheet microscopy, maximum intensity projection of an E11.5 embryo immunostained for SF1:eGFP (green) and PAX8 (red). (D) 3D optical coherence tomography image of the forebrain and heart region from an E8.5 embryo. (E) Micro-computed tomography, showing surface rendering (left) and mid-sagittal digital section (right) of E12.5 embryo. (F) Annular array high-frequency ultrasound mid-sagittal section of an E14.5 embryo. (G) Magnetic resonance image of horizontal digital section of adult mouse brain. (H) Optical projection tomography image of E16.5 lung, SOX9 progenitor cells (green), and SOX2 airway cells (magenta). (I) High-resolution episcopic microscopy, showing two resection planes of an E14.5 embryo showing a ventricular-septal defect. (A, Image courtesy of Sabrina Alam and Loydie Jerome-Majewska; B, image modified from Arora et al. 2016, with permission from The Company of Biologists; C, image courtesy of Corey Bunce and Blanche Capel; D, image modified, with permission from Wang et al. 2017, The Optical Society; E, Image modified from Hsu et al. 2016; F, image modified from Aristizábal et al. 2013, with permission from Elsevier; G, image modified from Kovacević et al. 2005, with permission from Oxford University Press; H, image courtesy of Jichao Chen; I, image courtesy of WJ Weninger and SH Geyer.)

produce accurate 3D reconstructions of fixed material but are not suitable for live tissue (Fig. 7I). Block face imaging, episcopic fluorescence image capturing (EFIC) (e.g., *Cited2*; Weninger et al. 2005), and high-resolution episcopic microscopy (HREM) maintain image alignment by photographing the block face before microtome sectioning, but currently cannot be used with fluorescent signals.

Each of these modalities has different and sometimes overlapping advantages and limitations. Often more than one type of imaging will be valuable for the full phenotypic analysis of mutations affecting the mouse embryo.

Cite this chapter as *Cold Spring Harb Protoc*; doi:10.1101/pdb.over107973

REFERENCES

Anderson MJ, Magidson V, Kageyama R, Lewandoski M. 2020. *Fgf4* maintains *Hes7* levels critical for normal somite segmentation clock function. *eLife* 9: e55608. doi:10.7554/eLife.55608

Aristizábal O, Mamou J, Ketterling JA, Turnbull DH. 2013. High-throughput, high-frequency 3-D ultrasound for in utero analysis of embryonic mouse brain development. *Ultrasound Med Biol* 39: 2321–2332. doi:10.1016/j.ultrasmedbio.2013.06.015

Armit C, Richardson L, Venkataraman S, Graham L, Burton N, Hill B, Yang Y, Baldock RA. 2017. eMouseAtlas: an atlas-based resource for understanding mammalian embryogenesis. *Dev Biol* 423: 1–11. doi:10.1016/j.ydbio.2017.01.023

Arora R, Fries A, Oelerich K, Marchuk K, Sabeur K, Giudice LC, Laird DJ. 2016. Insights from imaging the implanting embryo and the uterine environment in three dimensions. *Development* 143: 4749–4754. doi:10.1242/dev.144386

Bamforth SD, Schneider JE, Bhattacharya S. 2012. High-throughput analysis of mouse embryos by magnetic resonance imaging. *Cold Spring Harb Protoc* 2012: 93–101. doi:10.1101/pdb.prot067538

Baron MH. 2003. Embryonic origins of mammalian hematopoiesis. *Exp Hematol* 31: 1160–1169. doi:10.1016/j.exphem.2003.08.019

Baron MH, Isern J, Fraser ST. 2012. The embryonic origins of erythropoiesis in mammals. *Blood* 119: 4828–4837. doi:10.1182/blood-2012-01-153486

Behringer RR, Gertsenstein M, Nagy KV, Nagy A. 2014. *Manipulating the mouse embryo: a laboratory manual.* Cold Spring Harbor Laboratory Press, Cold Spring Harbor, NY.

Bunce C, McKey J, Capel B. 2021. Concerted morphogenesis of genital ridges and nephric ducts in the mouse captured through whole-embryo imaging. *Development* 148: dev199208. doi:10.1242/dev.199208

Chang H, Huylebroeck D, Verchueren K, Guo Q, Matzuk MM, Zwijsen A. 1999. Smad5 knockout mice die at mid-gestation due to multiple embryonic and extraembryonic defects. *Development* 126: 1631–1642. doi:10.1242/dev.126.8.1631

Colas JF, Sharpe J. 2009. Live optical projection tomography. *Organogenesis* 5: 211–216. doi:10.4161/org.5.4.10426

Davenport TG, Jerome-Majewska LA, Papaioannou VE. 2003. Mammary gland, limb, and yolk sac defects in mice lacking *Tbx3*, the gene mutated in human ulnar mammary syndrome. *Development* 130: 2263–2273. doi:10.1242/dev.00431

Degenhardt K, Wright AC, Horng D, Padmanabhan A, Epstein JA. 2010. Rapid 3D phenotyping of cardiovascular development in mouse embryos by micro-CT with iodine staining. *Circ Cardiovasc Imaging* 3: 314–322. doi:10.1161/CIRCIMAGING.109.918482

Foster FS, Brown AS. 2012. Microultrasound and its application to longitudinal studies of mouse eye development and disease. *Cold Spring Harb Protoc* 2012: 494–503. doi:10.1101/pdb.prot068544

Foster FS, Zhang M, Duckett AS, Cucevic V, Pavlin CJ. 2003. *In vivo* imaging of embryonic development in the mouse eye by ultrasound biomicroscopy. *Invest Ophthalmol Vis Sci* 44: 2361–2366. doi:10.1167/iovs.02-0911

Garcia MD, Udan RS, Hadjantonakis AK, Dickinson ME. 2011. Live imaging of mouse embryos. *Cold Spring Harb Protoc* 2011: pdb.top104. doi:10.1101/pdb.top104

Gnarra JR, Ward JM, Proter FD, Wagner JR, Devor DE, Grinberg A, Emmert-Buck MR, Westphal H, Klausner RD, Linehan WM. 1997. Defective placental vasculogenesis causes embryonic lethality in VHL-deficient mice. *Proc Natl Acad Sci* 94: 9102–9107. doi:10.1073/pnas.94.17.9102

Golden HB, Sunder S, Liu Y, Peng X, Dostal DE. 2012. In utero assessment of cardiovascular function in the embryonic mouse heart using high-resolution ultrasound biomicroscopy. *Methods Mol Biol* 843: 245–263. doi:10.1007/978-1-61779-523-7_23

Hadjantonakis A-K, Cox LL, Tam PPL, Nagy A. 2001. An X-linked GFP transgene reveals unexpected paternal X-chromosome activity in trophoblastic giant cells of the mouse placenta. *Genesis* 29: 133–140. doi:10.1002/gene.1016

Harrelson Z, Kelly RG, Goldin SN, Gibson-Brown JJ, Bollag RJ, Silver LM, Papaioannou VE. 2004. *Tbx2* is essential for patterning the atrioventricular canal and for morphogenesis of the outflow tract during heart development. *Development* 131: 5041–5052. doi:10.1242/dev.01378

Hsu CW, Wong L, Rasmussen TL, Kalaga S, McElwee ML, Keith LC, Bohat R, Seavitt JR, Beaudet AL, Dickinson ME. 2016. Three-dimensional microCT imaging of mouse development from early post-implantation to early postnatal stages. *Dev Biol* 419: 229–236. doi:10.1016/j.ydbio.2016.09.011

Jerome LA, Papaioannou VE. 2001. DiGeorge syndrome phenotype in mice mutant for the T-box gene, *Tbx1*. *Nat Genet* 27: 286–291. doi:10.1038/85845

Kovacević N, Henderson JT, Chan E, Lifshitz N, Bishop J, Evans AC, Henkelman RM, Chen XJ. 2005. A three-dimensional MRI atlas of the mouse brain with estimates of the average and variability. *Cereb Cortex* 15: 639–645. doi:10.1093/cercor/bhh165

Larina IV, Garcia MD, Vadakkan TJ, Larin KV, Dickinson ME. 2012. Imaging mouse embryonic cardiovascular development. *Cold Spring Harb Protoc* 2012: 1035–1043. doi:10.1101/pdb.top071498

Lee EY, Chang CY, Hu N, Wang YC, Lai CC, Herrup K, Lee WH, Bradley A. 1992. Mice deficient for Rb are nonviable and show defects in neurogenesis and haematopoiesis. *Nature* 359: 288–294. doi:10.1038/359288a0

Li Y, Behringer RR. 1998. *Esx1* is an X-chromosome-imprinted regulator of placental development and fetal growth. *Nat Genet* 20: 309–311. doi:10.1038/3129

Liu C, Martin J. 2010. Visualizing fetal mouse great blood vessels by plastic casting. *Cold Spring Harb Protoc* 2010: pdb prot5433. doi:10.1101/pdb.prot5433

McDole K, Guignard L, Amat F, Berger A, Malandain G, Royer LA, Turaga SC, Branson K, Keller PJ. 2018. In toto imaging and reconstruction of post-implantation mouse development at the single-cell level. *Cell* 175: 859–876.e833. doi:10.1016/j.cell.2018.09.031

Metscher BD. 2011. X-ray microtomographic imaging of intact vertebrate embryos. *Cold Spring Harb Protoc* 2011: 1462–1471. doi:10.1101/pdb.prot067033

Mikkola HK, Klintman J, Yang H, Hock H, Schlaeger TM, Fujiwara Y, Orkin SH. 2003. Haematopoietic stem cells retain long-term repopulating activity and multipotency in the absence of stem-cell leukaemia SCL/tal-1 gene. *Nature* 421: 547–551. doi:10.1038/nature01345

Mohun TJ, Weninger WJ. 2012. Generation of volume data by episcopic three-dimensional imaging of embryos. *Cold Spring Harb Protoc* 2012: 681–682. doi:10.1101/pdb.prot069591

Morasso MI, Grinberg A, Robinson G, Sargent TD, Mahon KA. 1999. Placental failure in mice lacking the homeobox gene *Dlx3*. *Proc Natl Acad Sci* 96: 162–167. doi:10.1073/pnas.96.1.162

Nagy A. 2010. Visualizing fetal mouse vasculature by India ink injection. *Cold Spring Harb Protoc* 2010: pdb prot5371. doi:10.1101/pdb.prot5371

Nagy A, Gertsenstein M, Vintersten K, Behringer R. 2008. Karyotyping mouse cells. *Cold Spring Harb Protoc* 2008: pdb.prot4706. doi:10.1101/pdb.prot4706

Nagy A, Gertsenstein M, Vintersten K, Behringer R. 2009. Alcian blue staining of the mouse fetal cartilaginous skeleton. *Cold Spring Harb Protoc* 2009: pdb prot5169. doi:10.1101/pdb.prot5169

Noctor SC. 2011. Time-lapse imaging of fluorescently labeled live cells in the embryonic mammalian forebrain. *Cold Spring Harb Protoc* 2011: 1350–1361. doi:10.1101/pdb.prot066605

Otaka K, Hiradate Y, Kobayashi N, Shirakata Y, Tanemura K. 2015. Distribution of the sex chromosome during mouse spermatogenesis in testis tissue sections. *J Reprod Dev* 61: 375–381. doi:10.1262/jrd.2015-013

Ovchinnikov D. 2009. Alcian blue/alizarin red staining of cartilage and bone in mouse. *Cold Spring Harb Protoc* 2009: pdb prot5170. doi:10.1101/pdb.prot5170

Papaioannou VE, Behringer RR. 2023a. Getting around an early lethal phenotpye in mice with chimeras. *Cold Spring Harb Protoc* doi:10.1011/pdb.over107979

Papaioannou VE, Behringer RR. 2023b. Tissue- and/or temporal-specific mutations in mice using conditional alleles. *Cold Spring Harb Protoc* doi:10.1011/pdb.over107980

Papaioannou VE, Behringer RR. 2023c. Phenotypic analysis of periimplantation to mid-gestation lethality in mice. *Cold Spring Harb Protoc* doi:10.1011/pdb.over107972

Papaioannou VE, Behringer RR. 2023d. Uncovering phenotypes in mutant mice by determining embryo, organ, tissue, and cell developmental potential. *Cold Spring Harb Protoc* doi:10.1011/pdb.over107974

Papaioannou VE, Behringer RR. 2023e. Phenotypic analysis of dominant mutant effects in mice. *Cold Spring Harb Protoc* doi:10.1011/pdb.over107978

Papaioannou VE, Behringer RR. 2023f. Orientation of implanted or dissected mouse embryos for histological sectioning. *Cold Spring Harb Protoc* doi:10.1011/pdb.prot108094

Papaioannou VE, Behringer RR. 2023g. Phenotypic analysis of perinatal lethality in mice. *Cold Spring Harb Protoc* doi:10.1011/pdb.over107975

Papaioannou VE, Behringer RR. 2023h. Phenotypic analysis of preimplantation lethality in mice. *Cold Spring Harb Protoc* doi:10.1011/pdb.over107971

Papaioannou VE, Behringer RR. 2023i. Analysis of mid- to late-gestation phenotypes in mice. *Cold Spring Harb Protoc* doi:10.1011/pdb.over107973

Papaioannou VE, Behringer RR. 2023j. Sex genotyping mice by polymerase chain reaction. *Cold Spring Harb Protoc* doi:10.1011/pdb.prot108062

Papaioannou VE, Behringer RR. 2023k. Recovering a targeted mutation in mice from embryonic stem cell chimeras or CRISPR–Cas founders. *Cold Spring Harb Protoc* doi:10.1011/pdb.over107959

Phoon CK, Ji RP, Aristizábal O, Worrad DM, Zhou B, Baldwin HS, Turnbull DH. 2004. Embryonic heart failure in $NFATc1^{-/-}$ mice: Novel mechanistic insights from in utero ultrasound biomicroscopy. *Circ Res* **95**: 92–99. doi:10.1161/01.RES.0000133681.99617.28

Quintana L, Sharpe J. 2011. Optical projection tomography of vertebrate embryo development. *Cold Spring Harb Protoc* 2011: 586–594. doi:10.1101/pdb.top116

Rivera-Peréz JA, Mallo M, Gendron-Maguire M, Behringer RR. 1995. Goosecoid is not an essential component of the mouse gastrula organizer but is required for craniofacial and rib development. *Development* **121**: 3005–3012. doi:10.1242/dev.121.9.3005

Roffler-Tarlov S, Gibson Brown JJ, Tarlov E, Stolarov J, Chapman DL, Alexiou M, Papaioannou VE. 1996. Programmed cell death in the absence of c-Fos and c-Jun. *Development* **122**: 1–9. doi:10.1242/dev.122.1.1

Ruffins SW, Jacobs RE. 2011. MRI in developmental biology and the construction of developmental atlases. *Cold Spring Harb Protoc* 2011: top100. doi:10.1101/pdb.top100

Sekine K, Ohuchi H, Fujiwara M, Yamasaki M, Yoshizawa T, Sato T, Yagishita N, Matsui D, Koga Y, Itoh N, et al. 1999. Fgf10 is essential for limb and lung formation. *Nat Genet* **21**: 138–141. doi:10.1038/5096

Sharpe J, Wong RO. 2011. *Imaging in developmental biology: a laboratory manual.* Cold Spring Harbor Laboratory Press, Cold Spring Harbor, NY.

Shawlot W, Behringer RR. 1995. Requirement for Lim1 in head-organizer function. *Nature* **374**: 425–430. doi:10.1038/374425a0

Speck N, Peeters M, Dzierzak E. 2002. Development of the vertebrate hematopoietic system. In *Mouse development: patterning, morphogenesis, and organogenesis* (ed. Rossant J, Tam PPL), pp. 191–210. Academic, San Diego.

Srinivas S. 2010. Imaging cell movements in egg-cylinder stage mouse embryos. *Cold Spring Harb Protoc* 2010: pdb.prot5539. doi:10.1101/pdb.prot5539

Sta Maria NS, Zhou C, Lee SJ, Valiulahi P, Li X, Choi J, Liu X, Jacobs R, Comai L, Reddy S. 2021. Mbnl1 and Mbnl2 regulate brain structural integrity in mice. *Commun Biol* **4**: 1342. doi:10.1038/s42003-021-02845-0

Udan RS, Piazza VG, Hsu CW, Hadjantonakis AK, Dickinson ME. 2014. Quantitative imaging of cell dynamics in mouse embryos using light-sheet microscopy. *Development* **141**: 4406–4414. doi:10.1242/dev.111021

Varlet I, Collignon J, Robertson EJ. 1997. Nodal expression in the primitive endoderm is required for specification of the anterior axis during mouse gastrulation. *Development* **124**: 1033–1044. doi:10.1242/dev.124.5.1033

Wang S, Garcia MD, Lopez AL III, Overbeek PA, Larin KV, Larina IV. 2017. Dynamic imaging and quantitative analysis of cranial neural tube closure in the mouse embryo using optical coherence tomography. *Biomed Opt Express* **8**: 407–419. doi:10.1364/BOE.8.000407

Wang S, Larina IV, Larin KV. 2020. Label-free optical imaging in developmental biology [Invited]. *Biomed Opt Express* **11**: 2017–2040. doi:10.1364/BOE.381359

Weninger WJ, Floro KL, Bennett MB, Withington SL, Preis JI, Barbera JPM, Mohun J, Dunwoodie SL. 2005. *Cited2* is required both for heart morphogenesis and establishment of the left–right axis in mouse development. *Development* **132**: 1337–1348. doi:10.1242/dev.01696

Williams PR, Morgan JL, Kerschensteiner D, Wong RO. 2013. *In vitro* imaging of retinal whole mounts. *Cold Spring Harb Protoc* 2013: pdb.prot072645. doi:10.1101/pdb.prot072645

Winnier G, Hargett L, Hogan BL. 1997. The winged helix transcription factor *MFH1* is required for proliferation and patterning of paraxial mesoderm in the mouse embryo. *Genes Dev* **11**: 926–940. doi:10.1101/gad.11.7.926

Wong MD, van Eede MC, Spring S, Jevtic S, Boughner JC, Lerch JP, Henkelman RM. 2015. 4D atlas of the mouse embryo for precise morphological staging. *Development* **142**: 3583–3591. doi:10.1242/dev.125872

Wu L, de Bruin A, Saavedra HI, Starovic M, Trimboli A, Yang Y, Opavska J, Wilson P, Thompson JC, Ostrowski MC, et al. 2003. Extra-embryonic function of Rb is essential for embryonic development and viability. *Nature* **421**: 942–947. doi:10.1038/nature01417

Yoder MC, Papaioannou VE, Breitfeld PP, Williams DA. 1994. Murine yolk sac endoderm- and mesoderm-derived cell lines support *in vitro* growth and differentiation of hematopoietic cells. *Blood* **83**: 2436–2443. doi:10.1182/blood.V83.9.2436.2436

Zhang J, Wu D, Turnbull DH. 2018. In utero MRI of mouse embryos. *Methods Mol Biol* **1718**: 285–296. doi:10.1007/978-1-4939-7531-0_17

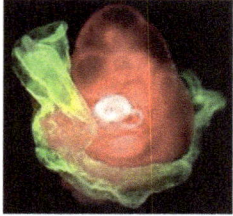

CHAPTER 12

Uncovering Phenotypes in Mutant Mice by Determining Embryo, Organ, Tissue, and Cell Developmental Potential

The death of an embryo during gestation does not necessarily preclude the study of the mutant embryo or the developmental potential of its individual cells, tissues, or organs. Whole-embryo in vitro culture prior to the time of death will allow real-time observation of living embryos and direct comparisons with controls. Organ anlage can be removed from embryos and cultured in vitro beyond the time of death of the whole embryo. In both whole embryos and organ anlage culture, fluorescent protein reporters may be used productively to follow cell types or specific gene expression changes. Some cells, such as hematopoietic cells, and organ anlage, may be suitable for transplantation to wild-type hosts for further analysis of their potential. Additionally, cell lines, including embryonic stem (ES) cells, trophoblast stem (TS) cells, extraembryonic endoderm (XEN) stem cells, and epiblast-derived stem cells (EpiSC), can be derived from mutant embryos to reveal the potential of the mutant cells outside the context of the whole organism. Mutant stem cells or even whole mutant embryos can be used to test potential in chimeras or in teratomas.

INTRODUCTION

DETERMINING EMBRYO/ORGAN/TISSUE/CELL DEVELOPMENTAL POTENTIAL

- POSTIMPLANTATION EMBRYO CULTURE
- TISSUE/ANLAGEN/ORGAN CULTURE
- CELL LINES
- CHIMERAS ⟶ Chapter 17
- CELL/ORGAN TRANSPLANTATION
- TERATOMAS

In a developing organism, the normality of the whole depends on the correct functioning of all of the parts. A defect in one cell type or tissue of an embryo will quickly compromise neighboring or dependent cells, setting up a chain reaction of problems that can encompass the entire organism.

Opening artwork: Courtesy of Sonja Nowotschin and Kat Hadjantonakis.

© 2024 Cold Spring Harbor Laboratory Press
Cite this chapter as *Cold Spring Harb Protoc*; doi:10.1101/pdb.over107974

This effect is particularly marked in defects that have their onset during early development. Our goal in previous chapters of determining the time of death and then trying to determine the earliest onset of the mutant phenotype is to identify the earliest role of the mutant gene that starts the catastrophic cascade. However, the effects of an early phenotype may be just the tip of the iceberg in terms of gene function. You may know that your favorite gene is expressed more widely in time or space than the earliest-identified defective tissue. Or you may not be able to distinguish a primary defect from a secondary defect due to the accumulation of problems in a developmental cascade. If you have an early embryonic lethal but your primary goal is to study the role of your gene in later embryonic stages or even in the adult, you may be able to entirely circumvent early lethality to get at later stages. If this is your situation, see Chapter 17: Getting around an Early Lethal Phenotype in Mice with Chimeras (Papaioannou and Behringer 2023a) and Chapter 18: Tissue- and/or Temporal-Specific Mutations in Mice Using Conditional Alleles (Papaioannou and Behringer 2023b).

On the other hand, much information can be gained by short-term embryo culture, allowing you to observe development directly. Alternatively, removing specific tissues from the detrimental effects of a dying embryo and testing developmental potential of tissues, either in isolation or recombined with wild-type tissue, are useful in addition to studying in utero development. These approaches are helpful in distinguishing primary and secondary effects of the mutation, and you may also be able to get at some of the later effects of the gene in cases for which a placental or other insufficiency is limiting embryo development. The potential of mutant tissue or organs to survive and differentiate when removed from developmental constraints, and/or from detrimental effects due to failure of other essential organs, may provide valuable clues to later gene function that would otherwise be obscured by early lethality. What follows are some of the varieties of ways to test the developmental potential of embryos or embryonic tissue. These should be selectively applied to your mutant phenotype depending on how far the mutant embryo develops before the lethal effect appears and on what you have already learned about the lethal effect. As always, the experiments should be guided by the expression pattern of the gene. However, even genes with ubiquitous expression can have tissue-specific defects.

IN VITRO CULTURE OF POSTIMPLANTATION EMBRYOS

Culture of preimplantation and periimplantation embryos was discussed in Chapter 9: Phenotypic Analysis of Preimplantation Lethality in Mice (Papaioannou and Behringer 2023c). For later-stage embryos, whole-embryo culture is feasible for periods of 24–72 h between E5.5 and E11.0 (Aguilera-Castrejon et al. 2021), during which gastrulation and early organogenesis can be directly observed and even manipulated. A variety of techniques for static or rotating culture has been developed, with conditions dependent on the age of the embryos at the time of explantation (see Chapter 12 in Behringer et al. 2014). Although the outermost layers of the fetal membranes (giant trophoblast cells and parietal endoderm) are removed in these protocols, the embryo is cultured with the placenta and yolk sac either intact or partially dissected so that the integrity of the circulation is retained, allowing observation of the development of these extraembryonic tissues as well.

In the context of mutation analysis, whole-embryo culture allows the direct comparison of development of homozygous mutant and wild-type embryos. The protocols are quite exacting and require a considerable investment of time and energy, but can be very informative for certain types of mutations, such as those affecting morphogenetic movements, particularly when combined with cell marking, such as the application of the fluorescent dye DiI (e.g., *amot*; Shimono and Behringer 2003). The advent and increasing availability of transgenic animals with fluorescent proteins localized to different subcellular organelles (Hadjantonakis and Papaioannou 2004), to specific cell types (Gu et al. 2018), or photoconvertible fluorescent proteins (Griswold et al. 2011) can be combined very usefully with whole-embryo culture and epifluorescent or confocal fluorescent microscopy so that developmental events can be observed in real time in live embryos. You may wish to cross your mutant with mice carrying such fluorescent markers, or, if you put such a fluorescent marker into your targeting construct as an expression reporter (Chapter 3: Mouse Gene-Targeting Strategies for Maximum Ease

TABLE 1. Some embryonic or postnatal tissues that can be cultured and/or transplanted with examples of protocols and/or use in mutational analysis of the indicated genes

	Gene analyzed	Reference(s)
A. Organs that can be cultured transiently in vitro		
Allantois	*Tbx4*	Arora et al. 2012a
Foregut/trachea bud	*Tbx4, Tbx5*	Arora et al. 2012b
Gonads	*Sox9*	Chaboissier et al. 2004
Heart		Kruithof et al. 2003
Hematopoietic tissue		Frisch and Calvi 2014
Kidney		Srinivas et al. 1999
Limb buds	*Tbx4*	Naiche and Papaioannou 2003
Lung	*Tbx4; Tbx5*	Arora et al. 2012b; Miura and Shiota 2000
Mammary glands		Heuberger et al. 1982
Palate	*Tgfb3*	Dudas et al. 2004
Teeth	*Msx1*	Bei et al. 2000
B. Tissues or cells that can be transplanted to heterotopic or orthotopic sites		
Hematopoietic tissues	*Ptpn2*	You-Ten et al. 1997
Liver fragments		Shiojiri et al. 1995
Liver cells		Rhim et al. 1994
Mammary glands	*Igfr1*	Bonnette and Hadsell 2001
Ovaries	*Gsc*	Rivera-Perez et al. 1995
Pituitary	*Ghrhr*	Hammer et al. 1984
Skin	*Pdgfa*	Karlsson et al. 1999
Spermatogonial stem cells		Brinster and Zimmermann 1994
Testes tubules		Schlatt et al. 2003

and Versatility) [Papaioannou and Behringer 2023d]), you can use whole-embryo culture to track the mutant (and wild-type) cells that express your gene.

IN VITRO CULTURE OF EMBRYONIC TISSUES, ANLAGEN, AND ORGANS

The long history of experimental embryology has provided us with techniques for isolating almost any embryonic germ layer, anlage, organ, or cell type, and a myriad of cell and tissue culture methods have been developed for specific tissues that can be used in mutation analysis (Table 1A; see Chapter 5 in Behringer et al. 2014). The separated tissues can be grown in isolation or recombined with other interacting tissues, either wild type or mutant. For example:

- the early germ layers of the embryo can be separated by a combination of mechanical and enzymatic dissection for culture in isolation (Nagy et al. 2006a);

- the layers of the yolk sac can be separated and cell lines derived from them (Yoder et al. 1994);

- the allantoic bud can be grown in organ culture (e.g., *Tbx4*; Downs et al. 2001; Arora et al. 2012a) or allantoic cells can be dispersed and grown as a monolayer;

- limb buds (Paradis et al. 2019; Arostegui and Underhill 2021), lung buds (e.g., *Tbx4* and *Tbx5*; Arora et al. 2012b), or mammary buds (Kogata and Howard 2015) can be isolated and will undergo development in vitro;

- ureteric buds can be isolated and recombined with metanephric mesenchyme and will undergo branching morphogenesis (Srinivas et al. 1999).

Organ Culture

For many in vitro organ culture systems, the organ or anlage of interest is cultured on top of a medium-permeable substrate (e.g., a filter) that sits above the medium (Fig. 1). The organ grows at the air–medium interface. Some tissues or organs can be cultured submerged in medium. The

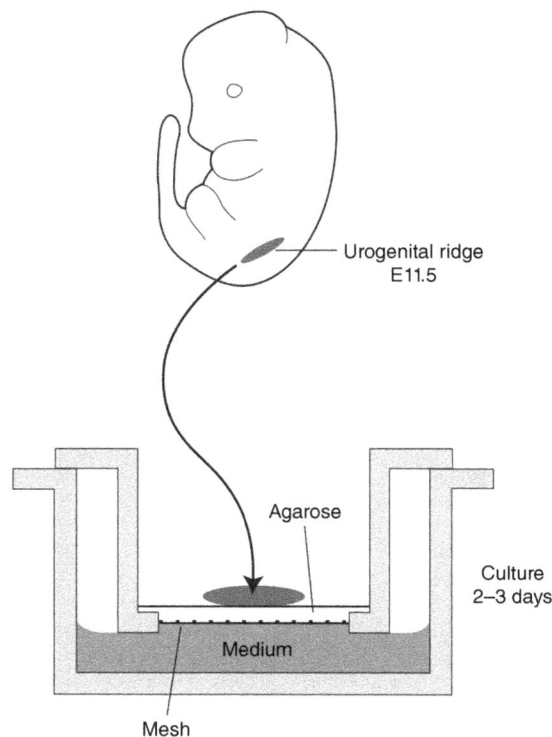

FIGURE 1. In vitro organ culture. In this example, the urogenital ridge containing the gonad and mesonephros is dissected from an E11.5 embryo and placed in a cassette on a medium-permeable filter, in this case with a layer of agarose. This cassette is then placed into a well of medium. The organ is cultured at the air–medium interface for 2–3 d. During this period of time, the bipotential gonads will differentiate into morphologically distinguishable ovaries or testes, depending on their genotype. (Diagram courtesy of Akio Kobayashi and Richard Behringer.)

cultured organs can then be analyzed for morphology, histology, molecular markers, cell proliferation, and death. In addition, the organs can be manipulated at the initiation of the in vitro culture to monitor cell fates and migration, for example by marking with vital dyes or by the inclusion of fluorescent protein reporters to dynamically image the development of the mutant organ in vitro.

In the context of mutant analysis, these techniques can be used to study an affected organ or tissue for several days beyond the lethal time point of the whole embryo. Because of the inherent variability associated with tissue and organ culture, the key is to explant the tissue, after carefully stage-matching the embryos, before the onset of the mutant effect. Then, compare the behavior of mutant and wild-type tissue. For the most part, only short-term cultures will be informative because the techniques do not support normal development in the long term. You might be able to gain some time by rescuing mutant tissue from an early placental (or cardiovascular) defect and thus watch the unfolding of mutant effects in vitro. Assays of cell proliferation, death, and migration, as well as the use of molecular markers, may all be useful in combination with in vitro culture.

Marking Cells in Organ Culture

Similar to whole-embryo culture, the use of fluorescent protein reporters in organ culture allows you to study dynamic cell behaviors of all mutant cells or of expressing mutant cells, if you have a reporter knock-in or cross your mutant with transgenic lines carrying fluorescent reporters. In addition, these in vitro culture methods allow the recombination of tissues, and, depending on the type of phenotypic effect observed, it may be useful to combine mutant and wild-type tissue in organ culture to test, for example, the integrity of inductive interactions in the mutant tissue (e.g., *Bmp4* [Furuta and Hogan 1998]; *Cerr1* [Shawlot et al. 2000]).

Defects in cell migration can be a cause of tissue-specific morphological defects. Cell migration can be inferred by collating static images of different embryos that have been fixed at sequential stages of development. However, only correlative conclusions can be drawn from such a study. To document migration rigorously, cells must be indelibly marked either chemically or genetically; they and their cellular progeny are then observed over time in the same embryo or tissue. In the absence of fluo-

Cite this chapter as *Cold Spring Harb Protoc*; doi:10.1101/pdb.over107974

rescent reporters, lipophilic dyes such as DiI or DiO can be used to mark cells of an organ (by exposure or intracellular injection). A drawback of chemical marking is that it becomes diluted as the marked cells proliferate.

A genetic strategy to mark cells exploits the DNA recombinase systems (i.e., Cre/*loxP* or Flp/*FRT*). For example, it is possible to generate mice that carry a Cre reporter allele that can be ubiquitously expressed (e.g., *Rosa26R*) and a CreER allele that is expressed in the tissue of interest (see Recombinase Mouse Strains in Chapter 18: Tissue- and/or Temporal-Specific Mutations in Mice Using Conditional Alleles [Papaiannou and Behringer 2023b]). CreER activity is induced by a pulse of tamoxifen, usually by intraperitoneal injection of the pregnant female at a specific time in development when the cells of interest are thought to be migrating. At a later stage of development, the embryo is examined for the presence and distribution of reporter-expressing cells. One difficulty with this technique is that you cannot determine exactly which cells give rise to the marked cells. An alternative would be to use a Cre reporter allele that uses a fluorescent protein reporter. If the starting tissue can be imaged at the initiation of the induction of Cre activity and subsequently during culture, then you can follow the migration of marked cells in real time.

MAKING CELL LINES FROM MUTANT EMBRYOS

Another way of revealing the potential of mutant embryonic cells outside the context of the complete organism is to derive cell lines directly from mutant embryos for cell proliferation, cell death, and biochemical analyses (see Chapter 8 in Behringer et al. 2014) (e.g., *Trp53*, *Mdm2*; de Rozieres et al. 2000). Homozygous mutant and control cells can be used to examine cell proliferation by simple cell counts, cell cycle characteristics by fluorescence-activated cell sorting (FACS), and cell death by terminal deoxynucleotidyl transferase dUTP nick end labeling (TUNEL), FACS analysis, or biochemical methods. The chromosomes of these cells can be examined to study genome integrity. Homozygous mutant cells can also be used for biochemical and molecular studies or as a substrate for in vitro genetic manipulation to determine the position of the gene of interest in a biochemical pathway.

You will already be familiar with the capabilities of one type of embryo-derived stem cell if your mutant mice were made using embryonic stem (ES) cell technology. Now, in the analysis of mutant mice derived from that technology, you can once again make use of the amazing properties of ES cells, but this time to inform on the mutant phenotype (e.g., *Twist*; Chen and Behringer 1995). If, for example, your homozygous mutant dies at embryonic day (E) 6.5 from a placental defect, but you know that the gene is expressed later in development, perhaps in yolk sac mesoderm or nerve cells, you may wish to try deriving ES cells from homozygous mutant embryos (Nichols and Jones 2017). You will then have a source of homozygous mutant cells that can be subjected to culture conditions that promote mesoderm and/or neuronal differentiation, allowing you to study these processes in mutant cells in vitro. In the course of making ES cell lines from homozygous mutant embryos, wild-type and heterozygous cell lines will also be generated that will serve as controls and, possibly, material for uncovering gene dosage effects.

An alternative to making new ES cell lines from homozygous mutant embryos is to use the original targeted ES cells, which are heterozygous, and make them homozygous by either (1) selecting highly resistant cells with a high selective drug dose (see Chapter 10 in Behringer et al. 2014), (2) retargeting the wild-type allele using the initial targeting vector (after removal of the selectable marker from the mutant allele), or (3) CRISPR–Cas gene editing. You could also generate induced pluripotent stem (iPS) cells for similar purposes from heterozygotes (see Chapter 9 in Behringer et al. 2014) and then mutate the remaining wild-type allele.

Although less common than ES cells, stem cells known as trophoblast stem (TS) cells can be derived from the precursors of trophoblast lineages (Rossant 2006). If your mutation affects the placenta or trophoblast, it may be worth deriving TS cells for in vitro study. Similarly, you might want to derive extraembryonic endoderm (XEN) stem cells if your mutation alters primitive endoderm development (Rugg-Gunn 2017b) or epiblast-derived stem cells (EpiSCs) if your mutation affects the epiblast or its derivatives (Rugg-Gunn 2017a).

If making ES, TS, XEN, or EpiSC cell lines from your homozygous mutant embryo seems like a good idea, stop for a moment and anticipate another potential use for these cells: making chimeras, as described in the next section and in Chapter 17: Getting around an Early Lethal Phenotype in Mice with Chimeras (Papaiannou and Behringer 2023a). Before deriving ES or TS cells from the mutant embryos, consider introducing a transgenic cell marker by breeding heterozygotes with a marker-bearing strain to make the best use of the cells in chimeras (see Distinguishing Cells in Chimeras in Chapter 17: Getting around an Early Lethal Phenotype in Mice with Chimeras [Papaiannou and Behringer 2023a]).

Finally, primary embryonic fibroblast cell lines, called mouse embryonic fibroblasts (MEFs), can be made from embryos. Traditionally, MEFs are generated from E13.5 embryos. However, it is possible to generate MEFs routinely from embryos as early as E8.5 (see Chapter 8 in Behringer et al. 2014). MEFs may be useful for comparing proliferation rates or ease of immortalization between homozygous mutant and wild-type fibroblast cells.

TESTING DEVELOPMENTAL POTENTIAL IN CHIMERAS

Even if your homozygous mutant embryos die by E9.5, it may still be possible to examine mutant cells in the context of a developing embryo at later stages using combinations of wild-type and mutant tissue. The idea is to combine mutant and wild-type cells in a chimera (e.g., homozygous mutant ES cells combined with a wild-type preimplantation embryo). Using a marker to distinguish the mutant and wild-type cells (see Distinguishing Cells in Chimeras in Chapter 17: Getting around an Early Lethal Phenotype in Mice with Chimeras [Papaiannou and Behringer 2023a]), you can follow the mutant cells during development to determine whether they behave normally or aberrantly. This procedure is sometimes referred to as chimeric rescue, although this is a bit of a misnomer. Chimeras might successfully circumvent lethality, leading to "rescue" of the mutant cells or even the embryo; however, the mutant cells may still suffer cell-autonomous detrimental mutant effects (e.g., *Tbx6*; Chapman et al. 2003). Nonetheless, much information can be obtained by examining the mutant cells' behavior in combination with wild-type cells.

Chimeras are useful tools for studying the actions of a mutation in vivo and also when combined with embryo or organ culture in vitro. In a chimera study, the presence and distribution of mutant and wild-type cells are determined for a particular tissue by terminal analysis (e.g., β-galactosidase staining or fluorescent reporter expression and histology) of individual chimeras. The collation of the data from these chimeras provides insights into the actions of the mutation in that tissue. However, one of the technical conundrums of a chimera study is that individual chimeras or their tissues are typically sacrificed at a single time point. Thus, one does not know the initial distribution of mutant and wild-type cells at the start of the experiment nor what the distribution of the mutant and wild-type cells would have been later. The development of fluorescent protein reporters can now be used to address this problem and obtain more information than previously possible from a chimera experiment. Accordingly, chimeras can be generated using mutant and wild-type embryonic cells. The mutant or wild-type cells can be marked with a ubiquitously expressed fluorescent reporter. At a specific stage of development, the organ of interest is explanted into in vitro culture and imaged to determine the presence, distribution, and behavior of the mutant or wild-type cells. The chimeric organ is then imaged during the culture period and the individual mutant or wild-type cells monitored. In this manner, the two genotypes of cells in the organs can be followed from start to finish (Leclerc and Costantini 2016).

Because making chimeras can entail a lot of work, think carefully about the questions that can be answered by this approach. You will only want to consider this approach once you have a very clear idea of the cause of lethality. Then, if the homozygous mutants die early and you know that the gene is expressed later in tissues that you wish to study, or if you want to know whether a specific mutant phenotype is cell autonomous or not, see Chapter 17: Getting around an Early Lethal Phenotype in Mice with Chimeras (Papaiannou and Behringer 2023a).

Cite this chapter as *Cold Spring Harb Protoc*; doi:10.1101/pdb.over107974

TRANSPLANTATION OF CELLS AND ORGANS

Cells or organs can also be transplanted into postnatal to adult wild-type hosts to examine mutant cells, tissues, or organs at heterotopic sites (ectopic transfer) or orthotopic locations (the usual place in the body). These types of somatic cell chimeras can be very useful to reveal mutant phenotypes in an in vivo context (Table 1B) and whether the mutation acts cell-autonomously or nonautonomously. Hematopoietic tissues form early during embryogenesis and can be transplanted by venous injection. Likewise, ovaries, skin, and spermatogonial stem cells can be transplanted to orthotopic locations. Other tissues may be transplanted at heterotopic locations because surgical manipulations may not be available (e.g., pituitary, testis tubules). The transplanted cells, tissues, or organs can then grow and differentiate in vivo for various periods of time for subsequent analysis.

DIFFERENTIATION OF EARLY EMBRYOS AND TISSUE FRAGMENTS INTO TERATOMAS

An alternative way of determining the differentiation and long-term histogenic potential of certain isolated embryonic tissues is to culture them in vivo in ectopic sites in a histocompatible host. The host organism, usually an adult mouse, serves as a culture chamber, supplying nutrients and trophic factors to the transplanted tissue. Tumors called teratomas or teratocarcinomas develop from these transplanted embryonic fragments, often containing a variety of differentiated cell and tissue types chaotically organized. Teratocarcinomas are distinguished by the presence of undifferentiated embryonal carcinoma cells, which are transplantable stem cells. The composition of the tumors indicates the potential for histogenesis of the transplanted embryonic cells (e.g., *Tbx6*; Chapman et al. 2003). When comparing mutant and wild-type tumors, you may, for example, discover a specific cell type missing from the mutant grafts, thus indicating a role for the gene in the differentiation of that cell type. One caveat in this type of analysis, however, is that the host organism could be potentially supplying a soluble factor missing from the mutant tissue as a result of the mutation. This might result in rescue of the differentiation capacity of mutant tissue but is not very informative with regard to the missing factor.

Teratomas or teratocarcinomas will only develop from transplanted embryonic tissues that contain cells with wide developmental potential, which include blastocysts, the embryonic portion of pregastrula- or gastrula-stage embryos, and the isolated tail bud or genital ridges of mid-gestation embryos. Histocompatible or immune deficient hosts are essential to avoid an immune reaction. The tissue is transplanted under either the testes or kidney capsule (Nagy et al. 2006b) and left for several weeks. Analysis is usually done histologically using Masson's trichrome stain (Fig. 2) or other special histochemical stains that more clearly delineate different tissue types than hematoxylin and eosin staining. Transcriptomes can also be generated for a global assessment of gene expression. The tumors can also be analyzed for neoplastic potential (i.e., transplantability or the ability to grow as stem cell

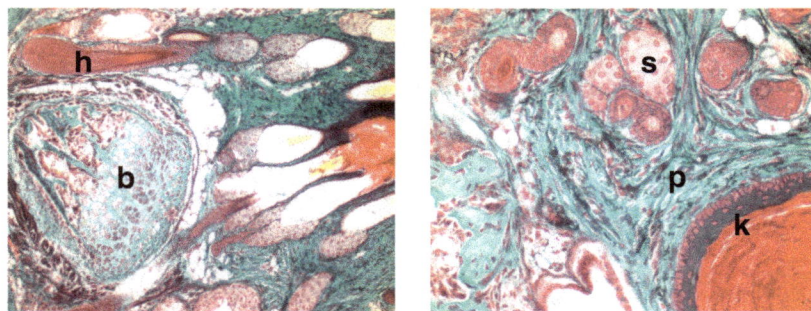

FIGURE 2. Example of a teratoma. Two sections through a teratoma show a multitude of cell types using Masson's trichrome stain. (b) Bone, (h) hair follicle, (k) keratinized epithelium, (p) pigment cells, (s) sebaceous gland.

lines in vitro) and for the expression of specific genes. Finally, teratomas can be used as the source for making fibroblast cell lines.

REFERENCES

Aguilera-Castrejon A, Oldak B, Shani T, Ghanem N, Itzkovich C, Slomovich S, Tarazi S, Bayerl J, Chugaeva V, Ayyash M, et al. 2021. Ex utero mouse embryogenesis from pre-gastrulation to late organogenesis. *Nature* 593: 119–124. doi:10.1038/s41586-021-03416-3

Arora R, del Alcazar CM, Morrisey EE, Naiche LA, Papaioannou VE. 2012a. Candidate gene approach identifies multiple genes and signaling pathways downstream of Tbx4 in the developing allantois. *PLoS ONE* 7: e43581. doi:10.1371/journal.pone.0043581

Arora R, Metzger R, Papaioannou VE. 2012b. Multiple roles and interactions of *Tbx4* and *Tbx5* in the development of the respiratory system. *PLoS Genet* 8: e1002866. doi:10.1371/journal.pgen.1002866

Arostegui M, Underhill TM. 2021. Murine limb bud organ cultures for studying musculoskeletal development. *Methods Mol Biol* 2230: 115–137. doi:10.1007/978-1-0716-1028-2_8

Behringer RR, Gertsenstein M, Nagy KV, Nagy A. 2014. *Manipulating the mouse embryo: a laboratory manual.* Cold Spring Harbor Laboratory Press, Cold Spring Harbor, NY.

Bei M, Kratochwil K, Maas RL. 2000. BMP4 rescues a non-cell-autonomous function of *Msx1* in tooth development. *Development* 127: 4711–4718. doi:10.1242/dev.127.21.4711

Bonnette SG, Hadsell DL. 2001. Targeted disruption of the IGF-I receptor gene decreases cellular proliferation in mammary terminal end buds. *Endocrinology* 142: 4937–4945. doi:10.1210/endo.142.11.8500

Brinster RL, Zimmermann JW. 1994. Spermatogenesis following male germ-cell transplantation. *Proc Natl Acad Sci* 91: 11298–11302. doi:10.1073/pnas.91.24.11298

Chaboissier MC, Kobayashi A, Vidal VI, Lutzkendorf S, van de Kant HJ, Wegner M, de Rooij DG, Behringer RR, Schedl A. 2004. Functional analysis of *Sox8* and *Sox9* during sex determination in the mouse. *Development* 131: 1891–1901. doi:10.1242/dev.01087

Chapman DL, Cooper-Morgan A, Harrelson Z, Papaioannou VE. 2003. Critical role for *Tbx6* in mesoderm specification in the mouse embryo. *Mech Dev* 120: 837–847. doi:10.1016/S0925-4773(03)00066-2

Chen ZF, Behringer RR. 1995. *twist* is required in head mesenchyme for cranial neural tube morphogenesis. *Genes Dev* 9: 686–699. doi:10.1101/gad.9.6.686

de Rozieres S, Maya R, Oren M, Lozano G. 2000. The loss of *mdm2* induces p53-mediated apoptosis. *Oncogene* 19: 1691–1697. doi:10.1038/sj.onc.1203468

Downs KM, Temkin R, Gifford S, McHugh J. 2001. Study of the murine allantois by allantoic explants. *Dev Biol* 233: 347–364. doi:10.1006/dbio.2001.0227

Dudas M, Nagy A, Laping NJ, Moustakas A, Kaartinen V. 2004. Tgf-β3-induced palatal fusion is mediated by Alk-5/Smad pathway. *Dev Biol* 266: 96–108. doi:10.1016/j.ydbio.2003.10.007

Frisch BJ, Calvi LM. 2014. Hematopoietic stem cell cultures and assays. *Methods Mol Biol* 1130: 315–324. doi:10.1007/978-1-62703-989-5_24

Furuta Y, Hogan BL. 1998. BMP4 is essential for lens induction in the mouse embryo. *Genes Dev* 12: 3764–3775. doi:10.1101/gad.12.23.3764

Griswold SL, Sajja KC, Jang CW, Behringer RR. 2011. Generation and characterization of *iUBC-KikGR* photoconvertible transgenic mice for live time-lapse imaging during development. *Genesis* 49: 591–598. doi:10.1002/dvg.20718

Gu B, Posfai E, Rossant J. 2018. Efficient generation of targeted large insertions by microinjection into two-cell-stage mouse embryos. *Nat Biotechnol* 36: 632–637. doi:10.1038/nbt.4166

Hadjantonakis A-K, Papaioannou VE. 2004. Dynamic *in vivo* imaging and cell tracking using a histone fluorescent protein fusion in mice. *BMC Biotechnol* 4: 33. doi:10.1186/1472-6750-4-33

Hammer RE, Palmiter RD, Brinster RL. 1984. Partial correction of murine hereditary growth disorder by germ-line incorporation of a new gene. *Nature* 311: 65–67. doi:10.1038/311065a0

Heuberger B, Fitzka I, Wasner G, Kratochwil K. 1982. Induction of androgen receptor formation by epithelium-mesenchyme interaction in embry-

onic mouse mammary gland. *Proc Natl Acad Sci* 79: 2957–2961. doi:10.1073/pnas.79.9.2957

Karlsson L, Bondjers C, Betsholtz C. 1999. Roles for PDGF-A and sonic hedgehog in development of mesenchymal components of the hair follicle. *Development* 126: 2611–2621. doi:10.1242/dev.126.12.2611

Kogata N, Howard BA. 2015. A protocol for studying embryonic mammary progenitor cells during mouse mammary primordial development in explant culture. In *Mammary stem cells: methods and protocols* (ed. Vivanco MdM), pp. 51–62. Springer, New York.

Kruithof BP, van den Hoff MJ, Wessels A, Moorman AF. 2003. Cardiac muscle cell formation after development of the linear heart tube. *Dev Dyn* 227: 1–13. doi:10.1002/dvdy.10269

Leclerc K, Costantini F. 2016. Mosaic analysis of cell rearrangements during ureteric bud branching in dissociated/reaggregated kidney cultures and in vivo. *Dev Dyn* 245: 483–496. doi:10.1002/dvdy.24387

Miura T, Shiota K. 2000. Time-lapse observation of branching morphogenesis of the lung bud epithelium in mesenchyme-free culture and its relationship with the localization of actin filaments. *Int J Dev Biol* 44: 899–902.

Nagy A, Gertsenstein M, Vintersten K, Behringer R. 2006a. Separating postimplantation germ layers. *CSH Protoc* doi:10.1101/pdb.prot4368

Nagy A, Gertsenstein M, Vintersten K, Behringer R. 2006b. Transplantation of tissues under the kidney capsule. *CSH Protoc* doi:10.1101/pdb.prot4382

Naiche LA, Papaioannou VE. 2003. Loss of *Tbx4* blocks hindlimb development and affects vascularization and fusion of the allantois. *Development* 130: 2681–2693. doi:10.1242/dev.00504

Nichols J, Jones K. 2017. Derivation of mouse embryonic stem (ES) cell lines using small-molecule inhibitors of Erk and Gsk3 signaling (2i). *Cold Spring Harb Protoc* doi:10.1101/pdb.prot094086

Papaioannou VE, Behringer RR. 2023a. Getting around an early lethal phenotype in mice with chimeras. *Cold Spring Harb Protoc* doi:10.1011/pdb.over107979

Papaioannou VE, Behringer RR. 2023b. Tissue- and/or temporal-specific mutations in mice using conditional alleles. *Cold Spring Harb Protoc* doi:10.1011/pdb.over107980

Papaioannou VE, Behringer RR. 2023c. Phenotypic analysis of preimplantation lethality in mice. *Cold Spring Harb Protoc* doi:10.1011/pdb.over107971

Papaioannou VE, Behringer RR. 2023d. Mouse gene-targeting strategies for maximum ease and versatility. *Cold Spring Harb Protoc* doi:10.1011/pdb.over107957

Paradis FH, Yan H, Huang C, Hales BF. 2019. The murine limb bud in culture as an in vitro teratogenicity test system. *Methods Mol Biol* 1965: 73–91. doi:10.1007/978-1-4939-9182-2_6

Rhim JA, Sandgren EP, Degen JL, Palmiter RD, Brinster RL. 1994. Replacement of diseased mouse liver by hepatic cell transplantation. *Science* 263: 1149–1152. doi:10.1126/science.8108734

Rivera-Perez JA, Mallo M, Gendron-Maguire M, Gridley T, Behringer RR. 1995. *Goosecoid* is not an essential component of the mouse gastrula organizer but is required for craniofacial and rib development. *Development* 121: 3005–3012. doi:10.1242/dev.121.9.3005

Rossant J. 2006. Derivation of trophoblast stem (TS) cell lines from blastocysts. *CSH Protoc* doi:10.1101/pdb.prot4407

Rugg-Gunn P. 2017a. Derivation and culture of epiblast stem cell (EpiSC) lines. *Cold Spring Harb Protoc* doi:10.1101/pdb.prot093971

Rugg-Gunn P. 2017b. Derivation and culture of extra-embryonic endoderm stem cell lines. *Cold Spring Harb Protoc* doi:10.1101/pdb.prot093963

Schlatt S, Honaramooz A, Boiani M, Scholer HR, Dobrinski I. 2003. Progeny from sperm obtained after ectopic grafting of neonatal mouse testes. *Biol Reprod* 68: 2331–2335. doi:10.1095/biolreprod.102.014894

Shawlot W, Min Deng J, Wakamiya M, Behringer RR. 2000. The cerberus-related gene, *Cerr1*, is not essential for mouse head formation. *Genesis*

Cite this chapter as *Cold Spring Harb Protoc*; doi:10.1101/pdb.over107974

26: 253–258. doi:10.1002/(SICI)1526-968X(200004)26:4<253::AID-GENE60>3.0.CO;2-D

Shimono A, Behringer RR. 2003. Angiomotin regulates visceral endoderm movements during mouse embryogenesis. *Curr Biol* **13:** 613–617. doi:10.1016/S0960-9822(03)00204-5

Shiojiri N, Wada JI, Tanaka T, Noguchi M, Ito M, Gebhardt R. 1995. Heterogeneous hepatocellular expression of glutamine synthetase in developing mouse liver and in testicular transplants of fetal liver. *Lab Invest* **72:** 740–747.

Srinivas S, Goldberg MR, Watanabe T, D'Agati V, al-Awqati Q, Costantini F. 1999. Expression of green fluorescent protein in the ureteric bud of transgenic mice: a new tool for the analysis of ureteric bud morphogenesis. *Dev Genet* **24:** 241–251. doi:10.1002/(SICI)1520-6408(1999)24:3/4<241::AID-DVG7>3.0.CO;2-R

Yoder MC, Papaioannou VE, Breitfeld PP, Williams DA. 1994. Murine yolk sac endoderm- and mesoderm-derived cell lines support in vitro growth and differentiation of hematopoietic cells. *Blood* **83:** 2436–2443. doi:10.1182/blood.V83.9.2436.2436

You-Ten KE, Muise ES, Itié A, Michaliszyn E, Wagner J, Jothy S, Lapp WS, Tremblay ML. 1997. Impaired bone marrow microenvironment and immune function in T cell protein tyrosine phosphatase-deficient mice. *J Exp Med* **186:** 683–693. doi:10.1084/jem.186.5.683

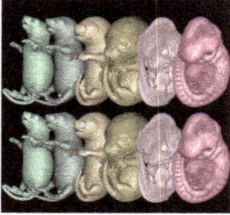

CHAPTER 13

Phenotypic Analysis of Perinatal Lethality in Mice

If homozygous mutants are found dead shortly after birth, further analysis of the phenotype will depend on knowing whether death occurred before or after parturition, which marks a major shift in physiological conditions of neonates. This chapter provides methods for determining the time of death and discusses possible causes of lethality during this period, including catastrophic morphological abnormalities or developmental delay of some or all organs. Attention is given to specific defects that could result in perinatal death, such as cranial nerve defects, cleft palate, diaphragm defects, and other problems that might not have been relevant during intrauterine development but become critical during the transition to extrauterine existence.

INTRODUCTION

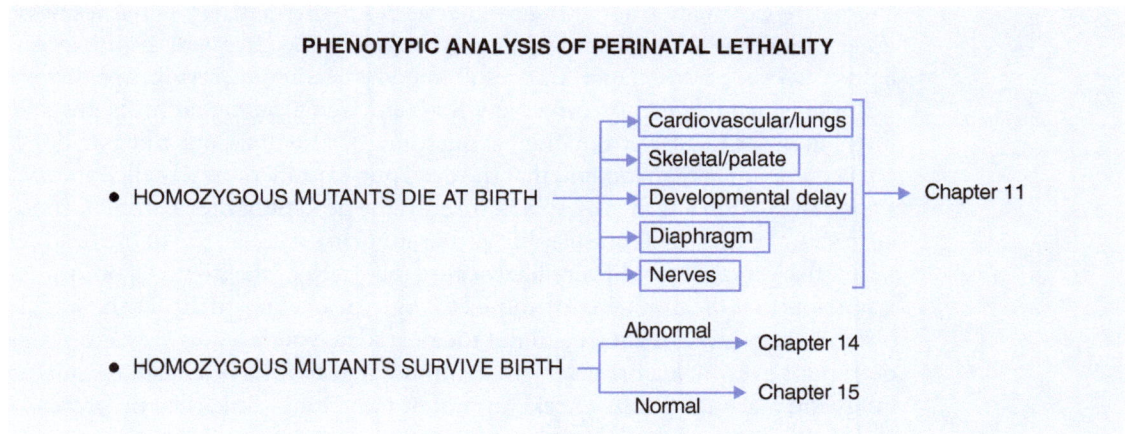

PHENOTYPIC ANALYSIS OF PERINATAL LETHALITY

- HOMOZYGOUS MUTANTS DIE AT BIRTH
  - Cardiovascular/lungs
  - Skeletal/palate
  - Developmental delay → Chapter 11
  - Diaphragm
  - Nerves

- HOMOZYGOUS MUTANTS SURVIVE BIRTH
  - Abnormal → Chapter 14
  - Normal → Chapter 15

Perinatal lethality is a common phenotype in mutational analysis. If you determine that homozygous mutants generated from heterozygous intercrosses are present at the appropriate frequency among the offspring at birth but are not viable, this chapter will guide you in their analysis. In determining the cause of perinatal lethality, as always, let the expression pattern of the gene mutated be a guide for the possible phenotypes to expect and therefore the most useful investigations to pursue. If you have

Opening artwork: Courtesy of Logan Hsu and Mary Dickinson.

© 2024 Cold Spring Harbor Laboratory Press
Cite this chapter as *Cold Spring Harb Protoc*; doi:10.1101/pdb.over107975

not previously investigated the perinatal expression of your favorite gene, either by experimentation or from published literature (e.g., Gene Expression Database; http://www.informatics.jax.org/expression.shtml), now is the time to do so in conjunction with the analysis of the mutant mice. If, on the other hand, the homozygous mutants are viable, the next two chapters deal with phenotypic analysis of obvious or visible abnormalities (see Chapter 14: Analysis of Postnatal Mutant Phenotypes in Mice [Papaioannou and Behringer 2023a]) or mice in which there is no obvious abnormality (see Chapter 15: The "No Phenotype" Challenge in Analyzing Mutant Mice [Papaioannou and Behringer 2023b]).

NEONATAL LOSSES—DETERMINING THE TIME OF DEATH

The first indication of perinatal lethality of mutant mice may be the discovery of dead offspring shortly after birth from heterozygous intercross matings. Instead of the anxiously awaited mutant mice, you may have found intact dead pups or body parts scattered around the cage, depending on whether and how much the mother cannibalized the remains. DNA is relatively robust, so save the bodies or pieces for genotyping. Determine the genotypes of all of the dead pups to determine whether the lethality is limited to the homozygous mutants or if heterozygous and wild-type pups are also among the dead.

There are two probable outcomes. In the first scenario, the dead pups are exclusively (or almost exclusively) homozygous mutants, suggesting that the lethality is limited to the homozygous mutant genotype. In this case you can proceed with the mutant analysis. However, in a second possible scenario, homozygous mutants as well as heterozygotes and/or wild-type pups may be among the dead. There are a couple of explanations for this outcome. It may fortuitously have been a difficult birth, causing the mother distress and leading to the loss of not only the homozygous mutants but also, nonspecifically, the pups of the other genotypes. It is still possible that homozygosity for the mutation causes lethality but you will need to examine additional litters to make this determination. An alternative explanation for the second scenario is that lethality is the result of poor husbandry or other environmental problems in the animal facility (e.g., frequent disturbances, noise, fumes, altered light cycle, temperature, diet) that result in poor breeding success independent of the genotype of the offspring. Investigate and correct any husbandry/environmental problems and examine additional litters in order to answer the original question: Are homozygous mutants born alive?

Let us assume that you find that the dead pups are all, or nearly all, homozygous mutants. This is a good indication to move on to determine if the dead pups died before or soon after birth. You may have an indication of a catastrophic visible abnormality (e.g., top of the head open and no brain or a herniated gut) that would very likely cause the death of the homozygous mutant pups. However, it is possible that the mother contributed to this type of abnormality as she was cleaning the pups after birth. If the bodies are intact and not too degraded, you can do a very simple test to determine if the dead pups ever took a breath of air, assuming that they have a head, mouth, and lungs. Dissect the lungs and place them into a beaker of water. If the lungs sink, the pup probably never took a breath after birth but might still have been born alive; if the lungs float, the mouse was alive at birth and took air into its lungs, but succumbed shortly thereafter. Do histology on the lungs to follow up on this initial conclusion, which might help determine the next steps in the analysis. If your mutated gene is expressed in the lungs, the lack of breathing could be a direct effect. On the other hand, a variety of other reasons may explain why a pup may fail to breathe at birth, including death before birth.

To determine whether a mutant was alive or dead at the time of birth, one could attempt to be present at the time of parturition. However, this is basically futile because mice usually give birth during the dark phase of the light cycle, and the length of gestation is variable, give or take a day. The quickest and most direct way to determine if full-term fetuses are alive or dead just before birth is to set up timed matings (i.e., check plugs with or without selecting females for estrus; see Box 1 in Chapter 5: Recovering a Targeted Mutation in Mice from Embryonic Stem Cell Chimeras or CRISPR–Cas

Cite this chapter as *Cold Spring Harb Protoc*; doi:10.1101/pdb.over107975

Founders [Papaioannou and Behringer 2023c]) and perform a cesarean section (C-section) (see Nagy et al. 2006) at embryonic day (E)18.5. This is also very useful because you can bypass any abnormalities that might have been caused by the mother as she cleans the pups at birth by determining if catastrophic abnormalities observed previously are also present upon C-section. Wild-type pups should all be recovered alive following C-sections performed at E18.5 and, thus, if the homozygous mutants die late in gestation, you will detect them when you perform the C-section. These pups will not move, you will not be able to revive them, and they may appear necrotic. If the homozygous mutants are alive just before birth, you may see them moving and struggling to breathe following C-section, but they may never turn pink and will probably die within minutes, possibly indicating that perinatal lethality was caused by an inability to breathe. Alternatively, the homozygous mutant pups might move, breathe, become pink, and squeak, suggesting that they can survive birth (or at least a C-section), but then die shortly thereafter. A wild-type pup delivered by C-section should move vigorously. Observe the pups obtained from the C-section to determine if any are lethargic or in distress in any way. If the pups have all survived the C-section, foster them onto a surrogate female (Nagy et al. 2006; see Protocol 4.2: Cross-Fostering of Newborn Mice from Embryonic Stem Cell Chimeras or CRISPR–Cas Founders [Papaioannou and Behringer 2023d]) and observe them periodically to discover if any of the pups die or lag behind in their postnatal development. Genotype any pups that die to determine if they are the homozygous mutants. It is to be expected that the homozygous mutants will die soon after C-section if they were destined to die during natural parturition. However, it is possible that C-section could rescue a perinatal lethality, indicating that the phenotype is sensitive to the stress of labor and birth. This would provide a way of recovering live homozygous mutants to investigate the phenotype further. For each of the scenarios described above, analyze sufficient numbers of timed matings to do a statistical analysis to support your conclusions (see Box 1 in Chapter 8: Phenotypic Analysis: Assessing Timing of Recessive Prenatal Lethality in Mice [Papaioannou and Behringer 2023e]).

If you determine that the homozygous mutants die before birth, see Chapter 11: Analysis of Mid- to Late-Gestation Phenotypes in Mice (Papaioannou and Behringer 2023f). If they survive at least the first day after birth by C-section, see Chapter 14: Analysis of Postnatal Mutant Phenotypes in Mice (Papaioannou and Behringer 2023a). If the homozygous mutants die at the time of birth, see the following section.

CAUSES OF LETHALITY AT BIRTH

There are various reasons why a mutant mouse might die at birth. As mentioned above, catastrophic abnormalities are clearly incompatible with life after birth. If the mutants are intact and appear morphologically normal following a gross anatomical assessment (see Chapter 14: Analysis of Post-natal Mutant Phenotypes in Mice [Papaioannou and Behringer 2023a]), then the most likely immediate causes of perinatal lethality are the inability to breathe or problems associated with the change in blood flow from the fetal to postnatal pattern that accompanies the first breaths. Mice can develop to term without a head or lungs, but they need to breathe to survive birth. Below we list some of the common causes of perinatal lethality, how to identify them, and how to study their development. Depending on the nature of the gene mutated and on its expression pattern, some of these causes may be more relevant than others in causing lethality of your particular mutant.

Catastrophic Abnormalities

A variety of catastrophic abnormalities resulting from the loss or abnormal development of an essential postnatal organ can lead to the immediate death of a homozygous mutant newborn pup (e.g., *Lhx1*; Shawlot and Behringer 1995) (Fig. 1). If there is no head, brain, mouth, or lungs, then it will be impossible for the animal to breathe and it will die within minutes after birth. Another major

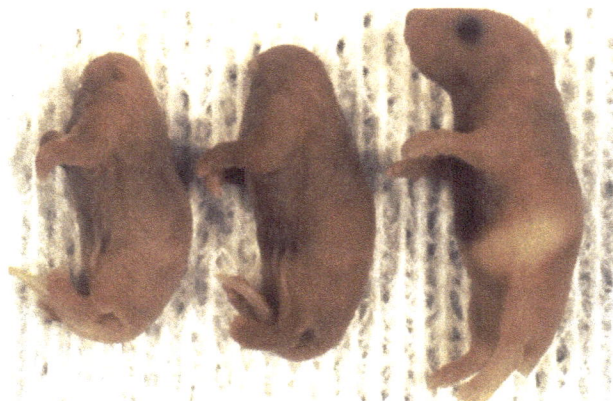

FIGURE 1. Catastrophic abnormalities. *Lhx1*-null mice (*left* and *center*) lack most of the head, a mouth, and kidneys. Control littermate (*right*). Because of these abnormalities the *Lhx1*-null pups die immediately after birth. Note the milk in the stomach of the control, wild-type littermate, indicating that it had been feeding properly. (Modified from Shawlot and Behringer 1995.)

defect that will lead to a quick death is herniation of the gut at the site of the umbilical attachment. If the guts are externalized, it is very possible that the mother may eat the externalized gut as she cleans the pup after birth. If the skin of homozygous mutant pups is not well developed, then the pups may die very quickly because of dehydration and the loss of barrier function (e.g., *Krt5*; Peters et al. 2001). Catastrophic phenotypes have their origins prior to birth even though they may not result in lethality until that time.

Cardiovascular Defects

In addition to the physical challenges of parturition, dramatic changes occur within the heart and circulation concomitant with a shift in the site of oxygenation of the blood from the placenta to the lungs. The ductus arteriosus, which during fetal development shunts most of the blood leaving the heart through the pulmonary artery into the aorta, closes, and the pulmonary arteries dilate, allowing greater perfusion of the newly inflated lungs. At the same time, the increase in blood pressure in the left atrium due to the increased pulmonary circulation closes the foramen ovale, the interatrial opening that, during fetal life, allows oxygenated blood from the placenta to enter the left side of the heart to be pumped out through the aorta. The closing of these two shunts with the onset of breathing at birth separates pulmonary from systemic circulation, establishing the postnatal blood flow pattern (Fig. 2). Failure of either of these shunts to close can cause respiratory and circulatory distress and result in neonatal lethality. Examination of the lungs, heart, and aortic arch arteries by gross and histological examination will reveal structural defects that can then be further explored during fetal life (see Chapter 11: Analysis of Mid- to Late-Gestation Phenotypes in Mice [Papaioannou and Behringer 2023f]) to determine their developmental origin (e.g., *Tbx1*; Jerome and Papaioannou 2001).

Micro computed tomography (microCT) can be used to reconstruct the complex three-dimensional structures of the heart and great vessels in fixed specimens (Hsu et al. 2016). The entire newborn is scanned, and the heart can then be computationally segmented from the body (digital dissection) and viewed whole or as planar sections in any orientation (Fig. 3C,D).

Developmental Delay

Developmental delay is a broad term indicating that a mouse is not fully mature when compared to age-matched controls. At birth, a pup with developmental delay may not have fully mature essential organ systems, such as the lungs, and would not be able to survive outside of the mother; it will likely be smaller than controls (Fig. 4; e.g., *Dph1*; Chen and Behringer 2004). Histological analysis of organs

Cite this chapter as *Cold Spring Harb Protoc*; doi:10.1101/pdb.over107975

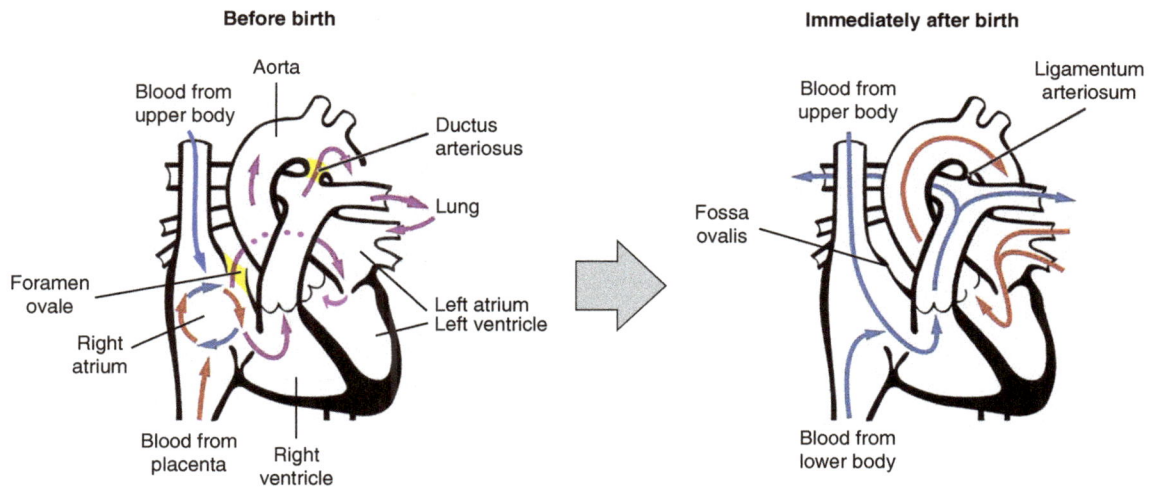

FIGURE 2. Diagram of cardiovascular changes at birth. During fetal life, oxygenated blood from the placenta (red) enters the heart through the right atrium where it mixes with deoxygenated blood (blue) from the body. Partially oxygenated blood (purple) is shunted through the interatrial septum via the foramen ovale to the left atrium. Blood is pumped into the ventricles through the right and left atrioventricular canals and is then pumped out through the aorta from the left ventricle and through the pulmonary artery from the right ventricle. However, because the lungs are not functional and resistance in the pulmonary arteries is high, most of the blood leaving the left ventricle through the pulmonary trunk is shunted through the ductus arteriosus into the aorta. At birth, with the shutdown of the umbilical circulation, the initiation of breathing, and the inflation of the lungs, the ductus arteriosus closes (becoming the ligamentum arteriosum), the pulmonary arteries dilate, and increased pressure in the left atrium closes the foramen ovale (forming the fossa ovalis), thus separating pulmonary from systemic circulation. (Adapted from Anatomy & Physiology. Provided by: OpenStax CNX. Located at: https://courses.lumenlearning.com/suny-ap2/chapter/adjustments-of-the-infant-at-birth-and-postnatal-stages/.)

should be performed as a simple screen to assess the maturation of the various tissues in comparison to controls. Determination of the timing of the developmental delay will require timed matings and a prenatal analysis (see Chapter 10: Phenotypic Analysis of Periimplantation to Mid-Gestation Lethality in Mice [Papaioannou and Behringer 2023g] and Chapter 11: Analysis of Mid- to Late-Gestation Phenotypes in Mice [Papaioannou and Behringer 2023f]).

Cranial Nerve Defects

Another reason why a newborn mouse may not survive the perinatal period is a defect in cranial nerve development (e.g., *Hoxa1*; Lufkin et al. 1991). The cranial nerves comprise 12 pairs of nerves that derive from the neural crest and are essential for sensory and motor functions (Table 1). The simplest way to visualize the initial stages of cranial nerve development for comparison of homozygous mutants and wild-type control littermates is to perform a whole-mount immunostain (see Chapter 17 in Behringer et al. 2014) at E10.5 using the 2H3 antineurofilament monoclonal antibody available from the Developmental Studies Hybridoma Bank (https://dshb.biology.uiowa.edu/) (Fig. 5).

Skeletal Defects

Abnormalities in skeleton development can also cause perinatal lethality (e.g., *Sp7*; Nakashima et al. 2002). For example, if the rib cage is too small or split, the mouse will have problems breathing (Fig. 6). It is probably a good idea to generate a skeleton preparation (see Ovchinnikov 2009) of any perinatal lethal mutant. The skeleton is a robust tissue. Thus, you can generate a skeleton preparation even for pups found dead in the cage. The skin or viscera can be used for genotyping. Compare the skeletons of the mutants with controls, paying particular attention to the size and structure of the rib cage and any other areas in which you know your gene is expressed.

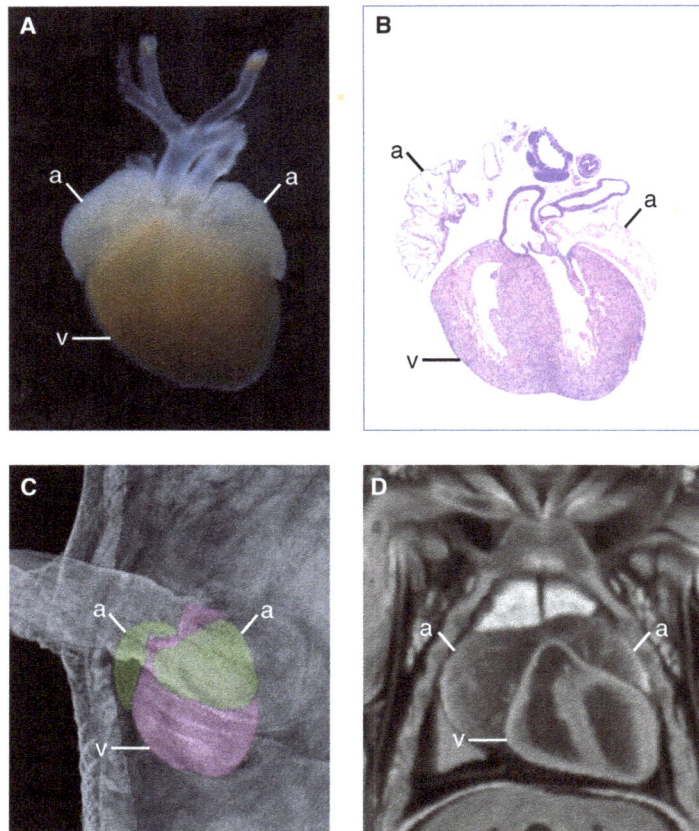

FIGURE 3. Volumetric imaging using micro computed tomography (microCT). (*A*) Freshly dissected E18.5 wild-type heart with great vessels. (*B*) Hematoxylin and eosin (H&E) stained frontal section of a wild-type heart for comparison showing atrial and ventricular chambers. (*C*) Surface rendering of wild-type newborn heart imaged by microCT in the intact newborn and digitally dissected (segmented) from the entire body. (*D*) Frontal digital section from microCT imaging of newborn wild-type heart showing atrial and ventricular chambers. (a) Atrium, (v) ventricle. (*A,B*, Courtesy of Shannon Erhardt and Jun Wang; *C,D*, courtesy of Logan Hsu and Mary Dickinson.)

Cleft Palate

Cleft palate is a relatively common phenotype of mutant mice that can cause death soon after birth (e.g., *Msx1*; (Satokata and Maas 1994). The mice may be born alive and breathe air, but much of the air ends up in the stomach, causing a bloated belly. To determine if the mutant neonate has cleft palate, it is best to remove the head from the body, remove the jaw, and view the upper part of the head from the ventral side to see the palate. If there is a cleft of the soft palate, it will be morphologically obvious (Fig. 7A). If you identify cleft palate, the next step is to generate a skeleton preparation (see Ovchinnikov 2009) to view the skull bones (Fig. 7B,C). The maxillary and palatine shelves form the hard palate of the skull. An easy way to observe these bones in a skeleton preparation is to remove the skull from the body and remove the lower jaw from the skull. You can then acquire images of the ventral side of the skull to examine the palate. The palate forms around E14.5 and E15.5 when the bilateral palatine shelves fuse along the midline. Therefore, analyze the mutants by histology between E11.5, as palatal shelf formation initiates, and E15.5, when they fuse. Frontal sections of the head provide the best view of the forming palatal shelves (Fig. 7D,E). A three-dimensional view of the forming palate can be obtained with scanning electron microscopy (SEM) or microCT imaging. Depending on what you observe in your histological and/or imaging analysis and on the nature and expression pattern of the gene, you may want to consider measuring cell proliferation and apoptosis during palate formation as well as analyzing relevant differentiation markers by in situ hybridization or immunostaining.

Cite this chapter as *Cold Spring Harb Protoc*; doi:10.1101/pdb.over107975

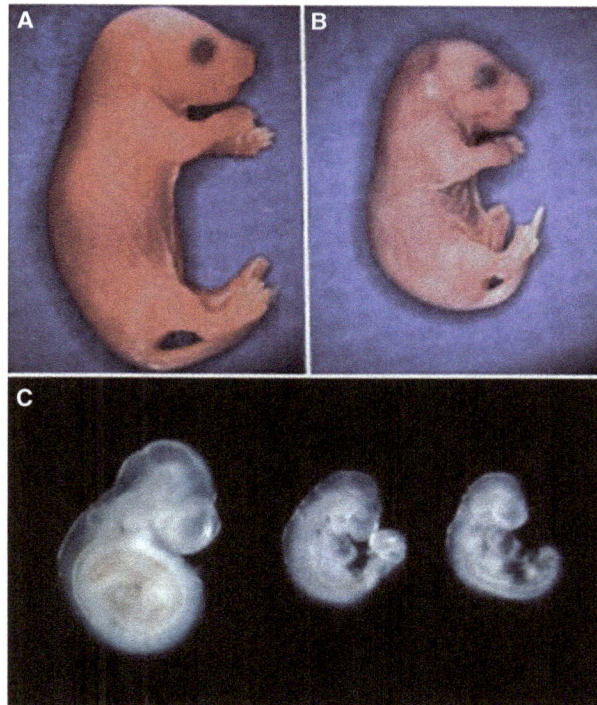

FIGURE 4. Developmental delay. (*A*) Littermate control and (*B*) *Dph1*-null neonates, showing developmental delay at birth. (*C*) E9.5 *Dph1*-null embryos (*middle* and *right*) with control littermate (*left*). (Reprinted, with permission, from Chen and Behringer 2004.)

Diaphragm Defects

Another cause of breathing problems is a diaphragm defect (e.g., *Myog*; Hasty et al. 1993). A hernia of the diaphragm can be detected grossly on careful dissection. You may find organs of the peritoneal cavity pushed up through the hernia into the chest cavity. Alternatively, the diaphragm may be too thin to support breathing. This is best assessed by histology in comparison to controls. MicroCT imaging could provide a 3D view of the diaphragm.

Other Causes of Perinatal Lethality

A defect in the breathing centers of the brain could also lead to perinatal lethality (e.g., *Bdnf*; Erickson et al. 1996). This is difficult to determine, but one could examine the neural circuitry

TABLE 1. Cranial nerves

Number	Name	Modality	Function
I	Olfactory	Sensory	Smell
II	Optic	Sensory	Vision
III	Oculomotor	Motor	Eye movement, pupil dilation
IV	Trochlear	Motor	Eye movement
V	Trigeminal	Motor and sensory	Head and face touch, pain; chewing
VI	Abducens	Motor	Eye movement
VII	Facial	Motor and sensory	Taste, ear touch, facial muscle movement
VIII	Vestibulocochlear	Sensory	Hearing, balance
IX	Glossopharyngeal	Motor and sensory	Taste; tongue, tonsil, pharynx touch; swallowing
X	Vagus	Motor and sensory	Glands, digestion, heart rate
XI	Spinal accessory	Motor	Head movement
XII	Hypoglossal	Motor	Tongue movement

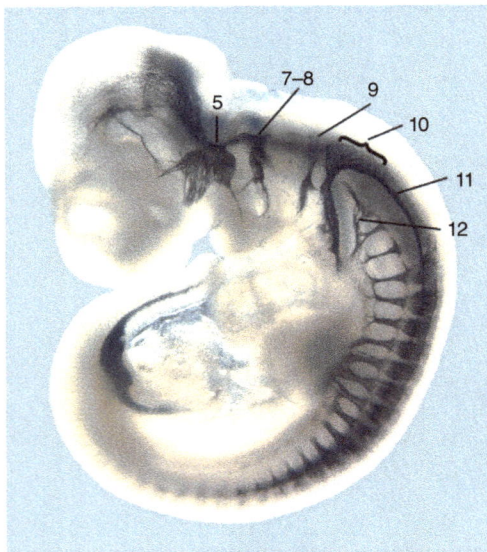

FIGURE 5. Whole-mount immunostaining for cranial nerves and spinal ganglia. Wild-type, E10.5 embryo immunostained using an antineurofilament antibody. Numbers indicate cranial nerves. (Image courtesy of Deborah L. Guris and Akira Imamoto.)

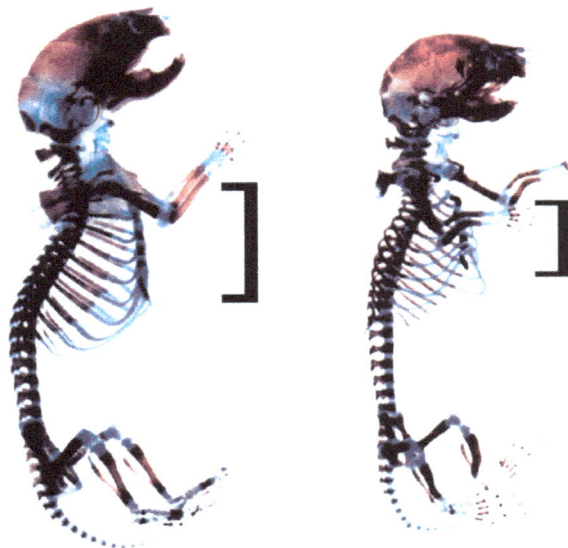

FIGURE 6. Skeleton preparation to visualize cartilage and bone. Wild-type (*left*) and *Sox9*$^{+/-}$ (*right*) newborn mice. Brackets highlight the difference in size of the rib cages between the mutant and control. (Image courtesy of Benoit de Crombrugghe.)

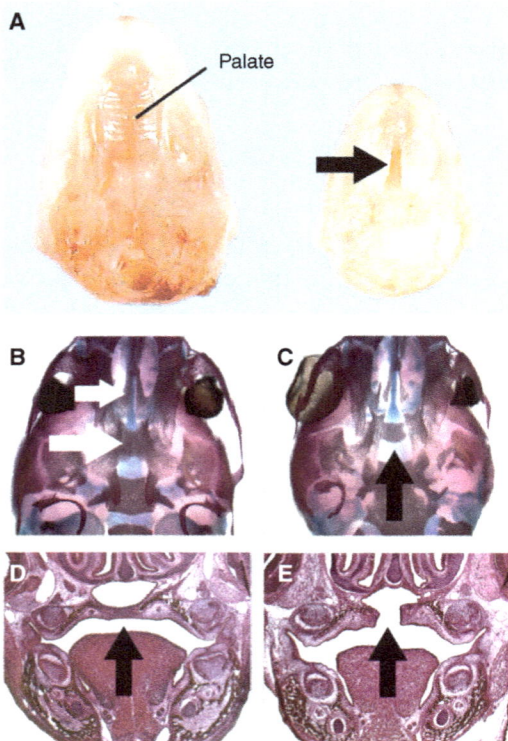

FIGURE 7. Morphological and histological analysis of cleft palate. (*A*) Ventral view of the upper portions of the heads of wild-type (*left*) and *Dph1* homozygous mutant (*right*) newborn mice, showing the roof of the mouth. The palate region of the control is intact, whereas the mutant has cleft palate (arrow). Note that in this case, the age-matched mutant is also smaller because of developmental delay. (*B, C*) Skeleton preparations showing ventral views of the skulls of wild-type (*B*) and *Tbx1* homozygous mutant (*C*) newborn mice. The jaws have been removed to more easily view the bones of the palate. (Large white arrow) Palatine shelf, (small white arrow) maxillary shelf, (black arrow) absence of bone regions. (*D,E*) Histology of palate formation in wild-type (*D*) and *Tbx1* homozygous mutant (*E*) E16.5 embryos. Frontal sections show the presence and absence, respectively of palate closure (arrows). (*A*, Image provided by Chun-Ming Chen and Richard Behringer; *B,C*, Adapted, with permission, from Jerome and Papaioannou 2001, ©Nature Publishing Group; *D,E* courtesy of Robert Kelly, Loydie Jerome-Majewska, and Virginia Papaioannou.)

Cite this chapter as *Cold Spring Harb Protoc*; doi:10.1101/pdb.over107975

that regulates breathing by histology or marker gene analysis, particularly in areas of expression of the mutated gene. This would at least provide correlative evidence for a central nervous system (CNS) defect.

Additionally, other defects unrelated to breathing may be to blame for perinatal death. Anemia could lead to perinatal death. Anemic pups will appear very pale in comparison to controls. An analysis of the blood (red blood cell counts, hemoglobin concentration, blood smears, etc.) would be a reasonable line of investigation. A bleeding disorder could also lead to perinatal lethality (e.g.; *F10*; Dewerchin et al. 2000). Blood in the gut or abdomen, or blood spots on the skin, would be indications of a bleeding disorder. In these cases, measure platelet counts and analyze blood clotting factors.

REFERENCES

Behringer RR, Gertsenstein M, Nagy KV, Nagy A. 2014. *Manipulating the mouse embryo: a laboratory manual.* Cold Spring Harbor Laboratory Press, Cold Spring Harbor, NY.

Chen CM, Behringer R. 2004. *Ovca1* regulates cell proliferation, embryonic development, and tumorigenesis. *Genes Dev* 18: 320–332. doi:10.1101/gad.1162204

Developmental Studies Hybridoma Bank. https://dshb.biology.uiowa.edu/ [Accessed December 15, 2022].

Dewerchin M, Liang Z, Moons L, Carmeliet P, Castellino FJ, Collen D, Rosen ED. 2000. Blood coagulation factor X deficiency causes partial embryonic lethality and fatal neonatal bleeding in mice. *Thromb Haemost* 83: 185–190. doi:10.1055/s-0037-1613783

Erickson JT, Conover JC, Borday V, Champagnat J, Barbacid M, Yancopoulos G, Katz DM. 1996. Mice lacking brain-derived neurotrophic factor exhibit visceral sensory neuron losses distinct from mice lacking NT4 and display a severe developmental deficit in control of breathing. *J Neurosci* 16: 5361–5371. doi:10.1523/JNEUROSCI.16-17-05361.1996

Hasty P, Bradley A, Morris JH, Edmondson DG, Venuti JM, Olson EN, Klein WH. 1993. Muscle deficiency and neonatal death in mice with a targeted mutation in the myogenin gene. *Nature* 364: 501–506. doi:10.1038/364501a0

Hsu CW, Wong L, Rasmussen TL, Kalaga S, McElwee ML, Keith LC, Bohat R, Seavitt JR, Beaudet AL, Dickinson ME. 2016. Three-dimensional microCT imaging of mouse development from early post-implantation to early postnatal stages. *Dev Biol* 419: 229–236. doi:10.1016/j.ydbio.2016.09.011

Jerome LA, Papaioannou VE. 2001. DiGeorge syndrome phenotype in mice mutant for the T-box gene, *Tbx1*. *Nat Genet* 27: 286–291. doi:10.1038/85845

Lufkin T, Dierich A, LeMeur M, Mark M, Chambon P. 1991. Disruption of the *Hox-1.6* homeobox gene results in defects in a region corresponding to its rostral domain of expression. *Cell* 66: 1105–1119. doi:10.1016/0092-8674(91)90034-V

Nagy A, Gertsenstein M, Vintersten K, Behringer R. 2006. Caesarean section and fostering. *Cold Spring Harb Protoc* doi:10.1101/pdb.prot4381

Nakashima K, Zhou X, Kunkel G, Zhang Z, Deng JM, Behringer RR, de Crombrugghe B. 2002. The novel zinc finger-containing transcription factor Osterix is required for osteoblast differentiation and bone formation. *Cell* 108: 17–29. doi:10.1016/S0092-8674(01)00622-5

Openstax. 2022. https://courses.lumenlearning.com/suny-ap2/chapter/adjustments-of-the-infant-at-birth-and-postnatal-stages [Accessed December 15, 2022].

Ovchinnikov D. 2009. Alcian blue/alizarin red staining of cartilage and bone in mouse. *Cold Spring Harb Protoc* doi:10.1101/pdb.prot5170

Papaioannou VE, Behringer RR. 2023a. Analysis of postnatal mutant phenotypes in mice. *Cold Spring Harb Protoc* doi:10.1011/pdb.over107976

Papaioannou VE, Behringer RR. 2023b. The "no phenotype" challenge in analyzing mutant mice. *Cold Spring Harb Protoc* doi:10.1011/pdb.over107977

Papaioannou VE, Behringer RR. 2023c. Recovering a targeted mutation from embryonic stem cell chimeras or CRISPR–Cas founders. *Cold Spring Harb Protoc* doi:10.1011/pdb.over107959

Papaioannou VE, Behringer RR. 2023d. Cross-fostering of newborn mice from embryonic stem cell chimeras or CRISPR–Cas founders. *Cold Spring Harb Protoc* doi:10.1011/pdb.prot107987

Papaioannou VE, Behringer RR. 2023e. Phenotypic analysis: assessing timing of recessive prenatal lethality in mice. *Cold Spring Harb Protoc* doi:10.1011/pdb.over107970

Papaioannou VE, Behringer RR. 2023f. Analysis of mid- to late-gestation phenotypes in mice. *Cold Spring Harb Protoc* doi:10.1011/pdb.over107973

Papaioannou VE, Behringer RR. 2023g. Phenotypic analysis of periimplantation to mid-gestation lethality in mice. *Cold Spring Harb Protoc* doi:10.1011/pdb.over107976

Peters B, Kirfel J, Büssow H, Vidal M, Magin TM. 2001. Complete cytolysis and neonatal lethality in keratin 5 knockout mice reveal its fundamental role in skin integrity and in epidermolysis bullosa simplex. *Mol Biol Cell* 12: 1775–1789. doi:10.1091/mbc.12.6.1775

Satokata I, Maas R. 1994. *Msx1* deficient mice exhibit cleft palate and abnormalities of craniofacial and tooth development. *Nat Genet* 6: 348–356. doi:10.1038/ng0494-348

Shawlot W, Behringer RR. 1995. Requirement for Lim1 in head-organizer function. *Nature* 374: 425–430. doi:10.1038/374425a0

CHAPTER 14

Analysis of Postnatal Mutant Phenotypes in Mice

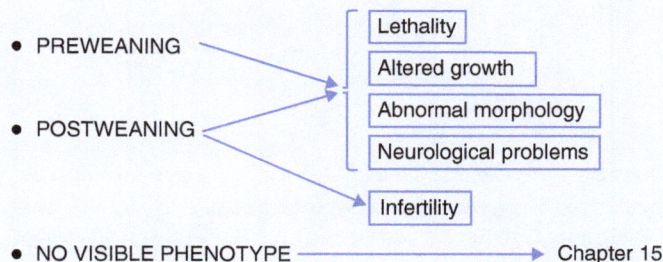

Viable homozygous mutant newborn mice may show effects of a mutation at any time during their development by exhibiting abnormal structure, function, or lethality. This chapter guides the analysis of postnatal mice through gross anatomical assessment and the detection of visible phenotypes prior to weaning such as altered growth patterns, neurological problems, or abnormalities in movement or coordination. Advice on marking pups for identification purposes and providing adequate nutrition in the event of eating problems is given. After weaning and at the onset of puberty, different phenotypes may become manifest, including compromised growth and vigor and reproductive problems in males and/or females. Assessing infertility in each sex is addressed.

INTRODUCTION

If, in the analysis of a spontaneous or induced mutation, you have ruled out prenatal and perinatal lethality of homozygous mutants (see Chapter 9: Phenotypic Analysis of Preimplantation Lethality in Mice [Papaioannou and Behringer 2023a]; Chapter 10: Phenotypic Analysis of Periimplantation to Mid-Gestation Lethality in Mice [Papaioannou and Behringer 2023b]; Chapter 11: Analysis of Mid- to Late-Gestation Phenotypes in Mice [Papaioannou and Behringer 2023c]; Chapter 13: Phenotypic Analysis of Perinatal Lethality in Mice [Papaioannou and Behringer 2023d]), you will be presented with live homozygous mutant pups from heterozygous matings on the day of birth. The next step is to observe their growth and

Opening artwork: Courtesy of Andy Salinger and Monica Justice.

© 2024 Cold Spring Harbor Laboratory Press
Cite this chapter as *Cold Spring Harb Protoc;* doi:10.1101/pdb.over107976

BOX 1. MARKING NEWBORN PUPS

You may wish to genotype newborn pups from a heterozygous intercross so that you can follow the development of a mutant phenotype starting at birth. It is easy to cut a small piece of tail from the newborn pups for polymerase chain reaction (PCR) genotyping, but how do you mark the pups to match them with their genotypes? One crude but effective method is to use a permanent marking pen to number each pup, usually by placing marks on the tail, but you will have to check periodically to touch up the marks if they begin to fade. Once the pups are old enough, use a traditional marking system (toe clip or ear punch) to keep track of the mice. Another strategy is to clip the toes to mark the neonates. However, the toes of mouse pups have not yet separated at birth, making toe clipping difficult. It is possible to use fine scissors to cut the most distal part of the toe to create an indelible mark. The small drops of blood from the toe and tail cuts do not usually seem to bother the mothers.

Before performing any of these procedures, check with your local animal care and use committee (IACUC) for appropriate approvals.

development to weaning and on into adulthood to determine when, where, or if abnormalities arise. The gene expression pattern will be a guide, but a comprehensive analysis will turn up any unexpected phenotypes to provide you with the full range of the phenotypic consequences of the mutation. In a surprising number of mutational studies, the homozygous mutant mice appear at first glance to be normal—the dreaded "no phenotype" mutation. If this is the case, after completing the assessments in this chapter, see Chapter 15: The "No Phenotype" Challenge in Analyzing Mutant Mice (Papaioannou and Behringer 2023e), which will guide you through a thorough analysis to reveal even the most cryptic phenotypes.

Throughout the analysis you will need a way of identifying individual mice by genotype to follow their fates, preferably from birth onward. If the pups survive 7–10 d, you will be able to take a tail sample for genotyping and mark them by distinguishing toe clips (if approved by your local animal care committee). Likewise, if they survive to weaning, you can number them by ear punching (see Chapter 3 in Behringer et al. 2014). These are indelible marks that allow you to identify each animal throughout its life. However, if the homozygous mutants survive birth but die before 7 d and you would like to identify the homozygous mutants, retrospectively, from the day of birth, you will need an alternative marking system for newborns (Box 1).

GROSS ANATOMICAL ASSESSMENT

One of the first steps to take with homozygous mutants is to examine them for gross abnormalities, focusing on known areas of gene expression from your studies or published information (e.g., Gene Expression Database, http://www.informatics.jax.org/expression.shtml). This examination can be performed at any time between birth and adulthood, depending on the characteristics of your gene and the effects of the mutation (e.g., when it causes lethality, if it is lethal). Examine the external features of both males and females using the checklist provided in Figure 1 as a guide. Next, sacrifice the homozygous mutants, as well as appropriate controls, both males and females, and perform an internal examination of the visceral organs. Use the checklist provided in Figure 2 as a guide. The male and female reproductive organs are shown in Figure 3. A guide to the distribution of the five pairs of mammary glands is shown in Figure 4. After performing this analysis, choose the appropriate section below for further analysis of visible or fertility abnormalities that appear before or after weaning or sexual maturity.

VISIBLE MUTANT PHENOTYPES BEFORE WEANING

Weaning is the transition a pup makes from feeding on milk from its mother to eating solid food on its own. Normally, this will occur when you remove the litter from the mother, but even if the offspring

☐ Body size: males bigger than females of same age

☐ Teeth: normal occlusion, incisors, and molars present
☐ Whiskers

☐ Eyes normal size and open by 14 days

☐ Head shape

☐ Ears

☐ Toes on forelimbs and hindlimbs (five on each); toenails on the dorsal side of each digit

☐ Hair expected color and normal texture

☐ Genitalia of male or female

☐ Tail length and straightness

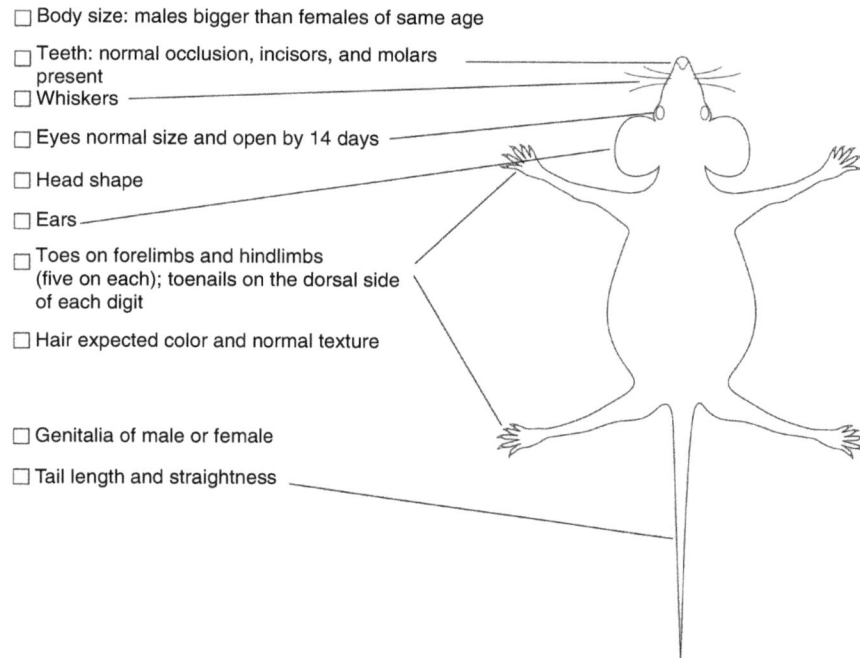

FIGURE 1. External examination checklist for postnatal mice. Check the indicated external organs and tissues to determine if they are present and grossly normal relative to wild-type age- and sex-matched littermates.

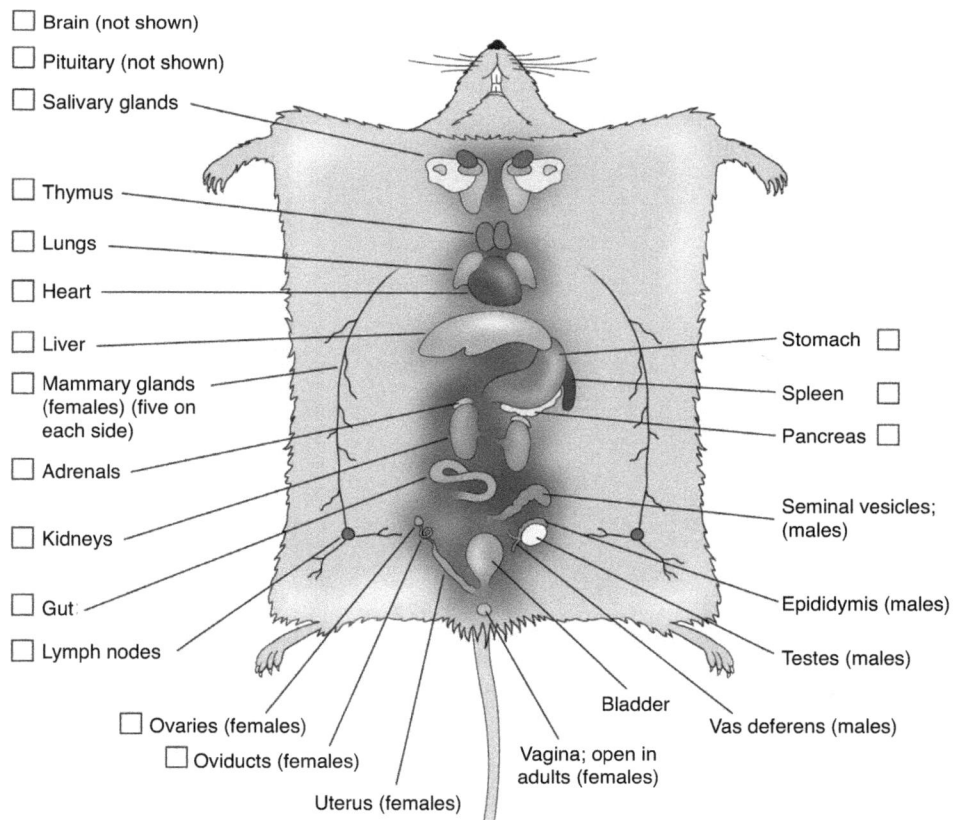

☐ Brain (not shown)

☐ Pituitary (not shown)

☐ Salivary glands

☐ Thymus

☐ Lungs

☐ Heart

☐ Liver

☐ Mammary glands (females) (five on each side)

☐ Adrenals

☐ Kidneys

☐ Gut

☐ Lymph nodes

☐ Ovaries (females)

☐ Oviducts (females)

Uterus (females)

Vagina; open in adults (females)

Bladder

Stomach ☐

Spleen ☐

Pancreas ☐

Seminal vesicles; (males)

Epididymis (males)

Testes (males)

Vas deferens (males)

FIGURE 2. Internal examination checklist for postnatal mice. Check the indicated internal organs and tissues to determine if they are present and grossly normal relative to wild-type age- and sex-matched littermates. (Adapted, with permission, from Green and Roscoe 1975, © The Jackson Laboratory.)

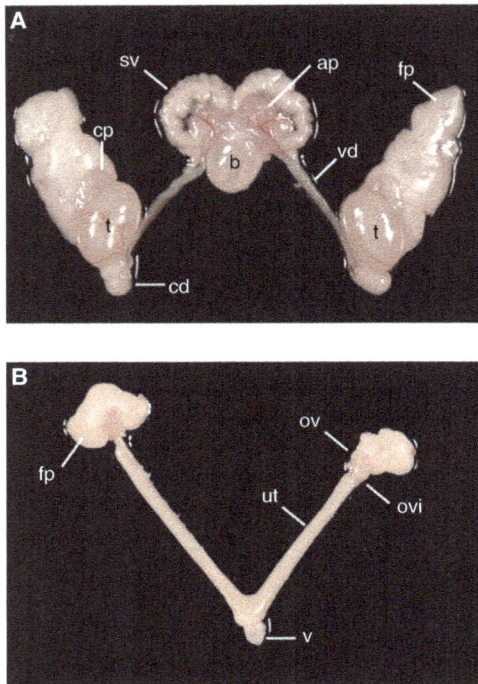

FIGURE 3. Male and female reproductive organs. Male (*A*) and female (*B*) reproductive tracts. (ap) Anterior prostate, (b) bladder, (cd) cauda epididymis, (cp) caput epididymis, (fp) fat pad, (ov) ovary, (ovi) oviduct, (sv) seminal vesicle, (t) testis, (ut) uterine horn, (v) vagina, (vd) vas deferens. (Reprinted from Jamin et al. 2002, ©Nature Publishing Group.)

are left with the mother, they will begin to depend more and more on solid food starting at ~3 wk postnatally. Thus, weaning is an important milestone in mouse development. Many requirements of life before weaning can be provided by maternal care and feeding. Once the pups are weaned, they are completely on their own and are thus challenged to survive. Accordingly, we have split the analysis of postnatal visible mutant phenotypes into before weaning and after weaning. Many phenotypes, such as morphological abnormalities, are evident throughout these periods, but some only appear after

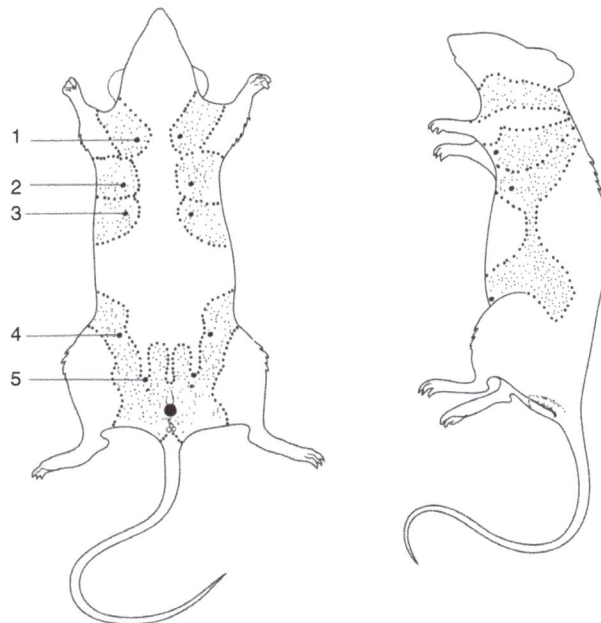

FIGURE 4. Mammary glands in the female mouse. Ventral (*left*) and lateral (*right*) views showing the positions of the five pairs of nipples and mammary glands located subcutaneously, attached to the dermal side of the skin. (Adapted, with permission, from Green and Roscoe 1975, © The Jackson Laboratory.)

 Cite this chapter as *Cold Spring Harb Protoc*; doi:10.1101/pdb.over107976

weaning, at the time of sexual maturity, or during old age. Phenotypes that appear in relative old age are considered in Chapter 15: The "No Phenotype" Challenge in Analyzing Mutant Mice (Papaioannou and Behringer 2023e).

Preweaning Lethality

Lethality will usually be presaged by the appearance of a sickly pup or one that is not thriving like its littermates. However, because you cannot watch your mice continuously, you may also simply find them dead in the cage. There is always a temptation to let a sickly pup continue to live, but more often than not it will die when you are not present and either be cannibalized or degrade, precluding some types of analysis. The best course of action is to sacrifice sickly pups for genotyping and phenotypic analysis as soon as they are detected.

One simple reason that a mouse might die, even if it initially appears active and vigorous, is that it does not or cannot feed properly. Complete lack of feeding would cause death very quickly, not only because of a lack of nourishment but also because of the lack of liquid normally obtained from the milk. You can easily determine if a pup is feeding during the first week postnatally by the presence or absence of milk in the stomach. A mouse that has been nursing will have its stomach full of milk that is white and this can be seen on the left side of the abdomen through the translucent skin (Fig. 5; see also Figure 1 in Chapter 13: Phenotypic Analysis of Perinatal Lethality in Mice [Papaioannou and Behringer 2023d]). By the way, if you find that the stomach is on the right side of the abdomen rather than the left, then you have identified a defect in left–right patterning. Other reasons for failure to feed and thus no milk in the stomach could be certain types of cleft palate or cranial nerve defects (see Chapter 13: Phenotypic Analysis of Perinatal Lethality in Mice [Papaioannou and Behringer 2023d]) or the inability to emit sounds. Mouse pups emit ultrasonic vocalizations that elicit maternal behavior such as retrieval of scattered pups to the nest. In the absence of vocalizations, the homozygous mutant pups are more likely to be left out of the nest and fail to nurse. Incidentally, if all of the pups, mutant and wild type, are scattered outside the nest, it could indicate a dominant hearing or olfaction problem on the part of the heterozygous mother because both of these senses are important for maternal recognition of the young.

Sudden death during the preweaning (or even postweaning) period that is not accompanied by eating problems could be attributable to defects in the conduction system of the heart that result in spontaneous ventricular arrhythmia or tachycardia. Conduction system defects may or may not be accompanied by structural defects, so functional tests such as an electrocardiogram (EKG) or a test that assesses heart rate may be called for.

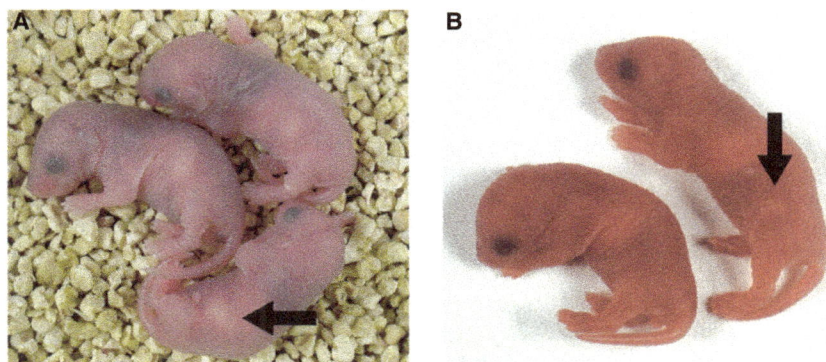

FIGURE 5. Detection of milk in the stomach of newborn mice and detection of the pancreas. (*A*) Three neonates are shown. A white milk spot (arrow) is seen on the left side of their abdomens through their translucent skin. (*B*) Wild-type (*right*) and elastase-diphtheria toxin transgenic (*left*) mice. These mice have not yet fed. In the wild-type mouse, a triangular white patch (arrow) is seen through the skin on the left side of the pup. This is the pancreas, not milk in the stomach. Elastase-directed expression of the toxin in the pancreas causes ablation and hence no white patch in the transgenic pup (Palmiter et al. 1987).

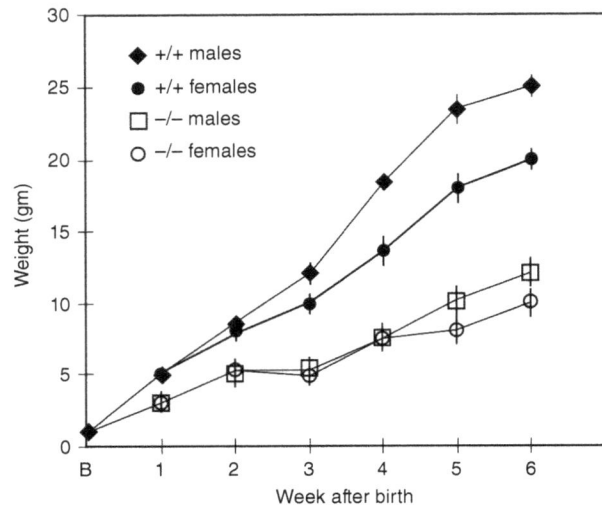

FIGURE 6. Growth curves of wild-type and growth-deficient mice. Pups generated from *Dmbx1* heterozygous intercrosses were marked at birth and weighed at 1-wk intervals. The weights of wild-type males and females diverge between 1 and 2 wk after birth. This sexual dimorphism is maintained as the mice age. The rate of growth of *Dmbx1* homozygous mutant males and females is always less than that of the controls. (Image courtesy of Akihira Ohtoshi and Richard Behringer.)

Altered Growth Pattern

One prevalent phenotype in mutant mice is a growth defect. Affected mice appear otherwise healthy but are smaller than controls. You can document a growth defect by generating a growth curve for the mutants and controls. Male and female mice are sexually dimorphic with respect to body size. Therefore, you will need to segregate your data by sex. Generate a sufficient number (>5) of each genotype and sex to make a statistically significant judgment on any growth differences. Weigh each mouse at regular intervals (e.g., daily, every other day, or weekly) and generate a growth curve. Be sure to include error bars for each genotype at each time point to indicate the observed variation. An example of growth curves from wild-type and mutant mice with a growth defect (e.g., *Dmbx1*; Ohtoshi and Behringer 2004) is shown in Figure 6. From birth to weaning, mice gain weight rapidly. There is usually a small decrease in the rate of growth of normal mice around weaning as the pups make the transition from milk to solid food. At ~6 wk of age, the rate of growth slows and eventually plateaus. A mutant with a growth deficiency will have a growth curve that diverges significantly from wild type.

A fundamental question to address is whether the growth deficiency manifests itself before or after birth. Accordingly, determine the weight of the mutants relative to controls obtained from a cesarean section at E18.5 (see Nagy et al. 2006). A cesarean section will bypass variation that might occur on the day of birth if the mice have fed and have varying amounts of milk in their stomachs. If you find that the mutants and controls are of similar weight at E18.5, then the growth deficiency developed postnatally. The neural regulation and pituitary control of somatic growth would be one avenue of subsequent investigation. Alternatively, a growth deficiency could be caused by a skeletal defect, metabolic defects, hyperactivity, or dental problems that interfere with feeding (e.g., *Fos*; Johnson et al. 1992).

Morphological Abnormalities

If your examination of the external and internal features of the homozygous mutants (Figs. 1 and 2) has identified an overt defect in a tissue or organ system, the tissue or organ may be absent, hypoplastic, or abnormal in morphology (Fig. 7; e.g., *En2* [Joyner et al. 1991]; *Nude (Foxn1)* [Militzer 2001]). You may have already anticipated such a phenotype because of the expression pattern of the gene mutated. First, document the gross abnormality in the intact animal, either awake or anesthetized, if the morphological abnormality is externally visible. If the defect is internal, sacrifice the mouse and perform a dissection to reveal the affected tissue or organ and acquire images for documentation. It is always important to acquire images with a size marker (e.g., a ruler). If the mutant organ is

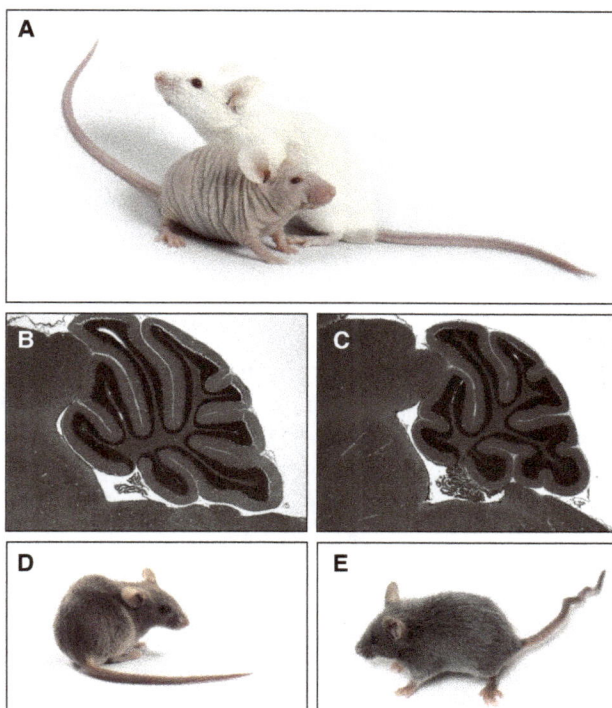

FIGURE 7. External and internal mutant postnatal phenotypes. (*A*) Nude (*Foxn1/Foxn1*) (*left*) mice lack fur and have thymus differentiation defects that result in reduced numbers of T cells. Control mouse (*right*). (*B,C*) Engrailed2 mutant mice (*En2/En2*) are viable with hypoplastic cerebellums. Hematoxylin and eosin (H&E)-stained midsagittal histological section of cerebellum of control (*B*) and homozygous mutant (*C*). (*D,E*) Tail-kink mutant mice (*Tk/Tk*) have numerous bends in their tails. Control (*D*), *Tk/Tk* (*E*). (*A,* Reprinted, with permission, from © The Jackson Laboratory; *B,C,* images courtesy of Alexandra Joyner; *D,* image courtesy of Charles River Laboratories; *E,* image courtesy of Mouse Genome Informatics: http://www.informatics.jax.org)

present, dissect it out and acquire images of the isolated organ. Remember to acquire images of controls also. The next step will depend on the type of tissue or organ that is affected. If you find a skeletal defect, then a skeleton preparation would be informative (see Chapter 17 in Behringer et al. 2014; see Nagy et al. 2009; Ovchinnikov 2009).

Generally, though, the next step would be to perform a histological analysis of the affected tissue or organ to understand which cell types are altered. Subsequently, gene expression studies comparing mutants and controls, including in situ hybridization and immunohistochemistry/immunofluorescence of molecular markers specific to the affected tissue organ, should provide insights into the molecular defects associated with the abnormal morphology. Cell proliferation or death assays will provide further insights into tissue defects. Finally, you will need to understand the development of the morphological abnormality and perform a prenatal analysis to determine the primary defect and its time of onset (see Chapter 10: Phenotypic Analysis of Periimplantation to Mid-Gestation Lethality in Mice [Papaioannou and Behringer 2023b]; Chapter 11: Analysis of Mid- to Late-Gestation Phenotypes in Mice [Papaioannou and Behringer 2023c]).

Neurological Problems

The Mouse Genome Informatics (MGI) Phenotype Ontology Browser (http://www.informatics.jax.org/vocab/mp_ontology) provides a useful classification of standard terms for annotating phenotypic data. Some of the terms associated with behavior/neurological phenotypes, many of which have additional subcategories, are the following:

- Abnormal circadian behavior
- Abnormal consumption behavior
- Abnormal emotion/affect behavior
- Abnormal grooming behavior

- Abnormal motor capabilities/coordination/movement
- Abnormal olfactory behavior
- Abnormal sensory capabilities/reflexes/nociception
- Abnormal social/conspecific interaction
- Seizures

These classifications provide standardized phenotype categories to consider as you assess your mutants for potential neurological disorders. Some of these phenotypes are more obvious than others to detect. Eating-behavior defects may become immediately obvious if such homozygous mutant pups are smaller than their littermate controls. However, overeating may not be immediately obvious because it may take longer to accumulate weight. Learning and memory alterations as well as sleep pattern/circadian rhythm defects are not obvious and require sophisticated tests using specialized equipment for assessment (e.g., *Fos*; Paylor et al. 1994; Honrado et al. 1996). If you suspect these types of phenotypes, then it may be wise to enlist a collaborator with the appropriate expertise. Abnormal social behavior, especially aggressive behavior, may become apparent because you will notice that heterozygous and wild-type mice in cages with homozygous mutant mice develop injuries (e.g., scars on their backs). Seizures, indicative of central nervous system (CNS) defects, will also be obvious, but only if you are present when they occur.

The SHIRPA (SmithKline Beecham, Harwell, Imperial College, Royal London Hospital, Phenotype Assessment) protocol (Rogers et al. 1997) includes a comprehensive quantitative assessment of behavioral aspects of mouse phenotypes and provides a useful framework for the analysis of mutants, particularly models of neurological disease. A primary screen uses standard methods to create a behavioral and functional profile by observational assessment, a secondary screen uses a battery of tests for a comprehensive behavioral assessment and pathological analysis, and a tertiary screen uses more specialized tests of specific neural pathways. Here, we discuss the analysis of a subset of neurological phenotypes. Abnormalities in motor capabilities, coordination, or movement are among the overt neurological phenotypes (e.g., *Mpz*; Giese et al. 1992). Thus, if you have a mutant phenotype of this class, it will likely be very obvious. Abnormalities in sensory capabilities, reflexes, or nociception are not overt phenotypes but can be assessed using a set of simple methods that are covered in Chapter 15: The "No Phenotype" Challenge in Analyzing Mutant Mice (Papaioannou and Behringer 2023e).

Abnormalities in Motor Capabilities/Coordination/Movement

There are many causes of an abnormal locomotor phenotype. The first thing to do is to document when the mutant behavior is first detected and whether it becomes progressively stronger or milder as the mice age. This will give you an indication of when the cellular, tissue, or organ defect first becomes detectable and the time frame of any changes. At birth, mouse pups can roll sideways. At 3–4 d after birth, they can pivot. By ~6–7 d, they can walk, and at 9 d, they can run unsteadily. By 15 d postpartum, the pups can move like adults (Wimer and Fuller 1966). Table 1 lists some of the phenotype ontology terms of abnormal locomotor activity. These are precise standardized definitions of locomotor behavioral defects. Document a neurological phenotype on video and remember to also document age- and sex-matched wild-type controls for comparison. When taking video recordings, set the stage so that the viewing area does not have anything in it that will distract the viewer from watching the behavior of the mouse. For example, have a platform or cage with only the test mouse. If you use a platform, cover it with a sheet of white filter paper. If you use a cage, use one with fresh bedding. You will quickly realize that mice urinate and defecate as a stress response to a new environment. Simply wait a few minutes until they have nothing else to produce, clean up the platform or cage, and then begin imaging. Use a white or light-colored backdrop if you are shooting the video from a lateral view, unless the mouse is albino, in which case a darker background might be better. Document sufficient

Cite this chapter as *Cold Spring Harb Protoc*; doi:10.1101/pdb.over107976

TABLE 1. Some abnormal motor capabilities/coordination/movement phenotype subcategories from the Mammalian Phenotype Browser[a] with standard definitions

Akinesia	Absence of movement or loss of the ability to move, such as temporary or prolonged paralysis or "freezing in place"
Ataxia	Inability to coordinate voluntary muscular movements
Bradykinesia	Decreased spontaneity and movement
Dystonia	State of abnormal tonicity (hypertonicity or hypotonicity) in any tissue resulting in impairment of voluntary movement
Hyperactivity	General restlessness or excessive movement; more frequent movement from one place to another
Hypoactivity	Reduced movement from one place to another
Impaired coordination	Reduced ability to execute integrated movements of muscle
Increased (or decreased) vertical activity	Greater (or less) than average time spent jumping or rearing
Jumpy	Marked by fitful, jerky movements
Lethargy	Mild impairment of consciousness resulting in reduced alertness and awareness
Negative geotaxis	Mice, when placed on a downward slanting grid, walk down without turning around, whereas wild-type mice will always turn around and walk upward
No spontaneous movement	Failure to make any change in position or posture
Paralysis	Loss of power of voluntary movement in a muscle through injury or disease of it or its nerve supply
Peculiar gait	Unusual or distinctive way of walking
Retropulsion	When placed in a new environment, mice will walk backward and then may walk forward, whereas wild-type mice will immediately walk forward, or freeze momentarily and then walk forward

[a]http://www.informatics.jax.org/vocab/mp_ontology/.

numbers of homozygous mutants and controls to come up with a consistent assessment of the mutant phenotype.

Myelination, the formation of myelin on axons, insulates axons for fine motor control and also functions to speed nerve conduction. Thus, defects in myelin can lead to abnormalities in motor control and nerve conduction. In the mouse, myelination of the neurons that regulate motor control occurs at ~10–14 d postpartum. Thus, locomotor defects caused by myelin dysfunction will probably not manifest themselves until 1–2 wk after birth (e.g., *Adam22*; Sagane et al. 2005). Ataxia or "shivering" phenotypes that become apparent at this stage suggest a myelination defect. Developmental defects in locomotor coordination could also suggest an abnormality in the cerebellum. This region of the brain functions to regulate movement and balance. In the mouse, a significant amount of cerebellar development and maturation occurs after birth. The major neuronal cell types in the cerebellum (granule and Purkinje cells) undergo migration, axon extension, dendrite formation, and synaptogenesis until the end of the third week after birth. Thus, abnormalities in movement and balance that become apparent after 14 d postpartum may indicate a cerebellar defect (e.g., *CerS1*; Ginkel et al. 2012).

Investigate whether the neurological phenotype is more likely caused by a defect of the CNS, peripheral nervous system (PNS), or the vestibular (inner ear) system. Again, let the expression pattern of the gene in question serve as a guide for which tissues to analyze. The first step would be to do a histological analysis of the brain and peripheral nerves (e.g., the sciatic nerve). standard hematoxylin and eosin (H&E) staining is sufficient for a superficial look at the brain. Stains that are useful for detecting defects in myelin include luxol fast blue or gold chloride. You can also do immunostaining for myelin basic protein (MBP) or myelin proteolipid protein (PLP) for the CNS, and MBP or peripheral myelin protein zero for the PNS. With a decrease in myelin, you will see normal numbers of axon bundles but a reduction in myelin around the bundles. If there is a neuronal defect, you may see fewer axons but the axons that are present will have normal amounts of myelin. Combinations of these two phenotypes might occur. It is also possible that the histology of the neural tissues will be uninformative because of a defect in a discrete set of neurons. Knowing the precise expression pattern of the gene of interest will be very important in this situation. However, if the gene is ubiquitously expressed, then the analysis will be challenging.

If this is the case, you will need to perform more detailed neurological tests to learn more about the phenotype.

VISIBLE AND FERTILITY MUTANT PHENOTYPES AFTER WEANING AND SEXUAL MATURITY

Compromised Growth and Vigor

Most abnormalities will be picked up before weaning with a thorough internal and external exam (see Figs. 1 and 2) and careful observation of the mutant mice. However, a growth defect or failure to thrive that only becomes evident after weaning may be due to problems with eating or digestion of solid food. One reason that a pup may not thrive around the time of weaning is an absence of teeth or malocclusion of the incisors (e.g., *Src* [Soriano et al. 1991]; *Fos* [Johnson et al. 1992]). Without teeth or with abnormal teeth, the pups will not be able to eat the hard food pellets that are the usual diet of laboratory mice. To circumvent this problem and keep the mice viable for further investigation, mice without teeth or with malocclusion can be fed a commercially available soft mouse chow. Alternatively, you can make your own soft mouse food (Box 2). The incisors of mice grow continuously, but because the upper and lower incisors meet and are worn down by gnawing, they are maintained at a constant length. If the incisors are misaligned, then they will become excessively long, preventing the mouse from eating properly (Fig. 8). If this is a feature of your mutant phenotype and you want the mice to survive, you will have to trim the teeth periodically with strong scissors and perhaps put them on a soft food diet.

Compromised growth and vigor could also be indications of problems with digestion and metabolism. If the teeth of the homozygous mutant appear normal, then perhaps there is a defect in the gastrointestinal tract (gut, pancreas, etc.). Another type of gastrointestinal defect that can cause abnormal growth and even death is megacolon. Mice with megacolon have defects in their ability to defecate. Alternatively, there may be abnormalities in the animals' metabolism. Clinical chemistry assays can be used to pinpoint the accumulation or deficiency of a metabolite, potentially guiding you to the primary defects in a metabolic pathway.

Infertility in Males

Between 4 and 6 wk, a previously cryptic phenotype, infertility, may become apparent. If your mutated gene is expressed in the gonads or reproductive tract during development or in the adult, this is an obvious phenotype to look for. However, many factors contribute to successful reproduction so that gonadal gene expression is not a prerequisite for effects on fertility. Here we present an overview of how to identify and analyze a male fertility defect. At some point, as the analysis progresses more deeply, you may want to seek the advice of an expert on male reproduction. As a starting point, you can simply breed the mutant males to determine if they are fertile. However, if you have generated extra homozygous mutant males and you already suspect a potential effect on spermatogenesis, you

BOX 2. SOFT MOUSE FOOD (COURTESY OF PHIL SORIANO)

Ingredients

10 g agar
100 g commercial (human) baby formula
300 g powdered mouse chow (grind it up or scrape the bottom of the barrel)
Dissolve the agar in 300 mL H_2O in a microwave oven, on high.
Dissolve the baby formula in 500 mL H_2O at room temperature.
Combine the melted agar and baby formula, stirring vigorously.
Quickly add powdered mouse chow while stirring.
Chill and serve.

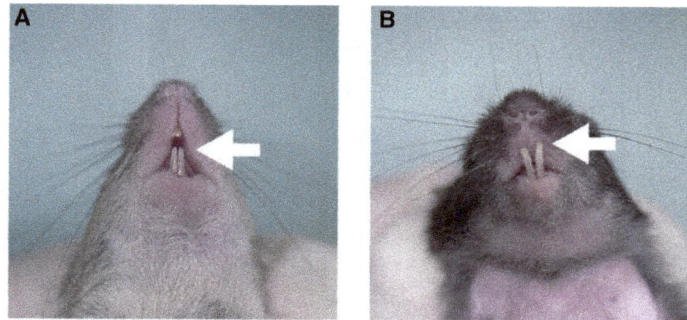

FIGURE 8. Mouse incisors. The top and bottom pair of incisors of a mouse grow continuously. Because they are aligned with one another and meet, they are constantly worn down and maintained at a certain length (A). If the incisors are misaligned (B), they will not meet and will continue to grow, causing the mouse difficulties in eating. The mice can be provided with a soft diet, or such teeth can be trimmed so that the mouse can feed. (Images courtesy of Kin Ming Kwan and Richard Behringer.)

can do a few simple tests without breeding that can potentially get you to a phenotype more quickly. First, determine if you have recovered the predicted proportion of homozygous mutant males from heterozygous intercross matings. Do a statistical analysis to reach a conclusion (see Box 1 in Chapter 8: Phenotypic Analysis: Assessing Timing of Recessive Perinatal Lethality in Mice [Papaioannou and Behringer 2023f]). If there are too many males, determine the genetic sex of all of the homozygous mutants (see Protocol 5.1: Sex Genotyping Mice by Polymerase Chain Reaction [Papaioannou and Behringer 2023g]). This will indicate whether you have phenotypic males that are genetically female (XX). XX males are invariably sterile because they lack the Y chromosome that carries essential spermatogenesis genes and they have too many X chromosomes that interfere with spermatogenesis. You should have already determined that the homozygous mutant males look like males and have all of the male-specific organs.

Next, check the size of the testes of adult mutants relative to age-matched controls. Some defects in spermatogenesis will cause a deficiency of spermatogenic cell production, resulting in smaller testes (e.g.; *Gnrhr*; Fig. 9; Pask et al. 2005). After checking testis size, examine the cauda epididymis (see Chapter 15 in Behringer et al. 2014). The cauda epididymis of an adult male should be white in appearance because of the mature sperm that it holds. A cauda epididymis that is small and translucent is an indication of a sperm defect. Release the sperm from the cauda epididymis into an appropriate medium as if you were performing an in vitro fertilization experiment (see Chapter 15 in Behringer et al. 2014). If there is sperm, is it motile? If there is no sperm or the sperm is immotile, you may have revealed a potential fertility phenotype. Check sufficient numbers of homozygous mutant males and sibling controls to determine if the phenotypes that you are observing are only observed in the homozygous mutants.

Although many fertility mutants can be identified using these methods, there are others that will appear to be wild type by those criteria. Thus, ultimately, you will have to perform a breeding test. Breed homozygous mutant males with wild-type females for a set period of time (8–12 wk). Count the number of litters produced and the number of pups in each litter. Wild-type males bred to wild-type females serve as controls. Test sufficient numbers of homozygous mutant and control males to establish statistical significance. If no pregnancies occur, then set up timed matings with wild-type females to determine if the mutant males can generate a plug (see Box 1 and Fig. 2 in Chapter 5: Recovering a Targeted Mutation in Mice from Embryonic Stem Cell Chimeras or CRISPR–Cas Founders [Papaioannou and Behringer 2023h]). A defect in reproductive behavior may prevent successful mating (generation of a plug) although spermatogenesis is otherwise normal (e.g., *Fos*; Baum et al. 1994). If a plug is generated, then females from timed matings can be sacrificed during gestation to determine if they are pregnant (i.e., you do not have to wait until birth). If you find embryos, then you can conclude that the male is fertile. To be thorough, it is probably a good idea to genotype the embryos to verify paternity. They should all be heterozygotes. This still does not verify

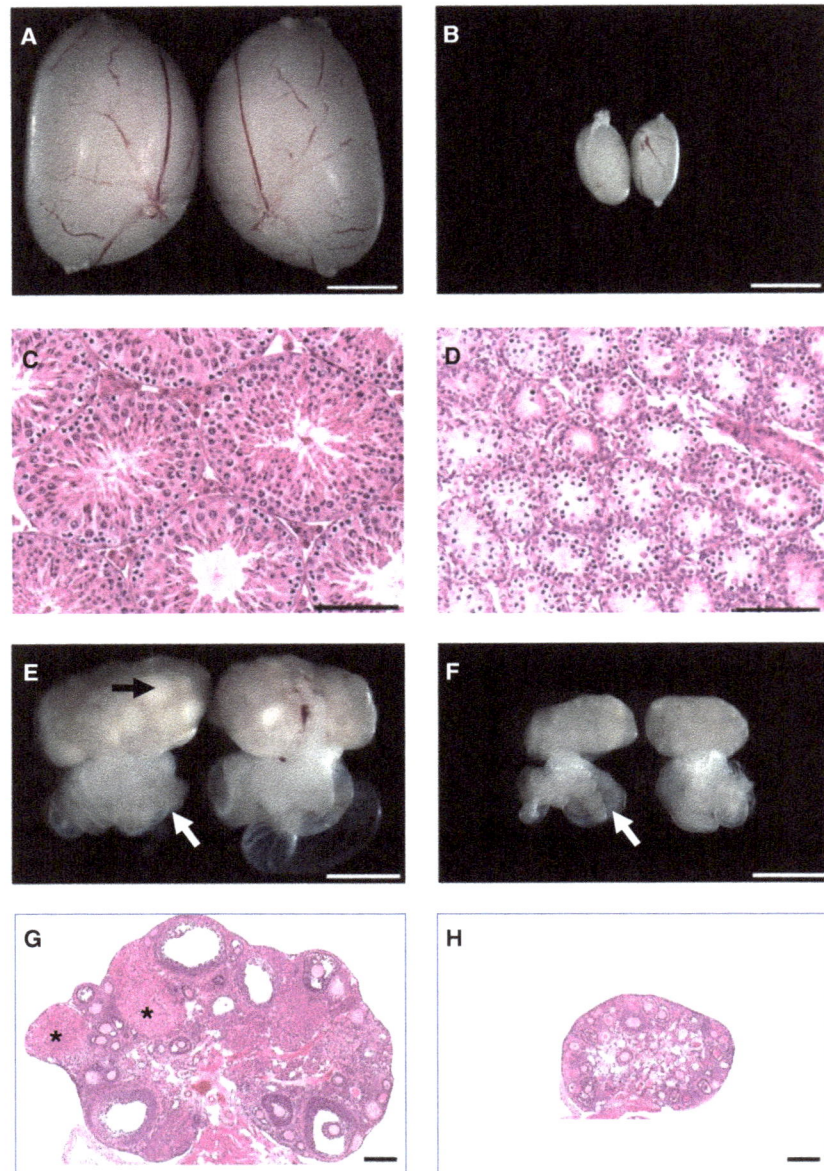

FIGURE 9. Gonadal mutant phenotypes. (*A,B*) Testes from wild-type (*A*) and *Gnrhr/Gnrhr* mutants (*B*). The mutant testes are greatly reduced in size relative to body weight. Scale bars, 2 mm. (*C,D*) H&E-stained histological sections of testes from wild-type (*C*) and *Gnrhr/Gnrhr* mutants (*D*), showing spermatogenesis in the seminiferous tubules (circular structures in these sections). Spermatogenesis in the *Gnrhr/Gnrhr* mutant testes is arrested in meiosis (see Fig. 10), resulting in smaller-diameter tubules. Scale bars, 0.1 mm. (*E,F*) Ovaries with attached oviducts (white arrows) from wild-type (*E*) and *Gnrhr/Gnrhr* mutants (*F*). Corpus luteum (black arrow in *E*). Scale bars, 1 mm. (*G,H*) H&E-stained histological sections of ovaries from wild-type (*G*) and *Gnrhr/Gnrhr* mutants (*H*) with an arrest at the secondary stage of folliculogenesis (see Fig. 10). Corpora lutea (asterisks), which are indicators of successful ovulation, are lacking in the *Gnrhr/Gnrhr* mutant ovaries. Scale bars, 0.2 mm. (Modified from Stewart et al. 2012.)

that the males are completely wild type with respect to spermatogenesis because as little as 10% sperm production is sufficient for fertility.

If no pregnancies are obtained from the homozygous mutant males, then the next step is to determine if their sperm can fertilize oocytes. Set up timed matings with wild-type females to obtain plugs. Collect the oocytes from the oviducts (see Chapter 4 in Behringer et al. 2014) and view them under the dissecting microscope to determine if they have been fertilized (i.e., are two pronuclei present? See Figure 1B in Chapter 9: Phenotype Analysis of Preimplantation Lethality in Mice [Papaioannou and Behringer 2023a]). If they are not fertilized, then the sperm from the homozygous mutant are likely to be defective. If there are two pronuclei, then culture the embryos and determine the percentage of fertilized oocytes that progress to the blastocyst stage (see Protocol 9.1: Phenotypic Analysis of Preimplantation Mouse Embryos in Culture [Papaioannou and Behringer 2023i]). Compare these numbers with wild-type controls. It is possible that the mutant sperm can fertilize the oocytes but then development fails. This would also suggest that the mutant sperm are defective.

At this point, you know if your homozygous mutant males can produce a plug, whether they have motile sperm, and whether the mutant sperm can fertilize oocytes. What do you do next? Spermatogenesis is a multistage process that is well-characterized. If no sperm are present in your mutant, it is likely that they lack germ cells or have a block in spermatogenesis (e.g., *Sys*; MacGregor et al. 1990). On the other hand, the production of defective sperm that is immotile or cannot fertilize oocytes is likely caused by abnormalities in spermiogenesis, the transformation of a round haploid cell called the spermatid into the morphologically distinct spermatozoa.

At birth, the testes only have spermatogonia, the stem cells that produce the spermatozoa. By 10 d after birth, a cohort of the spermatogonia has progressed into meiosis to generate spermatocytes. By 19 d, they have become haploid round spermatids. The first mature spermatozoa are subsequently formed at ~30 d after birth. This initial formation of sperm after birth is called the first wave of spermatogenesis (Fig. 10). Subsequent overlapping waves generate the variety of stages present in the adult testes. In adult males, it takes ~41 d for spermatogonial stem cells to form mature spermatozoa. Thus, you can histologically analyze homozygous mutant males at various times after birth to determine when spermatogenesis goes wrong. Alternatively, you can analyze adult testes histologically to determine which stages of spermatogenesis are abnormal (Fig. 9). Transmission electron microscopy (TEM) can be very useful for determining the precise lesions in spermatogenic cells.

Analyses of cell proliferation and programmed cell death are usually called for in the analysis of a sperm generation defect. When spermatogenesis begins to fail, it usually worsens with time, so analyze the homozygous mutant males at several time points. Finally, if you find a defect in sperm formation,

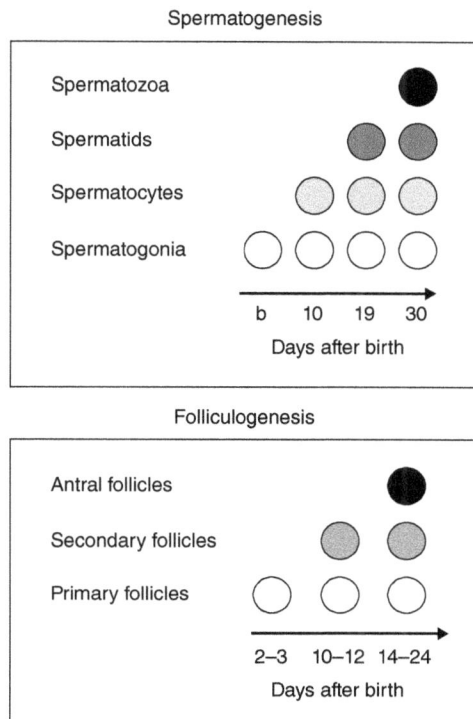

FIGURE 10. The first waves of spermatogenesis in males and folliculogenesis in females. (*Top*) Diagram showing the various types of sperm cells present at different days after birth (b), during the first wave of spermatogenesis. At birth, only spermatogonia are present in the testes. With time, more differentiated sperm cell types are present in the postnatal testis. Meiotic spermatocytes are present from day 10 onward and haploid spermatids are present from day 19 onward. The first mature spermatozoa are present at 30 d after birth. Subsequently, the length of time needed to generate spermatozoa increases to ~41 d. (*Bottom*) Diagram showing the various types of follicles present at different days after birth during the first wave of folliculogenesis. By 2–3 d after birth, all oocytes are in meiotic prophase in primary follicles. With time, more differentiated follicles are present in the postnatal ovary. Secondary follicles are present by 10–12 d after birth and antral follicles are present by 14–24 d after birth.

it may be caused by a defect that is intrinsic to the germ cells, the supporting somatic cells of the testis, or even the brain or pituitary. Insights into the primary defect can be obtained by knowing which tissues express the gene of interest. If the gene is expressed in both germ cells and other relevant cell types, a conditional knockout might help to identify the defective tissue (see Chapter 18: Tissue- and/ or Temporal-Specific Mutations in Mice Using Conditional Alleles [Papaioannou and Behringer 2023j]). Assisted reproductive technologies could also be used to determine which aspect of spermatogenesis is defective (e.g., *Zfy2*; Yamauchi et al. 2015).

Infertility in Females

Infertility in one sex does not necessarily predict infertility in the other. As always, the gene expression pattern of the gene mutated can be a guide. Genes expressed in primordial germ cells, the early indifferent gonad, the gonads, or other reproductive organs of both sexes could potentially affect the fertility of both sexes. Other genes may be exclusive to the gonads or other reproductive organs of one sex or the other.

As a starting point, breed the females to determine if they are fertile. However, if you have generated extra homozygous mutant females and already suspect a potential effect on female reproduction, you can do an initial examination of the ovaries and reproductive organs in parallel with the breeding that can potentially get you to a phenotype more quickly. You should have already determined that the homozygous mutant females look like females and have all of the female-specific organs (see Figs. 1–3). Examine the morphology and histology of the homozygous mutant ovaries, uterus, and oviducts and compare to age-matched controls (Fig. 9). In mice, an initial wave of folliculogenesis takes place in a typical pattern soon after birth (Fig. 10). By 3–5 d after birth, the ovaries contain oocytes that are all arrested in meiotic prophase. By 10–12 d after birth, secondary-stage follicles are present that contain oocytes surrounded by two or more layers of granulosa cells. Between 14 and 24 d, some follicles have progressed to antral stages. (Cells closest to the oocytes are called cumulus granulosa cells and cells lining the antrum wall are called mural granulosa cells.) By 5–6 wk, there may be evidence of ovulation (i.e., corpus lutea). Are all of these stages of folliculogenesis present in the homozygous mutant ovary? A block in a specific stage of folliculogenesis will point you in the right direction for finding the primary defect in oogenesis. Check sufficient numbers of homozygous mutant females and sibling controls to determine if the phenotypes that you are observing are only observed in the homozygous mutants.

Mate homozygous mutant females with wild-type males, preferably known-fertile males. Set up the matings and wait for a pregnancy or establish timed matings to check for plugs (see Box 1 and Fig. 2 in Chapter 5: Recovering a Targeted Mutation in Mice from Embryonic Stem Cell Chimeras or CRISPR–Cas Founders [Papaioannou and Behringer 2023h]). The advantage of setting up timed matings is that you can determine if the homozygous mutant females are capable of mating. If no plugs are found in the homozygous mutant females, then the chances of obtaining a productive pregnancy are very low. The lack of a plug could suggest a behavioral or hormonal problem and not necessarily a germ cell defect. Assisted reproduction techniques (e.g., in vitro fertilization) can be used to determine if the ovaries and germ cells are functional (see Chapter 6: Strategies for Maintaining Mouse Mutations [Papaioannou and Behringer 2023k]). If there are plugs but no pregnancies, then you will need to determine the basis for this infertility.

If you have infertile homozygous mutant females, determine if you have recovered the predicted proportion of homozygous mutant females from the heterozygous intercross matings. Do a statistical analysis to reach a conclusion (see Box 1 in Chapter 8: Phenotypic Analysis: Assessing Timing of Recessive Prenatal Lethality in Mice [Papaioannou and Behringer 2023f]). If there are too many females, determine the genetic sex of all of the homozygous mutants (see Protocol 5.1: Sex Genotyping Mice by Polymerase Chain Reaction [Papaioannou and Behringer 2023g]). This will indicate whether you have phenotypic females that are genetically male (XY). The presence of a Y chromosome will cause defects in oogenesis and hinder the production of functional oocytes.

Cite this chapter as *Cold Spring Harb Protoc*; doi:10.1101/pdb.over107976

If homozygous mutant females do not become pregnant, then use hormones to induce superovulation (see Chapter 3 in Behringer et al. 2014, 2018) in 3-wk-old homozygous mutant females and mate with wild-type male studs. If a plug is present the next morning, collect the oocytes and look under a dissection microscope to determine if they have been fertilized (i.e., they contain two pronuclei; see Fig. 1B in Chapter 9: Phenotypic Analysis of Preimplantation Lethality in Mice [Papaioannou and Behringer 2023a]). Culture them overnight to see if they progress to the two-cell stage and, if so, score the percentage of two-cell-stage embryos in comparison with controls. For a more stringent test, continue culturing the embryos to the blastocyst stage (see Protocol 9.1: Phenotypic Analysis of Preimplantation Mouse Embryos in Culture [Papaioannou and Behringer 2023i]). This can also be done using natural timed matings. If you are able to obtain fertilized oocytes that progress in culture to the blastocyst stage, then the fertility defect is most likely at the stage of implantation or later (e.g., *Lif*; Stewart et al. 1992). If the mutant oocytes cannot be fertilized or they do not progress, this indicates that the oocytes are in some way defective.

REFERENCES

Baum MJ, Brown JJG, Kica E, Rubin BS, Johnson RS, Papaioannou VE. 1994. Effect of a null mutation of the *c-fos* proto-oncogene on sexual behavior of male mice. *Biol Reprod* **50**: 1040–1048. doi:10.1095/biolreprod50.5.1040

Behringer RR, Gertsenstein M, Nagy KV, Nagy A. 2014. *Manipulating the mouse embryo: a laboratory manual.* Cold Spring Harbor Laboratory Press, Cold Spring Harbor, NY.

Behringer R, Gertsenstein M, Nagy KV, Nagy A. 2018. Administration of gonadotropins for superovulation in mice. *Cold Spring Harb Protoc* doi:10.1011/pdb.prot092403

Gene Expression Database. 2022. http://www.informatics.jax.org/expression.shtml [Accessed December 14, 2022].

Giese KP, Martini R, Lemke G, Soriano P, Schachner M. 1992. Mouse P_0 gene disruption leads to hypomyelination, abnormal expression of recognition molecules, and degeneration of myelin and axons. *Cell* **71**: 565–576. doi:10.1016/0092-8674(92)90591-Y

Ginkel C, Hartmann D, vom Dorp K, Zlomuzica A, Farwanah H, Eckhardt M, Sandhoff R, Degen J, Rabionet M, Dere E, et al. 2012. Ablation of neuronal ceramide synthase 1 in mice decreases ganglioside levels and expression of myelin-associated glycoprotein in oligodendrocytes. *J Biol Chem* **287**: 41888–41902. doi:10.1074/jbc.M112.413500

Green EL, Roscoe B. 1975. *Biology of the laboratory mouse*, 2nd ed. Dover, New York.

Honrado GI, Johnson RS, Golombek DA, Spiegelman BM, Papaioannou VE, Ralph MR. 1996. The circadian system of *c-fos* deficient mice. *J Comp Physiol A* **178**: 563–570. doi:10.1007/BF00190186

Jamin SP, Arango NA, Mishina Y, Hanks MC, Behringer RR. 2002. Requirement of Bmpr1a for Müllerian duct regression during male sexual development. *Nat Genet* **32**: 408–410. doi:10.1038/ng1003

Johnson RS, Spiegelman BM, Papaioannou VE. 1992. Pleiotropic effects of a null mutation in the *c-fos* proto-oncogene. *Cell* **71**: 577–586. doi:10.1016/0092-8674(92)90592-Z

Joyner AL, Herrup K, Auerbach BA, Davis CA, Rossant J. 1991. Subtle cerebellar phenotype in mice homozygous for a targeted deletion of the *En-2* homeobox. *Science* **251**: 1239–1243. doi:10.1126/science.1672471

MacGregor GR, Russell LD, Van Beek ME, Hanten GR, Kovac MJ, Kozak CA, Meistrich ML, Overbeek PA. 1990. Symplastic spermatids (sys): a recessive insertional mutation in mice causing a defect in spermatogenesis. *Proc Natl Acad Sci* **87**: 5016–5020. doi:10.1073/pnas.87.13.5016

Militzer K. 2001. Hair growth pattern in nude mice. *Cells Tissues Organs* **168**: 285–294. doi:10.1159/000047845

Mouse Genome Informatics (MGI) Phenotype Ontology Browser. 2022. http://www.informatics.jax.org/vocab/mp_ontology [Accessed December 14, 2022].

Mouse Genome Informatics. 2022. http://www.informatics.jax.org/mgihome/other/citation.shtml [Accessed December 14, 2022].

Nagy A, Gertsenstein M, Vintersten K, Behringer R. 2006. Caesarean section and fostering. *Cold Spring Harb Protoc* doi:10.1011/pdb.prot4381

Nagy A, Gertsenstein M, Vintersten K, Behringer R. 2009. Alizarin red staining of post-natal bone in mouse. *Cold Spring Harb Protoc* doi:10.1011/pdb.prot5171

Ohtoshi A, Behringer RR. 2004. Neonatal lethality, dwarfism, and abnormal brain development in *Dmbx1* mutant mice. *Mol Cell Biol* **24**: 7548–7558. doi:10.1128/MCB.24.17.7548-7558.2004

Ovchinnikov D. 2009. Alcian blue/alizarin red staining of cartilage and bone in mouse. *Cold Spring Harb Protoc* doi:10.1011/pdb.prot5170

Palmiter RD, Behringer RR, Quaife CJ, Maxwell F, Maxwell IH, Brinster RL. 1987. Cell lineage ablation in transgenic mice by cell-specific expression of a toxin gene. *Cell* **50**: 435–443. doi:10.1016/0092-8674(87)90497-1

Papaioannou VE, Behringer RR. 2023a. Phenotypic analysis of preimplantation lethality in mice. *Cold Spring Harb Protoc* doi:10.1011/pdb.over107971

Papaioannou VE, Behringer RR. 2023b. Phenotypic analysis of periimplantation to mid-gestation lethality in mice. *Cold Spring Harb Protoc* doi:10.1011/pdb.over107972

Papaioannou VE, Behringer RR. 2023c. Analysis of mid- to late-gestation phenotypes in mice. *Cold Spring Harb Protoc* doi:10.1011/pdb.over107973

Papaioannou VE, Behringer RR. 2023d. Phenotypic analysis of perinatal lethality in mice. *Cold Spring Harb Protoc* doi:10.1011/pdb.over107975

Papaioannou VE, Behringer RR. 2023e. The "no phenotype" challenge in analyzing mutant mice. *Cold Spring Harb Protoc* doi:10.1011/pdb.over107977

Papaioannou VE, Behringer RR. 2023f. Phenotypic analysis: assessing timing of recessive perinatal lethality in mice. *Cold Spring Harb Protoc* doi:10.1011/pdb.over107970

Papaioannou VE, Behringer RR. 2023g. Sex genotyping mice by polymerase chain reaction. *Cold Spring Harb Protoc* doi:10.1011/pdb.prot108062

Papaioannou VE, Behringer RR. 2023h. Recovering a targeted mutation in mice from embryonic stem cell chimeras or CRISPR–Cas founders. *Cold Spring Harb Protoc* doi:10.1011/pdb.over107959

Papaioannou VE, Behringer RR. 2023i. Phenotypic analysis of preimplantation mouse embryos in culture. *Cold Spring Harb Protoc* doi:10.1011/pdb.prot108090

Papaioannou VE, Behringer RR. 2023j. Tissue- and/or temporal-specific mutations in mice using conditional alleles. *Cold Spring Harb Protoc* doi:10.1011/pdb.over107980

Papaioannou VE, Behringer RR. 2023k. Strategies for maintaining mouse mutations. *Cold Spring Harb Protoc* doi:10.1011/pdb.over107960

Pask AJ, Kanasaki H, Kaiser UB, Conn PM, Janovick JA, Stockton DW, Hess DL, Justice MJ, Behringer RR. 2005. A novel mouse model of hypogonadotrophic hypogonadism: *N*-ethyl-*N*-nitrosourea-induced gonadotropin-releasing hormone receptor gene mutation. *Mol Endocrinol* **19**: 972–981. doi:10.1210/me.2004-0192

Paylor R, Johnson RS, Papaioannou VE, Speigelman BM, Wehner JM. 1994. Behavioral assessment of c-*fos* mutant mice. *Brain Research* **651**: 275-282. doi:10.1016/0006-8993(94)90707-2

Rogers DC, Fisher EMC, Brown SDM, Peters J, Hunter AJ, Martin JE. 1997. Behavioral and functional analysis of mouse phenotype: SHIRPA, a proposed protocol for comprehensive phenotype assessment. *Mamm Genome* **8**: 711–713. doi:10.1007/s003359900551

Sagane K, Hayakawa K, Kai J, Hirohashi T, Takahashi E, Miyamoto N, Ino M, Oki T, Yamazaki K, Nagasu T. 2005. Ataxia and peripheral nerve hypomyelination in ADAM22-deficient mice. *BMC Neurosci* **6**: 33. doi:10.1186/1471-2202-6-33

Soriano P, Montgomery C, Geske R, Bradley A. 1991. Targeted disruption of the c-src proto-oncogene leads to osteopetrosis in mice. *Cell* **64**: 693–702. doi:10.1016/0092-8674(91)90499-O

Stewart CL, Kaspar P, Brunet LJ, Bhatt H, Gadi I, Kontgen F, Abbondanzo SJ. 1992. Blastocyst implantation depends on maternal expression of leukaemia inhibitory factor. *Nature* **359**: 76–79. doi:10.1038/359076a0

Stewart MD, Deng JM, Stewart CA, Mullen RD, Wang Y, Lopez S, Serna MK, Huang CC, Janovick JA, Pask AJ, et al. 2012. Mice harboring *Gnrhr* E90K, a mutation that causes protein misfolding and hypogonadotropic hypogonadism in humans, exhibit testis size reduction and ovulation failure. *Mol Endocrinol* **26**: 1847–1856. doi:10.1210/me.2012-1072

Wimer RE, Fuller JL. 1966. Patterns of behavior. In *Biology of the laboratory mouse* (ed. Green EL), pp. 629–653. McGraw-Hill, New York.

Yamauchi Y, Riel JM, Ruthig V, Ward MA. 2015. Mouse Y-encoded transcription factor *Zfy2* is essential for sperm formation and function in assisted fertilization. *PLoS Genet* **11**: e1005476. doi:10.1371/journal.pgen.1005476

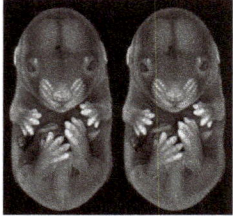

CHAPTER 15

The "No Phenotype" Challenge in Analyzing Mutant Mice

If homozygous mutant mice survive to adulthood, are fertile, and have no visible phenotypes attributable to mutation of the relevant gene, there are a number of possible reasons why an effect of the mutation is not evident. Technical errors that might have occurred during gene targeting or genotyping must first be eliminated. Variable penetrance of the mutation should be considered as well as the possibility of age-related or late-onset phenotypes, such as tumor formation or other pathologies. The gene expression pattern and nature of the protein product of the gene could provide clues. A number of simple tests can be applied to uncover cryptic phenotypes that are not easily seen on casual inspection (e.g., tests of the senses and of balance and coordination). Genetic and environmental challenges can be applied to overtly normal mutant mice to reveal deviations from normal.

INTRODUCTION

NO OVERT POSTNATAL PHENOTYPE

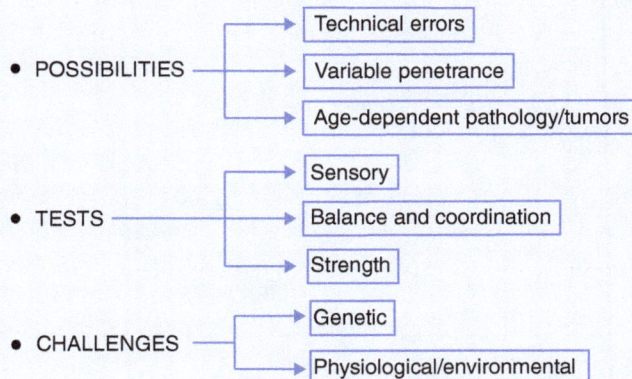

- POSSIBILITIES
  - Technical errors
  - Variable penetrance
  - Age-dependent pathology/tumors

- TESTS
  - Sensory
  - Balance and coordination
  - Strength

- CHALLENGES
  - Genetic
  - Physiological/environmental

It is highly unlikely that the mutation of a functional gene will be completely without effect. More likely, it is a matter of knowing how to look for the phenotype (e.g., *Hprt*; Kuehn et al. 1987). To arrive at the current chapter, you found that homozygous mutants, both male and female, are born at

Opening artwork: Courtesy of Liang Liang and Paul Trainor.

Mendelian frequency, survive to adulthood, are fully fertile, and have no obvious abnormalities. One possibility is a late-onset phenotype, but unless you have very strong reasons to believe that the mutation will only result in a phenotype in old age, you probably will not want to wait until the mice are 2 years old before exploring other options. You could set up a tumor watch if cancer is a possibility or set aside a cohort of mice to age, but at the same time, try some experiments that might reveal a cryptic mutant phenotype. You might also want to consider taking advantage of available mouse phenotyping services (e.g., https://ors.od.nih.gov/sr/dvr/drs/Pages/Mouse%20Phenotyping .aspx; https://www.mmpc.org/; https://www.mousephenotype.org/).

In the following sections, we delve deeply into less visible phenotypes, such as behavioral modifications, and also present the mutant mice with various challenges that might reveal a phenotype. In following up these lines of investigation, you can be selective depending on the expression pattern of the gene or what you know about the protein and its potential activity. Before proceeding, however, it is worth considering some trivial explanations for your mutant not having a visible phenotype. Perform a few control experiments to verify that your homozygous mutants really are null for the gene under study and that you are recovering the expected numbers of homozygous mutants.

POSSIBLE CAUSES FOR NO VISIBLE PHENOTYPE AMONG POSTNATAL MUTANTS

Genotyping Errors

Somehow, an error may have occurred in genotyping. Check your Southern blots or polymerase chain reaction (PCR) results. If you are using Southern analysis, make sure that you are using the correct probe (i.e., sequence it). If you are using PCR, check your primers and amplification sizes. Collect new samples from the putative homozygous mutants and re-genotype.

Mutant Gene Is Functional

If you designed a loss-of-function allele, you may not have completely inactivated the gene. Review your knockout strategy (see Chapter 3: Mouse Gene-Targeting Strategies for Maximum Ease and Versatility [Papaioannou and Behringer 2023a]). Because this is the first time that you will have a homozygous mutant mouse, you can perform a rigorous experimental test to determine if you have truly deleted the desired sequences. Perform a PCR that flanks the deleted region and sequence the amplified gene product to verify the deletion. A PCR strategy that detects a wild-type allele for the gene of interest should not amplify DNA from a homozygous mutant. You can also perform a Southern analysis of homozygous mutant DNA (tail or other tissue), using a probe within the region that has been deleted. The predicted wild-type band should be absent. Strip the membrane and rehybridize it with an irrelevant probe as a positive control for the presence of DNA. If you have not deleted all protein-coding exons, perform reverse transcriptase (RT)-PCR or other RNA analysis using probes that are 5′ and 3′ of the targeted deletion to determine if any transcripts are produced from the mutated locus. If you have antibodies to the gene product, perform a western blot or immunohisto-chemistry/immunofluorescence on homozygous mutant tissues to determine if any protein or partial protein is generated.

Variable Penetrance Phenotype

It is possible that you will recover homozygous mutants that are viable and fertile with no obvious defects but fewer of them than the predicted Mendelian frequency. This could indicate a variably penetrant mutant phenotype such that some homozygous mutants are viable with no obvious abnormalities but some die before birth and thus were not detected. Examine the ratios of homozygous mutants obtained relative to heterozygotes and wild type, using a larger sample than before (see Box 1 in Chapter 8: Phenotypic Analysis: Assessing Timing of Recessive Prenatal Lethality in Mice [Papaioannou and Behringer 2023b]). If you find a statistically significant underrepresentation of

 Cite this chapter as *Cold Spring Harb Protoc*; doi:10.1101/pdb.over107977

homozygous mutants after birth, establish timed matings between heterozygotes to determine when the homozygous mutants are being lost and analyze the prenatal phenotype (see Chapter 9: Phenotypic Analysis of Preimplantation Lethality in Mice [Papaioannou and Behringer 2023c]; Chapter 10: Phenotypic Analysis of Periimplantation to Mid-Gestation Lethality in Mice [Papaioannou and Behringer 2023d]; Chapter 11: Analysis of Mid- to Late-Gestation Phenotypes in Mice [Papaioannou and Behringer 2023e]; Chapter 12: Uncovering Phenotypes in Mutant Mice by Determining Embryo, Organ, Tissue, and Cell Development [Papaioannou and Behringer 2023f]).

No Overt Abnormality

The nontrivial possibility is that homozygosity for the null mutation results in no overt abnormality, at least at the age you are examining. At this point, think about the expected phenotype based on the expression pattern of the gene and its protein and use the following sections to analyze the mutants as seems appropriate.

AGE-DEPENDENT PHENOTYPES

Many genes act to maintain homeostasis after birth. Thus, phenotypes caused by mutations in these types of genes may not become apparent until the mouse ages sufficiently. Similarly, genes that affect the propensity to develop certain diseases may only be evident in older mice.

Tumor Formation

The type of gene or its expression pattern may already indicate that the mutation you have generated has a role in tumor formation (e.g., *Trp53*; Donehower et al. 1992). To determine if this is true, you will need to age your mutant mice and systematically analyze them for tumor formation. This involves generating sufficient numbers of experimental and control mice, allowing them to age under the same conditions, and determining if, when, and what types of tumors form. Ultimately, you will generate a tumor-free survival curve that plots the number of mice surviving tumor free as a function of time (Fig. 1). You will also be able to determine the mean latency of tumor detection. In addition, this type of study will provide information on the types of tumors that form (i.e., the tumor spectrum). Design the tumor watch study very carefully because it can take 2–3 yr to perform, complete, and analyze. If you do not have a sufficient number of mice with the appropriate genotypes at the beginning of your study, you may not obtain statistically significant (i.e., publishable) data, potentially wasting years of effort.

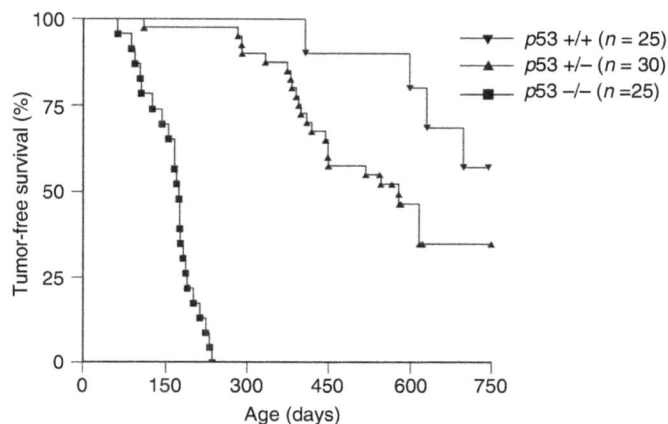

FIGURE 1. Tumor-free survival curves. A cohort of *p53* (*Trp53*) homozygous and heterozygous mutants and wild-type controls was followed over time for survival and tumor formation. (Image provided by Gigi Lozano.)

The genetic background of the mice in a tumor study is a very important consideration and should be as consistent as possible to minimize variation (see Chapter 6: Strategies for Maintaining Mouse Mutations [Papaioannou and Behringer 2023g]). The most straightforward choices are a coisogenic or congenic strain or an F_1 between two inbred strains. An initial tumor formation study could also be done on a mixed, F_2 genetic background, which can be informative because it contains the entire range of genetic variation between the two inbred strains used. F_3 or backcrosses to one of the inbred strains create genetic backgrounds that may be biased depending on the genetic mixture of the parents used to generate the cohort for the tumor watch. The choice of which inbred strains to use will be dependent on characteristics, such as known tumor susceptibility of different strains, and will be specific to your particular experimental design.

Once you have decided on a genetic background, generate a cohort of homozygous and heterozygous mutants as well as wild-type controls that are sex- and genetic background–matched. Decide on the number of mice for each genotype that you will need for the study. Publications that include survival curves generally have data for about 25 experimental and 25 control animals. We suggest that you generate 50 homozygous mutants, 50 heterozygotes, and 50 wild-type mice (equal numbers of males and females) because you will lose mice along the way (e.g., a mouse may die on a weekend and its tissues could degrade or be cannibalized before you can get to it). Fortunately, you do not have to generate the mice all at once. They can be accumulated over time. It is also possible to perform an "in vitro fertilization (IVF) expansion" to generate many mice at one time (see Takeo and Nakagata 2018). IVF is used to generate many zygotes that can be transferred into surrogates to yield plenty of mice to be used in the tumor-watch cohort.

Once mice are entered into the study at a specific age, monitor them on a regular basis (e.g., twice a week) for obvious tumor formation, moribundity, or death. Kill moribund mice for analysis. Ultimately, this will result in the generation of what is called a tumor-free survival curve (Fig. 1). For each tumor-bearing animal that is killed, perform a complete histopathological workup (see Protocol 15.1: Necropsy Guide for the Collection of Tissues from Mice with or without Tumors [Papaioannou and Behringer 2023h]). If the tumor is large, fix small pieces for histology, in situ hybridization, and immunohistochemistry. These tumor samples are precious because they take so long to generate. Thus, make sure that the fixative you use is compatible with the different types of analyses you might perform. If the tumor is large enough, snap freeze some of it (dry ice or liquid nitrogen) for protein and nucleic acid analyses (e.g., copy number variation or transcriptome analysis). If you are not a histopathologist, collaboration with one could be valuable. The tumor watch should be performed until the mice are at least 2 yr of age before concluding that there is no effect of a homozygous or heterozygous mutation. However, a strong effect on tumor formation could be detected in much less time.

Other Age-Related Pathologies

Many other types of pathologies can develop as mice age, including skeletal, cardiovascular, and neurological. The approach to studying these mutant phenotypes is generally the same. Sufficient numbers of age- and sex-matched mutants and controls on a consistent genetic background are set aside to age for various amounts of time and examined for the development of the pathology of interest. Here we discuss the analysis of age-related skeletal pathologies as a general example. Because of the long time frame, aging studies are generally only done when you have sufficient reason to suspect that an age-related pathology will develop based on the properties of the mutated gene.

Age-related skeletal pathologies include osteoporosis (bone loss) (e.g., *Notch2*; Fukushima et al. 2017), osteopetrosis (excess bone formation or lack of remodeling) (e.g., *ob*; Ducy et al. 2000), and arthritis (e.g., *Il1r1*; Horai et al. 2000) that can become progressively more severe as mutant mice age. Bone formation is a balance between osteoblast and osteoclast function. Accordingly, these are the two primary cell types to consider when presented with bone formation pathologies. Arthritis is a disease of the joints. Therefore, the joints would be the focus of an investigation of arthritis. You may notice these pathologies initially because they can lead to overt abnormalities such as kyphosis (curvature of the spine), reduced movements, or swelling of the joints.

Cite this chapter as *Cold Spring Harb Protoc*; doi:10.1101/pdb.over107977

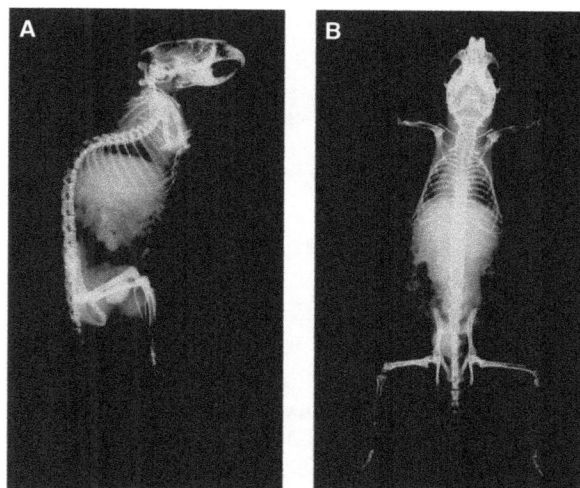

FIGURE 2. Contact X-ray images of skeletons. (A) Lateral view of wild-type 3-mo-old female mouse. (B) Dorsal view of same mouse. Repositioning the mouse is used to show specific regions of the skeleton. (Image courtesy of Yuji Mishina and Richard Behringer.)

The analysis begins with a documentation of the intact live animals. Photograph the homozygous and heterozygous mutants and age- and sex-matched wild-type controls while they are either awake or anesthetized. Make sure to include a size marker in the image (e.g., a ruler). To assess skeleton structure without sacrificing animals, take X-ray pictures of anesthetized mice (Fig. 2). These X-ray images can provide sufficient resolution to determine general bone density and other bone abnormalities (e.g., fractures). A next step might be to generate a skeleton preparation (see Chapter 17 in Behringer et al. 2014; see Nagy et al. 2009) to grossly examine the bones and joints in more detail. Bone histopathology will provide cellular and tissue organization details.

Once you have an initial indication that you are dealing with an age-dependent skeletal pathology, you will need to perform a comprehensive temporal study of the development of the pathology. Set aside cohorts of homozygous and heterozygous mutants and wild-type controls that will be analyzed at various ages. Remember that there may be differences in the phenotypes of males and females. Thus, it is necessary to use sex-matched controls. Because these types of studies can potentially take a long time to perform, make sure that you have also set aside sufficient numbers of animals to account for losses as they age to generate statistically significant data.

TESTS TO UNCOVER CRYPTIC PHENOTYPES

Tests of the Senses

Here we present some simple tests, which generally do not require highly specialized equipment, to test sensory perception to determine if your homozygous mutant mouse can perceive and process sensory information. If these simple tests reveal a defect, you can delve more deeply with more sophisticated measures of the specific senses, or you may wish to seek out an expert collaborator.

Vision

How can you determine if a mouse has normal vision? Unfortunately, there is no simple, definitive way to test for vision in mice. A first attempt at assessing the eyes of a mutant would simply be a gross dissection and histological analysis in comparison with wild-type controls. Gross assessments will indicate if the overall size of the eye is different from controls (e.g., *Pitx3*; Wada et al. 2014). Cataracts are also identifiable upon visual inspection (e.g., *Crybb2*; Zhang et al. 2008). Histological assessments will indicate any major abnormalities in the eye such as abnormal retina or abnormal differentiation of cell types (e.g., *Vsx1*; Ohtoshi et al. 2004). Optical coherence tomography (OCT) is a live, noninvasive imaging technology that can visualize the structure of the retina, including its various stereotypical layers but requires specialized equipment (Larina et al. 2012).

A simple test to assess response to light is by observing if the pupils constrict when exposed to light (e.g., from a small flashlight) and dilate when the light is removed, although this does not test visual function. Functional tests usually require specialized equipment and varying amounts of technical expertise. One method uses an apparatus called a visual tracking drum (Fig. 3; Thaung et al. 2002). The visual tracking drum is a cylinder approximately 1 foot in diameter and 20 inches tall that can rotate at two revolutions per minute. Cards with black and white vertical stripes are placed on the inside of the drum. The mouse is placed on a central, stationary, raised platform. The top of the drum is open to monitor the head movements of the mouse. If the mouse can see the pattern on the inside of the drum as it rotates, it will follow the movement of the pattern with its head. If its vision is severely compromised, it will not show this "head-tracking" behavior. This method does not require training the mouse because head-tracking behavior is a reflex response.

If you need more detailed measurements of vision, consider collaborating with a vision expert to generate an electroretinogram (ERG) for your mouse (e.g., *Vsx1*; Ohtoshi et al. 2004). ERG recordings measure the eye's electrical response to a flash of light. Responses are recorded using an electrode placed on the surface of the eye while the mouse is anesthetized. The ERG is usually generated by an instrument called a ganzfeld stimulator, which assesses the state of the entire retina.

Be aware that many of the common inbred mouse strains carry a recessive mutation called retinal degeneration 1 (*rd1*) in the *Pde6b* gene that causes postnatal loss of photoreceptors in the retina, resulting in vision loss. Some inbred strains that carry the *rd1* allele include C3H and CBA. Substrains of 129 and C57BL/6 mice are wild type for *Pde6b*. However, another recessive retinal degeneration allele (*rd8*) that causes vision impairment exists in the *Crb1* locus that is present in the widely used C57BL/6N substrain (Mattapallil et al. 2012). Thus, if you are studying genes that may influence the eye, be sure to check the *Pde6b* genotype of your genetic background or you may end up chasing a phenotype that is not specific to your engineered mutation.

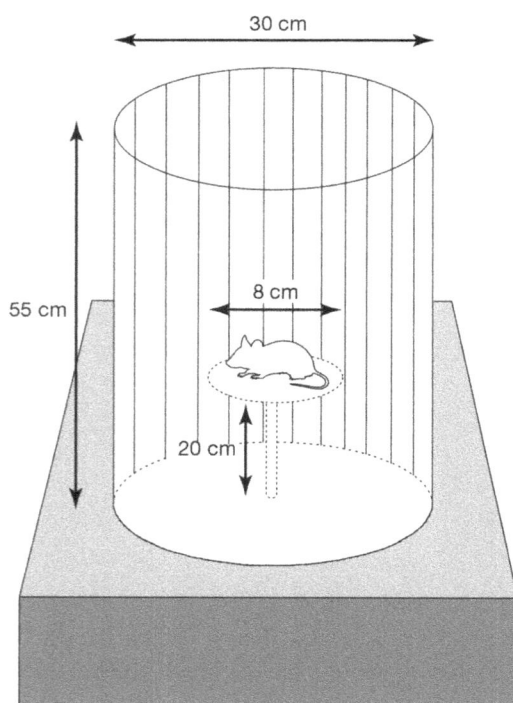

FIGURE 3. Diagram of visual tracking drum apparatus. (Redrawn, with permission from Elsevier, from Thaung et al. 2002.)

Hearing

A simple test of deafness or severe hearing loss is to use a click box to generate a sound of specific intensity to test for an acoustic startle response, the Preyer reflex, a physical response to an auditory stimulus. The click box generates a brief 20-kHz tone at a 90-db sound pressure level (SPL) when held 30 cm above the mouse. The Preyer reflex can be strong (the mouse jumps) or subtle (a backward flick of the ears). This test can identify both peripheral nervous system (PNS) and central nervous system (CNS) defects (e.g., *Ysb*; Dong et al. 2002). Because mice respond to the ultrasonic vocalizations of their offspring for such behaviors as recovering scattered pups, poor mothering skills among mutant females can also indicate deafness.

Smell and Taste

Tests for smell and taste are quite specialized. It is likely that you will not be concerned about these types of phenotypes unless your gene is expressed in the nose, tongue, or relevant regions of the brain and/or this is your primary area of interest. However, a simple test to detect the absence of a sense of smell is to bury a piece of smelly cheese or chocolate in the home cage and measure the time it takes homozygous mutants to discover it compared with control mice. As with hearing, mice depend to some extent on olfactory cues for the recognition of their offspring. The loss of the sense of smell could lead to bad parenting, as evidenced by pups scattered out of the nest or even cannibalized by a mutant mother.

Touch and Nociception

The perception of touch allows a mouse to interact with its environment and avoid injury. Touch perception is routinely assessed using the von Frey filament test (e.g., *Drg11*; Chen et al. 2001). The mouse is placed on a wire mesh platform and von Frey filaments (fibers of different diameters) are inserted through the mesh from below, touching the foot pads. The force of the filament on the skin increases as you continue to push the probe until the filament bends. Once the filament bends, the force will be maintained, not increased. Mice will withdraw their paws when they perceive the touch. A threshold for withdrawal is determined by using filaments of increasing diameter.

Nociception is the perception of pain. There are several different types of pain. Acute pain can be a "good" pain because it protects an organism from injury by outside noxious stimuli, such as fire, hot water, etc. In rodents, two simple tests for heat-induced pain are the tail flick test and the hot plate test. The tail-flick test measures reflex spinal nociception. The mouse is restrained in a tube and the tail is either exposed to a radiant heat source or it is immersed in water at 52°C. The amount of time it takes for the mouse to flick its tail away from the heat source or attempt to remove it from the water is measured. The cutoff time is 10 sec. The hot plate test measures supraspinal nociception. By placing the mouse on a hot plate set at 52°C, a wild-type mouse will either lick its hind paws or jump off of the hot plate within seconds, whereas a mutant mouse may have a delayed response or no response at all. The maximum exposure of the mouse to the heat is 30 sec.

Chronic pain perception is measured using an inflammation-induced paradigm, the injection of a dilute solution of formalin or Freund's adjuvant into the mouse's paw. Fifteen microliters of a 5% formalin solution is injected subcutaneously into the ventral surface of one hind foot. A wild-type mouse will feel pain induced by the formalin. The behavioral measurements of this type of pain include favoring, lifting, shaking, and licking the paw. These measurements are taken periodically over ~30 min. Mice with defects in nociception will not show these behaviors or will show these behaviors less frequently than controls.

Before performing any of the above tests, be sure to obtain the necessary approvals from your institutional animal care and use committee, which can provide specific guidelines for any tests of pain perception.

Neurological Tests of Balance and Coordination

Careful observation of mice in their home cage can sometimes reveal behavioral or neurological phenotypes. For example, circling or twirling behavior or a head tilt can indicate vestibular problems. Even though a mouse appears normal by visual inspection, it may still have a neurological defect that can be detected using some very simple tests (Kelly et al. 2003). Below are a few that can be easily performed without complex and expensive equipment. Learning and memory deficits require more sophisticated testing and may be required if your gene is implicated in brain function (https://med.stanford.edu/sbfnl/services/bm/lm.html).

Tail Suspension Test

The tail suspension test is a simple test that can reveal if a mouse has some kinds of neurological defects. It is indicative of a developmental defect, as the response changes with age, but it cannot distinguish between a CNS, PNS, or combined abnormality. Hold the mouse by the tail suspended over the cage. A wild-type mouse will spread its limbs outward, whereas mice with certain neurological defects will hold their limbs into their body, so-called "huggers" (Fig. 4). Wild-type mice at 12–14 d postpartum also show this hugging behavior but they eventually outgrow it.

Grip Strength Test

We have all had the experience of a mouse clinging to the wire lid of a cage as the lid is lifted, sometimes leading to concern about missing mice. This trait of mice can be used as a simple measure of grip strength. Place a mouse on top of a wire cage lid, gently shake the cage lid several times to make sure the mouse is gripping, then invert the lid over the cage at ~25 cm. Normal mice can hold this upside-down position for several minutes. Measuring the length of time it takes a mutant mouse to drop compared to wild-type mice could reveal a neuromuscular defect. Cutoff time is 60 sec.

Swim Test

The ability to swim is a common test of vestibular function of the inner ear. Mice with vestibular defects cannot maintain their orientation to gravity when not in contact with a solid surface. Fill a

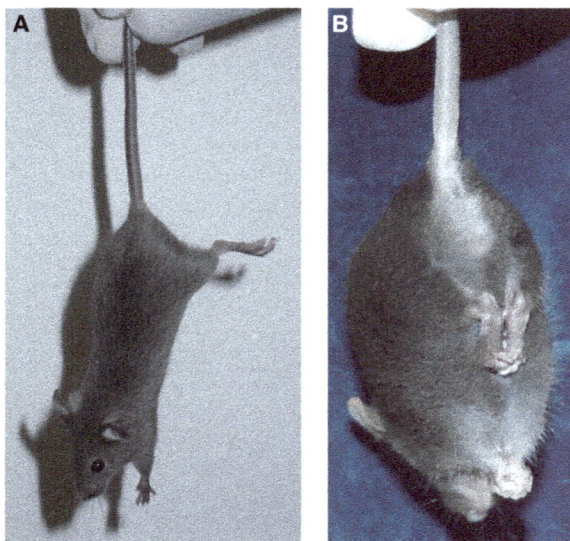

FIGURE 4. Tail suspension test. Wild-type mice will spread their limbs when suspended by their tails (A). However, mice with neurological defects such as Nur30/Nur30 mutants grasp their limbs to their bodies, so-called "huggers" (B). (B, Courtesy of Andy Salinger and Monica Justice; see also, Kile et al. 2003.)

Cite this chapter as Cold Spring Harb Protoc; doi:10.1101/pdb.over107977

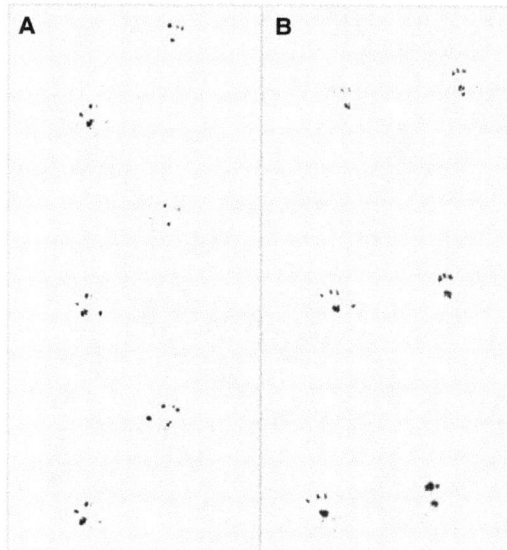

FIGURE 5. Footprints of mice to show gait. Ink was applied to the hindlimbs of wild-type and mutant mice. The mice then walked on white filter paper leaving their tracks. Wild-type mice (*A*) have an alternating gait. In this example, the mutant (*B*) hops.

large container (e.g., a mouse cage or aquarium) with tepid water. Hold the mouse by the tail and lower it into the water. A wild-type mouse will swim ("dog paddle") and keep their head above water. Mice with vestibular defects will be unable to swim or to remain at the surface of the water (e.g.; *Ysb*; Dong et al. 2002). The inability to swim could also be a result of defects in muscles, skeleton, limbs, and nerves. Be ready to rescue mice that cannot swim and dry all the mice off in a warm cage.

Gait Analysis

The gait of a mouse can be informative. Wild-type mice have a regular, alternating gait. Variation from this pattern can indicate ataxia or a morphological defect that causes an abnormal gait. First, watch how the homozygous mutant walks around the cage and compare this to wild-type mice. Are there any obvious differences? To document the gait of a mouse, dip the feet of your mouse in nontoxic ink and have the mouse walk on a large piece of filter paper before the ink dries. You could use different colors of ink for the forelimbs and hindlimbs (Wertman et al. 2019). We find it useful to create a chute through which the mouse can walk, which gives a nice visual record of the gait in one direction. If you identify an abnormal gait in the homozygous mutants (Fig. 5), a video record is useful for further analysis (Preisig et al. 2016).

Balance and Motor Coordination

A simple test of balance and motor coordination is the rotarod test. The rotarod is a machine with a dowel held in a horizontal position that rotates with increasing speed over time (Fig. 6). A wild-type mouse can be trained to stay on the rotarod for a certain period of time. However, it will eventually fall off as the speed is increased. This test is quantitative because you score the number of seconds the mouse can stay on the rotarod. Mice with defects in balance and coordination, possibly indicative of a cerebellar defect, will fall off the rotarod more quickly than wild-type mice (e.g., *En2*; Gerlai et al. 1996).

Genetic Challenges Applied to Viable, Overtly Normal Mutant Mice

A mouse homozygous for a null mutation is, by definition, deficient for a specific gene product. Accordingly, even if it is viable and appears normal, one can consider the genotype of the mouse to be sensitized. Thus, when challenged, the homozygous mutant or even a heterozygote may yield an abnormal phenotype. The complexity of the mammalian genome is likely the explanation for a

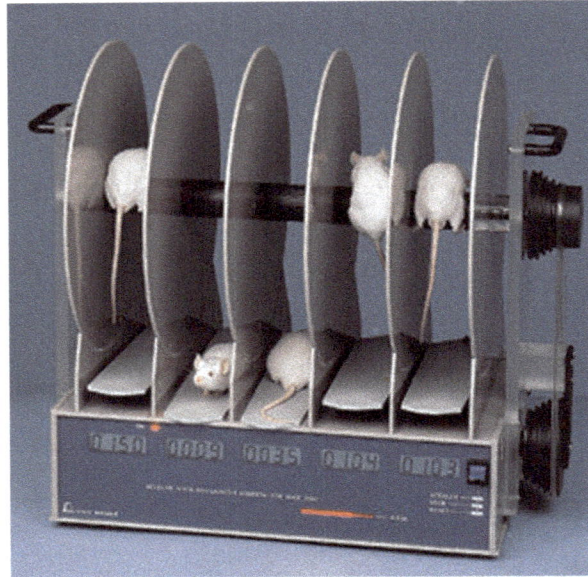

FIGURE 6. Rotarod apparatus. (Image courtesy of myNeuroLab.com, St. Louis, MO.)

high frequency of overtly normal homozygous mutant mice. Therefore, think about the types of challenge experiments to pursue, even at the beginning of each project, based on what you know about the gene and its protein.

Conceptually, two types of challenges can be used: genetic and environmental. Both types can yield information about the role of the mutated gene in development and homeostasis. Of course, the more you know about the gene in question, the easier it will be to decide what type of challenge to perform. A genetic challenge to a mutation is basically a genetic interaction experiment. The gene that is mutated may have redundant functions with another gene (e.g., *MyoD*, *Myf5*; Rudnicki et al. 1993). Alternatively, it may be compensated by the up-regulation of another gene. Adding an additional mutation within the same biochemical pathway may compromise that pathway, leading to a mutant phenotype. A full discussion of crosses to generate double mutants is provided in Chapter 7: Special Breeding Techniques for Use in Mouse Mutation Analysis [Papaioannou and Behringer 2023i]. The expression of transgenes on a homozygous mutant genetic background may also yield a phenotype specific to the targeted mutation.

Genetic Background

As we have mentioned before (Chapter 6: Strategies for Maintaining Mouse Mutations [Papaioannou and Behringer 2023g]), the genetic background on which a mutation is studied can have a great influence on the penetrance and expressivity of the mutant phenotype (e.g., *Egfr*; Threadgill et al. 1995). You can move the mutation onto any inbred genetic background by generating a congenic line (see Protocol 6.1: Backcrossing to Generate a Congenic Mouse Strain [Papaioannou and Behringer 2023j]). The choice of the genetic background can be based on the well-known characteristics of the specific inbred background. You can also move the mutation onto an outbred genetic background to expose the mutation to genetic diversity. An extension of this idea is to cross mice with the mutation to a different species or subspecies of mouse (e.g., *Mus spretus* or *Mus musculus castaneus*). The logic behind this strategy is that a different species or subspecies of mouse will have many more genetic differences relative to a different inbred or even outbred strain to modify the expression of a mutation. If you try this, keep in mind that *Mus musculus*/*M. spretus* F_1 interspecific hybrid males are sterile. Fortunately, the F_1 interspecific hybrid females are fertile. There are no sterility concerns with *M. musculus*/*M. musculus castaneus* F_1 intraspecific hybrid males and females.

Cite this chapter as *Cold Spring Harb Protoc*; doi:10.1101/pdb.over107977

Crosses with a Transgenic Strain

Transgenic mice generated by zygote injections or gene knock-ins express gene products in appropriate and/or ectopic tissues. These transgenic mice can be used to provide a genetic challenge through crosses to a homozygous mutant mouse that is otherwise phenotypically normal, causing it to yield a mutant phenotype. For example, consider a gene that is suspected of being involved in cancer that, when mutated, results in homozygous mutant mice that are viable and overtly normal. One way to challenge these mice genetically would be to use a transgenic mouse line that expresses an oncogene that induces tumor formation. In mice homozygous (or heterozygous) for the engineered mutation that also carry the transgene, tumor formation may be enhanced or suppressed, revealing a role for the engineered mutation in regulating tumorigenesis (Drosten et al. 2018).

Environmental Challenges Applied to Viable, Overtly Normal Mutant Mice

A challenge to a viable, overtly normal homozygous mutant does not necessarily have to be a genetic one. It is also possible, depending on the expression pattern and the putative role of the gene in question, to use environmental challenges to elicit a mutant phenotype from an otherwise normal homozygous mutant. The many types of environmental challenges that can be applied to a mutant include a variety of physiological stresses. Alterations in metabolism could be challenged by feeding the mice a special diet, such as a high-fat diet (e.g., *Fabp4*; Hotamisligil et al. 1996). A high-fat diet could also be used to stress the cardiovascular system by increasing the formation of fat deposits on arterial walls. Chemicals (e.g., carcinogens) and radiation are common challenges to induce tumor formation or reveal defects in DNA repair and genome stability, respectively. A challenge of the immune system can be performed by exposing the mutants to various pathogens. Remember that many of these types of environmental challenges can also be applied to cells derived from the homozygous mutant (e.g., mouse embryonic fibroblasts) to reveal cellular phenotypes. As always, heterozygous mutants might also be affected by such challenges and should be included in any experimental design.

REFERENCES

Behringer RR, Gertsenstein M, Nagy KV, Nagy A. 2014. *Manipulating the mouse embryo: a laboratory manual.* Cold Spring Harbor Laboratory Press, Cold Spring Harbor, NY.

Chen ZF, Rebelo S, White F, Malmberg AB, Baba H, Lima D, Woolf CJ, Basbaum AI, Anderson DJ. 2001. The paired homeodomain protein DRG11 is required for the projection of cutaneous sensory afferent fibers to the dorsal spinal cord. *Neuron* 31: 59–73. doi:10.1016/S0896-6273(01)00341-5

Donehower LA, Harvey M, Slagle BL, McArthur MJ, Montgomery CA Jr, Butel JS, Bradley A. 1992. Mice deficient for p53 are developmentally normal but susceptible to spontaneous tumours. *Nature* 356: 215–221. doi:10.1038/356215a0

Dong S, Leung KK, Pelling AL, Lee PY, Tang AS, Heng HH, Tsui LC, Tease C, Fisher G, Steel KP, et al. 2002. Circling, deafness, and yellow coat displayed by yellow submarine (ysb) and light coat and circling (lcc) mice with mutations on chromosome 3. *Genomics* 79: 777–784. doi:10.1006/geno.2002.6783

Drosten M, Guerra C, Barbacid M. 2018. Genetically engineered mouse models of K-Ras-driven lung and pancreatic tumors: validation of therapeutic targets. *Cold Spring Harb Perspect Med* 8: a031542. doi:10.1101/cshperspect.a031542

Ducy P, Amling M, Takeda S, Priemel M, Schilling AF, Beil FT, Shen J, Vinson C, Rueger JM, Karsenty G. 2000. Leptin inhibits bone formation through a hypothalamic relay: a central control of bone mass. *Cell* 100: 197–207. doi:10.1016/S0092-8674(00)81558-5

Fukushima H, Shimizu K, Watahiki A, Hoshikawa S, Kosho T, Oba D, Sakano S, Arakaki M, Yamada A, Nagashima K, et al. 2017. NOTCH2 Hajdu–Cheney mutations escape SCF[FBW7]-dependent proteolysis to promote osteoporosis. *Mol Cell* 68: 645–658.e645. doi:10.1016/j.molcel.2017.10.018

Gerlai R, Millen KJ, Herrup K, Fabien K, Joyner AL, Roder J. 1996. Impaired motor learning performance in cerebellar En-2 mutant mice. *Behav Neurosci* 110: 126–133. doi:10.1037/0735-7044.110.1.126

Horai R, Saijo S, Tanioka H, Nakae S, Sudo K, Okahara A, Ikuse T, Asano M, Iwakura Y. 2000. Development of chronic inflammatory arthropathy resembling rheumatoid arthritis in interleukin 1 receptor antagonist-deficient mice. *J Exp Med* 191: 313–320. doi:10.1084/jem.191.2.313

Hotamisligil GS, Johnson RS, Distel RJ, Ellis R, Papaioannou VE, Spiegelman BM. 1996. Uncoupling of obesity from insulin resistance through a targeted mutation in aP2, the adipocyte fatty acid binding protein. *Science* 274: 1377–1379. doi:10.1126/science.274.5291.1377

International Mouse Phenotyping Consortium. https://www.mousephenotype.org/ [Accessed October 3, 2022].

Kelly MA, Low MJ, Phillips TJ, Wakeland EK, Yanagisawa M. 2003. The mapping of quantitative trait loci underlying strain differences in locomotor activity between 129S6 and C57BL/6J mice. *Mamm Genome* 14: 692–702. doi:10.1007/s00335-003-2273-0

Kile BT, Hentges KE, Clark AT, Nakamura H, Salinger AP, Liu B, Box N, Stockton DW, Johnson RL, Behringer RR, et al. 2003. Functional genetic analysis of mouse chromosome 11. *Nature* 425: 81–86. doi:10.1038/nature01865

Kuehn MR, Bradley A, Robertson EJ, Evans MJ. 1987. A potential animal model for Lesch–Nyhan syndrome through introduction of HPRT mutations into mice. *Nature* 326: 295–298. doi:10.1038/326295a0

Larina IV, Syed SH, Sudheendran N, Overbeek PA, Dickinson ME, Larin KV. 2012. Optical coherence tomography for live phenotypic analysis of embryonic ocular structures in mouse models. *J Biomed Opt* 17: 081410-1. doi:10.1117/1.JBO.17.8.081410

Mattapallil MJ, Wawrousek EF, Chan CC, Zhao H, Roychoudhury J, Ferguson TA, Caspi RR. 2012. The *Rd8* mutation of the *Crb1* gene is present in vendor lines of C57BL/6N mice and embryonic stem cells, and confounds ocular induced mutant phenotypes. *Invest Ophthalmol Vis Sci* **53**: 2921–2927. doi:10.1167/iovs.12-9662

Nagy A, Gertsenstein M, Vintersten K, Behringer R. 2009. Alizarin red staining of post-natal bone in mouse. *Cold Spring Harb Protoc* doi:10.1011/pdb.prot5171

National Mouse Metabolic Phenotyping Center. 2017. https://www.mmpc.org/ [Accessed October 3, 2022].

NIH Mouse Phenotyping. https://ors.od.nih.gov/sr/dvr/drs/Pages/Mouse%20Phenotyping.aspx [Accessed October 3, 2022].

Ohtoshi A, Wang SW, Maeda H, Saszik SM, Frishman LJ, Klein WH, Behringer RR. 2004. Regulation of retinal cone bipolar cell differentiation and photopic vision by the CVC homeobox gene *Vsx1*. *Curr Biol* **14**: 530–536. doi:10.1016/j.cub.2004.02.027

Papaioannou VE, Behringer RR. 2023a. Mouse gene-targeting strategies for maximum ease and versatility. *Cold Spring Harb Protoc* doi:10.1011/pdb.over107957

Papaioannou VE, Behringer RR. 2023b. Phenotypic analysis: assessing timing of recessive prenatal lethality in mice. *Cold Spring Harb Protoc* doi:10.1011/pdb.over107970

Papaioannou VE, Behringer RR. 2023c. Phenotypic analysis of preimplantation lethality in mice. *Cold Spring Harb Protoc* doi:10.1011/pdb.over107971

Papaioannou VE, Behringer RR. 2023d. Phenotypic analysis of periimplantation to mid-gestation lethality in mice. *Cold Spring Harb Protoc* doi:10.1011/pdb.over107972

Papaioannou VE, Behringer RR. 2023e. Analysis of mid- to late-gestation phenotypes in mice. *Cold Spring Harb Protoc* doi:10.1011/pdb.over107973

Papaioannou VE, Behringer RR. 2023f. Uncovering phenotypes in mutant mice by determining embryo, organ, tissue, and cell development potential in mice. *Cold Spring Harb Protoc* doi:10.1011/pdb.over107974

Papaioannou VE, Behringer RR. 2023g. Strategies for maintaining mouse mutations. *Cold Spring Harb Protoc* doi:10.1011/pdb.over107960

Papaioannou VE, Behringer RR. 2023h. Necropsy guide for the collection of tissues from mice with or without tumors. *Cold Spring Harb Protoc* doi:10.1011/pdb.prot108097

Papaioannou VE, Behringer RR. 2023i. Special breeding techniques for use in mouse mutation analysis. *Cold Spring Harb Protoc* doi:10.1011/pdb.over107961

Papaioannou VE, Behringer RR. 2023j. Backcrossing to generate a congenic mouse strain. *Cold Spring Harb Protoc* doi:10.1011/pdb.prot108039

Preisig DF, Kulic L, Krüger M, Wirth F, McAfoose J, Späni C, Gantenbein P, Derungs R, Nitsch RM, Welt T. 2016. High-speed video gait analysis reveals early and characteristic locomotor phenotypes in mouse models of neurodegenerative movement disorders. *Behav Brain Res* **311**: 340–353. doi:10.1016/j.bbr.2016.04.044

Rudnicki MA, Schnegelsberg PN, Stead RH, Braun T, Arnold HH, Jaenisch R. 1993. MyoD or Myf-5 is required for the formation of skeletal muscle. *Cell* **75**: 1351–1359. doi:10.1016/0092-8674(93)90621-V

Stanford Medicine Behavioral and Functional Neuroscience Laboratory. 2022. https://med.stanford.edu/sbfnl/services/bm/lm.html [Accessed October 3, 2022].

Takeo T, Nakagata N. 2018. In vitro fertilization in mice. *Cold Spring Harb Protoc* doi:10.1011/pdb.prot094524

Thaung C, Arnold K, Jackson IJ, Coffey PJ. 2002. Presence of visual head tracking differentiates normal sighted from retinal degenerate mice. *Neurosci Lett* **325**: 21–24. doi:10.1016/S0304-3940(02)00223-9

Threadgill DW, Dlugosz AA, Hansen LA, Tennenbaum T, Lichti U, Yee D, LaMantia C, Mourton T, Herrup K, Harris RC, et al. 1995. Targeted disruption of mouse EGF receptor: effect of genetic background on mutant phenotype. *Science* **269**: 230–234. doi:10.1126/science.7618084

Wada K, Matsushima Y, Tada T, Hasegawa S, Obara Y, Yoshizawa Y, Takahashi G, Hiai H, Shimanuki M, Suzuki S, et al. 2014. Expression of truncated PITX3 in the developing lens leads to microphthalmia and aphakia in mice. *PLoS ONE* **9**: e111432. doi:10.1371/journal.pone.0111432

Wertman V, Gromova A, La Spada AR, Cortes CJ. 2019. Low-cost gait analysis for behavioral phenotyping of mouse models of neuromuscular disease. *J Vis Exp* **2019**: 10.3791/59878. doi:10.3791/59878

Zhang J, Li J, Huang C, Xue L, Peng Y, Fu Q, Gao L, Zhang J, Li W. 2008. Targeted knockout of the mouse βB2-crystallin gene (Crybb2) induces age-related cataract. *Invest Ophthalmol Vis Sci* **49**: 5476–5483. doi:10.1167/iovs.08-2179

Necropsy Guide for the Collection of Tissues from Mice with or without Tumors

> Mice that die at any stage of a mutational analysis, whether during early life or during ageing or longitudinal studies such as tumor survival studies, can yield important information. This protocol provides a necropsy guide for the collection and processing of tissue samples to provide material for complete histological or immunostaining analysis.

INTRODUCTION

During a longitudinal tumor study to determine a tumor-free survival curve, regular monitoring will identify sick or dead animals or mice with visible signs of tumors. Mice with tumors and moribund animals should be sacrificed for necropsy and collection of tumor and tissue samples for histopathology. Necropsy on dead animals can also yield useful data provided the animals have not degraded significantly. For each animal that is sacrificed or found dead, a complete histopathological workup, including normal tissue as well as any tumors, will provide valuable information as part of the tumor study.

MATERIALS

It is essential that you consult the appropriate Material Data Safety sheets and your institution's Environmental Health and Safety Office for proper handling of equipment and hazardous materials used in this protocol.

RECIPES: Please see the end of this protocol for recipes indicated by <R>. Additional recipes can be found online at http://cshprotocols.cshlp.org/site/recipes.

Reagents

Appropriate fixative for histology or immunostaining (e.g., Bouin's (Bouin's fixative <R>; 4% paraformaldehyde [PFA] <R>) or other special stains) (see Nagy et al. 2007).

Equipment

Dissection instruments

METHOD

1. Generate a list of tissues that will be collected from every mouse in addition to samples of any tumors detected. This may depend on the nature and expression of the gene under study or on information such as tissue susceptibility to tumor formation.

   You can never take too many samples because you cannot predict whether you will have to go back to a certain tissue for more analysis.

2. Perform necropsy as soon as the mouse is euthanized or is found dead. If the necropsy cannot be performed immediately, refrigerate the body. Never freeze the body because this will cause tissue distortion.

   However, specific tissues may be frozen for RNA (transcriptome) and DNA (loss of heterozygosity or copy number analysis) studies. In addition, specific tissues can be processed for frozen sections.

3. Fix tissues minimally at a 1:10 ratio of tissue to fixative at room temperature (Bouin's) or 4°C (PFA).

   Ideally, tissue samples should be no thicker than 5 mm for good penetration of the fixative. The length and width can be greater. Sometimes you may have no choice but to fix a large sample (e.g., the entire head).

4. Keep detailed notes as part of your permanent record during necropsy. Devise a numbering system for each animal and tissue to be used consistently during the entire duration of the tumor study, which may last for several years and involve multiple investigators. Describe the location, size, shape, pattern, amount, consistency, involvement, and color of each tumor. Take tissue samples from the tumor and nearby control tissues.

5. Acquire images of the tumor in situ and dissected away from the body. Establish an image numbering system to link images to a specific mouse and tissue sample (see Step 4).

RECIPES

Bouin's Fixative

2 g picric acid powder *or* saturated aqueous picric acid (see below)

   Picric acid is explosive in crystal form!

20 g paraformaldehyde
Formalin (40% w/v formaldehyde), as an alternative to paraformaldehyde (see below)
NaOH (1 N)
2× Phosphate-buffered saline (PBS)
Glacial acetic acid (to be used with saturated picric acid and formalin; see below)

To prepare 1 L of Bouin's fixative, dissolve 2 g of picric acid in 500 mL of H_2O. Filter through a Whatman No. 1, or equivalent. Add 20 g of paraformaldehyde, and heat to 60°C in a fume hood. Add a few drops of 1 N NaOH to dissolve. Cool and add 500 mL of 2× PBS.

Alternatively, combine 75 mL of saturated picric acid, 25 mL of formalin, and 5 mL of glacial acetic acid. Store at 4°C.

Paraformaldehyde (PFA; 4%)

Reagent	Quantity (for 100 mL)	Final concentration
Paraformaldehyde (PFA)	4 g	4%
Phosphate-buffered saline for immunohistochemistry (10×), diluted to 1× <R>	100 mL	

Slowly dissolve PFA in 1× phosphate-buffered saline over low heat until solution clears. Cool. Use immediately or aliquot and store for up to 1 yr at −20°C.

REFERENCE

Nagy A, Gertsenstein M, Vintersten K, Behringer R. 2007. Fixation of mouse embryos and tissues. *Cold Spring Harb Protoc* doi:10.1011/pdb.prot4702

CHAPTER 16

Phenotypic Analysis of Dominant Mutant Effects in Mice

Dominant effects of a mutation may show up at any time during a mutational analysis, including during the early stages of an embryonic stem (ES) cell gene targeting experiment. Here, we discuss the mechanisms of dominant and semidominant effects and how they might appear if they show up in heterozygous ES cells, in ES cell chimeras, or in heterozygous progeny of chimeras. Similarly, dominant effects may be seen in mice heterozygous for CRISPR–Cas-targeted, -induced, or spontaneous mutations. If the dominant effects prevent the germline transmission of ES cells or cause fertility problems in heterozygotes, they can severely limit further analysis of the mutation. Ways to circumvent such reproductive problems are presented. The special case of imprinted genes, which may be functionally hemizygous and present a different phenotype when inherited from the mother than when inherited from the father, is discussed.

INTRODUCTION

PHENOTYPIC ANALYSIS OF DOMINANT MUTATIONS

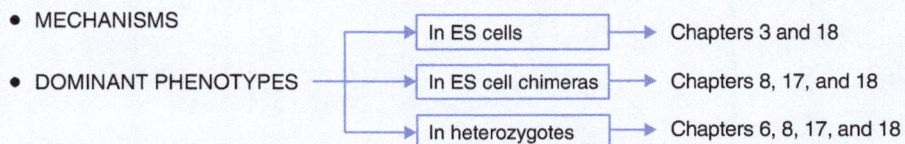

- MECHANISMS → In ES cells → Chapters 3 and 18
- DOMINANT PHENOTYPES → In ES cell chimeras → Chapters 8, 17, and 18
  → In heterozygotes → Chapters 6, 8, 17, and 18

Strictly speaking, a dominant mutation is one that produces a specific phenotype in heterozygous mice and the same phenotype in homozygous mutants. A semidominant mutation is one in which there is a specific phenotype in heterozygotes, but an even more severe version of the phenotype in homozygous mutants (e.g., *Brachyury*). Here, for the practical purpose of mutation analysis, we consider a dominant effect to be any phenotype seen in a heterozygous mouse. When more is known about how and why a phenotype appears and whether the homozygous effect is the same or more severe, the mutation can be classified as dominant or semidominant. Dominant effects of a mutation can appear at almost any stage in the process of making a targeted mutation or at any stage of development either in targeted, induced, or spontaneously occurring mutations. This chapter provides additional information about dominant effects and how to deal with them. If you have a mutant mouse with a dominant

Opening artwork: Courtesy of Virginia E. Papaioannou.

Cite this chapter as *Cold Spring Harb Protoc*; doi:10.1101/pdb.over107978

phenotype, you might want to take a detour now through Chapter 6: Strategies for Maintaining Mouse Mutations (Papaioannou and Behringer 2023a) for hints on breeding in general and for specific tips on breeding dominant fertility mutants.

MECHANISMS OF DOMINANT EFFECTS

The dominant effects of a mutation can occur in several ways depending on the nature of the genetic lesion (Fig. 1). If a mutation results in a null allele such that there is no gene product produced, a heterozygous phenotype could be the result of haploinsufficiency for the gene product. This pheno-

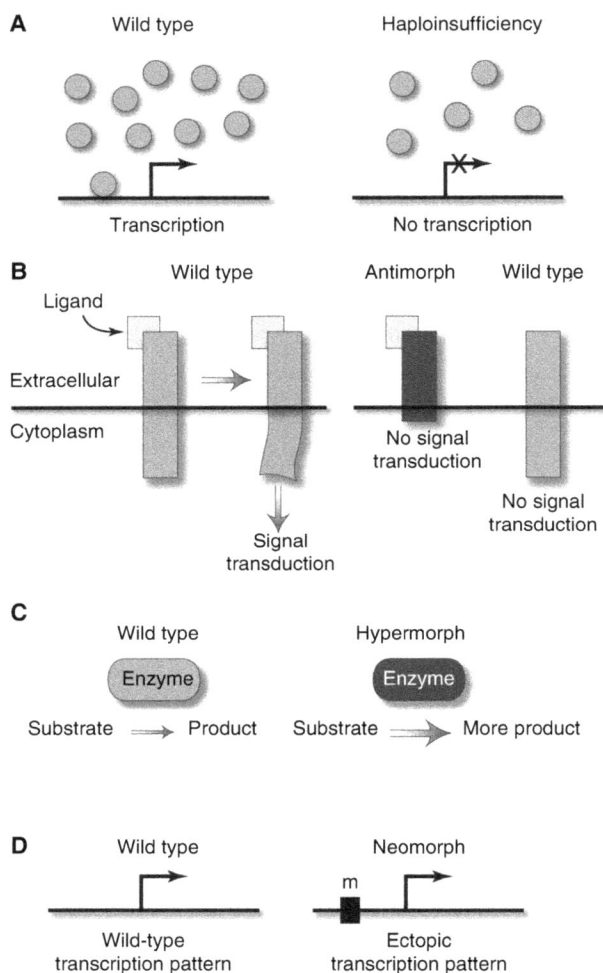

FIGURE 1. Mechanisms of action of dominant mutations. (A) Haploinsufficiency. In this example, wild-type levels of a transcription factor are required to achieve a critical concentration to activate transcription of a target gene. When there is only one dose of the gene, the level of protein is too low (i.e., the concentration is too low) to activate target gene transcription. (B) Dominant negative or antimorph. In this example, binding of a ligand to a wild-type transmembrane receptor induces a conformational change of its cytoplasmic domain that causes a signal to be transduced. A mutant version of the receptor (darker shading) can bind the ligand but because the mutation truncates the protein, it lacks the cytoplasmic domain and cannot transduce a signal. If the ligand is at limiting concentration, the mutant receptors could sequester it from the wild-type receptors resulting in reduced or no signal transduction. (C) Hypermorph. In this example, a wild-type enzyme acts on a substrate to catalyze the production of a product. The mutant protein (darker shading) can catalyze this reaction more efficiently, leading to a greater production of the product. (D) Neomorph. In this example, a wild-type allele is expressed in a specific pattern. The mutation (*black box* [m]) causes a change in the tissue specificity of transcription, leading to a new, ectopic site of expression.

Cite this chapter as *Cold Spring Harb Protoc*; doi:10.1101/pdb.over107978

type implies that there is a dose-sensitive requirement for the gene product and that the haploid amount is not sufficient to produce a wild-type phenotype. Other mechanisms for obtaining a dominant effect would be a mutation that results in the production of an altered protein that in some way interferes with the function of the wild-type protein (an antimorph or dominant negative), a mutation that results in a hyperactive protein (a hypermorph), or a mutation that encodes an altered protein or changes the expression pattern to produce a new activity, possibly unrelated to the function of the wild-type protein (a neomorph). In these cases, it is not the lowered level of wild-type protein that causes the phenotype, but rather the presence of an abnormal protein or an abnormal expression pattern.

CRISPR–Cas gene editing adds a further complication. It is very possible that the CRISPR–Cas founders you generated are mosaics—that is, an individual founder may have multiple populations of cells with various alleles that may have different activities. Some cells may be heterozygous and some cells may have both alleles mutated. Thus, it is possible to observe a phenotype in your CRISPR–Cas founders, but because of the genetic complexity that can occur in a CRISPR–Cas founder, it is challenging to determine if this phenotype is a dominant effect. It is most prudent to segregate the mutant alleles in progeny and then determine if the heterozygotes display a dominant phenotype.

For null alleles, there are several special cases, listed below, in which the loss of a single allele can result in the complete lack of a gene product rather than just haploinsufficiency, and in these situations, a mutation is quite likely to act as a dominant. This could become evident at any developmental stage.

Here are some situations in which loss of a single allele results in complete loss of a gene product.

1. *A mutation in an X-chromosome-linked gene.* In male mice (as well as male embryonic stem [ES] cells), which have only one X chromosome, a mutation in an X-linked gene results in the only copy of the gene being mutated (e.g., *Ar*; Yeh et al. 2002). In females, most tissues undergo random X-chromosome inactivation and, thus, in females heterozygous for an X-linked mutation, half of the cells will have only one functional wild-type allele and the other half will have no functional wild-type alleles, creating a mosaic of cells with and without the wild-type gene product (Fig. 2A,B). In certain embryonic tissues, such as the trophectoderm, primitive endoderm, and their derivatives, there is preferential inactivation of the paternally derived X chromosome. An X-linked mutation will result in an effective null in these tissues in heterozygous females—that is, a dominant effect if the mutant allele is inherited from the mother, but no effect if it is inherited from the father (Fig. 2C,D; see also Fig. 2 in Chapter 2: Obtaining or Generating Gene Mutations in Mice [Papaioannou and Behringer 2023b]).

2. *A mutation in a Y-chromosome-linked gene.* Just like mutations in X-chromosome-linked genes, Y-chromosome-linked gene mutations will show dominant effects in males (or male ES cells) because the only copy of the gene is mutated (e.g., *Ddx3y*; Rohozinski et al. 2002). A likely consequence of a Y-chromosome gene mutation is male infertility because many of the genes on the Y chromosome are involved in spermatogenesis.

3. *Imprinted autosomal genes resulting in an effective null with only a single copy of the mutated allele.* These are genes that are epigenetically inactivated depending on whether they are inherited through the male or female germline (i.e., from the father or the mother). For example, if a gene is inactivated by imprinting when it is derived from the father, a null mutation coming from the mother will result in an embryo with a null phenotype, even though it is genetically heterozygous (see Fig. 3 in Chapter 2: Obtaining or Generating Gene Mutations in Mice [Papaioannou and Behringer 2023b]) (e.g., *Igf2*; DeChiara et al. 1991).

If you know that your gene is X-linked or imprinted, these possibilities should have already been considered. If the gene is not X-linked and does not reside in a chromosomal region already known to be subject to imprinting, it could still be imprinted. Imprinting is easily checked by determining

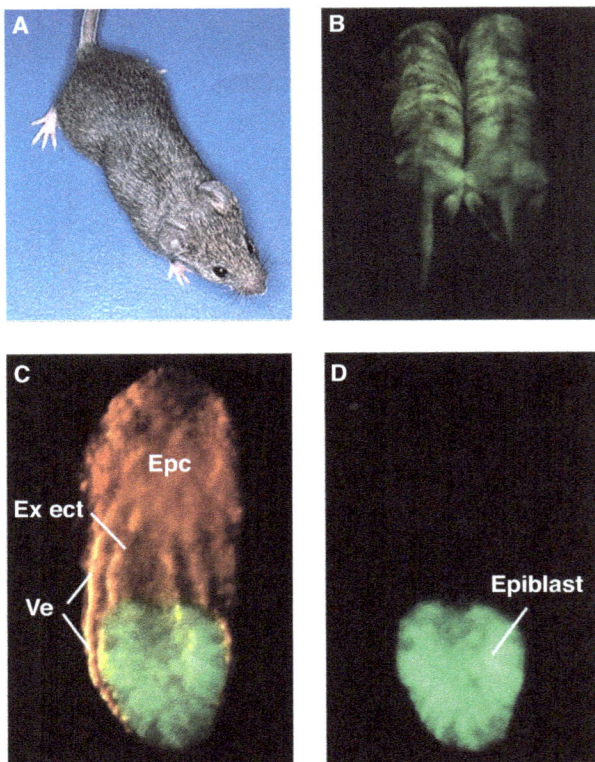

FIGURE 2. Examples of X-chromosome inactivation. (A) *Tabby* is an X-chromosome-linked mutation (*Eda^Ta*) that affects hair, tooth, and endocrine gland development. In heterozygous females, random X-chromosome inactivation on an agouti genetic background leads to coat color variegation. Note the fine transverse stripes of light-colored normal and dark tabby hair. (B) Newborn female mice carrying one copy of an X-linked green fluorescent protein (GFP) transgene. Random inactivation of one X chromosome results in a mosaic fluorescent/nonfluorescent pattern in the skin. (C,D) Nonrandom inactivation of the paternally derived X chromosome in extraembryonic tissues. This embryonic day 7.5 female embryo inherited an X-linked GFP transgene from its father. There is a mosaic of GFP-expressing and nonexpressing cells in the embryonic portion of the embryo, the epiblast (as seen by fluorescence microscopy in both *C* and *D*), indicating random X inactivation, whereas the extraembryonic region is nonfluorescent (seen as orange in *C*, which is an overlay of bright-field and fluorescence images) because of preferential inactivation of the paternally derived X chromosome in the extraembryonic tissues. (epc) Ectoplacental cone, (ex ect) extraembryonic ectoderm, (ve) visceral endoderm. (*A*, Courtesy of the Mouse Genome Informatics; http://www.informatics.jax .org [The Jackson Laboratory, Bar Harbor, Maine]; *B–D*, courtesy of Kat Hadjantonakis.)

whether the mutation has the same dominant effect when inherited from the mother as it does when inherited from the father.

DOMINANT EFFECTS AT DIFFERENT STAGES OF MUTATION ANALYSIS

The following sections present special considerations for dealing with dominant effects appearing in ES cells, in chimeras, in mice heterozygous for spontaneous or induced mutations, and in mice with mutations in imprinted genes.

Dominant Effects in ES Cells

We have pointed out the difficulty in determining if a heterozygous mutation causes the death of ES cells, due to a dominant mutant effect (see Chapter 4: Embryonic Stem Cell Gene Targeting and Chimera Production in Mice [Papaioannou and Behringer 2023c]), although it is a formal possibility. Because the outcome of such an effect would be the lack of targeted ES cell colonies, the same as if

Cite this chapter as *Cold Spring Harb Protoc*; doi:10.1101/pdb.over107978

there were no targeting events at all, it will be virtually impossible to determine a dominant effect because there will never be any cells to study. If the gene is X-chromosome-linked and you have been using an XY ES cell line without success, you might be able to circumvent the problem by using an XX ES cell line for gene targeting. If the gene is Y-linked, this will not be an option. Targeting a gene in an imprinted region should not create this type of problem because only one of the two alleles is likely to be imprinted, and furthermore, imprints appear to be lost altogether in some ES cell lines.

If the gene is neither sex-chromosome-linked nor imprinted, a genuine dominant effect that prevents the recovery of targeted ES cell clones would have to affect a basic cellular function, such as proliferation, metabolism, and adhesion, essential for ES cell survival in tissue culture. In the case of a null mutation, it is possible but unlikely that a truly basic cellular function would be subject to such acute dose sensitivity. If the mutation is a dominant negative (antimorph) or neomorph, the aberrant protein might be able to wreak havoc on critical basic cellular functions. In either case, a possibility for further study would be to make a different type of mutation, e.g., a conditional null or a hypomorph or to use a so-called rescuing transgene (Box 1).

Dominant Effects in ES Cell Chimeras

Another unlikely but formally possible situation is a dominant effect that does not affect ES cell growth, but causes the death of chimeric embryos. This could indicate a non-cell-autonomous effect such that the presence of the mutant cells is actually detrimental to the development of the embryo as a whole. It could also be the result of the ES cells initially making very large contributions to critical tissues in the chimera, but failing later because of the effects of the mutation, causing death of the chimera. However, a much more likely outcome of haploinsufficiency would be poor contribution of the heterozygous ES cells to chimeras or contributions only to specific (unaffected) cell and tissue types. On the other hand, for example, a neomorph or dominant-negative (antimorph) mutation of a critical signaling molecule might well disrupt the development of a chimera. In either case, further analysis could be accomplished by making a constant supply of chimeras and following the analytical steps outlined in the following associated chapters (Chapter 8: Phenotypic Analysis: Assessing Timing of Recessive Prenatal Lethality in Mice [Papaioannou and Behringer 2023d]; Chapter 9: Phenotypic Analysis of Preimplantation Lethality in Mice [Papaioannou and Behringer 2023e]; Chapter 10: Phenotypic Analysis of Periimplantation to Mid-Gestation Lethality in Mice [Papaioannou and Behringer 2023f]; Chapter 11: Analysis of Mid- to Late-Gestation Phenotypes in Mice [Papaioannou and Behringer 2023g]; Chapter 12: Uncovering Phenotypes in Mutant Mice by Determining Embryo, Organ, Tissue, and Cell Developmental Potential [Papaioannou and Behringer 2023h]) for homozygous mutants to determine the time and cause of death of the chimeras. For this analysis, the inclusion of a cell marker specific to either the ES cells or the embryonic cells is essential for recognizing chimeras and determining the tissue distribution of the heterozygous ES cells throughout development (see Chapter 17: Getting Around an Early Lethal Phenotype in Mice with Chimeras [Papaioannou and Behringer 2023i]).

Any number of genes might have dominant effects that could be seen in neonatal or adult chimeras (e.g., *Zic3*; Purandare et al. 2002). One such effect could even mask chimerism: If the mutation you made affects coat color, it is possible that a dominant effect would show up in chimeras. For example, in pigmented ES cell ⟷ albino embryo chimeras, a dominant mutation causing the death of melanoblasts could result in a cryptic chimera with an all-white coat. Similarly, in agouti ⟷ non-agouti chimeras, a dominant effect on the agouti signaling pathway might affect the agouti pattern of chimeras. The severity of a dominant phenotype in chimeras will usually be correlated with the level of contribution of ES cells to the affected tissue. If the effect is not too severe and does not affect the reproductive capacity of the chimeras, the mutation should still be recoverable from the germline of the chimeras and can then be studied in heterozygous or homozygous mutant mice, as outlined above and in Chapter 13: Phenotypic Analysis of Perinatal Lethality in Mice (Papaioannou and Behringer 2023j); Chapter 14: Analysis of Postnatal Mutant Phenotypes in Mice (Papaioannou and Behringer

BOX 1. TRANSGENE RESCUE OF MUTANT ES CELLS, CHIMERAS, OR EMBRYOS

A dominant mutation that causes an ES cell-lethal, male-sterile, or embryo-lethal phenotype can be hard to study because you will not recover targeted ES cell clones, generate germline-transmitting male chimeras, or obtain viable heterozygous mutants after birth, respectively. To study these types of dominant mutations, one option is to use a transgene (e.g., a wild-type copy of the gene being mutated) to rescue the mutant phenotype. However, hypermorphic and neomorphic mutations cannot be rescued with a wild-type transgene, and transgene rescue depends on wild-type ES cells being tolerant of extra copies of the gene in question. If you suspect a dominant, ES cell–lethal phenotype, you can introduce a wild-type copy of the gene of interest into wild-type ES cells (e.g., by electroporation) and then use gene targeting to mutate an endogenous allele to recover heterozygous mutant cell lines that are rescued by the wild-type transgene (Fig. 3A). If heterozygous cell lines are rescued, this provides formal proof of a dominant, ES cell-lethal phenotype and might encourage you to make a hypomorphic or conditional allele for further study.

If you suspect a dominant male-sterile or embryo-lethal phenotype, you can introduce a wild-type copy of the gene of interest into already targeted ES cells. Generate male chimeras using the transgene-containing targeted ES cells and breed them with wild-type females. The transgene may rescue a dominant male sterility in the chimeras, leading to germline transmission and the recovery of heterozygous progeny with or without the transgene (Fig. 3B). The male progeny without the transgene can be used to study the dominant effects of the mutation on spermatogenesis, whereas the mutation itself can be maintained by breeding heterozygous females or the heterozygous male progeny carrying the transgene. A transgene might also rescue a dominant embryo lethality. In this case, only progeny carrying the transgene will be viable (Fig. 3C). Use the heterozygous progeny with the rescuing transgene to breed with wild-type mice. Provided that the integration of the rescuing transgene is not linked tightly to the endogenous locus, segregation of the genes will provide nontransgenic heterozygous mice in the resulting litters that can be used for analysis of the dominant embryo-lethal mutant phenotype. Of course, you will have to genotype for both endogenous alleles and the rescuing transgene to find the desired embryos.

The rescuing transgene can be designed so that it is under the control of its endogenous regulatory sequences or heterologous sequences, depending on the situation. Although conceptually simple, in practical terms it is not always clear which endogenous or heterologous sequences will be most effective for transgene expression in the correct temporal and spatial pattern. Rarely are the endogenous sequences that regulate gene expression thoroughly characterized. In addition, a heterologous regulatory element may not be sufficient for relevant transgene expression, or it may lead to overexpression and a gain-of-function phenotype.

2023k); Chapter 15: The "No Phenotype" Challenge in Analyzing Mutant Mice (Papaioannou and Behringer 2023l).

If, however, the dominant effect prevents germline transmission of the ES-derived genotype, it will be harder to recover the mutation and establish it in the germline initially; it will be the same as if there were no germline transmission. Any dominant effects limited to gametogenesis in one sex or the other could be circumvented by breeding the mutation through the germline of the other sex and then producing sterile offspring of the affected sex for a gametogenesis study. However, in the case of a dominant effect on spermatogenesis, or a haploid effect on sperm function, the initial problem of lack of germline transmission (assuming that you are using an XY ES cell line) will require getting the mutation into the female germline either by breeding from female chimeras or by repeating the gene-targeting experiment using a female ES cell line, and hoping that the effect on gametogenesis does not extend to oogenesis. Alternatively, you could try a rescuing transgene (Box 1) or isolating an XO ES cell line from the targeted XY ES cells (see Protocol 16.1: Isolation of XO Subclones from XY Murine Embryonic Stem Cells [Papaioannou and Behringer 2023m]). Because most gene-targeting experiments make use of XY ES cells, any dominant effects on oogenesis will usually not be revealed until the mutation is safely in the germline, in which case it can be maintained by breeding through the male.

If a dominant gametogenesis effect produces sterility in both sexes, one way of getting a mutation in ES cells into the germline of mice is to produce a conditional mutation in which the gene remains functional throughout gametogenesis but can be mutated at will using, for example, an inducible Cre to study the gametogenesis phenotype in subsequent generations (see Chapter 18: Tissue- and/or

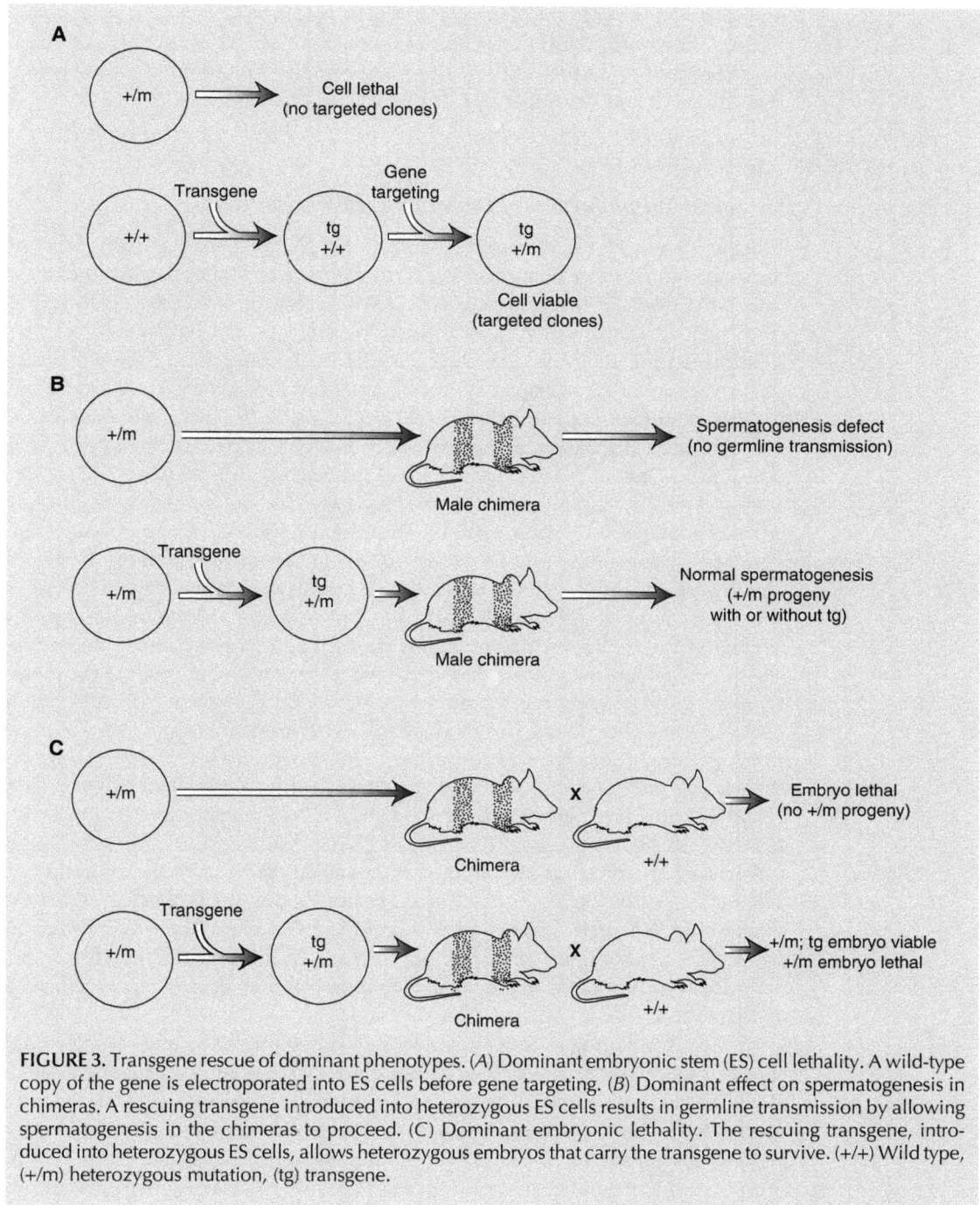

FIGURE 3. Transgene rescue of dominant phenotypes. (*A*) Dominant embryonic stem (ES) cell lethality. A wild-type copy of the gene is electroporated into ES cells before gene targeting. (*B*) Dominant effect on spermatogenesis in chimeras. A rescuing transgene introduced into heterozygous ES cells results in germline transmission by allowing spermatogenesis in the chimeras to proceed. (*C*) Dominant embryonic lethality. The rescuing transgene, introduced into heterozygous ES cells, allows heterozygous embryos that carry the transgene to survive. (+/+) Wild type, (+/m) heterozygous mutation, (tg) transgene.

Temporal-Specific Mutations in Mice Using Conditional Alleles [Papaioannou and Behringer 2023n]). Alternatively, you might try using a rescuing transgene (Box 1). This involves electroporating a wild-type transgene for the targeted allele into the targeted ES cells, selecting for random integrants, and then making chimeras. The transgene could rescue the mutant effect, allowing the targeted endogenous mutant allele to be transmitted through the germline.

Finally, a dominant mutant effect on sex determination might show up in chimeras and result in feminization, even if chimeras are made with XY ES cells. Alternatives for dealing with this problem would be to retarget an XX ES cell line, isolate XO ES cell lines from the targeted XY ES cells to get the mutation into the germline through the female (see Protocol 16.1: Isolation of XO Subclones from XY Murine Embryonic Stem Cells [Papaioannou and Behringer 2023m]), or make a conditional allele (see Chapter 18: Tissue- and/or Temporal-Specific Mutations in Mice Using Conditional Alleles [Papaioannou and Behringer 2023n]).

Dominant Effects in Heterozygous Mice

Dominant Morphological or Developmental Problems

Whether or not ES cell chimeras demonstrate a phenotype, a dominant effect of the mutation could show up in the heterozygous offspring of the chimeras any time after fertilization. A dominant effect that causes the death of heterozygous embryos such that no heterozygous offspring are recovered from chimeras that transmit the ES cell coat color genotype can be handled in one of several ways. One solution is to analyze the phenotype directly in the offspring of the chimeras using the diagnostic procedures outlined in Chapter 8: Phenotypic Analysis: Assessing Timing of Recessive Prenatal Lethality in Mice (Papaioannou and Behringer 2023d); Chapter 9: Phenotypic Analysis of Preimplantation Lethality in Mice (Papaioannou and Behringer 2023e); Chapter 10: Phenotypic Analysis of Periimplantation to Mid-Gestation Lethality in Mice (Papaioannou and Behringer 2023f); Chapter 11: Analysis of Mid- to Late-Gestation Phenotypes in Mice (Papaioannou and Behringer 2023g); Chapter 12: Uncovering Phenotypes in Mutant Mice by Determining Embryo, Organ, Tissue, and Cell Developmental Potential (Papaioannou and Behringer 2023h) for recessive mutations. If you have a good supply of chimeras that transmit ~100% of the ES cell genotype, then this might be feasible, although eventually you will be faced with replenishing the stock of chimeras to keep up the supply of embryos. Another alternative is to use a rescuing transgene (Box 1). Finally, you could consider producing a conditional allele to perpetuate the mutant allele in the germline of mice but produce the mutant effect at a time and or tissue of your choosing for further analysis (see Chapter 18: Tissue- and/or Temporal-Specific Mutations in Mice Using Conditional Alleles [Papaioannou and Behringer 2023n]). At the same time, you might also want to test-breed the chimeras with a different strain of mouse, just in case there is a strong genetic background effect and the mutants could survive on a different background.

Less severe developmental or physiological defects might also show up as dominant effects of the mutation in the heterozygous offspring of ES cell chimeras or in mice heterozygous for CRISPR–Cas, induced, or spontaneous mutations. For example, a skewing of the sex ratio in the heterozygous, ES cell–derived offspring of chimeras could indicate a dominant effect. If all heterozygous offspring are phenotypically female, you may have a dominant sex reversal effect. However disconcerting this may be, the XX heterozygous females should breed just fine and the mutation can be established in the germline. The dominant sex reversal phenotype can then be studied in the heterozygous XY offspring of these mice.

Other defects might be compatible with development to term or beyond, but still show up as dominant phenotypic effects (Fig. 4) (e.g., *Tbx1*, Jerome and Papaioannou 2001; *Sox9*, Bi et al. 2001). Be on the lookout for any characteristic phenotype associated with the heterozygous genotype and analyze it in the same way that you would a homozygous mutant phenotype, with the caveat that phenotypes caused by imprinted genes must be approached differently. If it is a readily detectable abnormality and is 100% penetrant in heterozygotes, you can even use the dominant trait to genotype the animals. Do not make the assumption, however, that you have a dominant, as opposed to semidominant, mutation. It may be that the homozygous mutant phenotype is even more severe (Fig. 4) and can provide further information regarding gene function. Thus, you will want to mate heterozygotes together to produce litters with all three genotypes for a complete assessment of the heterozygous and homozygous mutant phenotypes.

Dominant Reproductive Problems

Even if you had no trouble getting germline transmission from chimeras, you may not be home free from dominant reproductive problems (e.g., *Klhl10*, Yan et al. 2004). Dominant effects on the process

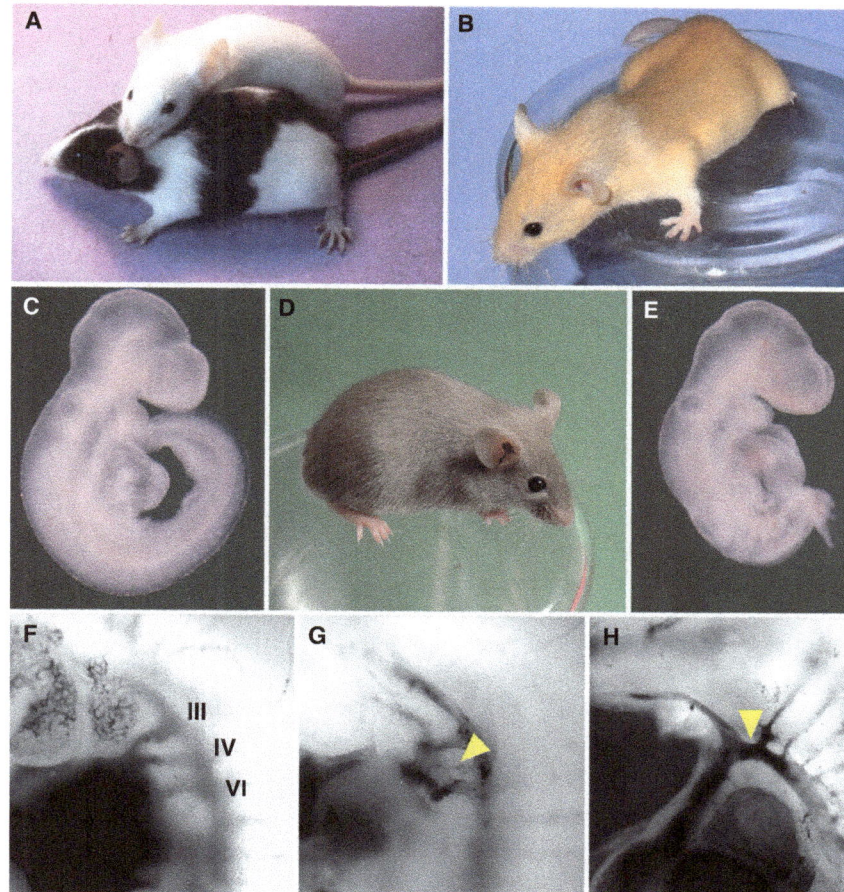

FIGURE 4. Examples of semidominant effects of mutations. (*A*) The gene *Kit*, formerly known as *Dominant White Spotting*, which is actually semidominant, encodes a transmembrane receptor tyrosine kinase. Animals heterozygous for the allele, *Kit^{W-sh}*, have large white spots (as well as other phenotypic effects), whereas homozygotes are black-eyed, all-white mice. (*B*) The *lethal yellow* allele of the *agouti* locus is embryonic lethal when homozygous but heterozygotes are viable, with a yellow coat. (*C–E*) The *Brachyury* or *T* locus encodes a transcription factor that affects mesoderm development. Heterozygotes for the *T^{Wis}* allele have no tails (*D*), whereas homozygous mutants have no posterior mesoderm beyond the lumbar region (*E*) and do not survive. A wild-type embryo (*C*) is shown for comparison. (*F–H*) The T-box transcription factor gene, *Tbx1*, has a number of phenotypic effects, including some that are recessive and some that show a semidominant pattern of inheritance. India ink injection into the heart (*bottom left* of each panel) of midgestation embryos outlines the developing aortic arch arteries. Normal embryos (*F*) have three aortic arches at this stage (III, IV, and VI), whereas heterozygous *Tbx1^{tm1Pa}* embryos (*G*) have a hypoplastic IVth arch artery (arrowhead) and homozygous mutant embryos (*H*) have only a single aortic arch artery (arrowhead). (*C,E,* Courtesy of Debbie Chapman; *F–H,* reprinted, with permission, from Jerome and Papaioannou 2001, ©Nature Publishing Group.)

of gametogenesis or other aspects of reproduction such as behavioral modifications in heterozygote males, females, or both sexes could cause sterility or subfertility, and may only show up in the next generation. You will be able to study this effect in heterozygous mice as outlined in Chapter 13: Phenotypic Analysis of Perinatal Lethality in Mice (Papaioannou and Behringer 2023j); Chapter 14: Analysis of Postnatal Mutant Phenotypes in Mice (Papaioannou and Behringer 2023k), but it will be more difficult to obtain the homozygous mutant animals, which could potentially have more severe or additional phenotypes. If only one sex is affected (e.g., as in dominant sex reversal), the mutation can be perpetuated by breeding from the unaffected sex, and then you can try assisted reproduction techniques for the affected sex when it comes time to produce homozygous mutants.

Special Considerations for Imprinted Genes

By definition, an imprinted gene will have different activity depending on the parent of origin. Thus, if a mutation in a paternally inactivated imprinted gene is inherited from the father, it has no effect. However, if that same mutation is inherited from the mother, it will be the only active allele and the mutant effect will appear to be dominant. Once you have established that you have an imprinted gene by demonstrating that the mutant effect is present in heterozygotes only when the mutant allele is inherited from one parent but not the other (Fig. 5A), you can study the mutant phenotype by breeding unaffected heterozygotes to wild-type mice to produce affected heterozygotes and then follow the procedures in Chapter 8: Phenotypic Analysis: Assessing Timing of Recessive Prenatal Lethality in Mice (Papaioannou and Behringer 2023d); Chapter 9: Phenotypic Analysis of Preimplantation Lethality in Mice (Papaioannou and Behringer 2023e); Chapter 10: Phenotypic Analysis of Periimplantation to Mid-Gestation Lethality in Mice (Papaioannou and Behringer 2023f); Chapter 11: Analysis of Mid- to Late-Gestation Phenotypes in Mice (Papaioannou and Behringer 2023g); Chapter 12: Uncovering Phenotypes in Mutant Mice by Determining Embryo, Organ, Tissue, and Cell Developmental Potential (Papaioannou and Behringer 2023h) to characterize the phenotype. However, you should also breed unaffected heterozygotes together to produce homozygous mutants because they may have a different or more severe phenotype, which could come about if imprinting is not complete

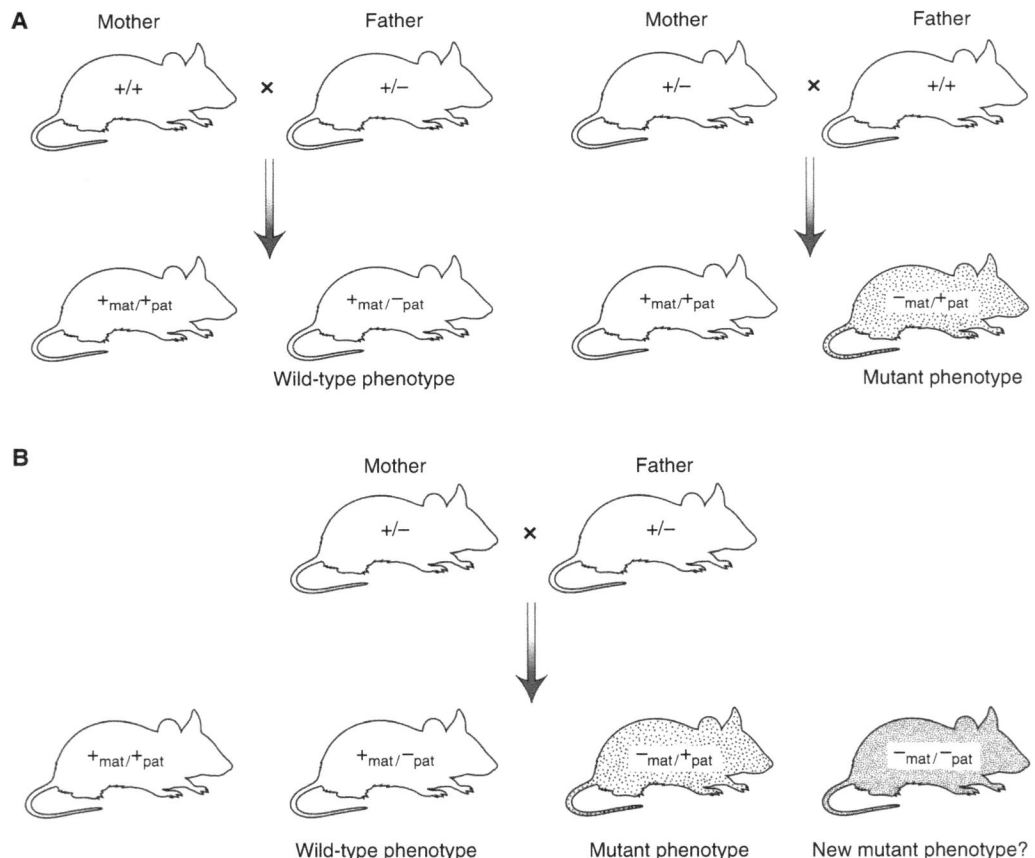

FIGURE 5. Breeding scheme for imprinted genes. (A) In this example of a paternally imprinted (inactivated) gene, reciprocal crosses establish that the heterozygous (+/−) phenotype is evident only when the mutant allele is inherited from the mother. To study this dominant effect, heterozygous females are crossed with wild-type (+/+) males to obtain heterozygotes with the mutant phenotype. (B) Heterozygotes can also be bred together to produce offspring of all possible genotype/phenotype combinations. If imprinting is ubiquitous, it is expected that the affected heterozygotes will have the same phenotype as homozygous mutants. However, if imprinting is tissue-specific such that some tissues escape imprinting, homozygous mutants could have a more severe mutant phenotype (darker stippling).

Cite this chapter as *Cold Spring Harb Protoc*; doi:10.1101/pdb.over107978

or if it is tissue-specific. In this cross, the litters will contain wild-type controls, both affected and unaffected heterozygotes, and homozygous mutants, providing all possible genotype/phenotype combinations for study (Fig. 5B).

Incidentally, a targeted mutation in an imprinted gene could be put to good use to provide information about the cell and tissue specificity of imprinting by detecting allele-specific expression patterns. In a cross in which the wild-type allele is inactivated, simply use a probe for a coding region deleted by the targeted mutation to see whether any cells or tissues express the wild-type allele. If so, it would be indicative of tissue-specific escape from imprinting in that tissue.

REFERENCES

Bi W, Huang W, Whitworth DJ, Deng JM, Zhang Z, Behringer RR, de Crombrugghe B. 2001. Haploinsufficiency of *Sox9* results in defective cartilage primordia and premature skeletal mineralization. *Proc Natl Acad Sci* **98:** 6698–6703. doi:10.1073/pnas.111092198

DeChiara TM, Robertson EJ, Efstratiadis A. 1991. Parental imprinting of the mouse insulin-like growth factor II gene. *Cell* **64:** 849–859. doi:10.1016/0092-8674(91)90513-X

Jerome LA, Papaioannou VE. 2001. DiGeorge syndrome phenotype in mice mutant for the T-box gene, *Tbx1*. *Nat Genet* **27:** 286–291. doi:10.1038/85845

Papaioannou VE, Behringer RR. 2023a. Strategies for maintaining mouse mutations. *Cold Spring Harb Protoc* doi:10.1011/pdb.over107960

Papaioannou VE, Behringer RR. 2023b. Obtaining or generating gene mutations in mice. *Cold Spring Harb Protoc* doi:10.1011/pdb.over107956

Papaioannou VE, Behringer RR. 2023c. Embryonic stem cell gene targeting and chimera production in mice. *Cold Spring Harb Protoc* doi: 10.1011/pdb.over107958

Papaioannou VE, Behringer RR. 2023d. Phenotypic analysis: assessing timing of recessive prenatal lethality in mice. *Cold Spring Harb Protoc* doi:10.1011/pdb.over107970

Papaioannou VE, Behringer RR. 2023e. Phenotypic analysis of preimplantation lethality in mice. *Cold Spring Harb Protoc* doi:10.1011/pdb.over107971

Papaioannou VE, Behringer RR. 2023f. Phenotypic analysis of periimplantation to mid-gestation lethality in mice. *Cold Spring Harb Protoc* doi:10.1011/pdb.over107972

Papaioannou VE, Behringer RR. 2023g. Analysis of mid- to late-gestation phenotypes in mice. *Cold Spring Harb Protoc* doi:10.1011/pdb.over107973

Papaioannou VE, Behringer RR. 2023h. Uncovering phenotypes in mutant mice by determining embryo, organ, tissue, and cell developmental potential. *Cold Spring Harb Protoc* doi:10.1011/pdb.over107974

Papaioannou VE, Behringer RR. 2023i. Getting around an early lethal phenotype in mice with chimeras. *Cold Spring Harb Protoc* doi:10.1011/pdb.over107979

Papaioannou VE, Behringer RR. 2023j. Phenotypic analysis of perinatal lethality in mice. *Cold Spring Harb Protoc* doi:10.1011/pdb.over107975

Papaioannou VE, Behringer RR. 2023k. Analysis of postnatal mutant phenotypes in mice. *Cold Spring Harb Protoc* doi:10.1011/pdb.over107976

Papaioannou VE, Behringer RR. 2023l. The "no phenotype" challenge in analyzing mutant mice. *Cold Spring Harb Protoc* doi:10.1011/pdb.over107977

Papaioannou VE, Behringer RR. 2023m. Isolation of XO subclones from XY murine embryonic stem cells. *Cold Spring Harb Protoc* doi:10.1011/pdb.prot108038

Papaioannou VE, Behringer RR. 2023n. Tissue- and/or temporal-specific mutations in mice using conditional alleles. *Cold Spring Harb Protoc* doi:10.1011/pdb.over107980

Purandare SM, Ware SM, Kwan KM, Gebbia M, Bassi MT, Deng JM, Vogel H, Behringer RR, Belmont JW, Casey B. 2002. A complex syndrome of left–right axis, central nervous system and axial skeleton defects in *Zic3* mutant mice. *Development* **129:** 2293–2302. doi:10.1242/dev.129.9.2293

Rohozinski J, Agoulnik AI, Boettger-Tong HL, Bishop CE. 2002. Successful targeting of mouse Y chromosome genes using a site-directed insertion vector. *Genesis* **32:** 1–7. doi:10.1002/gene.10020

Yan W, Ma L, Burns KH, Matzuk MM. 2004. Haploinsufficiency of kelch-like protein homolog 10 causes infertility in male mice. *Proc Natl Acad Sci* **101:** 7793–7798. doi:10.1073/pnas.0308025101

Yeh S, Tsai MY, Xu Q, Mu XM, Lardy H, Huang KE, Lin H, Yeh SD, Altuwaijri S, Zhou X, et al. 2002. Generation and characterization of androgen receptor knockout (ARKO) mice: an in vivo model for the study of androgen functions in selective tissues. *Proc Natl Acad Sci* **99:** 13498–13503. doi:10.1073/pnas.212474399

Protocol 16.1

Isolation of XO Subclones from XY Murine Embryonic Stem Cells

This is a simple procedure to isolate XO subclones from XY murine embryonic stem cells in situations that require transmission of a mutation through the female germline —for example, if the mutation adversely affects spermatogenesis. XY cells are plated at clonal density, and resulting colonies are genotyped by polymerase chain reaction for a Y-specific probe to identify clones that have spontaneously lost the Y chromosome.

INTRODUCTION

Most of the commonly used embryonic stem (ES) cell lines are genetically male (XY), and, consequently, most germline transmission of the ES cell genotype is through male chimeras. In some situations, you may require transmission through female chimeras but only have XY ES cells available. Similar to XX mice, XO mice develop as fertile females (Cattanach 1962). XY ES cells spontaneously lose the Y chromosome to become XO at about a 1% frequency in vitro (Eggan et al. 2002; Deng et al. 2011). Thus, an alternative to working with XX ES cells is to isolate XO ES cell subclones from an existing XY wild-type or mutant ES cell line.

MATERIALS

It is essential that you consult the appropriate Material Data Safety sheets and your institution's Environmental Health and Safety Office for proper handling of equipment and hazardous materials used in this protocol.

Reagents

Male (XY) murine ES cell line

Sry polymerase chain reaction (PCR) genotyping assay (see Protocol 15.1: Sex Genotyping Mice by Polymerase Chain Reaction [Papaioannou and Behringer 2023])

Standard ES cell culture conditions (see Chapter 8 in Behringer et al. 2014)

Equipment

Tissue culture plates, 96-well

METHOD

1. Plate XY ES cells at clonal density using culture conditions appropriate for the cell line.

2. After 7–10 d, pick 300–400 colonies into 96-well plates for expansion (see Nagy et al. 2006a).

3. When the ES cells are confluent, split the 96-well plates to freeze down master plates (see Nagy et al. 2006b) and generate duplicate plates for genotyping.

4. Isolate DNA from each well of the duplicate 96-well plates (see Nagy et al. 2006c) and genotype for the Y chromosome. The presence or absence of the Y chromosome can be assessed by PCR genotyping for the Y-chromosome-specific gene, *Sry*.

5. Karyotype your putative XO ES cell subclones (see Nagy et al. 2009). There should be 39 chromosomes.

6. Once you have identified multiple XO subclones, they can be used to generate female chimeras with a chance of germline transmission.

TROUBLESHOOTING

If you are using fibroblast feeders, you may want to check their Y-chromosome status, because if they are XY, they may contaminate the ES cells, especially if you are using PCR for genotyping. If this is a problem, ES cells can be purified by brief serial passages on tissue culture dishes because feeder cells adhere much more quickly than ES cells.

REFERENCES

Behringer RR, Gertsenstein M, Nagy KV, Nagy A. 2014. *Manipulating the mouse embryo: a laboratory manual.* Cold Spring Harbor Laboratory Press, Cold Spring Harbor, NY.

Cattanach BM. 1962. XO mice. *Genet Res* **3**: 487–490. doi:10.1017/S0016672300003335

Deng JM, Satoh K, Wang H, Chang H, Zhang Z, Stewart MD, Cooney AJ, Behringer RR. 2011. Generation of viable male and female mice from two fathers. *Biol Reprod* **84**: 613–618. doi:10.1095/biolreprod.110.088831

Eggan K, Rode A, Jentsch I, Samuel C, Hennek T, Tintrup H, Zevnik B, Erwin J, Loring J, Jackson-Grusby L, et al. 2002. Male and female mice derived from the same embryonic stem cell clone by tetraploid embryo complementation. *Nat Biotechnol* **20**: 455–459. doi:10.1038/nbt0502-455

Nagy A, Gertsenstein M, Vintersten K, Behringer R. 2006a. Isolating individual embryonic stem (ES) cell colonies by picking. *Cold Spring Harb Protoc* doi:10.1011/pdb.prot4410

Nagy A, Gertsenstein M, Vintersten K, Behringer R. 2006b. Freezing embryonic stem (ES) cells in 96-well plates. *Cold Spring Harb Protoc* doi:10.1011/pdb.prot4411

Nagy A, Gertsenstein M, Vintersten K, Behringer R. 2006c. Rapid preparation of DNA from cells in 96-well tissue culture dishes. *Cold Spring Harb Protoc* doi:10.1011/pdb.prot4413

Nagy A, Gertsenstein M, Vintersten K, Behringer R. 2009. Counting chromosomes in embryonic stem (ES) cells. *Cold Spring Harb Protoc* doi:10.1011/pdb.prot4404

Papaioannou VE, Behringer RR. 2023. Sex genotyping mice by polymerase chain reaction. *Cold Spring Harb Protoc* doi:10.1011/pdb.prot108062

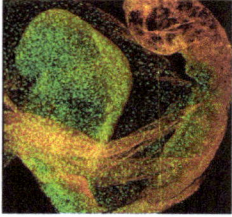

CHAPTER 17

Getting around an Early Lethal Phenotype in Mice with Chimeras

The same gene can have many different functions in different places in the body and/or at different times in development and adult life. Often only one organ or one developmental stage is of particular interest to an investigator. If, however, lethality or severe detrimental effects of a mutation prevent the study of the organ or stage of interest, there are a number of ways to circumvent an early effect. In this chapter, we discuss one way of getting around an early lethal phenotype by using chimeras, a method that is also useful for studying the mutant cells in the context of a wild-type host as part of the phenotypic analysis. The composition of chimeras with respect to embryonic cell lineages can be controlled to some extent to produce lineage-restricted chimeras with, for example, mutant cells restricted to certain lineages. Depending on the site of action of the mutant gene, this could result in chimeric "rescue." Details of how to distinguish mutant cells from wild type, an essential part of any chimera experiment, are discussed as well as methods to genotype the chimeras with respect to both component cell types.

INTRODUCTION

GETTING AROUND A LETHAL PHENOTYPE WITH CHIMERAS

- CONTROLLING CELL COMPOSITION
  - Mix in all tissues ⟶ Morula aggregation
  - Lineage-restricted ⟶ Blastocyst injection
- DISTINGUISHING MUTANT AND WILD-TYPE CELLS
- GENOTYPING CHIMERAS

When embarking on a mutational analysis, embryonic or fetal lethality may be the last thing you want or expect. But if that is the result, you may be looking for a way to circumvent early lethality so that the effects of the mutation can be studied at later stages of development, in specific tissues, or even in the adult mouse. The nature of development is such that the earliest effect of a mutation can potentially affect all subsequent stages, but the earliest time of a mutant effect may not correspond to the stage that you wanted to study or it may not be the only stage at which the gene is expressed.

Opening artwork: Courtesy of Guy Eakin, Kat Hadjantonakis, Virginia Papaioannou, and Richard Behringer.

© 2024 Cold Spring Harbor Laboratory Press
Cite this chapter as *Cold Spring Harb Protoc*; doi:10.1101/pdb.over107979

This chapter deals with ways to get around an early lethal phenotype or even a tissue-specific detrimental effect that may not be lethal, so that you can study later time points without worrying about whether the later effects are primary or secondary. For example, a gene may be expressed throughout the development of a particular organ, but if a mutation prevents the organ from initially forming, the role of the gene in the later development of the organ cannot be studied. We have already mentioned the possibilities offered by cell, tissue, or organ culture to study specific mutant effects that can essentially get around an embryo-lethal phenotype (see Chapter 12: Uncovering Phenotypes in Mutant Mice by Determining Embryo, Organ, Tissue, and Cell Developmental Potential [Papaioannou and Behringer 2023a] for an overview of the types of tissues that are amenable to study this way). In addition, assisted reproduction techniques (Chapter 6: Strategies for Maintaining Mouse Mutations [Papaioannou and Behringer 2023b]) can sometimes circumvent detrimental effects on fertility, and even something as simple as changing the genetic background by outcrossing to an outbred stock may ameliorate a mutant effect (e.g., *Ywhaz*; Yang et al. 2017). There are a number of other ways of getting around an early phenotype (lethal or not). One or more of these methods may be useful for the analysis of any gene that has a complex temporal and spatial pattern of expression. In this chapter, we discuss the use of chimeras, which can circumvent some lethal effects that are tissue specific and can also provide a tool for analysis of mutant effects (e.g., *Tbx6*; Chapman et al. 1996). In Chapter 18: Tissue- and/or Temporal-Specific Mutations in Mice Using Conditional Alleles (Papaioannou and Behringer 2023c), other methods will be discussed, including transgene rescue, regulatory mutations, and, in particular, conditional alleles.

USE OF CHIMERAS

Combining embryonic cells from two different embryos (or combining embryonic cells with stem cells) in a manner that will allow them to comingle and develop as a single organism (Fig. 1), commonly known as a chimera, is a time-honored embryology trick that was first used successfully in mammals more than half a century ago. The basic idea is that the developmental potential of the two contributing cell types is revealed by their behavior in combination in a single developing embryo. The method has been used to trace cell lineages, explore the range of developmental potential or plasticity of different cells, and test cells for totipotency.

For our purposes, chimeras can be used either to explore the potential of cells carrying a mutation in your favorite gene or for the practical purpose of circumventing lethality if viability is limited by the failure of a critical tissue early in development. With chimeric "rescue," you essentially combine mutant cells with wild-type cells and "ask" the wild-type cells to make up for the critical function missing in the mutant cells. If rescue works, you may bypass an early lethality, allowing the chimera to develop far enough for you to study later stages at which the gene might have additional roles. For exploring the potential of mutant cells, the behavior of cells in a chimera can tell you something about the nature of the mutant effect. For example, the mutant cells may function normally in the presence of wild-type cells if the wild-type cells supply a factor missing from the mutant cells. This would indicate a non-cell-autonomous effect of the mutation. Conversely, a cell-autonomous defect would still affect mutant cells whether or not they were developing in the presence of wild-type cells in a chimera (e.g., *Sox9*; Bi et al. 1999).

Before embarking on a chimera study, consider the series of questions in Box 1 to facilitate the design of your experiment. A common outcome of chimera experiments is that the distribution of mutant and wild-type cells is nonrandom or nonuniform when compared with chimeras made by the same method with two wild-type embryos (or a wild-type embryo and wild-type stem cells) (Fig. 2). Depending on where and when the gene is normally expressed and the nature of the nonrandom distribution, this might represent useful information bearing on the mutant effect. Exclusion of mutant cells from an organ or tissue indicates that the mutant cells, for some reason, cannot compete with the wild-type cells in the development of that organ or tissue. However, there are

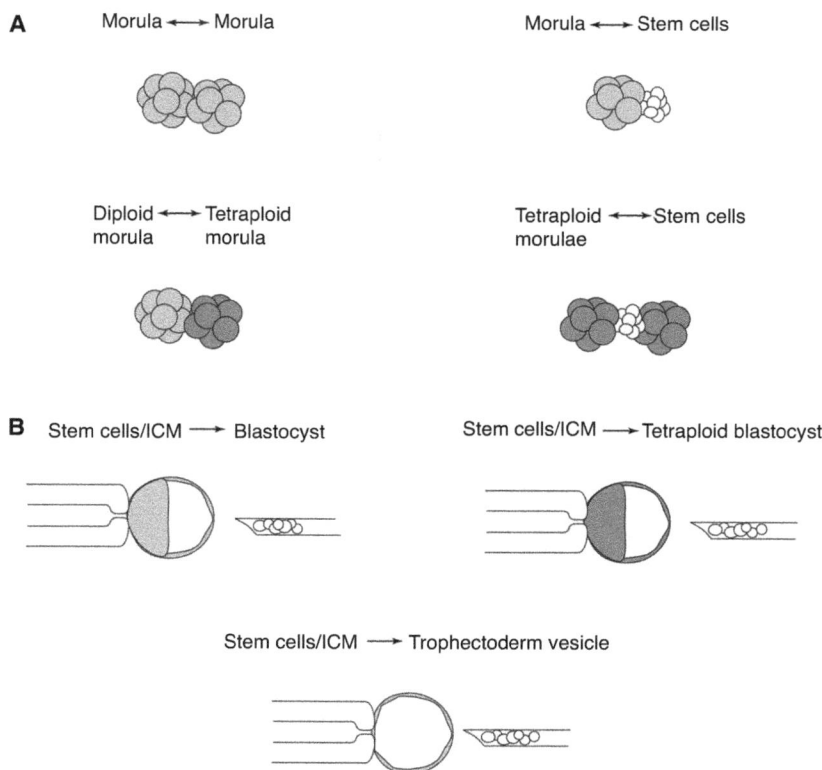

FIGURE 1. Different ways of making chimeras. (*A*) Aggregation at the morula stage. Embryos at the morula stage [embryonic day (E) 2.5] can be aggregated together or aggregated with stem cells (either embryonic stem [ES] or trophoblast stem [TS] cells). Tetraploid morulae can be used in combination with a diploid morula or diploid stem cells to produce tissue-restricted chimeras. In the case of aggregation of tetraploid morulae with stem cells, two morulae are used to increase the chance of successful development. (*B*) Injection of cells into the cavity of E3.5 blastocysts. Isolated inner cell mass (ICM) cells, whole ICMs, or stem cells can be injected into diploid or tetraploid blastocysts. In addition, an ICM can be injected into an isolated trophectoderm vesicle. All injection methods produce tissue-restricted chimeras. See Table 2 for the tissue composition of resulting chimeras.

many ways to arrive at a nonrandom mix of cells in a chimera, and the challenge is to figure out which applies; for example:

- The mutant cells may die before the chimera reaches a particular point in development.

- Mutant cells may have a lower proliferative rate than normal and are thus overgrown by wild-type cells.

- Mutant cells may not be able to respond to/emit normal inductive signals.

- Mutant cells may be inhibited in or incapable of cell movement or migration necessary to form an organ or tissue.

- Surface properties of mutant cells may render them incapable of interacting with wild-type cells.

- Mutant cells may be unable to complete a specific step in differentiation.

Any combination of altered properties could lead to segregation or sorting of cells in chimeras. It is necessary to examine a number of chimeras at different stages of development to narrow down these options. Better yet, visualization of the mutant cells in living chimeric embryos or tissue explants might be possible using short-term embryo or organ culture and vital fluorescent reporters to distinguish the mutant from wild-type cells (Figs. 2C,D, 3B–F). Of course, a prerequisite for making use of any chimeric combination is that you have some way of distinguishing between the two component cell types at the time of analysis. Similarly, it is unlikely that you will be able to distinguish the mutant

---

BOX 1. CHECKLIST FOR A CHIMERA EXPERIMENT

Answers to the following questions will facilitate the best design for your chimera experiment. Some of the questions are based on the supposition that you know the identity of the gene that is mutated, although this information is not essential for performing a chimera study.

- Which tissue(s) expresses the gene product in question?
- At which stage does expression initiate in the tissue?
- When does the tissue form during embryogenesis?
- When does the mutant phenotype first become apparent?
- Which chimera system will you use (morula aggregation, blastocyst injection, embryonic stem (ES) cells, trophoblast stem (TS) cells, tetraploid embryos, etc.)?
- What type of chimera will be your control? Usually, $+/+ \leftrightarrow +/+$ but ideally also $+/- \leftrightarrow +/+$.
- Have you controlled for genetic background differences?
- How will you genotype the chimeras?
- Which marking system will you use to follow the fate of mutant and/or wild-type cells (i.e., pigmentation, *lacZ*, fluorescent protein, etc.)?
- How many chimeras will you need for the study?
- What range of chimerism will be useful to address your specific question?
- Which stage(s) and tissue(s) will you analyze?
- What types of analyses will you perform (gross anatomy, static and time-lapse fluorescent imaging, histology, histochemistry, molecular markers)?

---

embryos from wild type at the time you make the chimeras, so you will need to have a strategy of genotyping the chimeras at the time of analysis that will distinguish heterozygous from homozygous mutant cells (see Genotyping Chimeras, below).

CONTROLLING CELL COMPOSITION IN CHIMERAS

Two basic methods for making chimeras are aggregation of embryos or cells at the morula (eight- to 16-cell) stage of development and injection of cells into blastocyst-stage embryos (Fig. 1; see Chapter 12 in Behringer et al. 2014). If you obtained your mutation through gene targeting in ES cells, then you will be familiar with one or the other of these methods for making ES cell chimeras. Both methods require basic skills in embryo recovery, handling, and transfer, whereas the injection method in addition requires specialized microinjection equipment and skills. Another variable in making chimeras is the nature of the cells used. There are few choices, because only embryonic cells and stem cells from embryos are able to make chimeras with preimplantation embryos. However, both diploid and tetraploid cells can be used, and the combinations of genotype, ploidy, and method provide plenty of variety. Each combination potentially provides different information because different types of cells have different developmental potential and/or some cells may offer a selective advantage over others. Thus, when making chimeras, you can take advantage of natural limitations in the developmental potential of different components to construct a chimera with specific tissue composition (Tables 1 and 2). Keep in mind that in addition to developmental limitations imposed on the two components of the chimeras by cell type and genotype with respect to the mutant allele, there may also be genetic background effects that skew the composition toward one component or the other. It is also worth considering using heterozygous as well as wild-type cells as controls, even if the mutation appears to be recessive, because heterozygous cells may not compete well when placed in competition with wild-type cells in a chimera, thus revealing a heterozygous phenotype.

 Cite this chapter as *Cold Spring Harb Protoc*; doi:10.1101/pdb.over107979

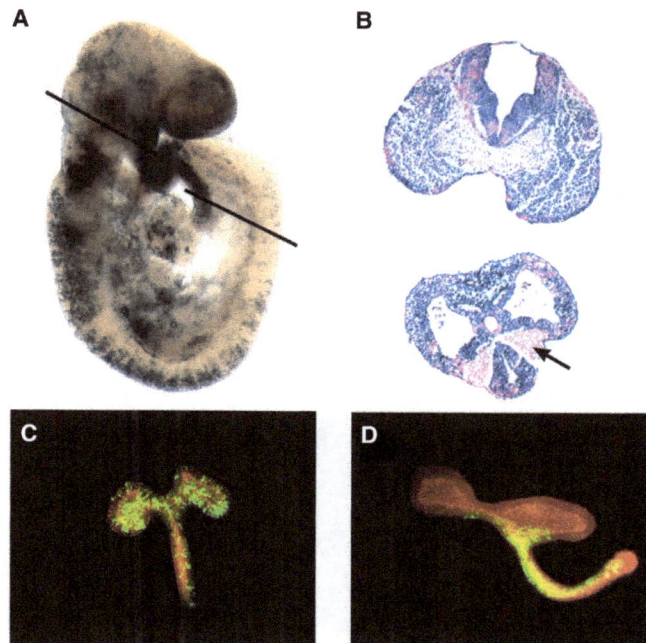

FIGURE 2. Nonrandom distribution of cells in chimeras due to a mutant gene effect. (A) Whole mount of a chimera made by blastocyst injection of *Tbx6* homozygous mutant embryonic stem (ES) cells carrying a ubiquitously expressed *lacZ* gene trap insertion (*ROSA26*) as a cell marker. The β-gal-positive mutant cells (blue) are distributed throughout the embryo but are much less dense in the posterior region, except for the tail bud, where the gene is normally expressed, which is composed mostly of mutant cells. (B) Section of a similar chimera counterstained with eosin. The lines in *A* represent the approximate level of the section. This chimera has a high level of contribution of mutant ES cells to all tissues with the notable exception of the somites, which are composed almost entirely of wild-type cells (pink, arrow). The mutation causes mutant cells to accumulate in the tail bud and prevents them from taking part in posterior somite formation. (C,D) Kidney primordia from chimeras made by injection of ES cells carrying a *Hoxb7/GFP* transgene into blastocysts. The ureteric bud epithelium was stained with anticytokeratin (red). The chimera in *C* was made with green fluorescent protein (GFP)-labeled, ES cells wild-type for *Ret* and shows a random contribution of GFP-positive cells to the ureteric bud. The chimera in *D* was made with GFP-labeled ES cells homozygous mutant for *Ret*. The mutant ES cells contribute only to the ureter and trunks of the ureteric bud branches but are excluded from the tips of the ureteric buds. (*A,* Courtesy of Debbie Chapman; *B,* adapted from Chapman et al. 2003, ©Elsevier; *C,D,* courtesy of Reena Shakya and Frank Costantini.)

If you are unsure about which kind of chimera will best suit your needs, read through the following three sections, which detail the limitations and advantages of each tissue combination and consult Tables 1 and 2. If your gene is expressed in extraembryonic tissues and/or you want to examine every possible tissue for defects, make chimeras with a mix of cells in all tissues. If you are not interested in looking at extraembryonic tissue but still want a chimeric mix in all tissues of the fetus, you can do morula aggregation. You also have the option of using blastocyst injection. If your mutant embryo has an extraembryonic defect and your goal is to rescue the lethal effect or determine the site of the extraembryonic defect, you can make chimeras with lineage-restricted contributions of cells from the two chimeric components.

Chimeras with a Mix of Cells in All Tissues—Morula Aggregation

Cleavage-stage embryos (morulae) can be aggregated together to make chimeras. Because of the pluripotency of cleavage-stage cells and because cell mixing is unimpeded at this time, the resulting chimeras have a mixture of cells in all tissues, both embryonic and extraembryonic (Table 2). Extraembryonic tissue refers to the fetal membranes, trophoblast, chorion, yolk sac, amnion, and allantois, in contrast to the embryonic tissue, which is the fetus or embryo proper. In fact, aggregation of morulae is the only way to get chimeras with a mix of the two components in every tissue.

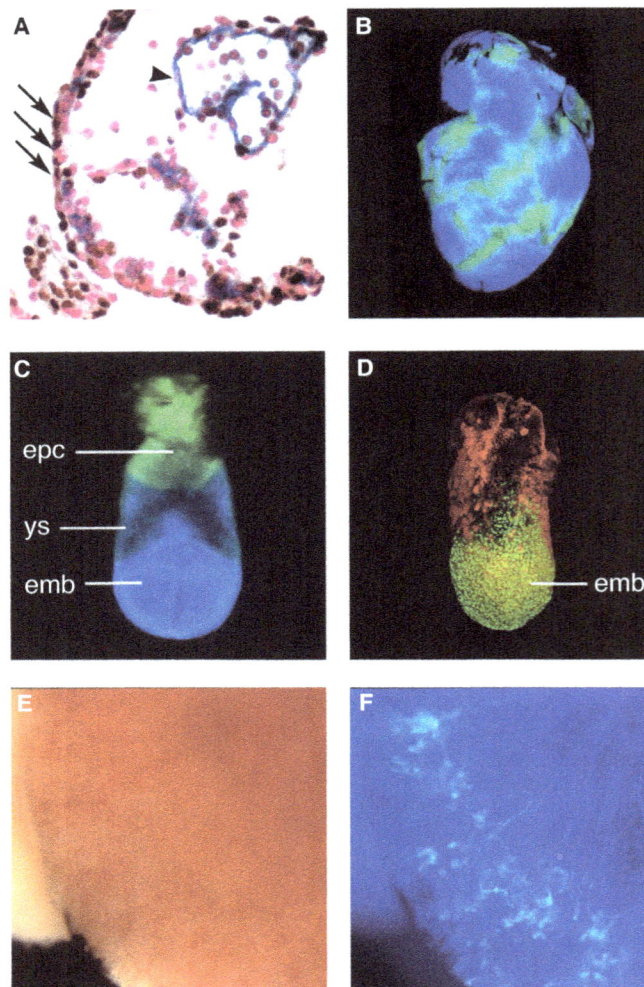

FIGURE 3. Distinguishing cells in a chimera using *GFP* transgenes. (*A*) Detection of chimerism and differentiation in the allantois of an E8.5 embryo using immunohistochemistry. This chimera was made with a wild-type embryo and a transgenic embryo carrying a ubiquitous, nuclear-localized GFP transgene. The embryo was subjected to sequential immunohistochemical labeling with anti-GFP and anti-PECAM (platelet endothelial cell adhesion molecule) antibodies with alkaline phosphatase- and peroxidase-conjugated secondary antibodies, respectively, sectioned, and counterstained with nuclear fast red. Anti-GFP immunohistochemistry and the counterstain distinguish cells as derived from either the wild-type embryo (pink nuclei) or GFP transgenic embryo (brown nuclei, arrows). Additionally, immunohistochemistry for PECAM, a molecular marker of endothelial differentiation, identifies cells that have begun to form the endothelium of blood vessels (blue cytoplasmic label, arrowhead), regardless of their origin. (*B*) Display of both chimeric components in an adult heart using two different fluorescent transgene markers. The chimera was made by aggregating a morula carrying an enhanced cyan fluorescent protein (ECFP) transgene with a morula carrying an enhanced yellow fluorescent protein (EYFP) transgene. The image is a double exposure using ECFP and EYFP filters, consecutively. (*C,D*) Comparison of cytoplasmic and nuclear-localized fluorescent markers in whole-mount gastrulation-stage chimeric embryos. (*C*) An E7.5 chimera made by aggregation of a tetraploid EYFP transgenic morula and a diploid ECFP transgenic morula. In both cases, the fluorescent protein is cytoplasmically localized. The diploid cells make up the entire embryonic region (emb) and the yolk sac (ys) mesoderm, whereas the tetraploid cells are confined to the extraembryonic tissues including the ectoplacental cone (epc) and yolk sac endoderm. The image is a double exposure using ECFP and EYFP filters, consecutively. (*D*) An E6.5 chimera made by injecting ES cells transgenic for a nuclear-localized EGFP into a wild-type blastocyst. The chimeric embryo has been counterstained with FM 4–64 (Molecular Probes, Oregon) to reveal cell boundaries (red). The progeny of the ES cells are in the embryonic region (emb). (*E,F*) Bright-field (*E*) and fluorescence (*F*) images of a 400-μm vibratome section of an adult kidney from a chimera made with wild-type and ECFP transgenic cells. The fluorescent cells are in the minority, making the structural details easy to see. (*A*, Courtesy of L.A. Naiche and Virginia Papaioannou; *B,C*, adapted from Hadjantonakis et al. 2002, licensee BioMed Central Ltd., https://doi.org/10.1186/1472-6750-2-11; *D*, courtesy of Guy Eakin, Kat Hadjantonakis, Virginia Papaioannou, and Richard Behringer; *E,F*, courtesy of Kat Hadjantonakis.)

TABLE 1. Limitations in the developmental potential of different embryonic components in chimeras

Tissue used to make the chimera	Fate of tissue in the chimera at a later stage of development			
	Trophectoderm and its placental derivatives	Yolk sac endoderm and parietal endoderm	Yolk sac mesoderm and other extra embryonic mesoderm	Embryo proper
Morula	✓	✓	✓	✓
Blastocyst	✓	✓	✓	✓
ICM		✓	✓	✓
ES cells			✓	✓
Tetraploid morula	✓	✓		
Tetraploid blastocyst	✓	✓		
TS cells	✓			
Trophectoderm vesicle	✓			

Possible fates of different kinds of cells in a chimera made by morula aggregation or by blastocyst injection are shown. Tissue is diploid unless otherwise stated. (ICM) Inner cell mass, (ES) embryonic stem, (TS) trophoblast stem.

If one component of the chimera is mutant, however, the composition of the chimeric mix may be affected by alterations in the developmental potential of the mutant cells, and this, of course, is what you will be looking for. The production of aggregation chimeras is straightforward (see Chapter 12 in Behringer et al. 2014) and consists of isolating embryos, removing the zona pellucida (Nagy et al. 2006f), culturing the embryos in contact with one another until they make a single aggregate (Nagy et al. 2006b), and then transferring the aggregates to a foster mother (Nagy et al. 2006g). Because in most cases the mutant genotype will not be evident at the time of aggregation, all of the embryos resulting from heterozygous matings are aggregated pairwise with wild-type embryos, and the genotypes of the chimeras are determined later during the analysis. Chimeras made with heterozygous or wild-type cells will serve as useful controls for technique and for possible genetic background effects. In addition to being able to figure out the genotype of the embryos, the two components must also be distinguishable through the use of a cellular marker to complete the analysis of the cellular distribution within the chimera. The resolution of the analysis will only be as good as the resolution of the cellular marker used, so a stable, cell-based, developmentally neutral marker is essential (see below).

The analysis of chimeras should begin around the time of lethality of the homozygous mutant. A thorough morphological assessment at this and later stages, similar to that done in determining the nature of the mutant effect in the first place (see Chapter 8: Phenotypic Analysis: Assessing Timing of Recessive Prenatal Lethality in Mice [Papaioannou and Behringer 2023d]; Chapter 9: Phenotypic

TABLE 2. Controlling the composition of chimeras using different methods and tissue combinations

Method	Components of chimera	Origin of trophectoderm in chimera	Origin of yolk sac and parietal endoderm in chimera	Origin of fetus and extraembryonic mesoderm in chimera
Aggregation	Morula ↔ morula	Mixed	Mixed	Mixed
	Morula ↔ ES cells	Morula	Morula	Mixed
	Morula ↔ TS cells	Mixed	Morula	Morula
	Morula ↔ tetraploid morula	Mixed	Mixed	Morula
	ES cells ↔ tetraploid morula	Tetraploid morula	Tetraploid morula	ES cells
Injection	ICM → blastocyst	Blastocyst	Mixed	Mixed
	ES → blastocyst	Blastocyst	Blastocyst	Mixed
	TS → blastocyst	Mixed	Blastocyst	Blastocyst
	ICM → tetraploid blastocyst	Tetraploid blastocyst	Mixed	ICM
	ES → tetraploid blastocyst	Tetraploid blastocyst	Tetraploid blastocyst	ES cells
	ICM → trophoblast vesicle	Trophoblast vesicle	ICM	ICM

Because of the limitation in tissue developmental potential, chimeras with different compositions of tissue can be made by choosing different tissue combinations and using either morula aggregation or blastocyst injection. Tissue is diploid unless otherwise stated. (ES) Embryonic stem, (TS) trophoblast stem, (ICM) inner cell mass.

Analysis of Preimplantation Lethality in Mice [Papaioannou and Behringer 2023e]; Chapter 10: Phenotypic Analysis of Periimplantation to Mid-Gestation Lethality in Mice [Papaioannou and Behringer 2023f]; Chapter 11: Analysis of Mid- to Late-Gestation Phenotypes in Mice [Papaioannou and Behringer 2023g]; Chapter 12: Uncovering Phenotypes in Mutant Mice by Determining Embryo, Organ, Tissue, and Cell Developmental Potential [Papaioannou and Behringer 2023a]; Chapter 13: Phenotypic Analysis of Perinatal Lethality in Mice [Papaioannou and Behringer 2023h]; Chapter 14: Phenotypic Analysis of Postnatal Mutant Phenotypes in Mice [Papaioannou and Behringer 2023i]), will indicate whether or not the phenotype is altered by the chimeric combination. Remember that there will be stochastic variation in the contribution of the different components to the chimeric tissues, independent of the genotypes, so you can expect to see a range in the overall percentage contribution of mutant and wild-type cells. A range in chimeric composition is useful because it allows you to study the behavior of mutant cells in combination with different proportions of wild-type cells. The results may look different when the mutant cells are in a majority or a minority. Although chimeras with a high proportion of mutant cells, so-called high-level chimeras, might look very similar to the homozygous mutants, low-level chimeras might have a mild phenotype and survive past the time of homozygous mutant lethality—in other words, a partial rescue. This will allow the examination of mutant cells in later-stage embryos. The type of analysis that you do will be similar to those that investigate the mutant alone, but in the case of chimeras, you will be looking at the phenotype in relation to the quantity and location of mutant cells.

If the phenotype of the mutant is an early lethality caused by a defect in extraembryonic tissues and it is rescued in chimeras, you will be able to study the role of the gene in embryonic tissues at later stages in chimeras. You might also consider making lineage-restricted chimeras that will completely rescue the extraembryonic phenotype and allow you to control the chimeric composition in different tissues of the chimera, for example, by producing a fetus derived exclusively of mutant cells with a mixture of mutant and wild-type cells in the placenta.

Lineage-Restricted Chimeras—Morula Aggregation with Stem Cells or Tetraploid Embryos

Morula ⟷ Embryonic Stem Cells

By using the relatively simple method of morula aggregation to make chimeras you can exploit the developmental restrictions of certain cell types to produce chimeras composed of nonrandom mixes of cells (Table 2). If you have homozygous mutant ES cells, you can aggregate the ES cells with wild-type morulae (Nagy et al. 2006c). Incidentally, this saves you the trouble of genotyping the chimeras because they are made with ES cells of known genotype, but be sure to make chimeras with wild-type and heterozygous ES cells as controls. The resulting chimeras will be a mix of components in the embryo proper and the mesoderm of the extraembryonic tissues. However, because of inherent restrictions in ES cell developmental potential (Table 1), the trophectoderm, its placental derivatives, and the extraembryonic endoderm will all be derived solely from the morula (Table 2). This combination might be useful if the mutant phenotype you have is limited to the placenta or yolk sac endoderm, in which case the mutant phenotype would be rescued in the chimeras. It can also help to distinguish a yolk sac endoderm defect from a yolk sac mesoderm defect by providing a yolk sac with the endoderm made up entirely of wild-type cells and the mesoderm made up of a mixture of wild-type and homozygous mutant cells. If you need to distinguish the affected layer for a yolk sac phenotype, mutant and wild-type cells can be completely segregated into one layer or the other in chimeras using tetraploid embryos and ES cells (see Chapter 12 in Behringer et al. 2014).

Morula ⟷ Trophoblast Stem Cells

If TS cells are available (Rossant 2006a,b), they can be used to make lineage-restricted chimeras. In this case, the derivatives of the TS cells are restricted to the trophectoderm, whereas the morula will contribute to the trophectoderm as well as make up the entire embryo proper, yolk sac, and parietal

Cite this chapter as *Cold Spring Harb Protoc*; doi:10.1101/pdb.over107979

endoderm (Table 2). A chimeric composition using wild-type TS cells and mutant morulae might be useful to rescue a trophectoderm defect if the mutant phenotype is limited to the trophectoderm in the early stages.

Diploid Morula ⟷ Tetraploid Morula

Another way of stacking the odds to control the composition of chimeras is to use tetraploid embryos. Tetraploid embryos can be made by electrofusion of the two blastomeres of a two-cell embryo (Nagy et al. 2006e; see also Chapter 12 in Behringer et al. 2014). By themselves, tetraploid embryos develop very poorly, and aggregated with diploid morulae, their cells are at a competitive disadvantage and only contribute well to certain cell lineages (Tables 1 and 2). In the context of mutant analysis, this method is sometimes called tetraploid rescue because it can be used to circumvent lethality caused by a defect in the trophectoderm, the placenta, or the yolk sac endoderm. If you know that this is the cause of lethality in your mutant and you want to study the role of the gene in other tissues in the later fetus, or even in adults, this method could provide a way to do so.

In a chimera, although they may be at a disadvantage, tetraploid cells are capable of making the trophectoderm of the blastocyst and its derivatives, as well as the primitive endoderm and its derivative, the yolk sac endoderm, but they are not so effective when it comes to making the inner cell mass (ICM) and its derivatives. That means that in a tetraploid ⟷ diploid chimera, the tetraploid cells are at a severe selective disadvantage in contributing to the ICM, but they can form a chimeric trophectoderm or yolk sac endoderm and possibly rescue a mutant effect in these tissues. A chimera made by aggregating a tetraploid embryo with a diploid embryo (Nagy et al. 2006a) will develop with a mixed population of cells in the trophectoderm and primitive endoderm but with primarily diploid cells in the ICM derivatives (the fetus and extraembryonic mesoderm). However, do not forget that the chimera starts out with a mix of diploid and tetraploid cells, and the cells sort out into separate compartments only gradually. This selection likely occurs between the blastocyst stage and gastrulation, but the precise timing is not known and could affect interpretation of chimeric patterns depending on the time of analysis and when your mutation acts.

Diploid ES Cell ⟷ Tetraploid Morula

For an even more extreme restriction of chimeric components, homozygous mutant ES cells can be used as the diploid component of the chimera (Nagy et al. 2006d). ES cells essentially have the reciprocal developmental potential compared with tetraploid cells in chimeras: They are capable of contributing to all ICM derivatives, but not to the trophectoderm or primitive endoderm lineages, so the resulting chimeras will have the components completely segregated (Table 2; e.g., *Foxa2*; Dufort et al. 1998). The nonmutant, tetraploid component will rescue any trophectoderm or endoderm defect, leaving the mutant ES-derived fetus to develop further (e.g., *Ascl2*; Guillemot et al. 1994). Its continuing development will thus allow you to search for developmental problems associated with the mutations in later-stage embryos. One complication is that an entirely ES-derived fetus will only survive to late gestation and birth if the ES cells have retained full developmental competence.

If your mutation results in a yolk sac defect, ES cell ⟷ tetraploid chimeras might also be an effective way to sort out whether the defect has its origin in the yolk sac endoderm or the yolk sac mesoderm, because this method results in a yolk sac with the two components restricted to one layer or the other of this bilaminar structure. The reciprocal combination of a mutant tetraploid embryo and wild-type ES cells provides the opposite distribution of mutant and wild-type cells in the yolk sac.

The disadvantages of these methods are that specialized equipment is required to make tetraploid embryos, and it may be an effort to derive homozygous mutant ES cells. However, if you think that there is a good chance that a wild-type trophectoderm and/or endoderm layer will get around the lethal problem or give you valuable information about the tissue restriction of a mutant effect, it may be worth the effort.

Lineage-Restricted Chimeras—Blastocyst Injection

Chimeras made using blastocyst injections (see Chapter 12 in Behringer et al. 2014) all have lineage restrictions because the trophectoderm and ICM have already been determined by the blastocyst stage and all cells that are successful at making chimeras at this stage have developmental restrictions. The ICM has the least restricted potential and, when injected into blastocysts, can contribute to the fetus as well as both layers of the yolk sac in a chimera; ES cells injected into blastocysts can contribute to the fetus and extraembryonic mesoderm, but not the extraembryonic endoderm. Injected TS cells are limited to contributing to the trophectoderm (although contribution levels will be very low), and tetraploid cells are limited to trophectoderm derivatives and extraembryonic endoderm (Tables 1 and 2). If you are interested in examining mutant cell behavior within the fetus and/or in the yolk sac and fetal membranes, but not the placenta, you can isolate ICMs from embryos of heterozygous matings and inject them into wild-type blastocysts. If you have homozygous mutant ES cells, you can inject these into wild-type blastocysts (or vice versa; e.g., *Nodal*; Varlet et al. 1997), and the only additional limitation will be that the yolk sac endoderm will not be chimeric. If your goal is to rescue a trophectoderm or yolk sac endoderm defect to let mutant embryos survive longer for study of later stages, you might try injecting mutant ICM cells or mutant ES cells into tetraploid blastocysts. These combinations give a completely mutant fetus in a wild-type trophectoderm with either a mixed or tetraploid extraembryonic endoderm, respectively.

For the sake of completeness, we should mention that chimeras can be made by injecting an ICM from one embryo into a mechanically dissected trophectoderm vesicle from another embryo resulting in complete separation of the two cell lineages (Table 2; e.g., *A*; Papaioannou and Gardner 1979), although the skills necessary for this procedure make it impractical for most purposes. Recently, ES, TS, and extraembryonic endoderm stem (XEN) cell lines have been combined to generate another type of chimera, completely in vitro–derived, called "synthetic embryos," which have the potential to be used in mutant phenotypic analysis (Amadei et al. 2022; Tarazi et al. 2022).

The analysis of any chimeric combination should begin around the time that a phenotype is first observed in the homozygous mutant embryos and continue as outlined in Chapter 8: Phenotypic Analysis: Assessing Timing of Recessive Prenatal Lethality in Mice (Papaioannou and Behringer 2023d); Chapter 9: Phenotypic Analysis of Preimplantation Lethality in Mice (Papaioannou and Behringer 2023e); Chapter 10: Phenotypic Analysis of Periimplantation to Mid-Gestation Lethality in Mice (Papaioannou and Behringer 2023f); Chapter 11: Analysis of Mid- to Late-Gestation Phenotypes in Mice (Papaioannou and Behringer 2023g); Chapter 12: Uncovering Phenotypes in Mutant Mice by Determining Embryo, Organ, Tissue, and Cell Developmental Potential (Papaioannou and Behringer 2023a); Chapter 13: Phenotypic Analysis of Perinatal Lethality in Mice (Papaioannou and Behringer 2023h); Chapter 14: Phenotypic Analysis of Postnatal Mutant Phenotypes in Mice (Papaioannou and Behringer 2023i) if a rescue or partial rescue is achieved. Even with no "rescue," the behavior of the mutant cells in combination with wild-type cells in the embryo provides valuable information about their developmental potential.

DISTINGUISHING CELLS IN CHIMERAS

Identifying the different cells making up a chimera is a simple matter of having a marker in one or the other cell components and detecting the marker at the time of analysis. Ideally, a marker should be stable, heritable, cell-localized, cell-autonomous, and developmentally neutral. It is also quite useful if the marker is ubiquitously expressed, easily detectable, and if several different variants are available.

Depending on the circumstances, a trade-off can sometimes be made with a less-than-ideal marker. One of the most common markers for detecting chimerism is coat color, but the genes affecting coat color meet only a few of the criteria of an ideal marker. The pigmentation pattern is stable, heritable, and easily detectable with several variants. However, pigmentation is highly tissue-specific, melanin granules are not always cell-localized (e.g., they are secreted into the hair shaft), and some pigment genes affect coat color in a non-cell-autonomous way (e.g., the protein of the agouti gene is a secreted signaling molecule). Nonetheless, if all you need is a rough approximation of the level of chimerism, as in assessing ES cell chimeras, this imperfect marker serves the purpose and is a noninvasive method of detection. Pigmentation differences can be detected in the coat postnatally and also in the embryonic retinal epithelium after it becomes pigmented at E10.5.

On the other hand, when investigating the behavior of mutant cells in combination with wild-type cells in a chimera, particularly in embryos, the most useful marker is one that allows positive identification of the origin of each and every cell either in histological sections, whole mounts, or tissue explants. It is also useful to be able to combine the assay for the chimera marker with assays for the molecular differentiation of tissues (Fig. 3A; e.g., *Krox20*; Voiculescu et al. 2001). It should be noted that if a knock-in reporter is part of your gene-targeting design, this reporter will not be ubiquitously expressed and, thus, if used in chimeras, does not mark all cells carrying the targeted mutation, only those expressing the reporter.

The most commonly used markers are introduced transgenes that have been demonstrated to be developmentally neutral and are more or less ubiquitously expressed. These transgenes are heritable, stable, and selected for their cell-autonomous, cell-localized properties. The Gt(ROSA) 26Sor gene trap insertion (called *Rosa26*) is a commonly used marker (Fig. 2A,B). This is a gene trap insertion of a *lacZ* transgene into a gene that is ubiquitously expressed in the embryo (but not in the extraembryonic tissues) and can be easily detected by X-Gal staining (Nagy et al. 2007). It appears to have no significant detrimental developmental effects even when homozygous, and it is thus a simple matter to breed the transgene onto the mutant stock of mice before making chimeras.

Increasingly, fluorescent protein genes are being used as transgenic reporters. For chimera studies, a transgenic, ubiquitously expressed, developmentally neutral, nuclear-localized green fluorescent protein (GFP) comes close to providing the ideal marker to identify every cell in a chimera (Figs. 2 and 3; Hadjantonakis and Papaioannou 2004). Fluorescent proteins have the advantages of the availability of several distinguishable spectral variants (Fig. 3B,C), the possibility of creating fusion proteins to localize the marker to different subcellular compartments (Fig. 3C,D; Kiyonari et al. 2019), and the feasibility of detection in living or fixed tissues either by fluorescence microscopy or by immunolocalization (Fig. 3A). In addition, embryos can be imaged vitally with time-lapse microscopy for a dynamic picture of development. Fluorescent protein genes can be targeted to specific loci, such as the *Rosa26* locus, or random transgene insertions can be selected for ubiquitous expression. Many mouse strains carrying such markers are available (e.g., mice.jax.org; Abe and Fujimori 2013).

For tracking mutant cells in a chimera, it is preferable to have the mutant cells as the marked population. If you want to perform a chimera experiment using an existing homozygous mutant ES cell line that does not have a marker, one possibility would be to target the *Rosa26* locus or another ubiquitously expressed locus of the mutant ES cells with a knock-in construct such as *lacZ*, or a fluorescent protein gene. A technical consideration is to use a selectable marker for this second gene targeting that is not already present in the ES cell line. For example, you cannot use G418 selection if the ES cell line is already neo resistant. If the selectable marker was floxed or flrted, then you could remove the selectable marker by CRE or FLP, allowing you to reuse the original selection scheme (Abuin and Bradley 1996).

Ideally, the marked population would be in the minority of cells in the chimera as it is easier to detect a few positive cells in a background of negative cells than vice versa. If the mutant cells are at a disadvantage, they may be in the minority. However, apart from the gross lineage restrictions detailed in previous sections, it is not possible to reliably control the quantitative contribution of chimeric components. One way around this limitation would be to mark both components of the chimeras with different, distinguishable reporters (Fig. 3B; Hadjantonakis et al. 2002; Lizier et al. 2016).

GENOTYPING CHIMERAS

If you are making chimeras between wild-type embryos and homozygous mutant ES cells, there will be no need to genotype the chimeras because they will all be +/+ ⟷ −/−. If the mutant ES cells also contain an independent marker (which could be produced by targeting a reporter to a ubiquitously expressed locus in the mutant ES cells), detecting the marked cells is the same as detecting the mutant cells. However, if you are making chimeras with embryos rather than ES cells, you will necessarily be making chimeras from +/+, +/−, and −/− embryos if you are breeding from heterozygotes because the mutation is lethal. Furthermore, the usual polymerase chain reaction (PCR) genotyping protocols that depend on detection of a mutant and wild-type allele will not help, because +/+ ⟷ −/− chimeras have both alleles and look the same on a PCR gel as +/+ ⟷ +/− chimeras. There are several possibilities for solving this problem.

1. Use two different mutant alleles that you can distinguish by PCR (Fig. 4; e.g., *Gsc* [Rivera-Perez et al. 1995]; *Tbx4* [Naiche and Papaioannou 2003]). This is not as difficult as it sounds because you will often have two alleles that act the same—that is, alleles before and after excision of a selection cassette that both act as null alleles, or alleles with two different selectable markers. Detection of both null alleles ensures that the chimera is +/+ ⟷ −/−.

2. Make use of tissue restriction in chimeric combinations to provide "pure" tissue for genotyping. For example, if you inject wild-type ICMs into blastocysts from heterozygous mutant matings, the trophectoderm derivatives and primitive endoderm will be derived from the blastocyst (Table 2). Thus, these tissues can be isolated from the chimeras for PCR genotyping without fear of contamination from the +/+ ICM derivatives. However, beware of maternal contamination when dissecting trophectoderm derivatives for PCR.

3. Use laser capture microdissection to isolate a sample composed of pure marked or unmarked cells, depending on which component has the marker, for PCR genotyping.

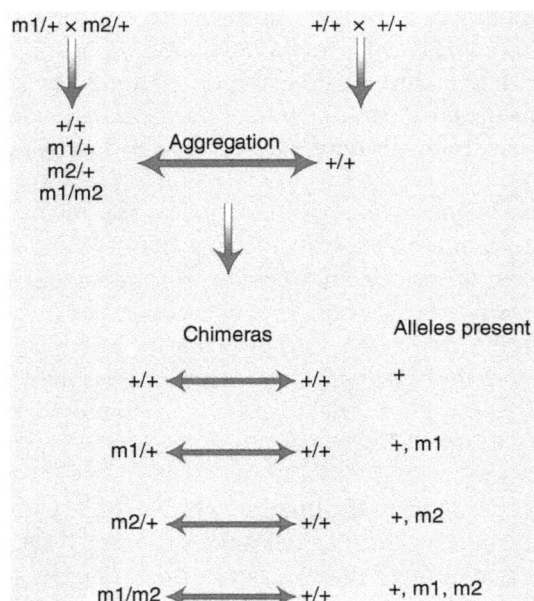

FIGURE 4. Genotyping chimeras using two different null alleles. Two different null alleles, m1 and m2, which can be distinguished by Southern blotting or polymerase chain reaction (PCR), are used. In one cross, m1 heterozygotes are bred with m2 heterozygotes, yielding embryos with four unique genotypes. Wild-type embryos are generated in the second cross. Morulae from the two crosses are aggregated pairwise to generate chimeras. The four different genotypic combinations of chimeras that are generated can be identified by Southern or PCR analysis of tissues. The chimeras with homozygous mutant and wild-type cells will have all three alleles (m1, m2, and +). The other chimeras can serve as controls.

Cite this chapter as *Cold Spring Harb Protoc*; doi:10.1101/pdb.over107979

4. Use statistics. This is probably the least satisfactory method, because it requires a large number of chimeras to be analyzed and depends on the existence of a clear, consistent difference in cell behavior between +/+ ⟷ −/− chimeras and +/+ ⟷ +/− or +/+ ⟷ +/+ chimeras. There will always be doubt as to the interpretation but you may have no other choice. Be sure to analyze a sufficient number of chimeras for statistical significance (see Box 1. Worked Examples of a Chi Square Test of Mendelian Segregation in Chapter 8: Phenotypic Analysis: Assessing Timing of Recessive Prenatal Lethality in Mice [Papaioannou and Behringer 2023d]).

REFERENCES

Abe T, Fujimori T. 2013. Reporter mouse lines for fluorescence imaging. *Dev Growth Differ* **55**: 390–405. doi:10.1111/dgd.12062

Abuin A, Bradley A. 1996. Recycling selectable markers in mouse embryonic stem cells. *Mol Cell Biol* **16**: 1851–1856. doi:10.1128/MCB.16.4.1851

Amadei G, Handford CE, Qiu C, De Jonghe J, Greenfeld H, Tran M, Martin BK, Chen DY, Aguilera-Castrejon A, Hanna JH, et al. 2022. Embryo model completes gastrulation to neurulation and organogenesis. *Nature* **610**: 143–153. doi:10.1038/s41586-022-05246-3

Behringer RR, Gertsenstein M, Nagy KV, Nagy A. 2014. *Manipulating the mouse embryo: a laboratory manual.* Cold Spring Harbor Laboratory Press, Cold Spring Harbor, NY.

Bi W, Deng JM, Zhang Z, Behringer RR, de Crombrugghe B. 1999. *Sox9* is required for cartilage formation. *Nat Genet* **22**: 85–89. doi:10.1038/8792

Chapman DL, Agulnik I, Hancock S, Silver LM, Papaioannou VE. 1996. *Tbx6*, a mouse T-box gene implicated in paraxial mesoderm formation at gastrulation. *Dev Biol* **180**: 534–542. doi:10.1006/dbio.1996.0326

Chapman DL, Cooper-Morgan A, Harrelson Z, Papaioannou VE. 2003. Critical role for *Tbx6* in mesoderm specification in the mouse embryo. *Mech Dev* **120**: 837–847. doi:10.1016/S0925-4773(03)00066-2

Dufort D, Schwartz L, Harpal K, Rossant J. 1998. The transcription factor HNF3β is required in visceral endoderm for normal primitive streak morphogenesis. *Development* **125**: 3015–3025. doi:10.1242/dev.125.16.3015

Guillemot F, Nagy A, Auerbach A, Rossant J, Joyner AL. 1994. Essential role of *Mash-2* in extraembryonic development. *Nature* **371**: 333–336. doi:10.1038/371333a0

Hadjantonakis A-K, Macmaster S, Nagy A. 2002. Embryonic stem cells and mice expressing different GFP variants for multiple non-invasive reporter usage within a single animal. *BMC Biotechnol* **2**: 11. doi:10.1186/1472-6750-2-11

Hadjantonakis A-K, Papaioannou VE. 2004. Dynamic *in vivo* imaging and cell tracking using a histone fluorescent protein fusion in mice. *BMC Biotechnol* **4**: 33. doi:10.1186/1472-6750-4-33

Kiyonari H, Kaneko M, Abe T, Shioi G, Aizawa S, Furuta Y, Fujimori T. 2019. Dynamic organelle localization and cytoskeletal reorganization during preimplantation mouse embryo development revealed by live imaging of genetically encoded fluorescent fusion proteins. *Genesis* **57**: e23277. doi:10.1002/dvg.23277

Lizier M, Anselmo A, Mantero S, Ficara F, Paulis M, Vezzoni P, Lucchini F, Pacchiana G. 2016. Fusion between cancer cells and macrophages occurs in a murine model of spontaneous *neu*⁺ breast cancer without increasing its metastatic potential. *Oncotarget* **7**: 60793–60806. doi:10.18632/oncotarget.11508

Nagy A, Gertsenstein M, Vintersten K, Behringer R. 2006a. Assembling aggregates between diploid and tetraploid embryos. *Cold Spring Harb Protoc* doi:10.1011/pdb.prot4425

Nagy A, Gertsenstein M, Vintersten K, Behringer R. 2006b. Assembling aggregates between diploid embryos. *Cold Spring Harb Protoc* doi:10.1011/pdb.prot4423

Nagy A, Gertsenstein M, Vintersten K, Behringer R. 2006c. Assembling aggregates between embryonic stem (ES) cells and diploid embryos. *Cold Spring Harb Protoc* doi:10.1011/pdb.prot4424

Nagy A, Gertsenstein M, Vintersten K, Behringer R. 2006d. Assembling aggregates between embryonic stem (ES) cells and tetraploid embryos. *Cold Spring Harb Protoc* doi:10.1011/pdb.prot4426

Nagy A, Gertsenstein M, Vintersten K, Behringer R. 2006e. Production of tetraploid embryos. *Cold Spring Harb Protoc* doi:10.1011/pdb.prot4422

Nagy A, Gertsenstein M, Vintersten K, Behringer R. 2006f. Removal of zona pellucida. *Cold Spring Harb Protoc* doi:10.1011/pdb.prot4421

Nagy A, Gertsenstein M, Vintersten K, Behringer R. 2006g. Uterine transfer. *Cold Spring Harb Protoc* doi:10.1011/pdb.prot4380

Nagy A, Gertsenstein M, Vintersten K, Behringer R. 2007. Staining whole mouse embryos for β-galactosidase (*lacZ*) activity. *Cold Spring Harb Protoc* doi:10.1011/pdb.prot4725

Naiche LA, Papaioannou VE. 2003. Loss of *Tbx4* blocks hindlimb development and affects vascularization and fusion of the allantois. *Development* **130**: 2681–2693. doi:10.1242/dev.00504

Papaioannou VE, Behringer RR. 2023a. Uncovering phenotypes in mutant mice by determining embryo, organ, tissue, and cell developmental potential. *Cold Spring Harb Protoc* doi:10.1011/pdb.over107974

Papaioannou VE, Behringer RR. 2023b. Strategies for maintaining mouse mutations. *Cold Spring Harb Protoc* doi:10.1011/pdb.over107960

Papaioannou VE, Behringer RR. 2023c. Tissue- and/or temporal-specific mutations in mice using conditional alleles. *Cold Spring Harb Protoc* doi:10.1011/pdb.over107980

Papaioannou VE, Behringer RR. 2023d. Phenotypic analysis: assessing timing of recessive prenatal lethality in mice. *Cold Spring Harb Protoc* doi:10.1011/pdb.over107970

Papaioannou VE, Behringer RR. 2023e. Phenotypic analysis of preimplantation lethality in mice. *Cold Spring Harb Protoc* doi:10.1011/pdb.over107971

Papaioannou VE, Behringer RR. 2023f. Phenotypic analysis of periimplantation to mid-gestation lethality in mice. *Cold Spring Harb Protoc* doi:10.1011/pdb.over107972

Papaioannou VE, Behringer RR. 2023g. Analysis of mid- to late-gestation phenotypes in mice. *Cold Spring Harb Protoc* doi:10.1011/pdb.over107973

Papaioannou VE, Behringer RR. 2023h. Phenotypic analysis of perinatal lethality in mice. *Cold Spring Harb Protoc* doi:10.1011/pdb.over107975

Papaioannou VE, Behringer RR. 2023i. Analysis of postnatal mutant phenotypes in mice. *Cold Spring Harb Protoc* doi:10.1011/pdb.over107976

Papaioannou VE, Gardner RL. 1979. Investigation of the lethal yellow Ay/Ay embryo using mouse chimaeras. *J Embryol Exp Morphol* **52**: 153–163.

Rivera-Perez JA, Mallo M, Gendron-Maguire M, Gridley T, Behringer RR. 1995. Goosecoid is not an essential component of the mouse gastrula organizer but is required for craniofacial and rib development. *Development* **121**: 3005–3012. doi:10.1242/dev.121.9.3005

Rossant J. 2006a. Culturing trophoblast stem (TS) cell lines. *Cold Spring Harb Protoc* doi:10.1011/pdb.prot4406

Rossant J. 2006b. Derivation of trophoblast stem (TS) cell lines from blastocysts. *Cold Spring Harb Protoc* doi:10.1011/pdb.prot4407

Tarazi S, Aguilera-Castrejon A, Joubran C, Ghanem N, Ashouokhi S, Roncato F, Wildschutz E, Haddad M, Oldak B, Gomez-Cesar E, et al. 2022. Post-gastrulation synthetic embryos generated ex utero from mouse naive ESCs. *Cell* **185**: 3290–3306.e25. doi:10.1016/j.cell.2022.07.028

Varlet I, Collignon J, Robertson EJ. 1997. Nodal expression in the primitive endoderm is required for specification of the anterior axis during mouse gastrulation. *Development* 124: 1033–1044. doi:10.1242/dev.124.5.1033

Voiculescu O, Taillebourg E, Pujades C, Kress C, Buart S, Charnay P, Schneider-Maunoury S. 2001. Hindbrain patterning: *Krox20* couples segmentation and specification of regional identity. *Development* 128: 4967–4978. doi:10.1242/dev.128.24.4967

Yang J, Joshi S, Wang Q, Li P, Wang H, Xiong Y, Xiao Y, Wang J, Parker-Thornburg J, Behringer RR, et al. 2017. 14-3-3ζ loss leads to neonatal lethality by microRNA-126 downregulation-mediated developmental defects in lung vasculature. *Cell Biosci* 7: 58. doi:10.1186/s13578-017-0186-y

Cite this chapter as *Cold Spring Harb Protoc*; doi:10.1101/pdb.over107979

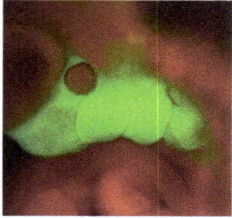

CHAPTER 18

Tissue- and/or Temporal-Specific Mutations in Mice Using Conditional Alleles

Although many existing mutations are null alleles, multipurpose conditional alleles that can be used to delete gene function in a tissue- and/or temporal-specific manner are increasingly the alleles of choice. There are two distinct but related advantages: first, early lethal effects of the mutation can be bypassed by leaving the gene intact until later stages in development; second, indirect or secondary effects on an organ of interest can be eliminated by tissue- or organ-specific gene deletion. In this chapter, we cover aspects of testing and using conditional alleles to ensure that the desired effect is obtained, including how to test the engineered conditional allele to ensure it functions as planned, and how to test any recombinase mouse strain used, including inducible transgenic or knock-in lines. Finally, we discuss how to use a conditional allele for maximum value in a phenotypic analysis.

INTRODUCTION

TISSUE/TEMPORAL-SPECIFIC MUTATIONS

- CONDITIONAL ALLELES
  - Testing conditional allele
  - Characterizing recombinase strains → Tissue-specific / Inducible
  - Using conditional alleles
- BEYOND CONDITIONAL ALLELES
  - Transgene rescue
  - Regulatory mutations

Individual genes can have a multitude of roles depending on the timing and location of their expression. A gene active during early stages of embryonic development may also be active during later organogenesis (e.g., *Tbx4*; Naiche and Papaioannou 2003) or play a role in homeostasis during adult life (e.g.; *Brca1*; Ludwig et al. 1997). Your area of interest may be in the development of a specific organ or in the formation of a specific type of tumor in the adult. Thus, an early lethal phenotype resulting from a null mutation could prevent the study of the roles of that gene in later development. Knowing

Opening artwork: Courtesy of Ayan Ray and Philippe Soriano.

the full expression pattern of a gene before you embark on a gene targeting experiment can provide an indication of potential cells and tissues where the gene might have an effect, which would allow you to anticipate an early lethal effect and inform your gene targeting strategy (see Chapter 3: Mouse Gene Targeting Strategies for Maximum Ease and Versatility [Papaioannou and Behringer 2023a]). However, if you find yourself in the position in which an early lethal effect of a null mutation prevents the study of the area of interest, there are a number of ways to circumvent the problem. We covered the use of organ culture (see Chapter 12: Uncovering Phenotypes in Mutant Mice by Determining Embryo, Organ, Tissue, and Cell Developmental Potential [Papaioannou and Behringer 2023b]) and assisted reproductive techniques (see Chapter 6: Strategies for Maintaining Mouse Mutations [Papaioannou and Behringer 2023c]) and the use of chimeras (see Chapter 17: Getting around an Early Lethal Phenotype in Mice with Chimeras [Papaioannou and Behringer 2023d]). Here we present additional strategies to get around an early effect that involve making additional genetic alterations. Conditional alleles are the predominant tools that allow you to confine the mutant effect to specific times, cell types, or cell lineages, effectively circumventing a mutant problem in some part of the embryo to reveal a mutant effect in others, and they have wide application for making different types of mutations. Transgene rescue can be used to supply a missing protein to allow further development of a mutant animal. Finally, creating regulatory mutations is a way of producing tissue-specific ablation or quantitative reduction of a gene product.

TYPES OF CONDITIONAL ALLELES

The incorporation of the bacteriophage-derived Cre/*loxP* DNA recombination system into the mouse genome, and subsequently the yeast-derived FLP/*frt* system and other bacteriophage-derived DNA recombination systems Dre/*rox* and Vika/*vox* (Anastassiadis et al. 2009; Karimova et al. 2018), have opened up conditional genetic experiments that were not previously possible (Gu et al. 1994). In the mouse, there are broadly two types of conditional alleles: (1) conditional gain-of-function and (2) conditional null. Both types function like wild-type alleles but are converted to gain-of-function or null alleles, respectively, upon the action of a DNA recombinase. Conditional gain-of-function alleles are typically transgene knock-ins at a locus that is ubiquitously expressed (e.g., *ROSA26*) and that are silent until activated by a DNA recombinase (Soriano 1999). They are used to express reporter genes but can also be used to ectopically express a gene of interest. Conditional null alleles are our focus here, as they are widely used in mutational analysis. In combination with recombinase mouse strains, they can be used to inactivate a gene of interest in a specific cell or tissue type, at a specific time in fetal development, or after birth (Lewandoski 2001).

TESTING CONDITIONAL NULL ALLELES

Making a conditional allele, as described in Chapter 3: Mouse Gene-Targeting Strategies for Maximum Ease and Versatility (Papaioannou and Behringer 2023a), gives you the option of leaving gene function intact until the allele is exposed to the appropriate site-specific recombinase, at which time a critical region of the gene is excised resulting in a loss of function. Conditional alleles can be generated by homologous recombination in embryonic stem (ES) cells or by CRISPR–Cas manipulations of ES cells or zygotes. Although a conditional allele with a positive selectable marker left in place, if generated in ES cells, may function as a wild-type allele, we recommend that you remove the selection cassette from the allele in ES cells before making chimeras (see Box 2 in Chapter 3: Mouse Gene-Targeting Strategies for Maximum Ease and Versatility [Papaioannou and Behringer 2023a]). This will circumvent any potential interference with gene function by the promoter of the selectable marker.

Once the conditional allele is in the germline of mice, whether the selectable marker is left in the locus or taken out, several tests must be performed to verify that the conditional allele is functioning as planned. First, you must ensure that the conditional allele actually functions like a wild-type allele rather than a null or hypomorph. If the conditional allele does function as a hypomorph, it can still be a useful allele (see below). Second, you must show that the site-specific recombinase produces the desired effect (i.e., the removal of the floxed or flrted exons), resulting in a functionally null allele. Only then can you make use of the conditional allele for time- or tissue-specific ablation.

A first step in testing a conditional allele is to show that it has no effect on gene function. To establish this, examine heterozygotes for the conditional allele and then breed the allele to homozygosity to determine that there is no mutant phenotype (e.g., *Bmpr1a* [Mishina et al. 2002]; *Lhx1* [Kwan and Behringer 2002]). If you have a recessive null allele at the locus available, you should also produce mice that are compound heterozygotes for the conditional allele and the null allele to show that the conditional allele is fully functional. If the homozygous conditional and compound heterozygous mice are normal and fertile, you probably have a good conditional allele.

On the other hand, if a mutant phenotype is detected in mice with the conditional allele, the most likely culprit would be the positive selection cassette if it was left in the locus. Remove the cassette by breeding as outlined in Chapter 6: Strategies for Maintaining Mouse Mutations (Papaioannou and Behringer 2023c) and then breed to homozygosity again. Keep the original allele, however, because it may prove to be a useful hypomorphic allele. If the selection cassette has already been removed, there is less chance of a mutant phenotype, but it is still possible if the remaining recombinase sites (*loxP* or *frt*) are fortuitously located in regulatory regions. Thus, even a conditional allele without a selection cassette must be tested for phenotypic effects. If a conditional allele without a selection cassette has a mutant phenotype, you can still make use of it if it proves to be a hypomorphic rather than a null allele, as mentioned above. You will be able to ablate the function of this hypomorphic allele conditionally, provided it allows development of the tissue of interest (e.g., *Nodal* [Lowe et al. 2001]; *Bmp4* [Kulessa and Hogan 2002]). However, be aware that a hypomorphic mutation could well have an effect on the tissue of interest, so that when the null allele is produced by recombinase, the development of the tissue might already be compromised.

The second step in testing a conditional allele is to show that site-specific recombinases have the desired effect on the allele. This should have already been tested in ES cells (Chapter 3: Mouse Gene-Targeting Strategies for Maximum Ease and Versatility [Papaioannou and Behringer 2023a]) but now that the allele is in mice, removing the floxed or flrted exon(s) serves the dual purpose of establishing the utility of the allele and producing a null mutation. Breed the mice carrying a conditional null allele with deleter mice as outlined in Chapter 6: Strategies for Maintaining Mouse Mutations (Papaioannou and Behringer 2023c) and recover mice heterozygous for the null allele. In addition to analyzing these mice for the effects of a null mutation (see Chapter 8: Phenotypic Analysis: Assessing Timing of Recessive Prenatal Lethality in Mice [Papaioannou and Behringer 2023e]; Chapter 9: Phenotypic Analysis of Preimplantation Lethality in Mice [Papaioannou and Behringer 2023f]; Chapter 10: Phenotypic Analysis of Periimplantation to Mid-Gestation Lethality in Mice [Papaioannou and Behringer 2023g]; Chapter 11: Analysis of Mid- to Late-Gestation Phenotypes in Mice [Papaioannou and Behringer 2023h]; Chapter 12: Uncovering Phenotypes in Mutant Mice by Determining Embryo, Organ, Tissue, and Cell Developmental Potential [Papaioannou and Behringer 2023b]; Chapter 13: Phenotypic Analysis of Perinatal Lethality in Mice [Papaioannou and Behringer 2023i]; Chapter 14: Analysis of Postnatal Mutant Phenotypes in Mice [Papaioannou and Behringer 2023j]), you can also use them in the breeding scheme to produce a tissue-specific mutation and to test whether the conditional allele is wild type or hypomorphic. First, go to Chapter 7: Special Breeding Techniques for Use in Mouse Mutation Analysis (Papaioannou and Behringer 2023k) and follow the breeding scheme to test for a hypomorphic or null allele. Then, proceed to characterize the Cre-expressing or other recombinase expressing mouse line you intend to use with the conditional allele.

RECOMBINASE MOUSE STRAINS

Conditional alleles are characterized by the inclusion of recombinase recognition sites flanking a genomic region to be excised or inverted by site-specific recombination upon exposure to a specific recombinase (Cre, Flp, Dre, Vika), usually by breeding with a mouse strain transgenic for a recombinase gene. There is an ever-increasing number of recombinase-expressing transgenic mouse lines available (e.g., https://www.jax.org/research-and-faculty/resources/cre-repository) or you may have generated transgenic or knock-in mouse lines yourself to express Cre (or the relevant recombinase) in specific tissues of interest. Whatever the source of the mouse lines, it is essential that they are fully characterized with respect to the recombinase activity pattern before use. To be absolutely sure, you should test it yourself to verify the tissue-specific pattern and timing of recombinase activity.

Characterizing Tissue-Specific Recombinase Transgenic or Knock-in Mouse Lines

Here is a guide for the characterization of *cre*-transgenic or knock-in mouse lines (the same strategy can be used to characterize *flp* or other recombinase transgenic or knock-in mouse lines). If you have received the mice from another laboratory, first genotype the mice using an allele-specific polymerase chain reaction (PCR) strategy to make sure that you have received the correct animals. The next step is to examine Cre activity. Typically, a Cre reporter mouse line such as *R26R* is used for this purpose (Fig. 1; Soriano 1999). This line contains a conditional *lacZ* reporter knock-in at the ubiquitously expressed *ROSA26* locus such that it can report Cre activity in all tissues. The Cre-expressing mouse line is crossed with the Cre reporter mouse line to generate double heterozygotes, and the reporter is used to visually assess the temporal and spatial patterns of Cre activity. Examine reporter expression in double heterozygotes at various embryonic stages, either in whole mounts or histological sections. It is very important to determine the first stages of development in which you detect Cre reporter activity because this indicates when and in which tissues a conditional knockout might first show its effects using this Cre line. Keep in mind that the expression of the reporter is the result of Cre-mediated recombination in a given cell at a specific time, and is heritable, and that all of the cellular progeny of

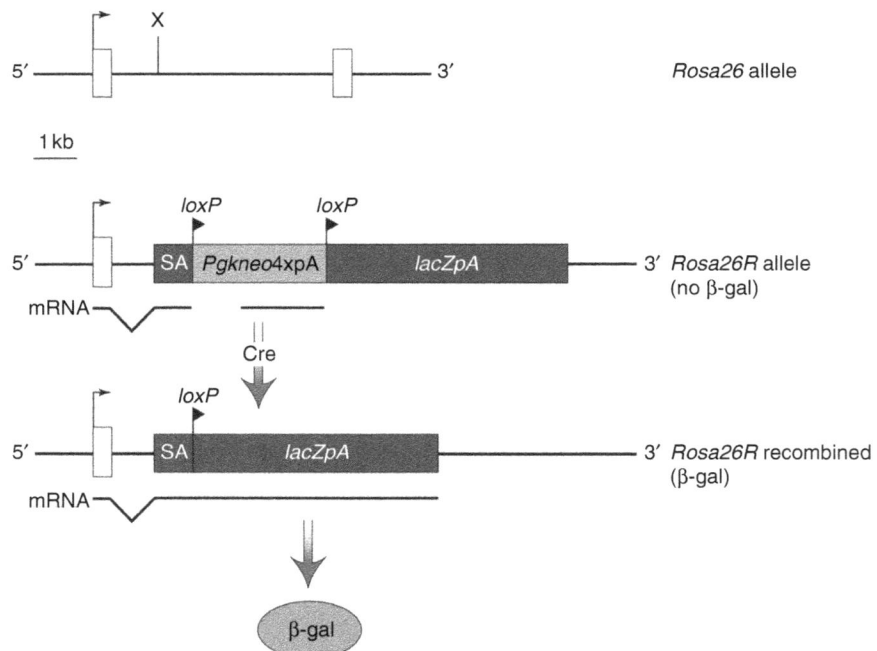

FIGURE 1. Generation and use of the *R26R* Cre reporter mouse line. The *Rosa26* locus encodes transcripts that lack an open reading frame. The *R26R* allele was generated by targeting a cassette into the XbaI (X) site located within the first intron. A *neo* expression cassette with 4xpA sequences blocks the transcription of the *lacZ* gene. Cre recombinase will delete the floxed *neo* cassette, allowing *lacZ* transcription. The splice acceptor (SA) sequence leads to the generation of a chimeric *R26-lacZ* transcript that results in the production of β-galactosidase (β-gal). All of the cellular progeny derived from a cell that has undergone this Cre-mediated recombination will stain positively for β-gal activity (Soriano 1999).

Cite this chapter as *Cold Spring Harb Protoc*; doi:10.1101/pdb.over107980

that cell will continue to express the reporter (effectively fate mapping the progeny of that cell). Thus, not all reporter-positive cells necessarily express the recombinase at a given time. It should be noted that some Cre reporters might be more sensitive to Cre activity than others. Therefore, you may want to characterize your Cre mouse lines with more than one type of Cre reporter mouse (Srinivas et al. 2001; Muzumdar et al. 2007).

Mosaic expression of a *cre* transgene or knock-in is relatively common. Accessing the amount and variation of mosaicism will be important for anticipating potential mutant phenotypes in a conditional knockout experiment. Genetic background can influence, either positively or negatively, the expression of a *cre* transgene or knock-in. Therefore, it is useful to characterize the expression of your *cre* transgene or knock-in on different genetic backgrounds or at least on the background you intend to use.

It is possible that a putative tissue-specific *cre* transgene or knock-in will also express Cre in the male or female germline (i.e., in the developing or mature germ cells). This "off-target" expression will preclude its use in a tissue-specific strategy. To determine if this is the case, generate mice that are heterozygous for the *cre* transgene or knock-in and also heterozygous for a Cre reporter allele. Then breed these double heterozygous mice to wild-type mice and examine Cre reporter expression in the resulting embryos. If Cre is active in the germline, ubiquitous expression of the Cre reporter will be present in the progeny. If you do find germline Cre activity in a Cre line, it may be limited to only the male or female germline, in which case you could adjust your crosses accordingly using the germline that does not have Cre activity (Liput 2018). Finally, remember that the timing of Cre activity and activation of a Cre reporter will likely be different from the timing of loss of expression of the endogenous gene by a conditional knockout (Fig. 2).

Characterizing Inducible Recombinase Transgenic or Knock-in Mouse Lines

In addition to using a conditional allele to create a tissue-specific mutation, it is also possible to control the timing of recombination of a conditional allele using mouse strains carrying Cre (or another recombinase) fusion protein genes that are inducible upon the application of specific drugs, providing temporal control of gene manipulation (Feil et al. 1996; Kellendonk et al. 1996). The Cre estrogen receptor (ER) or Cre progesterone receptor (PR) and their modified versions (e.g., CreER; Hayashi and McMahon 2002) are fusion proteins composed of Cre and mutant ligand binding domains of the ER and PR, respectively. These fusion proteins do not bind estrogen or progesterone but rather the artificial ligands tamoxifen and RU-486, respectively. In the absence of these artificial ligands, the Cre fusion proteins reside in the cytoplasm. However, in their presence, the Cre fusion proteins enter the nucleus and become active.

The characterization of an inducible CreER or CrePR transgenic or knock-in mouse line is similar to that described for a standard Cre mouse line. However, in addition to crosses with Cre reporter mice to characterize Cre expression, you also need to test the inducibility and extent of Cre activity at different times after administration of the inducing drug. This can require a lot of time, so if you have generated new transgenic mouse lines with tissue-specific inducible Cre expression, it may be more practical to first assay for Cre expression at the level of RNA or protein by in situ hybridization or immunostaining, respectively. Once you have identified expressing transgenic lines, then generate double heterozygotes with the inducible *cre* transgene and a Cre reporter mouse to test for inducibility.

The analysis of embryonic stages requires treatment of a pregnant female carrying the *cre* transgene and Cre reporter double-heterozygous embryos with tamoxifen (for CreER) or RU-486 or its analogs (for CrePR) at different stages of gestation. Thus, you will need to determine the sufficient drug dose and the number and timing of injections needed during gestation for the excision of the floxed allele and loss or activation of gene product. These drugs are light sensitive, so take precautions. In addition, it is very important to have the drugs completely dissolved in the diluent, which may require overnight stirring. For drug injections into mice, use vegetable oil as the diluent (Whitfield et al. 2015). Daily injections of the drug may be necessary to achieve sufficient recombination for your purposes

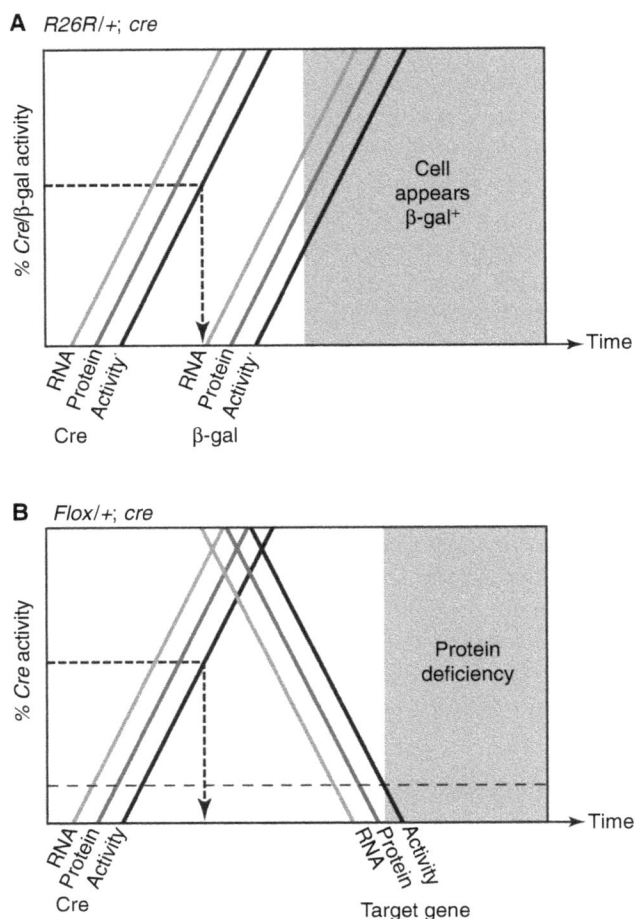

FIGURE 2. Comparison of Cre reporter activity with Cre-mediated loss of function at the cellular level. (*A*) Detection of β-gal activity in *R26R/+* x *cre* mice. As a *cre* transgene is transcribed and translated, Cre activity increases. At a certain threshold of Cre activity, the floxed DNA that blocks β-gal expression is deleted (dashed arrow). *lacZ* is then transcribed and translated into β-gal. After a certain time, there will be sufficient β-gal activity and individual cells will stain blue with X-gal (shaded area). (*B*) Time line for a deficiency of protein activity caused by Cre-mediated recombination of a conditional null allele (*flox*). As in *A* above, as the cre transgene is transcribed and translated, Cre activity increases. At a certain threshold of Cre activity, the floxed DNA is deleted (dashed arrow). Because the cell genotype is *flox/−*, no further gene product can be generated from the target gene. Residual target gene mRNA and protein will decay, gradually reducing residual levels of target gene protein activity. Once protein activity levels drop below a threshold (dashed line), the cell will be functionally deficient for that protein (shaded area). Thus, while a Cre reporter provides information about when Cre activity first initiates, it does not necessarily tell you when your target gene protein activity will be sufficiently reduced to cause a functional consequence.

(Hayashi and McMahon 2002). You can also have tamoxifen or RU-486 commercially encapsulated. Although expensive, these capsules can be implanted under the skin of the animal for constant slow release of the drug. Some inducible Cre lines express in the skin, in which case, tamoxifen or RU-486 can be applied topically (e.g., *IKK2*; Vasioukhin et al. 1999; Stratis et al. 2006).

Below is a guide for suggested doses of tamoxifen and RU-486 for induction of Cre activity. You will have to optimize these empirically for your particular study to balance recombination efficiency and embryo viability.

- Tamoxifen dose for injection of pregnant females:1.5–10 mg/pregnant female (Joyner and Zervas 2006; Concepcion et al. 2017).

- Tamoxifen dose for injection of postnatal animals: 0.75–1 mg for five consecutive days.

Cite this chapter as *Cold Spring Harb Protoc*; doi:10.1101/pdb.over107980

- RU-486 (24 h half-life) dose for injection of pregnant females is 50–100 µg/kg body weight. To prevent abortion, RU-486-treated mice must also be injected daily with progesterone at 0.5 mg/mouse to maintain the pregnancy (Rose et al. 2009).

- RU-486 dose for injection of postnatal animals: 100–150 µg/kg body weight, probably multiple (three) daily injections.

These doses can serve as a starting point for your particular study. Examine the fetuses for the expression of the Cre reporter at different times after the treatment. The analysis of postnatal stages requires treatment of the double heterozygotes after birth. Again, you will need to determine the drug dose and the number and timing of injections empirically to elicit sufficient recombination for your studies. After the induction of Cre activity, analyze the tissue of interest for expression of the Cre reporter at different times after the treatment to assess the timing and extent of Cre activity. These data can then be used as a guide for analysis of your intended inducible conditional knockout.

USING CONDITIONAL NULL ALLELES

Once tested and shown to have no mutant phenotype on its own and to produce a null allele upon exposure to recombinase, a conditional allele can be used to ablate gene function either in specific tissues and/or at specific stages of development by crossing mice carrying the conditional allele with mice carrying a well-characterized *cre* (or other appropriate recombinase) transgene. Either tissue- or stage-specific ablation approaches can be used to allow an embryo to develop beyond an early lethal effect caused by a homozygous null mutation. Additionally, a tissue-specific ablation using a recombinase transgene or knock-in expressed only in that tissue is useful if you are mainly interested in a certain organ or tissue or want to distinguish primary from secondary effects. A ubiquitous, stage-specific ablation approach, accomplished by using an inducible, ubiquitously expressed recombinase, allows you to study the role of the gene in multiple tissues later in development and could reveal later time points in development in which the lack of gene function causes lethality. For both approaches, the limiting factors, apart from space to breed the animals, are the number and variety of different recombinase transgenic or knock-in mouse lines available and your ability to verify the excision event in specific cells or tissues.

Cre (or Flp) can also be introduced in vivo both in embryos and adults using a variety of DNA, RNA, or protein transfer methods, including virally mediated gene transduction (Lee et al. 2008), electroporation of tissues and organs (Bland et al. 2017), and exposure to cell-permeable recombinases (Ryder et al. 2014). The advantages include the ability to choose the time and place to apply the recombinase; disadvantages include potential limitations of access to specific tissues and organs.

With respect to verifying which cells have excised the floxed/flrted exons, unless you have built a reporter into the allele that reports when the exons are excised (see Fig. 10 in Chapter 3: Mouse Gene-Targeting Strategies for Maximum Ease and Versatility [Papaioannou and Behringer 2023a]), you will have to rely on assumptions based on the known pattern of recombinase expression or on direct measurements of either Cre/Flp expression or the absence of the floxed/flrted exons. Because the level of recombinase expression may vary for different transgenes and the time of onset of recombinase expression may be gradual (Fig. 2), you will have to consider the possibility that tissues are mosaic for excision of the exons in the offspring of the cross and interpret the results accordingly. Similarly, the efficiency of excision may be variable for different recombinase transgenes or different targeted alleles, possibly leading to incomplete excision resulting in a variable phenotype (e.g., *Nr5a1*; Zhao et al. 2001).

Once you have chosen—and tested—a suitable recombinase transgenic line, tissue- or stage-specific mutations can be generated using several breeding schemes (e.g., *Fgf8*; Meyers et al. 1998). A very efficient scheme is to generate males heterozygous for a constitutive null allele of your gene and heterozygous or hemizygous for the targeted knock-in or transgene that directs tissue-specific recombinase expression—Cre in this example, +/−; *cre*/+ (Fig. 3). The constitutive null allele can either be an

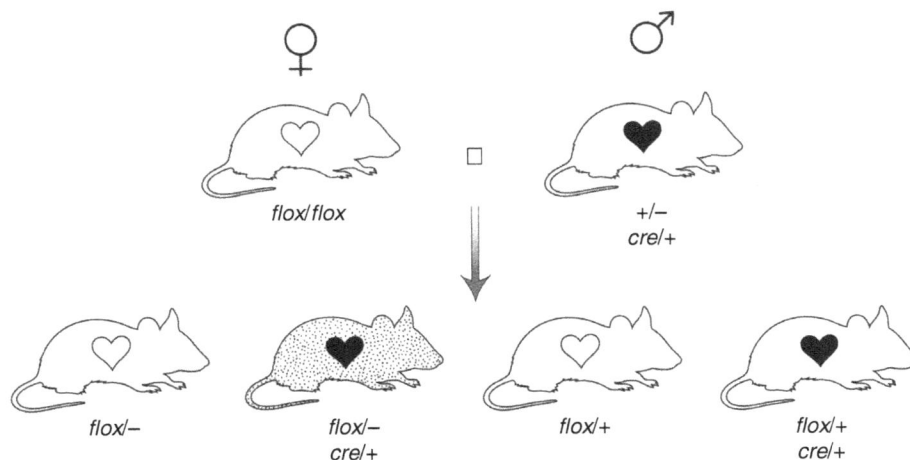

FIGURE 3. Breeding scheme for producing mice or embryos with a tissue-specific mutation. Females homozygous for a conditional allele (*flox*) are mated with males heterozygous or hemizygous for a gene-specific *cre* knock-in or a tissue-specific *cre* transgene (*cre/+*)—in this case, a heart-specific *cre* (solid heart) and heterozygous for a null allele of the gene (*+/−*). One in four of the resulting offspring (stippled) will carry both mutant alleles (*flox/−*) and also contain the *cre* transgene (*cre/+*), in which case the floxed exons will be deleted from the conditional allele in the tissues expressing Cre, namely, the heart, resulting in a heart-specific null.

independent null allele or a recombined null allele generated from the conditional null allele. Only a few *+/−; cre/+* males are needed for crosses with females homozygous for the conditional null allele (*flox/flox*). It is likely that you will need to sacrifice the *flox/flox* females to analyze embryonic time points, but they are easy to generate from a homozygous breeding stock, which obviates the need for genotyping. Using this cross, one in four of the resulting progeny will be *flox/−; cre/+*. With this strategy, Cre only has to act on one allele because the other is already null. If Cre is required to act on two floxed alleles per cell, it might not be as efficient as expected and may create mosaicism for the tissue-specific knockout. This, in turn, may generate a less severe mutant phenotype than a null.

When you generate *+/−; cre/+* stud males for crosses with *flox/flox* females in a tissue-specific knockout experiment (see Fig. 3), prescreen the stud males by crossing them with Cre reporter females (e.g., *R26R*) to verify that they can produce progeny with the correct pattern of Cre activity (Furuta and Behringer 2005). This is especially important when you know that the tissue-specific pattern of Cre activity for the particular *cre* transgenic line that you are using is sensitive to genetic background, position effects, or stochastic variation.

One note of caution: If you are using a recombinase "knock-in" allele for tissue-specific ablation (i.e., a recombinase gene that is inserted into an endogenous locus under the control of an endogenous promoter), it is also likely to be a loss-of-function allele for the gene into which the recombinase is inserted and could potentially influence the expressivity and penetrance of your tissue-specific mutant phenotype. The only ways around this are fairly drastic (e.g., recreating the recombinase-expressing mouse without ablating the endogenous gene function or using a comparable recombinase transgenic strain if one is available), but you should keep this in mind when interpreting your results. Similarly, you may have little choice of genetic background when it comes to selecting a recombinase-expressing mouse to breed with your conditional allele. You can backcross the mice to initiate the generation of a congenic strain (see Protocol 6.1: Backcrossing to Make a Congenic Mouse Strain [Papaioannou and Behringer 2023]), or accept that you are dealing with a mixed genetic background in the analysis.

Ideally, you should be able to document the loss of protein from your recombined conditional allele in the recombinase-expressing tissue. In many cases, however, antibodies may not be available. Documenting the loss of mRNA for the gene is the next best strategy and could be done on a cell-by-cell basis with in situ hybridization using an exon-specific probe within the floxed exons or on a whole-tissue basis using reverse transcription (RT)-PCR and exon-specific primers. It is important to realize that the timing of recombinase expression and the loss of mRNA, protein, and eventually function may be gradual. Timing will depend on the rate of turnover of the RNA and protein in question and, for inducible recombinase expression, the timing and kinetics of induction (Fig. 2).

Finally, the analysis. Knowing when and where the specific recombinase is expressed, you can extrapolate to a projected time of effect of gene ablation on a given tissue. Now, adapt the methods outlined in Chapter 8: Phenotypic Analysis: Assessing Timing of Recessive Prenatal Lethality in Mice (Papaioannou and Behringer 2023e); Chapter 9: Phenotypic Analysis of Preimplantation Lethality in Mice (Papaioannou and Behringer 2023f); Chapter 10: Phenotypic Analysis of Periimplantation to Mid-Gestation Lethality in Mice (Papaioannou and Behringer 2023g); Chapter 11: Analysis of Mid- to Late-Gestation Phenotypes in Mice (Papaioannou and Behringer 2023h); Chapter 12: Uncovering Phenotypes in Mutant Mice by Determining Embryo, Organ, Tissue, and Cell Developmental Potential (Papaioannou and Behringer 2023b); Chapter 13: Phenotypic Analysis of Perinatal Lethality in Mice (Papaioannou and Behringer 2023i); and Chapter 14: Analysis of Postnatal Mutant Phenotypes in Mice (Papaioannou and Behringer 2023j) to study the phenotype produced by the tissue- or stage-specific ablation of gene function.

BEYOND CONDITIONAL ALLELES

Transgene Rescue to Get around a Lethal Phenotype

It is possible to bypass an embryonic lethality caused by a mutation or to correct a specific defect by providing expression from a transgene to critical tissues (see Box 1 in Chapter 16: Phenotypic Analysis of Dominant Mutant Effects in Mice [Papaioannou and Behringer 2023m], in which a transgene was electroporated into ES cells to rescue a dominant embryo-lethal phenotype). Transgene expression can be supplied by an endogenous or heterologous regulatory element. Because transgene expression derived from zygote injections (vs. targeted knock-ins) is influenced by the site of integration, each transgenic mouse line will potentially provide a different level of expression and thus a different extent of rescue of the mutant phenotype, leading to the equivalent of an allelic series of phenotypes. On a wild-type background, the expression of the rescuing transgene may result in a gain-of-function phenotype that could also be informative (*Lhx1*; Kobayashi et al. 2005)

To use transgenic rescue, generate transgenic mice on a wild-type genetic background by zygote injection of your rescuing gene construct (see Chapter 7 in Behringer et al. 2014) and then backcross the transgene onto the mutant background to obtain homozygous mutant, transgenic mice for analysis. If you are generating bacterial artificial chromosome (BAC) transgenic mice (e.g., *Myo15*; Probst et al. 1998), you can use the entire BAC containing the wild-type allele, allowing you to use the plasmid backbone as a tag to detect the presence of the BAC in mice from this cross. This does not ensure that the entire BAC has integrated intact, but it is a quick and simple way to detect the presence of the foreign DNA. Alternatively, if the inbred strain from which the BAC is derived differs from the strain of the zygotes that you are using to make the transgenic, then you can use DNA polymorphisms between the strains to detect the presence of the BAC. You will find that genotyping a homozygous mutant with a transgene is much simpler in the case of a BAC transgene if you have two different mutant alleles (e.g., *neo* and *lacZ* knockout alleles). If you only have one mutant allele, you will have to use a DNA polymorphism to track the BAC on the mutant genetic background at the same time as genotyping for the mutant allele. This is because the genotyping strategies for the wild-type allele will yield the same result if you have a heterozygous or homozygous mutant along with the wild-type allele on the BAC. Fortunately, if you are using a BAC derived from a species other than the mouse, you can use species-specific probes or primers to detect the presence of the BAC in a mouse. When designing a transgene rescue experiment, carefully think through how you will follow transgene expression (northern, RT-PCR, in situ hybridization, etc.) on the wild-type and mutant genetic backgrounds.

There are a few caveats to the transgene rescue approach. First, it may be difficult to generate transgenic mice carrying a rescuing gene construct if overexpression of the gene is detrimental. If so, you may find that you only recover lines that express the transgene at low levels. This can be useful because these lines might lead to hypomorphic rescue of the mutant phenotype. If you find that the transgene is lethal on a wild-type genetic background, it would be possible, though not very practical, to inject the rescuing transgene construct directly into heterozygous and homozygous mutant zygotes from heterozygous crosses. In addition, if you are using a BAC transgene, remember that there may be other genes on the BAC that could influence the rescue and that expression of transgenes can be mosaic, which may influence the outcome of the experiment.

Regulatory Mutations for Tissue-Specific Knockout

The mouse genome has been sequenced, assembled, and annotated, and we now know the identity of most, if not all, coding regions. By comparison, less is known about the specific sequences that regulate gene transcription and other gene regulatory processes. Empirical studies of sequences located near coding regions have traditionally been used to identify tissue-specific transcriptional enhancers (Hammer et al. 1987a,b; Zhou et al. 1995). Because of the availability of multiple assembled animal genomes, cross-species comparisons of noncoding regions have begun to identify conserved sequences as candidate transcriptional regulatory elements that are then being tested using transgenic assays. Chromatin landscape analyses of organs, tissues, and cell types can provide a genome-wide assessment of candidate *cis*-regulatory sequences (Noordermeer and Duboule 2013; Lu et al. 2016; Wu et al. 2016; Gonen et al. 2018). Thus, more and more candidate tissue-specific transcriptional enhancers are being identified. This opens up the possibility of creating tissue-specific loss-of-function alleles by targeted mutagenesis (Fig. 4). These regulatory mutations are not limited to transcriptional enhancers; they could also include mutations that affect RNA splicing, polyadenylation sequences, RNA/protein binding sites in untranslated regions, etc. Just like a Cre/*loxP*-based conditional knockout, a regulatory mutation can allow you to bypass an early lethality caused by a standard knockout.

Regulatory mutations can generally be used for genes that are expressed in multiple tissues but not ubiquitously (e.g., *Sox9*; Gonen et al. 2018). This implies that the gene in question is regulated by multiple essential transcriptional enhancers. If the regulatory sequence acts redundantly with another *cis* element, then it may contribute only partially to the overall levels of tissue-specific transcription. Therefore, mutation of a redundant element may result in a hypomorphic allele that can also be very useful. Regulatory mutations can be made using a variety of gene-targeting or gene editing methods

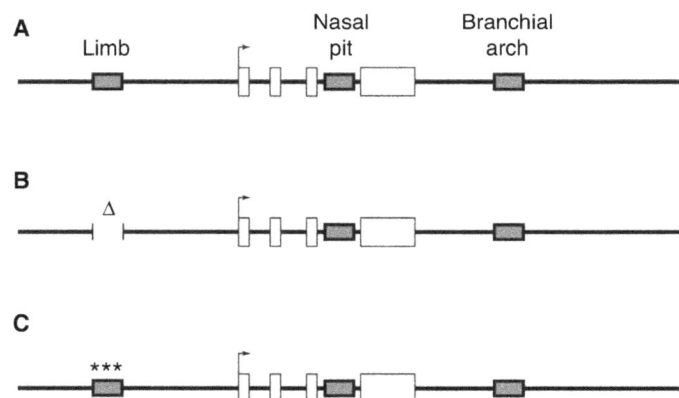

FIGURE 4. Tissue-specific knockouts caused by mutations in tissue-specific enhancers. (A) Structure of a gene with tissue-specific enhancers (shaded boxes) for limb, nasal pit, and branchial arch expression distributed 5′, within, and 3′ of the exons (open boxes), respectively. (B) Targeted deletion (Δ) of the limb enhancer. (C) Targeted base-pair changes (***) in a transcription factor binding site of the limb enhancer that eliminates binding. The alleles shown in B and C should result in loss of transcription of the gene only in the limb resulting in a tissue-specific knockout, unless there is a redundant limb enhancer.

 Cite this chapter as *Cold Spring Harb Protoc*; doi:10.1101/pdb.over107980

(see Chapter 2: Obtaining or Generating Gene Mutations in Mice [Papaioannou and Behringer 2023n]; Chapter 3: Mouse Gene-Targeting Strategies for Maximum Ease and Versatility [Papaioannou and Behringer 2023a]). Candidate *cis*-regulatory sequence can be deleted (Sagai et al. 2005; Gonen et al. 2018; Hojo et al. 2022). Alternatively, base-pair changes can be engineered to disrupt the enhancer activity of a regulatory sequence (e.g., *Amh*; Arango et al. 1999). These base-pair changes prevent the binding of *trans*-acting factors that regulate tissue-specific transcription. As with any mutation, regulatory mutations can be made homozygous or combined with a null allele, if available (Arango et al. 1999).

A tissue-specific knockout caused by a regulatory mutation differs fundamentally from a traditional Cre/*loxP*-based tissue-specific knockout in a number of ways. First, only one mutant allele (i.e., one mouse strain) is required for a regulatory mutation approach in comparison to at least two (*cre* and *flox*) but more likely three (*cre*, *flox*, *null*) alleles required for a Cre/*loxP*-based tissue-specific knockout. In addition, with a Cre/*loxP*-based system, the loss of the target gene product depends upon the timing, action, and extent of Cre expression in the tissue of interest. With a regulatory mutation, transcription will not be initiated in the tissue of interest, and, thus, a constitutive deficiency of the gene product results.

In recent years, new technologies have been developed to reduce or eliminate gene expression in specific tissues or at specific times, including new CRISPR–Cas systems to knock down specific mRNAs (Kushawah et al. 2020; Colognori et al. 2022) and inducible protein degradation systems (degrons) to deplete specific proteins (Yunusova et al. 2021; Abuhashem et al. 2022; Kanemaki 2022). These new methods are promising approaches to complement the current genetic tool kit to analyze mutant mouse phenotypes in the quest to understand gene function.

REFERENCES

Abuhashem A, Lee AS, Joyner AL, Hadjantonakis AK. 2022. Rapid and efficient degradation of endogenous proteins in vivo identifies stage-specific roles of RNA Pol II pausing in mammalian development. *Dev Cell* **57:** 1068–1080.e1066. doi:10.1016/j.devcel.2022.03.013

Anastassiadis K, Fu J, Patsch C, Hu S, Weidlich S, Duerschke K, Buchholz F, Edenhofer F, Stewart AF. 2009. Dre recombinase, like Cre, is a highly efficient site-specific recombinase in *E. coli*, mammalian cells and mice. *Dis Model Mech* **2:** 508–515. doi:10.1242/dmm.003087

Arango NA, Lovell-Badge R, Behringer RR. 1999. Targeted mutagenesis of the endogenous mouse *Mis* gene promoter: in vivo definition of genetic pathways of vertebrate sexual development. *Cell* **99:** 409–419. doi:10.1016/S0092-8674(00)81527-5

Behringer RR, Gertsenstein M, Nagy KV, Nagy A. 2014. *Manipulating the mouse embryo: a laboratory manual.* Cold Spring Harbor Laboratory Press, Cold Spring Harbor, NY.

Bland KM, Casey ZO, Handwerk CJ, Holley ZL, Vidal GS. 2017. Inducing Cre-lox recombination in mouse cerebral cortex through in utero electroporation. *J Vis Exp* **2017:** 56675. doi:10.3791/56675

Colognori D, Trinidad M, Doudna JA. 2022. Precise transcript targeting by CRISPR–Csm complexes. bioRxiv doi:10.1101/2022.06.20.496908

Concepcion D, Washkowitz AJ, DeSantis A, Ogea P, Yang JI, Douglas NC, Papaioannou VE. 2017. Cell lineage of timed cohorts of *Tbx6*-expressing cells in wild-type and *Tbx6* mutant embryos. *Biol Open* **6:** 1065–1073.

Feil R, Brocard J, Mascrez B, LeMeur M, Metzger D, Chambon P. 1996. Ligand-activated site-specific recombination in mice. *Proc Natl Acad Sci* **93:** 10887–10890. doi:10.1073/pnas.93.20.10887

Furuta Y, Behringer RR. 2005. Recent innovations in tissue-specific gene modifications in the mouse. *Birth Defects Res C Embryo Today* **75:** 43–57. doi:10.1002/bdrc.20036

Gonen N, Futtner CR, Wood S, Garcia-Moreno SA, Salamone IM, Samson SC, Sekido R, Poulat F, Maatouk DM, Lovell-Badge R. 2018. Sex reversal following deletion of a single distal enhancer of *Sox9*. *Science* **360:** 1469–1473. doi:10.1126/science.aas9408

Gu H, Marth JD, Orban PC, Mossmann H, Rajewsky K. 1994. Deletion of a DNA polymerase β gene segment in T cells using cell type–specific gene targeting. *Science* **265:** 103–106. doi:10.1126/science.8016642

Hammer RE, Krumlauf R, Camper SA, Brinster RL, Tilghman SM. 1987a. Diversity of alpha-fetoprotein gene expression in mice is generated by a combination of separate enhancer elements. *Science* **235:** 53–58. doi:10.1126/science.2432657

Hammer RE, Swift GH, Ornitz DM, Quaife CJ, Palmiter RD, Brinster RL, MacDonald RJ. 1987b. The rat elastase I regulatory element is an enhancer that directs correct cell specificity and developmental onset of expression in transgenic mice. *Mol Cell Biol* **7:** 2956–2967. doi:10.1128/MCB.7.8.2956

Hayashi S, McMahon AP. 2002. Efficient recombination in diverse tissues by a tamoxifen-inducible form of Cre: a tool for temporally regulated gene activation/inactivation in the mouse. *Dev Biol* **244:** 305–318. doi:10.1006/dbio.2002.0597

Hojo H, Saito T, He X, Guo Q, Onodera S, Azuma T, Koebis M, Nakao K, Aiba A, Seki M, et al. 2022. Rnx2 regulates chromatin accessibility to direct the osteoblast program at neonatal stages. *Cell Rep* **40:** 111315. doi:10.1016/j.celrep.2022.111315

Joyner AL, Zervas M. 2006. Genetic inducible fate mapping in mouse: establishing genetic lineages and defining genetic neuroanatomy in the nervous system. *Dev Dyn* **235:** 2376–2385. doi:10.1002/dvdy.20884

Kanemaki MT. 2022. Ligand-induced degrons for studying nuclear functions. *Curr Opin Cell Biol* **74:** 29–36. doi:10.1016/j.ceb.2021.12.006

Karimova M, Baker O, Camgoz A, Naumann R, Buchholz F, Anastassiadis K. 2018. A single reporter mouse line for Vika, Flp, Dre, and Cre-recombination. *Sci Rep* **8:** 14453. doi:10.1038/s41598-018-32802-7

Kellendonk C, Tronche F, Monaghan AP, Angrand PO, Stewart F, Schütz G. 1996. Regulation of Cre recombinase activity by the synthetic steroid RU 486. *Nucleic Acids Res* **24:** 1404–1411. doi:10.1093/nar/24.8.1404

Kobayashi A, Kwan KM, Carroll TJ, McMahon AP, Mendelsohn CL, Behringer RR. 2005. Distinct and sequential tissue-specific activities of the LIM-class homeobox gene *Lim1* for tubular morphogenesis during kidney development. *Development* **132:** 2809–2823. doi:10.1242/dev.01858

Kulessa H, Hogan BL. 2002. Generation of a *loxP* flanked *bmp4^loxP-lacZ* allele marked by conditional *lacZ* expression. *Genesis* **32:** 66–68. doi:10.1002/gene.10032

Kushawah G, Hernandez-Huertas L, Abugattas-Nuñez Del Prado J, Martinez-Morales JR, DeVore ML, Hassan H, Moreno-Sanchez I, Tomas-Gallardo L, Diaz-Moscoso A, Monges DE, et al. 2020. CRISPR–Cas13d induces efficient mRNA knockdown in animal embryos. *Dev Cell* **54**: 805–817.e807. doi:10.1016/j.devcel.2020.07.013

Kwan KM, Behringer RR. 2002. Conditional inactivation of *Lim1* function. *Genesis* **32**: 118–120. doi:10.1002/gene.10074

Lee N, Robitz R, Zurbrugg RJ, Karpman AM, Mahler AM, Cronier SA, Vesey R, Spearry RP, Zolotukhin S, Maclennan AJ. 2008. Conditional, genetic disruption of ciliary neurotrophic factor receptors reveals a role in adult motor neuron survival. *Eur J Neurosci* **27**: 2830–2837. doi:10.1111/j.1460-9568.2008.06298.x

Lewandoski M. 2001. Conditional control of gene expression in the mouse. *Nat Rev Genet* **2**: 743–755. doi:10.1038/35093537

Liput DJ. 2018. Cre-recombinase dependent germline deletion of a conditional allele in the Rgs9cre mouse line. *Front Neural Circuits* **12**: 68. doi:10.3389/fncir.2018.00068

Lowe LA, Yamada S, Kuehn MR. 2001. Genetic dissection of *nodal* function in patterning the mouse embryo. *Development* **128**: 1831–1843. doi:10.1242/dev.128.10.1831

Lu F, Liu Y, Inoue A, Suzuki T, Zhao K, Zhang Y. 2016. Establishing chromatin regulatory landscape during mouse preimplantation development. *Cell* **165**: 1375–1388. doi:10.1016/j.cell.2016.05.050

Ludwig T, Chapman DL, Papaioannou VE, Efstratiadis A. 1997. Targeted mutations of breast cancer susceptibility gene homologs in mice: lethal phenotypes of *Brca1*, *Brca2*, *Brca1/p53*, and *Brca2/p53* nullizygous embryos. *Genes Dev* **11**: 1226–1241. doi:10.1101/gad.11.10.1226

Meyers EN, Lewandoski M, Martin GR. 1998. An *Fgf8* mutant allelic series generated by Cre- and Flp-mediated recombination. *Nat Genet* **18**: 136–141. doi:10.1038/ng0298-136

Mishina Y, Hanks MC, Miura S, Tallquist MD, Behringer RR. 2002. Generation of *Bmpr/Alk3* conditional knockout mice. *Genesis* **32**: 69–72. doi:10.1002/gene.10038

Muzumdar MD, Tasic B, Miyamichi K, Li L, Luo L. 2007. A global double-fluorescent Cre reporter mouse. *Genesis* **45**: 593–605. doi:10.1002/dvg.20335

Naiche LA, Papaioannou VE. 2003. Loss of *Tbx4* blocks hindlimb development and affects vascularization and fusion of the allantois. *Development* **130**: 2681–2693. doi:10.1242/dev.00504

Noordermeer D, Duboule D. 2013. Chromatin architectures and *Hox* gene collinearity. *Curr Top Dev Biol* **104**: 113–148. doi:10.1016/B978-0-12-416027-9.00004-8

Papaioannou VE, Behringer RR. 2023a. Mouse gene-targeting strategies for maximum ease and versatility. *Cold Spring Harb Protoc* doi:10.1011/pdb.over107957

Papaioannou VE, Behringer RR. 2023b. Uncovering phenotypes in mutant mice by determining embryo, organ, tissue, and cell developmental potential. *Cold Spring Harb Protoc* doi:10.1011/pdb.over107974

Papaioannou VE, Behringer RR. 2023c. Strategies for maintaining mouse mutations. *Cold Spring Harb Protoc* doi:10.1011/pdb.over107960

Papaioannou VE, Behringer RR. 2023d. Getting around an early lethal phenotype in mice with chimeras. *Cold Spring Harb Protoc* doi:10.1011/pdb.over107979

Papaioannou VE, Behringer RR. 2023e. Phenotypic analysis: assessing timing of recessive prenatal lethality in mice. *Cold Spring Harb Protoc* doi:10.1011/pdb.over107970

Papaioannou VE, Behringer RR. 2023f. Phenotypic analysis of preimplantation lethality in mice. *Cold Spring Harb Protoc* doi:10.1011/pdb.over107971

Papaioannou VE, Behringer RR. 2023g. Phenotypic analysis of periimplantation to mid-gestation lethality in mice. *Cold Spring Harb Protoc* doi:10.1011/pdb.over107972

Papaioannou VE, Behringer RR. 2023h. Analysis of mid- to late-gestation phenotypes in mice. *Cold Spring Harb Protoc* doi:10.1011/pdb.over107973

Papaioannou VE, Behringer RR. 2023i. Phenotypic analysis of perinatal lethality in mice. *Cold Spring Harb Protoc* doi:10.1011/pdb.over107975

Papaioannou VE, Behringer RR. 2023j. Analysis of postnatal mutant phenotypes in mice. *Cold Spring Harb Protoc* doi:10.1011/pdb.over107976

Papaioannou VE, Behringer RR. 2023k. Special breeding techniques for use in mouse mutation analysis. *Cold Spring Harb Protoc* doi:10.1011/pdb.over107961

Papaioannou VE, Behringer RR. 2023l. Backcrossing to make a congenic mouse strain. *Cold Spring Harb Protoc* doi:10.1011/pdb.prot108039

Papaioannou VE, Behringer RR. 2023m. Phenotypic analysis of dominant mutant effects in mice. *Cold Spring Harb Protoc* doi:10.1011/pdb.over107978

Papaioannou VE, Behringer RR. 2023n. Obtaining or generating gene mutations in mice. *Cold Spring Harb Protoc* doi:10.1011/pdb.over107956

Probst F, Fridell R, Raphael Y, Saunders T, Wang A, Liang Y, Morell R, Touchman J, Lyons R, Noben-Trauth K, et al. 1998. Correction of deafness in *shaker-2* mice by an unconventional myosin in a BAC transgene. *Science* **280**: 1444–1447.

Rose MF, Ahmad KA, Thaller C, Zoghbi HY. 2009. Excitatory neurons of the proprioceptive, interoceptive, and arousal hindbrain networks share a developmental requirement for *Math1*. *Proc Natl Acad Sci* **106**: 22462–22467. doi:10.1073/pnas.0911579106

Ryder E, Doe B, Gleeson D, Houghton R, Dalvi P, Grau E, Habib B, Miklejewska E, Newman S, Sethi D, et al. 2014. Rapid conversion of EUCOMM/KOMP-CSD alleles in mouse embryos using a cell-permeable Cre recombinase. *Transgenic Res* **23**: 177–185. doi:10.1007/s11248-013-9764-x

Sagai T, Hosoya M, Mizushina Y, Tamura M, Shiroishi T. 2005. Elimination of a long-range *cis*-regulatory module causes complete loss of limb-specific Shh expression and truncation of the mouse limb. *Development* **132**: 797–803. doi:10.1242/dev.01613

Soriano P. 1999. Generalized *lacZ* expression with the ROSA26 Cre reporter strain. *Nat Genet* **21**: 70–71. doi:10.1038/5007

Srinivas S, Watanabe T, Lin CS, William CM, Tanabe Y, Jessell TM, Costantini F. 2001. Cre reporter strains produced by targeted insertion of *EYFP* and *ECFP* into the *ROSA26* locus. *BMC Dev Biol* **1**: 4. doi:10.1186/1471-213X-1-4

Stratis A, Pasparakis M, Markur D, Knaup R, Pofahl R, Metzger D, Chambon P, Krieg T, Haase I. 2006. Localized inflammatory skin disease following inducible ablation of I kappa B alpha in murine epidermis. *J Invest Dermatol* **126**: 614–620. doi:10.1038/sj.jid.5700092

The Jackson Laboratory Cre Portal. 2022. https://www.jax.org/research-and-faculty/resources/cre-repository [Accessed December 14, 2022].

Vasioukhin V, Degenstein L, Wise B, Fuchs E. 1999. The magical touch: genome targeting in epidermal stem cells induced by tamoxifen application to mouse skin. *Proc Natl Acad Sci* **96**: 8551–8556. doi:10.1073/pnas.96.15.8551

Whitfield J, Littlewood T, Soucek L. 2015. Tamoxifen administration to mice. *Cold Spring Harb Protoc* doi:10.1101/pdb.prot077966

Wu J, Huang B, Chen H, Yin Q, Liu Y, Xiang Y, Zhang B, Liu B, Wang Q, Xia W, et al. 2016. The landscape of accessible chromatin in mammalian preimplantation embryos. *Nature* **534**: 652–657. doi:10.1038/nature18606

Yunusova A, Smirnov A, Shnaider T, Lukyanchikova V, Afonnikova S, Battulin N. 2021. Evaluation of the OsTIR1 and AtAFB2 AID systems for genome architectural protein degradation in mammalian cells. *Front Mol Biosci* **8**: 757394. doi:10.3389/fmolb.2021.757394

Zhao L, Bakke M, Krimkevich Y, Cushman LJ, Parlow AF, Camper SA, Parker KL. 2001. Hypomorphic phenotype in mice with pituitary-specific knockout of steroidogenic factor 1. *Genesis* **30**: 65–69. doi:10.1002/gene.1034

Zhou G, Garofalo S, Mukhopadhyay K, Lefebvre V, Smith CN, Eberspaecher H, de Crombrugghe B. 1995. A 182 bp fragment of the mouse pro alpha 1(II) collagen gene is sufficient to direct chondrocyte expression in transgenic mice. *J Cell Sci* **108**: 3677–3684. doi:10.1242/jcs.108.12.3677

Useful Books and Other Resources

ABSOLUTE ESSENTIALS

Behringer R, Gertsenstein M, Vintersten Nagy K, Nagy A, eds. 2014. *Manipulating the mouse embryo: a laboratory manual.* Cold Spring Harbor Laboratory Press, Cold Spring Harbor, New York.

> The premier source of technical and theoretical guidance for mouse developmental biologists and geneticists; contains all the commonly used protocols for embryo and gene manipulation.

Joyner AL, ed. 2000. *Gene targeting: a practical approach*, 2nd ed. Oxford University Press, Oxford.

> An indispensable practical tool and information resource for scientists generating mouse mutants using embryonic stem (ES) cell technology. The gene targeting strategies can be modified to generate CRISPR–Cas alleles.

Kaufman MH. 1992. *The atlas of mouse development.* Academic, New York.

> The standard reference work and the definitive account of mouse embryology and development including serial histological sections of embryos throughout development, scanning electron micrographs of the external appearance of embryos, and special sections on the development of specific organ systems such as heart, tongue, palate, gonads, and kidney. Available online at https://www.emouseatlas.org/emap/about/what_is_eHistology.html.

Rossant J, Tam PPL, eds. 2002. *Mouse development: patterning, morphogenesis, and organogenesis.* Academic, San Diego, CA.

> A valuable collection of highly readable, informative reviews of mouse development from fertilization through axis formation and gastrulation, to lineage specification and differentiation and detailed organogenesis of many of the major organ systems.

Silver LM. 1995. *Mouse genetics: concepts and applications.* Oxford University Press, Oxford.

> A comprehensive practical guide to mouse breeding and genetics and an introduction to the mouse as a model system for genetic analysis; contains a complete description of the laboratory mouse, the tools used in analysis, procedures for carrying out genetic studies, and a detailed guide for performing breeding studies and interpreting experimental results. Available online at http://www.informatics.jax.org/silver/.

MOUSE GENETICS AND HUSBANDRY

Fox J, Barthold S, Davisson M, Newcomer C, Quimby F, Smith A, eds. 2006. *The mouse in biomedical research. Normative biology, husbandry, and models*, 2nd ed. Academic, New York.

> Extensive information on mouse biology, husbandry, and disease models.

Opening artwork: Kawanabe Kyōsai/Freer Gallery of Art, Smithsonian Institution, Washington, D.C.: Purchase—Charles Lang Freer Endowment, F1975.29.9.

Green EL, ed. 1998. *Biology of the laboratory mouse, by the staff of the Jackson Laboratory*, 2nd revised ed. Dover, New York.

A golden oldie on many aspects of mouse biology. This title emphasizes genetic variations of mice and their exploitation for the solution of biological problems. Includes practical chapters on husbandry, breeding systems, and record keeping, as well as chapters on development, physiology, life span, susceptibility to disease and tumors, and much more. Available online at http://www.informatics.jax.org/greenbook/.

Hedrich H., ed. 2012. *The laboratory mouse*, 2nd ed. Academic, New York.

Extensive information about mouse husbandry, biology, and experimentation.

Jackson Laboratory. 2009. *Handbook on genetically standardized JAX mice*, 6th ed. JAX Mice Literature, The Jackson Laboratory, Bar Harbor, ME.

A highly detailed compendium of information on all aspects of growth, reproduction, life span, and disease susceptibility of the various inbred, hybrid, recombinant inbred, congenic, and mutant strains and stocks, and mice with chromosomal aberrations, available from JAX. Also contains general husbandry procedures. Available online at https://resources.jax.org/therapeutic-area-essentials/jax-handbook-genetically-standardized-mice.

Lyon MF, Rastan S, Brown SDM, eds. 1995. *Genetic variants and strains of the laboratory mouse*, 3rd ed. International Committee on Standardized Genetic Nomenclature for Mice, Oxford University Press, New York.

A two-volume reference on the fundamentals of mouse genetics for researchers and advanced students; catalogs and describes known genetic variants and inbred strains of the mouse. Much of this information is now available through Mouse Genome Informatics (http://www.informatics.jax.org/).

Silvers WK. 1979. *The coat colors of mice: a model for mammalian gene action and interaction*. Springer-Verlag, Heidelberg.

Describes the many genetic factors that influence the coat colors of the house mouse and illustrates various multidisciplinary approaches that have been used to explore genotype–phenotype pathways; contains information on a large number of specific alleles at different loci and how they affect pigmentation. Available online at http://www.informatics.jax.org/wksilvers/.

Suckow MA, Danneman P, Brayton C. 2001. *The laboratory mouse*. A volume in the *Laboratory animal pocket reference series*. CRC Press, Boca Raton, FL.

Contains the basics of management, husbandry, equipment, and regulations involved in maintaining an animal facility, plus useful information on diseases and veterinary care of mice.

DEVELOPMENT, DEVELOPMENTAL LANDMARKS, AND STAGING SYSTEMS

Downs KM, Davies T. 1993. Staging of gastrulating mouse embryos by morphological landmarks in the dissecting microscope. *Development* **118**: 1255–1266.

Fujinaga M, Brown NA, Baden JM. 1992. Comparison of staging systems for the gastrulation and early neurulation period in rodents: a proposed new system. *Teratology* **46**: 183–190.

Kaufman MH, Bard JBL. 1999. *The anatomical basis of mouse development*. Academic, New York.

A companion to Kaufman's *The Atlas of Mouse Development*, which details the developmental anatomy of the early embryo, the transitional tissues, and all the major organ systems. It includes extensive reference indexes detailing developmental stage criteria, when tissues first appear, and the constituent tissues of embryos at each stage of development. Also covered are comparisons with normal and abnormal human development. Useful for understanding developmental anatomy in normal and mutant mice.

Le Garrec J-F, Dominguez JN, Desgrange A, Ivanovich KD, Raphaël E, Bangham JA, Torres M, Coen E, Mohun TJ, Meilhac SM. 2017. A predictive model of asymmetric morphogenesis from 3D reconstructions of mouse heart looping dynamics. *eLife* **6**: e28951.

Ruberte J, Carretero A, Navarro M. 2017. *Morphological mouse phenotyping: anatomy, histology and imaging*. Academic, New York.

> Normal mouse morphology at the anatomical, histological, and imaging levels.

Rugh R. 1990. *The mouse: its reproduction and development*. Oxford University Press, Oxford.

> A well-illustrated text on mouse development including reproduction, chronology of normal development, and organogenesis; includes histological sections and gross anatomy with many useful morphological markers of different stages of development.

Tam PPL, Nelson WJ, Rossant J, eds. 2013. *Mouse development: networks, switches, and morphogenetic processes*. Cold Spring Harbor Laboratory Press, Cold Spring Harbor, NY.

> Chapters on the molecular and cellular mechanisms of mammalian development with an emphasis on transcriptional and epigenetic switches that regulate cell differentiation and tissue morphogenesis.

Theiler K. 1989. *The house mouse: atlas of embryonic development*, 2nd ed. Springer-Verlag, New York.

> In this classic compendium, the author presents each stage of mouse development (the commonly used Theiler stages) with photographs and micrographs. Organ systems are systematically reconstructed from fertilization until after birth. Available online at https://www.emouseatlas.org/emap/ema/theiler_stages/house_mouse/book.html.

Wanek N, Muneoka K, Holler-Dinsmore G, Burton R, Bryant SV. 1989. A staging system for mouse limb development. *J Exp Zool* **249**: 41–49.

METHODS

Copp AJ, Cockroft DL, eds. 1990. *Postimplantation mammalian embryos: a practical approach*. Oxford University Press, Oxford.

> Methods of observing, manipulating, and analyzing implanted embryos. The topics include exo utero surgery, the morphological stages of postimplantation embryonic development, culturing postimplantation embryos, extracting macromolecules, and teratogen testing.

Doudna J, Mali P, eds. 2016. *CRISPR–Cas: a laboratory manual*. Cold Spring Harbor Laboratory Press, Cold Spring Harbor, NY.

> Protocols using CRISPR–Cas methods in various model systems including mouse.

Jackson II, Abbott CM, eds. 2001. *Mouse genetics and transgenics: a practical approach*. Oxford University Press, Oxford.

> Covers all aspects of using the mouse as a genetic model organism: care and husbandry; archiving stocks as frozen embryos or sperm; making new mutations by chemical mutagenesis; transgenesis; gene targeting; mapping mutations and polygenic traits by cytogenetic, genetic, and physical means; and disseminating and researching information via the Internet.

McLaren A, ed. 1976. *Mammalian chimaeras*. Developmental and cell biology series. Cambridge University Press, Cambridge.

> A classic monograph on what constitutes a chimera, how to make them, and how to use them.

Pease S, Saunders TL, eds. 2011. *Advanced protocols for animal transgenesis: an ISTT manual*. Springer, Berlin.

> This manual, generated by the International Society for Transgenic Technology, contains methods used to generate genetically modified animals.

Pinkert CA, ed. 2002. *Transgenic animal technology: a laboratory handbook.* Academic, Boston.

Covers the technical aspects of gene transfer—from molecular methods to whole-animal considerations—for important laboratory and domestic animal species.

Robertson EJ, ed. 1987. *Teratocarcinomas and embryonic stem cells: a practical approach.* Oxford University Press, Oxford.

Contains the definitive work on derivation and culture of ES cells. A classic.

Singh SR, Hoffman RM, Singh A. 2021. *Mouse genetics. Methods and protocols. Methods in molecular biology.* Humana, New York.

Numerous methods and protocols to generate and analyze genetically modified mice, including CRISPR–Cas.

Stern CD, Holland PWH, eds. 1993. *Essential developmental biology: a practical approach.* Oxford University Press, Oxford.

Comprehensive, easy-to-follow protocols and practical instructions for techniques, from traditional embryology to cellular and molecular methods. Includes complete reprints of commonly used stage tables for the main laboratory species.

Tuan RS, Lo CW, eds. 2000. *Developmental biology protocols*, Vols. I–III. *Methods in molecular biology*, Vols. 135–137. Humana, Totowa, NJ.

Three-volume laboratory manual providing principles, background, rationale, and step-by-step instructions for studying and analyzing the events of embryonic development. Covers developmental pattern and morphogenesis, embryo structure and function, cell lineage analysis, chimeras, experimental manipulation of embryos, application of viral vectors, organogenesis, abnormal development and teratology, screening and mapping of novel genes and mutations, transgenesis production and gene knockout, manipulation of developmental gene expression and function, analysis of gene expression, models of morphogenesis and development, and in vitro models and analysis of differentiation and development.

Tymms MJ, Kola I, eds. 2001. *Gene knockout protocols.* Humana, Totowa, NJ.

Describes techniques for the design of gene-targeting constructs and analysis of the mouse phenotype, covering techniques of embryo transplantation, in vitro embryonic stem cell differentiation, creation of aggregation chimeras, mouse pathology, embryo cryopreservation, and transplantation.

Wassarman PM, DePamphilis ML, eds. 1993. *Guide to techniques in mouse development. Methods in Enzymology*, Vol. 225. Academic, New York.

An extremely valuable, comprehensive compendium of techniques used in the study of mouse development, including record keeping for mouse colonies, in vitro fertilization, preimplantation and postimplantation embryo culture, and analysis of gene expression including in situ hybridization, gene targeting, and making transgenics. See Vols. 226 and 227 for updates and expansion of this series.

Wassarman PM, Soriano P. 2010. *Guide to techniques in mouse development, Part A: mice, embryos, and cells. Methods in enzymology.* Academic, New York.

Techniques and methodologies to study mouse development.

Wassarman PM, Soriano P. 2010. *Guide to techniques in mouse development, Part B: mouse molecular genetics. Methods in enzymology.* Academic, New York.

Techniques and methodologies to study mouse development.

Wilkinson DG, ed. 1998. *In situ hybridization: a practical approach*, 2nd ed. Oxford University Press, Oxford.

Another classic, containing protocols detailing the major techniques of in situ hybridization, including in situ hybridization to mRNA with oligonucleotide and RNA probes; analysis using light and electron microscopes; whole-mount in situ hybridization; double detection of RNAs and RNA plus protein; and fluorescent in situ hybridization to detect chromosomal sequences.

PHENOTYPIC ANALYSIS

Abate-Shen C, Politi K, Chodosh LA, Olive KP, eds. 2014. *Mouse models of cancer: a laboratory manual.* Cold Spring Harbor Laboratory Press, Cold Spring Harbor, NY.

> Reviews and protocols for all aspects of using the mouse in cancer research, including transgenic and genetically engineered mouse models and analysis of mouse cancer phenotypes and their application in clinical research.

Conway SJ, Kruzynska-Frejtag A, Kneer PL, Machnicki M, Koushik SV. 2003. What cardiovascular defect does my prenatal mouse mutant have, and why? *Genesis* **35:** 1–21.

> Review of phenotypic analysis of embryos with heart problems.

Crawley JN. 2000. *What's wrong with my mouse? Behavioral phenotyping of transgenic and knockout mice.* Wiley-Liss, New York.

> A molecular geneticist's guide to the best tests for analyzing behavioral phenotypes in transgenic and knockout mice.

Maronpot R, Boorman GA, Gaul BW, eds. 1999. *Pathology of the mouse: reference and atlas.* Cache River Press, Vienna, IL.

> Provides a comprehensive survey of mouse pathology; a heavily illustrated reference and atlas that covers all systems of the body.

Rogers DC, Fisher EMC, Brown SDM, Peters J, Hunter AJ, Martin JE. 1997. Behavioral and functional analysis of mouse phenotype: SHIRPA, a proposed protocol for comprehensive phenotype assessment. *Mam. Genome* **8:** 711–713.

> A procedure to characterize the phenotype of mice in three stages: a primary screen utilizing standard methods to provide a behavioral and functional profile by observational assessment, a secondary screen involving a comprehensive behavioral assessment battery and pathological analysis, and a tertiary screen tailored to the assessment of existing or potential models of neurological disease as well as assessment of phenotypic variability.

Sharpe J, Wong RO, eds. 2011. *Imaging in developmental biology.* Cold Spring Harbor Laboratory Press, Cold Spring Harbor, NY.

> An indispensable guide to imaging in model organisms, including mouse. Comprehensive reviews containing protocols for all types of imaging, including whole embryos, 3D, live imaging, and time lapse, and examples of their use in phenotyping studies.

Sundberg JP. 1994. *Handbook of mouse mutations with skin and hair abnormalities: animal models and biomedical tools.* CRC Press, Boca Raton, FL.

> Overview of mutant mouse resources and use, major features of cutaneous biology, and comparison of phenotypes of mouse mutations with human diseases.

Sundberg JP, Boggess D, eds. 2000. *Systematic approach to evaluation of mouse mutations.* CRC Press, Boca Raton, FL.

> Geared toward evaluating spontaneous mutations in adult mice using scaled-down versions of standard operating procedures for diagnostic medicine and biomedical research. Includes chapters on necropsy, photography, colony establishment, and management, and JAX mouse resources for mutant mice.

Ward JM, Mahler JF, Maronpot RR, Sundberg JP, Frederickson RM, eds. 2000. *Pathology of genetically engineered mice.* Iowa State University Press, Ames, IA.

Approaches to organ-specific evaluations, with examples of many currently available genetically engineered and naturally mutant mice to illustrate how specific defects can be studied. Topics include using confocal microscopy to study transgenic and knockout mice, inducing mutations, genetic background effects on the interpretation of phenotypes in induced mutant mice, the pathologic characterization of neurological mutants, and hepatic pathology.

ADDITIONAL WEB-BASED RESOURCES

Anatomy and Normal Biology

eMouseAtlas, the Edinburgh Mouse Atlas Project (eMAP)
https://www.emouseatlas.org/emap/home.html

It includes histological sections of mouse embryos throughout gestation (eHistology), as well as an atlas of 3D volume and surface rendered anatomy models with spatially mapped gene expression patterns derived from histological sections (eMAGE).

Mouse Nomenclature

Mouse Nomenclature Home Page
http://www.informatics.jax.org/mgihome/nomen/index.shtml

The authoritative source of official names for mouse genes, alleles, and strains. Nomenclature follows the rules and guidelines established by the International Committee on Standardized Genetic Nomenclature for Mice and is implemented through the Mouse Genomic Nomenclature Committee (MGNC).

New nomenclature for strain 129 mice.
http://www.informatics.jax.org/mgihome/nomen/strain_129.shtml

Finding and/or Archiving Mutants

Center for Animal Resources and Development (CARD)
https://cardmice.com/rbase/

Source and repository for mutant mouse strains in Japan.

European Mouse Mutant Archive (EMMA)
https://www.infrafrontier.eu/emma/

Source and repository for mutant mouse strains in Europe.

International Gene Trap Consortium (IGTC)
https://igtc.org/

An international resource of publicly available embryonic stem cells with gene trap insertions.

International Mouse Strain Resource (IMSR)
http://www.findmice.org

A database of available mouse resources, including mutant and normal strains and genetically engineered embryonic stem cell lines.

Knockout Mouse Project (KOMP)/International Mouse Phenotyping Consortium (IMPC)
https://mousephenotype.org

A resource of knockout mice and phenotypes with the stated aim of identifying the function of every protein-coding gene in the genome.

Mouse Mutant Resource (MMR)
https://www.jax.org/research-and-faculty/resources/mouse-mutant-resource

Unique disease model resource of mice with characterized spontaneous genetic mutations.

National Cancer Institute (NCI) Mouse Repository
https://frederick.cancer.gov/resources/repositories/nci-mouse-repository/MouseModels/

> A repository for mouse cancer models.

NIH Mutant Mouse Regional Resource Centers (MMRRC)
https://www.mmrrc.org

> The National Institutes of Health source and repository of mouse mutants and strains in the United States.

RIKEN Bioresource Research Center (RBRC)
https://mus.brc.riken.jp/en

> Source and repository for mutant mouse strains in Japan.

Genome, Genes, and Expression Resources

Allen Brain Map
https://portal.brain-map.org/

> Extensive atlases and data on the human and mouse brain, including anatomy, neural projections, genomic data, and developmental gene expression.

eEnsembl
http://useast.ensembl.org/index.html

> A genome browser for vertebrate genomes, including mouse, with tools for data-mining, BLAST searches, and prediction of variant effects.

Gene Expression Omnibus (GEO)
www.ncbi.nlm.nih.gov/geo

> A database containing MIAME (Minimum Information About a Microarray Experiment)-compliant high-throughput genomics data, including microarray- and sequencing-based data.

Mouse Genome Informatics (MGI)
http://www.informatics.jax.org/

> Provides integrated access to data on the genetics, genomics, and biology of the laboratory mouse, including but not limited to the following:
>
> - Genes, including gene and genome information, maps, gene expression, and disease models.
> - Phenotypes and mutant alleles curates spontaneous, induced, and genetically engineered mutations and the resulting phenotypes.
> - Gene Expression Database (GXD) integrates different types of gene expression information from the mouse and provides a searchable index of published experiments on endogenous gene expression during development.
> - Recombinase (cre) Activity, a collection and annotation of expression and activity of recombinase transgenic and knock-in mouse lines.
> - Mouse Models of Human Cancer Database (MMHCdb), a curated resource of mouse models of human cancer, including spontaneous and induced tumors in mice, genetically defined mouse models of cancer, and patient-derived xenografts (PDX) mouse models.
> - Gene Ontology (GO). The Mouse Genome Informatics (MGI) group is a founding member of the Gene Ontology Consortium (www.geneontology.org). MGI fully incorporates GO in the database and provides a GO browser.

Conditional Genetic Resources

Cre Portal, the Jackson Laboratory
https://www.jax.org/research-and-faculty/resources/cre-repository

Compendium of transgenic Cre driver lines including inducible, floxed, and Cre-excision reporter lines.

Mouse Phenotyping

IMPC, International Mouse Phenotyping Consortium
https://mousephenotype.org

A catalog of phenotypes of knockout mice with the aim of compiling a comprehensive catalogue of mammalian gene function.

National Mouse Metabolic Phenotyping Center
https://www.mmpc.org/

A National Institutes of Health-sponsored testing service for metabolic diseases in mice.

Examples of Phenotype Analysis

Examples and key references for specific types of phenotypic analyses or analyses using specific techniques are keyed to chapters in the book. The references are not meant to be comprehensive but rather provide a cogent example of different situations and how we or other investigators have dealt with them.

Chapter	Situation or phenotype	Example(s)	Reference(s)
2	ENU mutagenesis screen	Dominant screen	Hrabe de Angelis et al. 2000
		Recessive screen	Kasarskis et al. 1998
		Balancer screen	Kile et al. 2003
2	X-ray mutagenesis screen	Albino deletions	Holdener and Magnuson 1994
2	Insertional mutagenesis	Viral	Friedrich and Soriano 1991
		Gene trap	Gossler et al. 1989
		Transposon	Horie et al. 2003
2	Transgenic expression	*Gh* transgene	Palmiter et al. 1982
2, Box 1	Two classes of heterozygous females depending on maternal or paternal inheritance of mutant allele because of nonrandom X inactivation in placenta	*Esx1*	Li and Behringer 1998
3	Selectable marker cassette affecting gene expression at targeted locus	*Myf5*	Kaul et al. 2000
3	Marker recycling to reuse a targeting vector for the second allele		Abuin and Bradley 1996
3	Regulatory element for a gene in an intron of a neighboring gene	*Shh*	Lettice et al. 2003
3	An expression reporter producing a hypermorphic allele	*Vegfa*	Miquerol et al. 2000
3	Conditional allele that acts as a null or hypomorphic allele	*Fgf8*	Meyers et al. 1998
3, Box 3	HSV *tk* and male infertility		Braun et al. 1990
4	Mutant ES cells causing the death of chimeric embryos (X-linked gene)	*Gata1*	Pevny et al. 1991
4	A coat color phenotype that results in good chimeras not showing strong coat color chimerism	*Pdgfra*	Soriano 1997
5	No germline transmission from male chimeras due to a dominant effect of the mutation on gametogenesis	*Klhl10*	Yan et al. 2004
5	Male germline chimeras produce no heterozygous offspring due to dominant effect on embryos	*Vegfa*	Carmeliet et al. 1996
5	Perinatal lethality due to a dominant effect in the heterozygous offspring of chimeras	*Sox9*	Bi et al. 2001
5	Dominant effect on gametogenesis resulting in infertility	*Klhl10*	Yan et al. 2004
5	Sex-limited dominant effect on fertility	*Odsex*	Bishop et al. 2000
5	Dominant sex reversal	*Odsex*	Bishop et al. 2000
6	Deletion of a selection cassette in vivo	*Sox2Cre*	Hayashi et al. 2003
		CmvCre	Arango et al. 1999
6	Maintaining a mutant with a balancer chromosome	*Dph2l1*	Chen and Behringer 2004
6	Dealing with male infertility: intracytoplasmic sperm injection	*Tnp1, Tnp2*	Zhao et al. 2004
6	Dealing with female infertility: nonreceptive reproductive tract	*Lif*	Stewart et al. 1992
6, Box 1	Different phenotype on different genetic backgrounds	*Egfr*	Threadgill et al. 1995
7	Complementation testing to determine allelism of two mutations	*Tbx6*	Watabe-Rudolph et al. 2002

(*continued*)

Opening artwork: Katsushika Hokusai (1760–1849). Wikipedia Commons.

Continued

Chapter	Situation or phenotype	Example(s)	Reference(s)
7	Detection of a hypomorphic allele	*Tbx6*	Watabe-Rudolph et al. 2002
7	Functional redundancy or compensation by genes in the same gene family	*Myod1, Myf5*	Rudnicki et al. 1993
7	Similar mutant phenotypes but with no genetic interaction	*Amh, Amhr2*	Behringer et al. 1994; Mishina et al. 1996
7	Mutant phenotype caused by overexpression of a downstream gene	*Leftb, Cer1, Nodal*	Perea-Gomez et al. 2002
7	"Cheating Mendel"	*Sox9*	Akiyama et al. 2002
9	No mutant preimplantation embryos due to a sperm/oocyte incompatibility	DDK syndrome	Babinet et al. 1990
9	Arrest or delay during cleavage	*Pes1*	Lerch-Gaggl et al. 2002
9	Abnormal compaction of preimplantation embryos	*Cdh1*	Ohsugi et al. 1997
9	Abnormal blastocyst morphology (other than compaction problem)	*Nup214*	van Deursen et al. 1996
9	Failure to hatch	*Smarc1b*	Guidi et al. 2001
9	Failure to attach	*Mcl1*	Rinkenberger et al. 2000
9	Failure of blastocyst outgrowth	*Trrap*	Herceg et al. 2001
9	Failure of ICM in blastocyst outgrowth	*Fgf4*	Feldman et al. 1995
9	Analysis of isolated inner cell mass	*Mcl1*	Rinkenberger et al. 2000
9	Analysis of embryos in implantation delay	*Lif*	Nichols et al. 2001
9	TUNEL on blastocysts	*Mdm2*	Chavez-Reyes et al. 2003
9	In situ hybridization on preimplantation embryos	*Mcl1*	Rinkenberger et al. 2000
9	Two-dimensional gels on preimplantation embryos		Shi et al. 1994
9	Gene expression in whole preimplantation embryos: in situ	*Eomes*	Hancock et al. 1999
9	Whole-mount immunohistochemistry on preimplantation embryos	*Pou5f1*	Liu et al. 2004
		Fgfr1	Kurowski et al. 2019
9	RT-PCR on preimplantation embryos	*Fgf4*	Goldin and Papaioannou 2003
9, Protocol 2	Cell counting in ICM and trophectoderm of preimplantation embryos	*Fgf4*	Goldin and Papaioannou 2003
10	Periimplantation death E4.5–5.5	*Mdm2*	Montes de Oca Luna et al. 1995
		Fgf4	Feldman et al. 1995
10	Lethality E6.5–9.5: failure of gastrulation	*Wnt3*	Liu et al. 1999
10	Lethality E6.5–9.5: early growth retardation with normal developmental stage	*Myc*	Davis et al. 1993
10	Lethality E6.5–9.5: failure of extra-embryonic membrane/placenta to function properly	*Dlx3*	Morasso et al. 1999
10	Lethality E6.5–9.5: failure of vasculogenesis or hematopoiesis	*Jag1*	Xue et al. 1999
10	Lethality E6.5–9.5: failure of allantoic fusion	*Tbx4*	Naiche and Papaioannou 2003
10	Lethality E6.5–9.5: cardiac problems	*Mef2c*	Lin et al. 1997
10	Lethality E6.5–9.5: heart looping defects	*Mef2c*	Lin et al. 1997
10	Laterality defects	*Invs*	Morgan et al. 1998
		Tbx6	Hadjantonakis et al. 2008
10	Neural tube closure defects	*Twist1*	Chen and Behringer 1995
10		*Cart1*	Zhao et al. 1996
10		*Shroom3*	Hildebrand and Soriano 1999
10	Molecular characterization of mutant phenotypes	*Tbx6*	Chapman and Papaioannou 1998
10	Genotyping embryos from histological sections	*Hdh*	Zeitlin et al. 1995
10	Genetic interactions in early gestation lethals	*Brca1, Brca2, Trp53*	Ludwig et al. 1997
		Mdm2, Trp53	Montes de Oca Luna et al. 1995
10	Alterations in proliferation and/or apoptosis	*Hdh*	Zeitlin et al. 1995
11	Lethality E9.5–term: vasculogenesis and hematopoietic defects	*Tal1*	Mikkola et al. 2003
		Smad5	Chang et al. 1999
11	Lethality E9.5–term: cardiovascular defects	*Tbx1*	Jerome and Papaioannou 2001
		Tbx2	Harrelson et al. 2004
		Nfatc1	Phoon et al. 2004

(continued)

Continued

Chapter	Situation or phenotype	Example(s)	Reference(s)
11	Lethality E9.5–term: yolk sac defect	*Tbx3*	Davenport et al. 2003
11	Lethality E9.5–term: placental defects	*Dlx3*	Morasso et al. 1999
		Vhlh	Gnarra et al. 1997
		Rb	Lee et al. 1992; Wu et al. 2003
11	Separation of yolk sac germ layers		Yoder et al. 1994
		Nodal	Varlet et al. 1997
11	Large placenta, small fetus	*Esx1*	Li and Behringer 1998
11	E9.5–term: specific morphological defects:		
	Skeleton	*Tbx1*	Jerome and Papaioannou 2001
		Tbx3	Davenport et al. 2003
	Kidney	*Lhx1*	Shawlot and Behringer 1995
	Lung and limbs	*Fgf10*	Sekine et al. 1999
	Heart	*Tbx2*	Harrelson et al. 2004
11	E9.5–term: cell death	*Fos*	Roffler-Tarlov et al. 1996
11	Lethality E9.5–term: cell proliferation	*Foxc2*	Winnier et al. 1997
11	Imaging: confocal optical sectioning	*Wnt5a*	Arora et al. 2016
11	Imaging: optical coherence tomography	*Wdr19*	Wang et al. 2017
11	Imaging: X-ray micro-computed tomography	*Plxnd1*	Degenhardt et al. 2010
11	Imaging: microscopic magnetic resonance imaging	*Mbnl1/Mbnl2*	Sta Maria et al. 2021
11	Imaging: episcopic fluorescence image capturing	*Cited2*	Weninger et al. 2005
12	Whole-embryo culture	*Amot*	Shimono and Behringer 2003
12	Testing inductive interactions	*Cer1*	Shawlot et al. 2000
12	E9.5–term organ culture:		
	Allantoic bud	*Tbx4*	Arora et al. 2012a
	Eye	*Bmp4*	Furuta and Hogan 1998
	Foregut/trachea bud	*Tbx4, Tbx5*	Arora et al. 2012b
	Gonad	*Sox9*	Chaboissier et al. 2004
	Heart		Kruithof et al. 2003
	Hematopoietic tissue		Frisch and Calvi 2014
	Kidney primordia		Srinivas et al. 1999
	Limb buds	*Tbx4*	Naiche and Papaioannou 2003
		Tbx5	Arora et al. 2012b
	Lung buds	*Tbx4; Tbx5*	Arora et al. 2012b; Miura and Shiota 2000
	Mammary glands		Heuberger et al. 1982
	Palate	*Tgfb3*	Dudas et al. 2004
	Tooth buds	*Msx1*	Bei et al. 2000
12	E9.5–term—transplantation of cells and organs:		
	Hematopoietic tissues	*Ptpn2*	You-Ten et al. 1997
	Liver		Shiojiri et al. 1995
	Mammary buds	*Igfr1*	Bonnette and Hadsell 2001
	Ovaries	*Gsc*	Rivera-Perez et al. 1995
	Pituitary	*Ghrhr*	Hammer et al. 1984
	Skin	*Pdgfa*	Karlsson et al. 1999
	Spermatogonial stem cells		Brinster and Zimmermann 1994
	Testis tubules		Schlatt et al. 2003
12	Derivation and use of mouse embryonic fibroblasts	*Trp53, Mdm2*	de Rozieres et al. 2000
12	Using ES cells to investigate a mutant phenotype	*Twist*	Chen and Behringer 1995
12	Testing developmental potential of mutant cells in chimeras	*Tbx6*	Chapman et al. 2003
12	Use of teratomas to investigate a mutant phenotype	*Tbx6*	Chapman et al. 2003
13	Perinatal lethality: catastrophic	*Lhx1*	Shawlot and Behringer 1995
		Krt5	Peters et al. 2001
13	Perinatal lethality: cardiovascular	*Tbx1*	Jerome and Papaioannou 2001
13	Perinatal lethality: developmental delay	*Dph2l1*	Chen and Behringer 2004

(continued)

Continued

Chapter	Situation or phenotype	Example(s)	Reference(s)
13	Perinatal lethality: cranial nerve defects	*Hoxa1*	Lufkin et al. 1991
13	Perinatal lethality: skeletal defects	*Sp7*	Nakashima et al. 2002
13	Perinatal lethality: cleft palate	*Msx1*	Satokata and Maas 1994
13	Perinatal lethality: diaphragm	*Myog*	Hasty et al. 1993
13	Perinatal lethality: breathing	*Bdnf*	Erickson et al. 1996
13	Perinatal lethality: bleeding disorder	*F10*	Dewerchin et al. 2000
14	Postnatal altered growth	*Dmbx1*	Ohtoshi and Behringer 2004
		Fos	Johnson et al. 1992
14	Postnatal morphology	*Foxn1*	Militzer 2001
		En2	Joyner et al. 1991
14	Circadian rhythm alterations	*Fos*	Honrado et al. 1996
14	Postnatal neurological/behavioral problems	*Fos*	Paylor et al. 1994
14	Postnatal coordination/movement	*Mpz*	Giese et al. 1992
		Adam22	Sagane et al 2005
		CerS1	Ginkel et al. 2012
14	Post weaning: growth and vigor	*Src*	Soriano et al. 1991
		Fos	Johnson et al. 1992
14	Male infertility: small testes	*Gnrhr*	Pask et al. 2005
14	Abnormal reproductive behavior in males	*Fos*	Baum et al. 1994
14	Male infertility	*Sys*	MacGregor et al. 1990
	Male infertility: spermatogenesis	*Zfy2*	Yamauchi et al. 2015
14	Female infertility	*Lif*	Stewart et al. 1992
15	No visible mutant phenotype	*Hprt*	Kuehn et al. 1987
15	Age-dependent phenotypes: tumors	*Trp53*	Donehower et al. 1992
15	Age-dependent phenotypes: osteoporosis	*Notch2*	Fukushima et al. 2017
	Age-dependent phenotypes: osteopetrosis	*Lep*	Ducy et al. 2000
15	Age-dependent phenotypes: rheumatoid arthritis	*Il1r1*	Horai et al. 2000
15	Eye morphology	*Pitx3*	Wada et al. 2014
15	Cataracts	*Crybb*	Zhang et al. 2008
15	Test of vision	*Vsx1*	Ohtoshi et al. 2004
15	Test of hearing	*Ysb*	Dong et al. 2002
15	Test of touch	*Drg11*	Chen et al. 2001
15	Neurological tests of balance and coordination	C57BL6 and 129 strains	Kelly et al. 2003
	Swim test	*Ysb*	Dong et al. 2002
	Rotorod	*En-2*	Gerlai et al. 1996
15	Genetic challenge of overtly normal mutants	*MyoD, Myf5*	Rudnicki et al. 1993
		Egfr	Threadgill et al. 1995
15	Environmental challenge of overtly normal mutants	*Fabp4*	Hotamisligil et al. 1996
16	Mutation in an X-linked gene	*Ar*	Yeh et al. 2002
16	Mutation in a Y-linked gene	*Ddx3y*	Rohozinski et al. 2002
16	Mutation in an imprinted autosomal gene	*Igf2*	DeChiara et al. 1991
16	Dominant effects in chimeras	*Zic3*	Purandare et al. 2002
16	Dominant effects in heterozygous mice	*Tbx1*	Jerome and Papaioannou 2001
16	Perinatal dominant lethal	*Sox9*	Bi et al. 2001
16	Dominant reproductive problems	*Klhl10*	Yan et al. 2004
17	Using chimeras to analyze a phenotype	*Tbx6*	Chapman et al. 2003
17	Cell autonomous defect in chimeras	*Sox9*	Bi et al. 1999
17	ES cell–tetraploid embryo chimeras	*Foxa2*	Dufort et al. 1998
17	Tetraploid rescue of placental phenotype	*Ascl2*	Guillemot et al. 1994
17	ES cell–diploid embryo chimeras	*Nodal*	Varlet et al. 1997
17	Chimeras made with trophoblast vesicles	*A^y*	Papaioannou and Gardner 1979
17	Chimera analysis with a tissue-specific reporter	*Egr2*	Voiculescu et al. 2001
17	Two different alleles for genotyping chimeras	*Gsc*	Rivera-Perez et al. 1995
18	Different roles at different stages	*Tbx4*	Naiche and Papaioannou 2003
		Brca1	Ludwig et al. 1997

(continued)

Continued

Chapter	Situation or phenotype	Example(s)	Reference(s)
18	Testing a conditional allele	*Lhx1*	Kwan and Behringer 2002
		Bmpr1a	Mishina et al. 2002
18	Floxed allele that is a hypomorph but can still be used as a conditional allele	*Bmp4*	Kulessa and Hogan 2002
		Nodal	Lowe et al. 2001
18	Inducible recombination	CreER™	Hayashi et al. 2001
18	Cre reporter		Soriano 1999
18	Inducible recombination in epidermis	*Ikbkb*	Stratis et al. 2006
18	Incomplete excision by Cre resulting in hypomorphic phenotype	*Nr5a1*	Zhao et al. 2001
18	Conditional and tissue-specific mutation	*Fgf8*	Meyers et al. 1998
18	Gain of function of rescuing transgene	*Lhx1*	Kobayashi et al. 2005
18	Transgene rescue of a mutant phenotype	*Myo15*	Probst et al. 1998
18	Regulatory mutations	*Amh*	Arango et al. 1999
		Sox9	Gonen et al. 2018

REFERENCES

Abuin A, Bradley A. 1996. Recycling selectable markers in mouse embryonic stem cells. *Mol Cell Biol* **16**: 1851–1856.

Akiyama H, Chaboissier MC, Martin JF, Schedl A, de Crombrugghe B. 2002. The transcription factor Sox9 has essential roles in successive steps of the chondrocyte differentiation pathway and is required for expression of *Sox5* and *Sox6*. *Genes Dev* **16**: 2813–2828.

Arango NA, Lovell-Badge R, Behringer RR. 1999. Targeted mutagenesis of the endogenous mouse *Mis* gene promoter: in vivo definition of genetic pathways of vertebrate sexual development. *Cell* **99**: 409–419.

Arora R, del Alcazar CM, Morrisey EE, Naiche LA, Papaioannou VE. 2012a. Candidate gene approach identifies multiple genes and signaling pathways downstream of Tbx4 in the developing allantois. *PLoS ONE* **7**: e43581.

Arora R, Metzger R, Papaioannou VE. 2012b. Multiple roles and interactions of *Tbx4* and *Tbx5* in the development of the respiratory system. *PLoS Genet* **8**: e1002866.

Arora R, Fries A, Oelerich K, Marchuk K, Sabeur K, Giudice LC, Laird DJ. 2016. Insights from imaging the implanting embryo and the uterine environment in three dimensions. *Development* **143**: 4749–4754.

Babinet C, Richoux V, Guénet JL, Renard JP. 1990. The DDK inbred strain as a model for the study of interactions between parental genomes and egg cytoplasm in mouse preimplantation development. *Development (Suppl.)* **1990**: 81–87.

Baum MJ, Brown JJG, Kica E, Rubin BS, Johnson RS, Papaioannou VE. 1994. Effect of a null mutation of the c-*fos* proto-oncogene on sexual behavior of male mice. *Biol Reprod* **50**: 1040–1048.

Behringer RR, Finegold MJ, Cate RL. 1994. Mullerian-inhibiting substance function during mammalian sexual development. *Cell* **79**: 415–425.

Bei M, Kratochwil K, Maas RL. 2000. BMP4 rescues a non-cell-autonomous function of *Msx1* in tooth development. *Development* **127**: 4711–4718.

Bi W, Deng JM, Zhang Z, Behringer RR, de Crombrugghe B. 1999. *Sox9* is required for cartilage formation. *Nat Genet* **22**: 85–89.

Bi W, Huang W, Whitworth DJ, Deng JM, Zhang Z, Behringer RR, de Crombrugghe B. 2001. Haploinsufficiency of *Sox9* results in defective cartilage primordia and premature skeletal mineralization. *Proc Natl Acad Sci* **98**: 6698–6703.

Bishop CE, Whitworth DJ, Qin Y, Agoulnik AI, Agoulnik IU, Harrison WR, Behringer RR, Overbeek PA. 2000. A transgenic insertion upstream of *sox9* is associated with dominant XX sex reversal in the mouse. *Nat Genet* **26**: 490–494.

Bonnette SG, Hadsell DL. 2001. Targeted disruption of the IGF-I receptor gene decreases cellular proliferation in mammary terminal end buds. *Endocrinology* **142**: 4937–4945.

Braun RE, Lo D, Pinkert CA, Widera G, Flavell RA, Palmiter RD, Brinster RL. 1990. Infertility in male transgenic mice: disruption of sperm development by HSV-tk expression in postmeiotic germ cells. *Biol Reprod* **43**: 684–693.

Brinster RL, Zimmermann JW. 1994. Spermatogenesis following male germ-cell transplantation. *Proc Natl Acad Sci* **91**: 11298–11302.

Carmeliet P, Ferreira V, Breier G, Pollefeyt S, Kieckens L, Gertsenstein M, Fahrig M, Vandenhoeck A, Harpal K, Eberhardt C, et al. 1996. Abnormal blood vessel development and lethality in embryos lacking a single VEGF allele. *Nature* **380**: 435–439.

Chaboissier MC, Kobayashi A, Vidal VI, Lutzkendorf S, van de Kant HJ, Wegner M, de Rooij DG, Behringer RR, Schedl A. 2004. Functional analysis of *Sox8* and *Sox9* during sex determination in the mouse. *Development* **131**: 1891–1901.

Chang H, Huylebroeck D, Verchueren K, Guo Q, Matzuk MM, Zwijsen A. 1999. Smad5 knockout mice die at mid-gestation due to multiple embryonic and extraembryonic defects. *Development* **126**: 1631–1642.

Chapman DL, Papaioannou VE. 1998. Three neural tubes in mouse embryos with mutations in the T-box gene, *Tbx6*. *Nature* **391**: 695–697.

Chapman DL, Cooper-Morgan A, Harrelson Z, Papaioannou VE. 2003. Critical role for *Tbx6* in mesoderm specification in the mouse embryo. *Mech Dev* **120**: 837–847.

Chavez-Reyes A, Parant JM, Amelse LL, de Oca Luna RM, Korsmeyer SJ, Lozano G. 2003. Switching mechanisms of cell death in *mdm2*- and *mdm4*-null mice by deletion of p53 downstream targets. *Cancer Res* **63**: 8664–8669.

Chen CM, Behringer R. 2004. *Ovca1* regulates cell proliferation, embryonic development, and tumorigenesis. *Genes Dev* **18**: 320–332.

Chen ZF, Behringer RR. 1995. *twist* is required in head mesenchyme for cranial neural tube morphogenesis. *Genes Dev* **9**: 686–699.

Chen ZF, Rebelo S, White F, Malmberg AB, Baba H, Lima D, Woolf CJ, Basbaum AI, Anderson DJ. 2001. The paired homeodomain protein DRG11 is required for the projection of cutaneous sensory afferent fibers to the dorsal spinal cord. *Neuron* **31**: 59–73.

Davenport TG, Jerome-Majewska LA, Papaioannou VE. 2003. Mammary gland, limb, and yolk sac defects in mice lacking *Tbx3*, the gene mutated in human ulnar mammary syndrome. *Development* **130**: 2263–2273.

Davis AC, Wims M, Spotts GD, Hann SR, Bradley A. 1993. A null c-*myc* mutation causes lethality before 10.5 days of gestation in homozygotes and reduced fertility in heterozygous female mice. *Genes Dev* **7**: 671–682.

DeChiara TM, Robertson EJ, Efstratiadis A. 1991. Parental imprinting of the mouse insulin-like growth factor II gene. *Cell* **64**: 849–859.

Degenhardt K, Wright AC, Horng D, Padmanabhan A, Epstein JA. 2010. Rapid 3D phenotyping of cardiovascular development in mouse embryos by micro-CT with iodine staining. *Circ Cardiovasc Imaging* **3**: 314–322.

de Rozieres S, Maya R, Oren M, Lozano G. 2000. The loss of *mdm2* induces pp53-mediated apoptosis. *Oncogene* **19**: 1691–1697.

Dewerchin M, Liang Z, Moons L, Carmeliet P, Castellino FJ, Collen D, Rosen ED. 2000. Blood coagulation factor X deficiency causes partial embryonic lethality and fatal neonatal bleeding in mice. *Thromb Haemost* **83**: 185–190.

Donehower LA, Harvey M, Slagle BL, McArthur MJ, Montgomery CA Jr, Butel JS, Bradley A. 1992. Mice deficient for p53 are developmentally normal but susceptible to spontaneous tumours. *Nature* **356**: 215–221.

Dong S, Leung KK, Pelling AL, Lee PY, Tang AS, Heng HH, Tsui LC, Tease C, Fisher G, Steel KP, Cheah KS. 2002. Circling, deafness, and yellow coat displayed by yellow submarine (ysb) and light coat and circling (lcc) mice with mutations on chromosome 3. *Genomics* **79**: 777–784.

Downs KM, Temkin R, Gifford S, McHugh J. 2000. Study of the murine allantois by allantoic explants. *Dev Biol* **233**: 347–364.

Ducy P, Amling M, Takeda S, Priemel M, Schilling AF, Beil FT, Shen J, Vinson C, Rueger JM, Karsenty G. 2000. Leptin inhibits bone formation through a hypothalamic relay: a central control of bone mass. *Cell* **100**: 197–207.

Dudas M, Nagy A, Laping NJ, Moustakas A, Kaartinen V. 2004. Tgf-β3-induced palatal fusion is mediated by Alk-5/Smad pathway. *Dev Biol* **266**: 96–108.

Dufort D, Schwartz L, Harpal K, Rossant J. 1998. The transcription factor HNF3β is required in visceral endoderm for normal primitive streak morphogenesis. *Development* **125**: 3015–3025.

Erickson JT, Conover JC, Borday V, Champagnat J, Barbacid M, Yancopoulos G, Katz DM. 1996. Mice lacking brain-derived neurotrophic factor exhibit visceral sensory neuron losses distinct from mice lacking NT4 and display a severe developmental deficit in control of breathing. *J Neurosci* **16**: 5361–5371.

Feldman B, Poueymirou W, Papaioannou VE, DeChiara TM, Goldfarb M. 1995. Requirement of FGF-4 for postimplantation mouse development. *Science* **267**: 246–249.

Fitch KR, McGowan KA, van Raamsdonk CD, Fuchs H, Lee D, Puech A, Herault Y, Threadgill DW, Hrabe de Angelis M, Barsh GS. 2003. Genetics of dark skin in mice. *Genes Dev* **17**: 214–228.

Friedrich G, Soriano P. 1991. Promoter traps in embryonic stem cells: AA genetic screen to identify and mutate developmental genes in mice. *Genes Dev* **5**: 1513–1523.

Frisch BJ, Calvi LM. 2014. Hematopoietic stem cell cultures and assays. *Meth Mol Biol (Clifton, NJ)* **1130**: 315–324.

Fukushima H, Shimizu K, Watahiki A, Hoshikawa S, Kosho T, Oba D, Sakano S, Arakaki M, Yamada A, Nagashima K, et al. 2017. NOTCH2 Hajdu–Cheney mutations escape SCF^{FBW7}-dependent proteolysis to promote osteoporosis. *Mol Cell* **68**: 645–658.e645.

Furuta Y, Hogan BL. 1998. BMP4 is essential for lens induction in the mouse embryo. *Genes Dev* **12**: 3764–3775.

Gerlai R, Millen KJ, Herrup K, Fabien K, Joyner AL, Roder J. 1996. Impaired motor learning performance in cerebellar En-2 mutant mice. *Behav Neurosci* **110**: 126–133.

Giese KP, Martini R, Lemke G, Soriano P, Schachner M. 1992. Mouse P_0 gene disruption leads to hypomyelination, abnormal expression of recognition molecules, and degeneration of myelin and axons. *Cell* **71**: 565–576.

Ginkel C, Hartmann D, vom Dorp K, Zlomuzica A, Farwanah H, Eckhardt M, Sandhoff R, Degen J, Rabionet M, Dere E, et al. 2012. Ablation of neuronal ceramide synthase 1 in mice decreases ganglioside levels and expression of myelin-associated glycoprotein in oligodendrocytes. *J Biol Chem* **287**: 41888–41902.

Gnarra JR, Ward JM, Porter FD, Wagner JR, Devor DE, Grinberg A, Emmert-Buck MR, Westphal H, Klausner RD, Linehan WM. 1997. Defective placental vasculogenesis causes embryonic lethality in VHL-deficient mice. *Proc Natl Acad Sci* **94**: 102–107.

Goldin SN, Papaioannou VE. 2003. Paracrine action of FGF4 during peri-implantation development maintains trophectoderm and primitive endoderm. *Genesis* **36**: 40–47.

Gonen N, Futtner CR, Wood S, Garcia-Moreno SA, Salamone IM, Samson SC, Sekido R, Poulat F, Maatouk DM, Lovell-Badge R. 2018. Sex reversal following deletion of a single distal enhancer of Sox9. *Science* **360**: 1469–1473.

Gossler A, Joyner AL, Rossant J, Skarnes WC. 1989. Mouse embryonic stem cells and reporter constructs to detect developmentally regulated genes. *Science* **244**: 463–465.

Guidi CJ, Sands AT, Zambrowicz BP, Turner TK, Demers DA, Webster W, Smith TW, Imbalzano AN, Jones SN. 2001. Disruption of Ini1 leads to peri-implantation lethality and tumorigenesis in mice. *Mol Cell Biol* **21**: 3598–3603.

Guillemot F, Nagy A, Auerbach A, Rossant J, Joyner AL. 1994. Essential role of Mash-2 in extraembryonic development. *Nature* **371**: 333–336.

Hadjantonakis A-K, Pisano E, Papaioannou VE. 2008. Tbx6 regulates left/right patterning in mouse embryos through effects on nodal cilia and perinodal signaling. *PloS ONE* **3**: e2511.

Hancock SN, Agulnik SI, Silver LM, Papaioannou V.E. 1999. Mapping and expression analysis the mouse ortholog of Xenopus Eomesodermin. *Mech Dev* **81**: 205–208.

Harrelson Z, Kelly RG, Goldin SN, Gibson-Brown JJ, Bollag RJ, Silver LM, Papaioannou VE. 2004. Tbx2 is essential for patterning the atrioventricular canal and for morphogenesis of the outflow tract during heart development. *Development* **131**: 5041–5052.

Hasty P, Bradley A, Morris JH, Edmondson DG, Venuti JM, Olson EN, Klein WH. 1993. Muscle deficiency and neonatal death in mice with a targeted mutation in the myogenin gene. *Nature* **364**: 501–506.

Hayashi S, Tenzen T, McMahon AP. 2003. Maternal inheritance of Cre activity in a Sox2Cre deleter strain. *Genesis* **37**: 51–53.

Hayashi S, Lewis P, Pevney L, McMahon AP. 2002. Efficient gene modulation in mouse epiblast using a Sox2Cre transgenic mouse strain. *Mech Dev (Suppl. 1)* **119**: S97–S101.

Herceg Z, Hulla W, Gell D, Cuenin C, Lleonart M, Jackson S, Wang ZQ. 2001. Disruption of Trrap causes early embryonic lethality and defects in cell cycle progression. *Nat Genet* **29**: 206–211.

Heuberger B, Fitzka I, Wasner G, Kratochwil K. 1982. Induction of androgen receptor formation by epithelium–mesenchyme interaction in embryonic mouse mammary gland. *Proc Natl Acad Sci* **79**: 2957–2961.

Hildebrand JD, Soriano P. 1999. Shroom, a PDZ domain-containing actin-binding protein, is required for neural tube morphogenesis in mice. *Cell* **99**: 485–497.

Holdener BC, Magnuson T. 1994. A mouse model for human hereditary tyrosinemia I. *BioEssays* **16**: 85–87.

Honrado GI, Johnson RS, Golombek DA, Spiegelman BM, Papaioannou VE, Ralph MR. 1996. The circadian system of c-fos deficient mice. *J Comp Physiol A* **178**: 563–570.

Horai R, Saijo S, Tanioka H, Nakae S, Sudo K, Okahara A, Ikuse T, Asano M, Iwakura Y. 2000. Development of chronic inflammatory arthropathy resembling rheumatoid arthritis in interleukin 1 receptor antagonist-deficient mice. *J Exp Med* **191**: 313–320.

Horie K, Yusa K, Yae K, Odajima J, Fischer SE, Keng VW, Hayakawa T, Mizuno S, Kondoh G, Ijiri T, Matsuda Y, Plasterk RH, Takeda J. 2003. Characterization of Sleeping Beauty transposition and its application to genetic screening in mice. *Mol Cell Biol* **23**: 9189–9207.

Hotamisligil GS, Johnson RS, Distel RJ, Ellis R, Papaioannou VE, Spiegelman BM. 1996. Uncoupling of obesity from insulin resistance through a targeted mutation in aP2, the adipocyte fatty acid binding protein. *Science* **274**: 1377–1379.

Hrabe de Angelis MH, Flaswinkel H, Fuchs H, Rathkolb B, Soewarto D, Marschall S, Heffner S, Pargent W, Wuensch K, Jung M, et al. 2000. Genome-wide, large-scale production of mutant mice by ENU mutagenesis. *Nat Genet* **25**: 444–447.

Jerome LA, Papaioannou VE. 2001. DiGeorge syndrome phenotype in mice mutant for the T-box gene, Tbx1. *Nat Genet* **27**: 286–291.

Johnson RS, Spiegelman BM, Papaioannou VE. 1992. Pleiotropic effects of a null mutation in the c-fos proto-oncogene. *Cell* **71**: 577–586.

Joyner AL, Herrup K, Auerbach BA, Davis CA, Rossant J. 1991. Subtle cerebellar phenotype in mice homozygous for a targeted deletion of the En-2 homeobox. *Science* **251**: 1239–1243.

Karlsson L, Bondjers C, Betsholtz C. 1999. Roles for PDGF-A and sonic hedgehog in development of mesenchymal components of the hair follicle. *Development* **126**: 2611–2621.

Kasarskis A, Manova K, Anderson KV. 1998. A phenotype-based screen for embryonic lethal mutations in the mouse. *Proc Natl Acad Sci.* **95**: 7485–7490.

Kaul A, Köster M, Neuhaus H, Braun T. 2000. Myf-5 revisited: loss of early myotome formation does not lead to a rib phenotype in homozygous Myf-5 mutant mice. *Cell* **102**: 17–19.

Kelly MA, Low MJ, Phillips TJ, Wakeland EK, Yanagisawa M. 2003. The mapping of quantitative trait loci underlying strain differences in locomotor activity between 129S6 and C57BL/6J mice. *Mamm Genome* **14**: 692–702.

Kile BT, Hentges KE, Clark AT, Nakamura H, Salinger AP, Liu B, Box N, Stockton DW, Johnson RL, Behringer RR, et al. 2003. Functional genetic analysis of mouse chromosome 11. *Nature* **425**: 81–86.

Kruithof BP, van den Hoff MJ, Wessels A, Moorman AF. 2003. Cardiac muscle cell formation after development of the linear heart tube. *Dev Dynamics* **227**: 1–13.

Kuehn MR, Bradley A, Robertson EJ, Evans MJ. 1987. A potential animal model for Lesch–Nyhan syndrome through introduction of HPRT mutations into mice. *Nature* **326**: 295–298.

Kulessa H, Hogan BL. 2002. Generation of a *loxP* flanked *bmp4^{loxP-lacZ}* allele marked by conditional *lacZ* expression. *Genesis* **32**: 66–68.

Kurowski A, Molotkov A, Soriano P. 2019. FGFR1 regulates trophectoderm development and facilitates blastocyst implantation. *Dev Biol* **446**: 94–101.

Kwan KM, Behringer RR. 2002. Conditional inactivation of *Lim1* function. *Genesis* **32**: 118–120.

Lee EY, Chang CY, Hu N, Wang YC, Lai CC, Herrup K, Lee WH, Bradley A. 1992. Mice deficient for Rb are nonviable and show defects in neurogenesis and haematopoiesis. *Nature* **359**: 288–294.

Lerch-Gaggl A, Haque J, Li J, Ning G, Traktman P, Duncan SA. 2002. Pescadillo is essential for nucleolar assembly, ribosome biogenesis, and mammalian cell proliferation. *J Biol Chem* **277**: 45347–45355.

Lettice LA, Heaney SJ, Purdie LA, Li L, de Beer P, Oostra BA, Good D, Elgar G, Hill RE, de Graaf E. 2003. A long-range Shh enhancer regulates expression in the developing limb and fin and is associated with preaxial polydactyly. *Hum Mol Genet* **12**: 1725–1735.

Li Y, Behringer RR. 1998. *Esx1* is an X-chromosome-imprinted regulator of placental development and fetal growth. *Nat Genet* **20**: 309–311.

Lin Q, Schwarz J, Bucana C, Olson EN. 1997. Control of mouse cardiac morphogenesis and myogenesis by transcription factor MEF2C. *Science* **276**: 1404–1407.

Liu L, Czerwiec E, Keefe DL. 2004. Effect of ploidy and parental genome composition on expression of Oct-4 protein in mouse embryos. *Gene Expr Patterns* **4**: 433–441.

Liu P, Wakamiya M, Shea MJ, Albrecht U, Behringer RR, Bradley A. 1999. Requirement for *Wnt3* in vertebrate axis formation. *Nat Genet* **22**: 361–365.

Lowe LA, Yamada S, Kuehn MR. 2001. Genetic dissection of *nodal* function in patterning the mouse embryo. *Development* **128**: 1831–1843.

Ludwig T, Chapman DL, Papaioannou VE, Efstratiadis A. 1997. Targeted mutations of breast cancer susceptibility gene homologs in mice: lethal phenotypes of *Brca1*, *Brca2*, *Brca1/Brca2*, *Brca1/p53*, and *Brca2/p53* nullizygous embryos. *Genes Dev* **11**: 1226–1241.

Lufkin T, Dierich A, LeMeur M, Mark M, Chambon P. 1991. Disruption of the *Hox-1.6* homeobox gene results in defects in a region corresponding to its rostral domain of expression. *Cell* **66**: 1105–1119.

MacGregor GR, Russell LD, Van Beek ME, Hanten GR, Kovac MJ, Kozak CA, Meistrich ML, Overbeek PA. 1990. Symplastic spermatids (sys): a recessive insertional mutation in mice causing a defect in spermatogenesis. *Proc Natl Acad Sci* **87**: 5016–5020.

Meyers E, Lewandoski M, Martin G.R. 1998. An *Fgf8* mutant allelic series generated by Cre- and Flp-mediated recombination. *Nat Gene.* **18**: 136–141.

Miquerol L, Langille BL, Nagy A. 2000. Embryonic development is disrupted by modest increases in vascular endothelial growth factor gene expression. *Development* **127**: 3941–3946.

Mikkola HK, Klintman J, Yang H, Hock H, Schlaeger TM, Fujiwara Y, Orkin SH. 2003. Haematopoietic stem cells retain long-term repopulating activity and multipotency in the absence of stem-cell leukaemia SCL/tal-1 gene. *Nature* **421**: 547–551.

Militzer K. 2001. Hair growth pattern in nude mice. *Cells Tissues Organs* **168**: 285–294.

Mishina Y, Hanks MC, Miura S, Tallquist MD, Behringer RR. 2002. Generation of *Bmpr/Alk3* conditional knockout mice. *Genesis* **32**: 69–72.

Mishina Y, Rey R, Finegold MJ, Matzuk MM, Josso N, Cate RL, Behringer RR. 1996. Genetic analysis of the Mullerian-inhibiting substance signal transduction pathway in mammalian sexual differentiation. *Genes Dev* **10**: 2577–2587.

Miura T, Shiota K. 2000. Time-lapse observation of branching morphogenesis of the lung bud epithelium in mesenchyme-free culture and its relationship with the localization of actin filaments. *Int J Dev Biol* **44**: 899–902.

Montes de Oca Luna R, Wagner DS, Lozano G. 1995. Rescue of early embryonic lethality in *mdm2*-deficient mice by deletion of *p53*. *Nature* **378**: 203–206.

Morasso MI, Grinberg A, Robinson G, Sargent TD, Mahon KA. 1999. Placental failure in mice lacking the homeobox gene *Dlx3*. *Proc Natl Acat Sci* **96**: 162–167.

Morgan D, Turnpenny L, Goodship J, Dai W, Majumder K, Matthews L, Gardner A, Schuster G, Vien L, Harrison W, Elder FF, Penman-Splitt M, Overbeek P, Strachan T. 1998. *Inversin*, a novel gene in the vertebrate left–right axis pathway, is partially deleted in the *inv* mouse. *Nat Genet* **20**: 149–156.

Naiche LA, Papaioannou VE. 2003. Loss of *Tbx4* blocks hindlimb development and affects vascularization and fusion of the allantois. *Development* **130**: 2681–2693.

Nakashima K, Zhou X, Kunkel G, Zhang Z, Deng JM, Behringer RR, de Crombrugghe B. 2002. The novel zinc finger–containing transcription factor osterix is required for osteoblast differentiation and bone formation. *Cell* **108**: 17–29.

Nichols J, Chambers I, Taga T, Smith A. 2001. Physiological rationale for responsiveness of mouse embryonic stem cells to gp130 cytokines. *Development* **128**: 2333–2339.

Ohsugi M, Larue L, Schwarz H, Kemler R. 1997. Cell-junctional and cytoskeletal organization in mouse blastocysts lacking E-cadherin. *Dev Biol* **185**: 261–271.

Ohtoshi A, Behringer RR. 2004. Neonatal lethality, dwarfism, and abnormal brain development in *Dmbx1* mutant mice. *Mol Cell Biol* **24**: 7548–7558.

Ohtoshi A, Wang SW, Maeda H, Saszik S, Frishman L, Klein WH, Behringer RR. 2004. Regulation of retinal cone bipolar cell differentiation and photopic vision by the CVC homeobox gene *Vsx1*. *Curr Biol* **14**: 530–536.

Palmiter RD, Brinster RL, Hammer RE, Trumbauer ME, Rosenfeld MG, Birnberg NC, Evans RM. 1982. Dramatic growth of mice that develop from eggs microinjected with metallothionein-growth hormone fusion genes. *Nature* **300**: 611–615.

Papaioannou VE, Gardner RL. 1979. Investigation of the lethal yellow Ay/Ay embryo using mouse chimaeras. *J Embryol Exp Morphol* **52**: 153–163.

Pask AJ, Kanasaki H, Kaiser UB, Conn PM, Janovick JA, Stockton DW, Hess DL, Justice MJ, Behringer RR. 2005. A novel mouse model of hypogonadotrophic hypogonadism: *N*-ethyl-*N*-nitrosourea-induced gonadotropin-releasing hormone receptor gene mutation. *Mol Endocrinol (Baltimore, Md)* **19**: 972–981.

Paylor R, Johnson RS, Papaioannou VE, Speigelman BM, Wehner JM. 1994. Behavioral assessment of c-*fos* mutant mice. *Brain Res* **651**: 275–282.

Perea-Gomez A, Vella FD, Shawlot W, Oulad-Abdelghani M, Chazaud C, Meno C, Pfister V, Chen L, Robertson E, Hamada H, et al. 2002. Nodal antagonists in the anterior visceral endoderm prevent the formation of multiple primitive streaks. *De. Cell* **3**: 745–756.

Peters B, Kirfel J, Büssow H, Vidal M, Magin TM. 2001. Complete cytolysis and neonatal lethality in keratin 5 knockout mice reveal its fundamental role in skin integrity and in epidermolysis bullosa simplex. *Mol Biol Cell* **12**: 1775–1789.

Pevny L, Simon MC., Robertson E, Klein W, Tsai SF, D'Agati V, Orkin SH, Costantini F. 1991. Erythroid differentiation in chimaeric mice blocked by a targeted mutation in the gene for transcription factor GATA-1. *Nature* **349**: 257–260.

Phoon CK, Ji RP, Aristizábal O, Worrad DM, Zhou B, Baldwin HS, Turnbull DH. 2004. Embryonic heart failure in *NFATc1^{-/-}* mice: novel mechanistic insights from in utero ultrasound biomicroscopy. *Circ Res* **95**: 92–99.

Probst FJ, Fridell RA, Raphael Y, Saunders TL, Wang A, Liang Y, Morell RJ, Touchman JW, Lyons RH, Noben-Trauth K, et al. 1998. Correction of deafness in *shaker-2* mice by an unconventional myosin in a BAC transgene. *Science* **280**: 1444–1447.

Purandare SM, Ware SM, Kwan KM, Gebbia M, Bassi MT, Deng JM, Vogel H, Behringer RR, Belmont JW, Casey B. 2002. A complex syndrome of left–right axis, central nervous system and axial skeleton defects in *Zic3* mutant mice. *Development* **129**: 2293–2302.

Rinkenberger JL, Horning S, Klocke B, Roth K, Korsmeyer SJ. 2000. Mcl-1 deficiency results in peri-implantation embryonic lethality. *Genes Dev* **14**: 23–27.

Rivera-Perez JA, Mallo M, Gendron-Maguire M, Gridley T, Behringer RR. 1995. *goosecoid* is not an essential component of the mouse gastrula organizer but is required for craniofacial and rib development. *Development* **121**: 3005–3012.

Roffler-Tarlov S, Gibson Brown JJ, Tarlov E, Stolarov J, Chapman DL, Alexiou M, Papaioannou VE. 1996. Programmed cell death in the absence of c-Fos and c-Jun. *Development* 122: 1–9.

Rohozinski J, Agoulnik AI, Boettger-Tong HL, Bishop CE. 2002. Successful targeting of mouse Y chromosome genes using a site-directed insertion vector. *Genesis* 32: 1–7.

Rudnicki MA, Schnegelsberg PN, Stead RH, Braun T, Arnold HH, Jaenisch R. 1993. MyoD or Myf-5 is required for the formation of skeletal muscle. *Cell* 75: 1351–1359.

Sagane K, Hayakawa K, Kai J, Hirohashi T, Takahashi E, Miyamoto N, Ino M, Oki T, Yamazaki K, Nagasu T. 2005. Ataxia and peripheral nerve hypomyelination in ADAM22-deficient mice. *BMC Neurosci* 6: 33.

Satokata I, Maas R. 1994. *Msx1* deficient mice exhibit cleft palate and abnormalities of craniofacial and tooth development. *Nat Genet* 6: 348–356.

Schlatt S, Honaramooz A, Boiani M, Scholer HR, Dobrinski I. 2003. Progeny from sperm obtained after ectopic grafting of neonatal mouse testes. *Biol Reprod* 68: 2331–2335.

Sekine K, Ohuchi H, Fujiwara M, Yamasaki M, Yoshizawa T, Sato T, Yagishita N, Matsui D, Koga Y, Itoh N, Kato S. 1999. Fgf10 is essential for limb and lung formation. *Nat Genet* 21: 138–141.

Shawlot W, Behringer RR. 1995. Requirement for Lim1 in head-organizer function. *Nature* 374: 425–430.

Shawlot W, Min Deng J, Wakamiya M, Behringer RR. 2000. The cerberus-related gene, *Cerr1*, is not essential for mouse head formation. *Genesis* 26: 253–258.

Shi CZ, Collins HW, Buettger CW, Garside WT, Matschinsky FM, Heyner S. 1994. Insulin family growth factors have specific effects on protein synthesis in preimplantation mouse embryos. *Mol Reprod Dev* 37: 398–406.

Shimono A, Behringer RR. 2003. Angiomotin regulates visceral endoderm movements during mouse embryogenesis. *Curr Biol* 13: 613–617.

Shiojiri N, Wada JI, Tanaka T, Noguchi M, Ito M, Gebhardt R. 1995. Heterogeneous hepatocellular expression of glutamine synthetase in developing mouse liver and in testicular transplants of fetal liver. *Lab Invest* 72: 740–747.

Soriano P. 1997. The PDGF alpha receptor is required for neural crest cell development and for normal patterning of the somites. *Development* 124: 2691–2700.

Soriano P. 1999. Generalized *lacZ* expression with the *ROSA26* Cre reporter. *Nat Genet* 21: 70–71.

Soriano P, Montgomery C, Geske R, Bradley A. 1991. Targeted disruption of the c-src proto-oncogene leads to osteopetrosis in mice. *Cell* 64: 693–702.

Srinivas S, Goldberg MR, Watanabe T, D'Agati V, al-Awqati Q, Costantini F. 1999. Expression of green fluorescent protein in the ureteric bud of transgenic mice: a new tool for the analysis of ureteric bud morphogenesis. *Dev Genet* 24: 241–251.

Sta Maria NS, Zhou C, Lee SJ, Valiulahi P, Li X, Choi J, Liu X, Jacobs R, Comai L, Reddy S. 2021. Mbnl1 and Mbnl2 regulate brain structural integrity in mice. *Commun Biol* 4: 1342.

Stewart CL, Kaspar P, Brunet LJ, Bhatt H, Gadi I, Kontgen F, Abbondanzo SJ. 1992. Blastocyst implantation depends on maternal expression of leukaemia inhibitory factor. *Nature* 359: 76–79.

Stratis A, Pasparakis M, Markur D, Knaup R, Pofahl R, Metzger D, Chambon P, Krieg T, Haase I. 2006. Localized inflammatory skin disease following inducible ablation of I kappa B kinase 2 in murine epidermis. *J Invest Dermatol* 126: 614–620.

Threadgill DW, Dlugosz AA, Hansen LA, Tennenbaum T, Lichti U, Yee D, LaMantia C, Mourton T, Herrup K, Harris RC, et al. 1995. Targeted disruption of mouse EGF receptor: effect of genetic background on mutant phenotype. *Science* 269: 230–234.

van Deursen J, Boer J, Kasper L, Grosveld G. 1996. G2 arrest and impaired nucleocytoplasmic transport in mouse embryos lacking the proto-oncogene CAN/Nup214. *EMBO J* 15: 5574–5583.

Varlet I, Collignon J, Robertson EJ. 1997. Nodal expression in the primitive endoderm is required for specification of the anterior axis during mouse gastrulation. *Development* 124: 1033–1044.

Voiculescu O, Taillebourg E, Pujades C, Kress C, Buart S, Charnay P, Schneider-Maunoury S. 2001. Hindbrain patterning: *Krox20* couples segmentation and specification of regional identity. *Development* 128: 4967–4978.

Wada K, Matsushima Y, Tada T, Hasegawa S, Obara Y, Yoshizawa Y, Takahashi G, Hiai H, Shimanuki M, Suzuki S, et al. 2014. Expression of truncated PITX3 in the developing lens leads to microphthalmia and aphakia in mice. *PLoS ONE* 9: e111432.

Watabe-Rudolph M, Schlautmann N, Papaioannou VE, Gossler A. 2002. The mouse rib-vertebrae mutation is a hypomorphic *Tbx6* allele. *Mech Dev* 119: 251–256.

Weninger WJ, Floro KL, Bennett MB, Withington SL, Preis JI, Barbera JPM, Mohun tJ, Dunwoodie SL. 2005. *Cited2* is required both for heart morphogenesis and establishment of the left–right axis in mouse development. *Development* 132: 1337–1348.

Winnier G, Hargett L, Hogan BL. 1997. The winged helix transcription factor *MFH1* is required for proliferation and patterning of paraxial mesoderm in the mouse embryo. *Genes Dev* 11: 926–940.

Wu L, de Bruin A, Saavedra HI, Starovic M, Trimboli A, Yang Y, Opavska J, Wilson P, Thompson JC, Ostrowski MC, et al. 2003. Extra-embryonic function of Rb is essential for embryonic development and viability. *Nature* 421: 942–947.

Xue Y, Gao X, Lindsell CE, Norton CR, Chang B, Hicks C, Grendron-Maguire M, Rand EB, Weinmaster G, Gridley T. 1999. Embryonic lethality and vascular defects in mice lacking the Notch ligand Jagged1. *Hum Mol Genet* 8: 723–730.

Yamauchi Y, Riel JM, Ruthig V, Ward MA. 2015. Mouse Y-encoded transcription factor *Zfy2* is essential for sperm formation and function in assisted fertilization. *PLoS Genet* 11: e1005476.

Yan W, Ma L, Burns KH, Matzuk MM. 2004. Haploinsufficiency of kelch-like protein homolog 10 causes infertility in male mice. *Proc Natl Acad Sci* 101: 7793–7798.

Yeh S, Tsai MY, Xu Q, Mu XM, Lardy H, Huang KE, Lin H, Yeh SD, Altuwaijri S, Zhou X, et al. 2002. Generation and characterization of androgen receptor knockout (ARKO) mice: an in vivo model for the study of androgen functions in selective tissues. *Proc Natl Acad Sci* 99: 13498–13503.

Yoder MC, Papaioannou VE, Breitfeld PP, Williams DA. 1994. Murine yolk sac endoderm- and mesoderm-derived cell lines support in vitro growth and differentiation of hematopoietic cells. *Blood* 83: 2436–2443.

You-Ten KE, Muise ES, Itié A, Michaliszyn E, Wagner J, Jothy S, Lapp WS, Tremblay ML. 1997. Impaired bone marrow microenvironment and immune function in T cell protein tyrosine phosphatase-deficient mice. *J Exper Med* 186: 683–693.

Zeitlin S, Liu J-P, Chapman DL, Papaioannou VE, Efstratiadis A. 1995. Increased apoptosis and early embryonic lethality in mice nullizygous for the Huntington's disease gene homologue. *Nat Genet* 11: 155–163.

Zhang J, Li J, Huang C, Xue L, Peng Y, Fu Q, Gao L, Zhang J, Li W. 2008. Targeted knockout of the mouse βB2-crystallin gene (*Crybb2*) induces age-related cataract. *Invest Ophthalmol Vis Sci* 49: 5476–5483.

Zhao L, Bakke M, Krimkevich Y, Cushman LJ, Parlow AF, Camper SA, Parker KL. 2001. Hypomorphic phenotype in mice with pituitary-specific knockout of steroidogenic factor 1. *Genesis* 30: 65–69.

Zhao M, Shirley CR, Hayashi S, Marcon L, Mohapatra B, Suganuma R, Behringer RR, Boissonneault G, Yanagimachi R, Meistrich ML. 2004. Transition nuclear proteins are required for normal chromatin condensation and functional sperm development. *Genesis* 38: 200–213.

Zhao Q, Behringer RR, de Crombrugghe B. 1996. Prenatal folic acid treatment suppresses acrania and meroanencephaly in mice mutant for the *Cart1* homeobox gene. *Nat Genet* 13: 275–283.

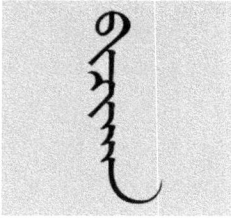

Glossary

AGM, aorta–gonad–mesonephros, a blood cell–forming region of the embryo.

AI, artificial insemination, experimental introduction of a viable sperm preparation into the female reproductive tract, usually delivered directly through the cervix. This procedure can be used to overcome infertility.

allantois, a mesoderm-derived extraembryonic structure that forms at the posterior pole of the embryo, grows through the extraembryonic coelom to attach to the chorion, and differentiates into the umbilical cord.

alleles, alternative or variant forms of the same gene occurring at the same locus on homologous chromosomes. Except in cases of duplication or deletion, an animal carries two alleles of autosomal genes, one derived from each parent.

allelic series, more than two different alleles of a given gene or locus. The phenotypic effects of an allelic series can sometimes be arranged in a dominance hierarchy with graded phenotypic effects.

alternative splicing, joining of different or alternative exons from primary transcripts from a single gene to produce different mRNA and protein isoforms.

anemia, the condition of a reduced number of circulating red blood cells relative to wild type.

aneuploid, a deviation from the standard karyotype, which in the mouse consists of 20 pairs of chromosomes including 19 autosomal pairs and the X and Y sex chromosomes. Such deviations can involve duplications, deletions, or translocations of whole or partial chromosomes.

anlagen, tissue primordium.

antimorph, a dominant-negative mutation encoding a mutant protein that interferes with the function of the wild-type protein.

ART (assisted reproductive technology), generally, in vitro technologies used to circumvent infertility, including IVF and ICSI.

autosomal, pertaining to chromosomes other than the X and Y sex chromosomes.

BAC, bacterial artificial chromosome.

backcross, breeding a mouse to one of its parental strains. This approach is used with forced heterozygosity to transfer an allele onto a different strain background by repeated backcrossing.

background genotype, *see genetic background.*

balancer chromosome, a chromosome with an inversion marked by a dominant visible marker (e.g., coat color). The inversion may or may not have a linked recessive lethal mutation. Balancer chromosomes can be created by chromosome engineering and are used in mutagenesis screens and for maintaining recessive lethal or sterile mutations.

β-gal, beta-galactosidase, protein product of the *Escherichia coli lacZ* gene.

Opening artwork: Mongolian word for "mouse."

bicistronic, an mRNA transcript encoding two different proteins; may be engineered using an internal ribosomal entry site.

blastocoel, the cavity of the blastocyst, also called the blastocyst cavity.

blastocyst, preimplantation-stage embryo between approximately E3.5 and E4.5, consisting of approximately 32–128 cells arranged as a single layer of trophoblast cells surrounding an inner cell mass, with an asymmetrically located cavity, the blastocyst cavity.

blastomeres, cells of early preimplantation embryos between the two-cell and approximately 16- to 32-cell stage. They are the products of cleavage in which cell division is not accompanied by cell growth.

Cas9, CRISPR-associated endonuclease 9.

cell autonomous, activity of a gene (or protein) that has its effect on the cell in which it is produced.

Chi squared (χ^2), test of probability to assess whether significant differences exist between an observed number of events and an expected number based on the null hypothesis. This method is commonly used to test for deviations from expected Mendelian segregation ratios.

chimera, a name appropriated from a mythical creature with the head and body of a lion, the head of a goat, and the tail of a serpent, now used to describe experimental mice made by the combination of two or more embryos or the combination of embryonic stem cells and an embryo. Genetic differences among the components of the chimera provide a means of identifying the cellular progeny of each component.

chorioallantoic placenta, definitive placenta in mice created by fusion of the allantois to the chorion, often referred to simply as placenta.

chorion, an extraembryonic membrane in the early postimplantation embryo composed of extraembryonic ectoderm and mesoderm. The allantois fuses with this membrane to form part of the chorioallantoic placenta.

CL, corpus luteum (plural: **corpora lutea**), the ovarian progesterone-secreting organ that forms from a collapsed follicle following ovulation.

cleavage, the process of cell division during preimplantation development, when cell division is not accompanied by cell growth and thus the size of the cells, known as blastomeres, decreases.

CNS, central nervous system, comprised of the brain and spinal cord.

codominant, the nature of inheritance of phenotype. A codominant phenotype exists when the effects of both alleles can be simultaneously detected.

co-isogenic strains, strains that differ from one another at a single locus, usually through the occurrence of a spontaneous mutation in an inbred strain or through targeted mutagenesis in an inbred strain.

compaction, process whereby the blastomeres of the preimplantation embryo pack into a solid ball of cells, cell–cell adhesion increases, and blastomeres flatten onto one another as a result of increased cell–cell contacts.

complementation test, cross to determine whether two different mutations are within the same gene.

compensation, loss of one gene product by mutation leading to the up-regulation of a nonallelic gene that ameliorates the severity of the phenotype caused by the mutation.

compound heterozygote, an animal carrying two different mutant alleles at a particular locus.

conditional allele, a genetically engineered allele designed to function as wild type but which can be induced to produce a mutant allele in a tissue- or time-specific manner.

congenic strains, strains that differ from one another at a single locus and at closely linked loci as well. Congenic strains are produced by repeated backcrossing of mice carrying a mutation to an inbred strain for at least 10 generations while maintaining heterozygosity for the mutation. Congenic strains differ from co-isogenic strains in that they are produced by breeding. They also differ by a small chromosomal segment, rather than by a single allele.

construct, DNA fragment(s) used to modify the mouse genome (e.g., a targeting construct).

copulation plug, a solid coagulation of the mouse male ejaculate that usually remains in the vagina of females for ~24 hr after mating, called simply a plug and used as an indication that mating has taken place.

corpora lutea, *see CL.*

Cre, causes recombination; DNA recombinase from *E. coli* bacteriophage P1 that recognizes the 34-bp *loxP* sequence; used to catalyze the deletion or inversion of floxed DNA sequences, depending on the orientation of two loxP sites.

CreER, chimeric protein composed of Cre fused to a mutated version of the estrogen receptor ligand binding domain that does not bind estrogen but can bind tamoxifen. Administration of tamoxifen induces the translocation of cytoplasmic CreER into the nucleus to act on floxed alleles.

CrePR, chimeric protein composed of Cre fused to a mutated version of the progesterone receptor ligand binding domain that does not bind progesterone but can bind RU-486. Administration of RU-486 induces the translocation of cytoplasmic CrePR into the nucleus to act on floxed alleles.

CRISPR, clustered regularly interspersed short palindromic repeats.

CRISPR–Cas, RNA guided endonuclease system for gene editing.

C-section, cesarean section, transabdominal delivery of fetal mice, usually done by sacrificing the mother.

cytospin, cell suspension centrifuged onto a microscope slide.

decidua, uterine tissue that differentiates around the implanting blastocyst. It is shed at parturition, hence the name.

decidualization, the reaction of the maternal uterine epithelium to an implanting embryo (or to certain other types of stimuli) that includes the development of decidual tissue surrounding the implantation site, some of which will take part in the formation of the definitive, chorioallantoic placenta.

deleter mice, genetically engineered mice that express Cre or Flp recombinase in the germline or in the embryonic cells that give rise to the germline. These mice are typically used to remove drug-selection cassettes in all embryonic cells.

DIC, differential interference contrast microscopy.

diploid, containing two homologous sets of chromosomes, referred to as 2N. Most somatic cells in mammals are diploid and divide by mitosis.

DNA recombinase systems, various systems used to genetically modify the mouse genome, including the bacteriophage-derived Cre/*loxP* system, the yeast-derived Flp/*FRT* system and other bacteriophage-derived systems such as Dre/*rox* and Vika/*vox*.

dominant, refers to the expression of phenotypes or the relationship of one allele to a second at the same locus. A dominant phenotype is detectable when only one variant allele is present. However, a single mutation can be dominant for some phenotypic traits and recessive for others.

dominant negative, a mutant protein that interferes with the function of the corresponding wild-type protein.

double heterozygote, an animal carrying one mutant allele at each of two different loci.

dpc, days postcoitus, the number of days after mating, used to indicate the age of embryos in some staging systems, the day of detection of a copulation plug is 0.5 dpc, largely replaced by E (embryonic day).

E, embryonic day, indicating the age of embryos; E0.5 is the midday of the day after mating, or the day of detection of a copulation plug.

EFIC, episcopic fluorescence image capturing.

EpiSC cells, epiblast-derived pluripotent stem cells.

electrofusion, electrical pulse that disrupts plasma membranes between cells, leading to fusion of the two cells. This method is used to fuse blastomeres of two-cell-stage embryos to generate tetraploid embryos for chimera production.

embryo proper, the body of the fetus excluding fetal membranes (extraembryonic components) and placenta.

endochondral ossification, bone formation utilizing a cartilage template.

endoreduplication, repeated rounds of DNA replication in the absence of intervening cell division. This process is characteristic of the trophoblastic giant cells and leads to polyploidy.

ENU, N-ethyl-N-nitrosourea, a chemical mutagen that usually causes single base-pair changes.

epiblast, tissue of the blastocyst that gives rise to the ectoderm, mesoderm, and endoderm of the embryo proper, as well as the extraembryonic mesoderm.

epistasis, the hierarchical relationship between two genes in a genetic pathway such that the effects of one mutation mask the effects of another.

erythrocyte, red blood cell.

ES cells, embryonic stem cells, pluripotent stem cells derived from the inner cell mass of blastocysts that can be perpetuated in vitro as undifferentiated cells, but that have a developmental potential similar to epiblast cells—that is, in chimeras they can contribute to all tissues of the fetus and the extraembryonic mesoderm but not extraembryonic endoderm or trophectoderm.

EST, expressed sequence tag, partial sequence from a cDNA clone.

euploid, the standard karyotype that in mice consists of 20 pairs of chromosomes including 19 autosomal pairs and the X and Y sex chromosomes.

expressivity, the severity of a mutant phenotype within an individual. This can vary among individuals with the same mutation.

extraembryonic ectoderm, a trophoblast-derived layer in the early postimplantation embryo that gives rise to part of the placenta.

extraembryonic, embryo-derived tissues, such as the amnion, chorion, and yolk sac, that will not make any cellular contribution to the body of the fetus. The term refers to tissue that is not part of the embryo proper.

FACS, fluorescence-activated cell sorting.

floxed, flanked by loxP sites.

Flp, flippase; DNA recombinase from yeast that recognizes the 34-bp FRT sequence.

flrted, flanked by FRT sites.

forced heterozygosity, selection of heterozygous mice as breeding stock for the next generation. This approach is used in backcrossing to transfer an allele to a different strain background or during inbreeding to maintain a mutation in the stock.

FRT sites, flippase recognition target, 34-bp sequence recognized by Flp recombinase. Contains two 13-bp inverted repeats and an 8-bp asymmetric spacer region in between, see DNA recombinase systems.

gametes, the haploid sex cells, sperm (spermatozoa), or oocytes that are the products of meiotic divisions and come together during fertilization to form the zygote.

gametogenesis, the process of formation of the gametes that includes meiosis and the cellular differentiation of the mature sperm and oocyte.

gastrulation, the process in early embryogenesis that results in the formation of mesoderm.

gene targeting, homologous recombination between DNA sequences residing in the chromosome and introduced DNA sequences. This technology allows for directed mutagenesis of a specific locus.

gene trap, an insertional mutagen usually containing a splice acceptor sequence and reporter; chimeric cDNAs that can facilitate cloning are generated.

genetic background, usually referring to the strain or stock on which a mutant allele is maintained (e.g., an inbred C57BL/6J background or an outbred background).

genetic modifier, a locus that enhances or suppresses the expression of another locus.

genotype, the genetic makeup of an organism as opposed to the physical appearance or phenotype; the combination of alleles at a specific locus.

germ layers, the three classic layers of a triploblastic embryo: the ectoderm, mesoderm, and endoderm.

germline transmission, in the context of embryonic stem cell–embryo chimeras, the inheritance of an ES cell–derived allele by its offspring.

gestation, period of embryogenesis, which takes place within the maternal reproductive tract.

GFP, green fluorescent protein. eGFP is enhanced green fluorescent, a variant with enhanced spectral qualities. The gene is incorporated into the mammalian genome mainly for the purpose of live imaging.

giant cells, trophoblast cells with very large nuclei resulting from endoreduplication of DNA. These cells surround the embryo at the maternal–embryo interface and form the outside fetal layer of the chorioallantoic placenta.

guide RNA, synthetic RNA with a Cas-binding domain and a sequence that recognizes the genomic target region for CRISPR–Cas gene editing.

haploid, containing a single set of chromosomes, referred to as N. Haploid cells result from meiotic division.

haploinsufficiency, a phenotype resulting from the presence of half the normal amount of gene product due to, for example, a null mutation in one allele, a deletion of one allele, imprinting of an allele, or X-chromosome inactivation. Contrast this phenotype with haplosufficiency, in which the amount of gene product generated by one allele is sufficient for normal function.

hatching, the process whereby the blastocyst escapes from the zona pellucida before implantation.

HCR, hairpin chain reaction or hybridization chain reaction RNA in situ hybridization using fluorescent probes. Multiple probes with different fluorescent colors can be used on individual sections or wholemount specimens.

hematopoiesis, the process of blood cell formation.

hemizygous, a locus with no corresponding locus on the homologous chromosome. This term usually refers to an insertional transgene and to X- and Y-chromosome-linked genes in males.

heterozygote, an animal that has two distinguishable alleles at a particular locus. The locus or the animal is considered heterozygous.

histocompatible, during transplantation, donor tissue is recognized by the recipient as self.

homolog/homologous, (1) member of a chromosome pair in diploid organisms; (2) genes in one species that have a common evolutionary origin and function as genes from another species, or genes within a species that have a common origin and function. (*See also ortholog.* Note that all orthologs are homologs but not all homologs are orthologs.)

homologous DNA, genetically identical DNA.

homologous recombination, genetic exchange between homologous DNA (e.g., between homologous chromosomes or between a chromosome and a gene-targeting construct containing homologous genomic DNA). (*See also gene targeting.*)

homology-directed repair, CRISPR–Cas-mediated homologous recombination.

homozygote, an organism that has two identical alleles at a particular locus. The locus or animal is considered homozygous. A double homozygote is an animal that is homozygous at two particular loci.

HREM, high-resolution episcopic microscopy.

hybrid mice, first filial (F_1) generation of mice from a cross between two inbred strains. These mice are all genetically uniform (i.e., identical to each other). However, when hybrid mice are mated together, the offspring, the F_2 generation, will be genetically nonuniform because these offspring will be segregating for all loci at which the parental strains differ; thus, each F_2 animal will have a unique, recombinant genetic makeup.

hybrid vigor, generally increased vigor and fecundity associated with heterozygosity at a large number of loci. This term is frequently used to refer to the offspring of a cross between two inbred strains that will be more vigorous than either parent strain.

hypermorph, a mutation that causes an increase in gene expression or protein activity compared to wild type.

hypomorph, a mutation that retains some wild-type function of the gene and results in a less severe phenotype than a null, or loss-of-function, allele.

ICM, inner cell mass, the cells enclosed within the trophectoderm layer at the blastocyst stage of development.

ICSI, intracytoplasmic sperm injection, sperm or sperm heads injected directly into oocytes. This method can be used to overcome certain types of infertility.

immunosurgery, a procedure for isolation of the ICM of the blastocyst by exposure of the embryo to antibodies, followed by complement-mediated lysis of the outer trophoblast layer.

implantation delay, also called diapause, a physiological delay in the implantation process accompanied by a delay in embryonic development. In mice, implantation delay can be brought about naturally by lactation and nursing of a previous litter, or experimentally by ovariectomy of a recently pregnant female prior to implantation.

implantation, the process whereby the blastocyst attaches to and invades the maternal uterine epithelium, elicits a decidual response in the mother, and establishes maternal–fetal exchange through placental structures. In the mouse, this process begins at E4.5.

imprinting, a process of epigenetic silencing of one allele of a homologous pair that takes place at certain loci during gametogenesis. Some loci are maternally imprinted and others are paternally imprinted.

inbred strains, for all practical purposes, the mice of an inbred strain are genetically identical to one another except for sex chromosome differences. Inbred strains are derived by many generations of close inbreeding, minimally 20 generations of brother–sister matings. However, if separate breeding colonies of inbred strains are maintained after the initial derivation, they are called substrains after 18 generations of separation in recognition of the probable accumulation of random genetic mutations that render them nonidentical.

inbreeding depression, general loss of vigor and fertility as a result of the loss of heterozygosity or hybrid vigor during the inbreeding process.

inbreeding, breeding scheme that results in increased homozygosity and is usually accomplished by brother–sister or father–daughter matings. A strain is considered inbred after 20 generations of brother–sister matings, when the probability of heterozygosity at any given locus is less than 2%. Inbreeding will occur in any small closed breeding colony.

indel mutation, insertion or deletion allele.

insertional mutation, a mutation in an endogenous gene brought about by the insertion of a segment of DNA.

intercross, a cross between two animals that have the same heterozygous genotype at a specific locus.

iPS cells, induced pluripotent stem cells.

IRES, internal ribosome entry site, used to create bicistronic mRNAs.

isogenic, genetically identical, as in DNA derived from different individuals of an inbred strain.

IVF, in vitro fertilization, combining sperm and oocytes in vitro for fertilization; usually followed by transfer of the resulting embryos to the oviduct or uterus for continued development.

karyotype, the number of chromosomes and the specific microscopic banding pattern seen upon staining of condensed chromatin.

knock-in allele, insertion of heterologous sequence into a specific locus generally used to produce expression in the pattern of the target locus

knockout mice, jargon that originally referred to mice with a nonfunctional allele produced by gene targeting in embryonic stem cells, but which has been generalized for use with all mutations made by the method. Because of this imprecision, the term should be well defined or avoided altogether.

labyrinth, region of the placenta in which maternal and fetal tissues are tightly juxtaposed and nutrient and waste exchange takes place.

locus (plural: **loci**), the position or location of a gene on a chromosome.

loss of function, referring to a mutation in which the protein is absent or rendered nonfunctional; also called a null or null mutation.

loxP, locus of crossing-over, 34-bp sequence recognized by Cre recombinase. Contains two 13-bp inverted repeats and an 8-bp asymmetric spacer region in between, *see DNA recombinase systems.*

lssDNA, long single-stranded DNA, used with CRISPR–Cas to knock in heterologous sequences to a specific locus.

marker recycling, method for creating homozygous mutant embryonic stem cell lines. A gene-targeting vector with a floxed drug-selection cassette is targeted to one allele and Cre recombinase is used to remove the drug-selection cassette. The heterozygous *neo*⁻ embryonic stem cells are then used to target the remaining wild-type allele, using the same gene-targeting vector and G418 selection.

MEFs, mouse embryonic fibroblasts.

meiosis, the nuclear reduction division process by which diploid germ cell precursors segregate chromosomes into haploid nuclei of sperm and oocytes; through this process, one diploid germ cell produces four haploid daughter cells.

mesonephros (plural: **mesonephroi**), one of the three pairs of embryonic renal organs that develop in mammals. The others are pronephros and metanephros.

MI, mitotic index, a measure of cell division determined by dividing the number of cells in mitosis by the total number of cells examined.

microCT, X-ray microcomputed tomography.

microMRI, microscopic magnetic resonance imaging.

missense mutation, mutation that results in a change in a specific amino acid.

moribund, near death.

morula, a preimplantation embryo of between eight and 32 cells. The name derives from the mulberry appearance of the blastomeres before compaction. A compacted morula describes the postcompaction embryo before blastocyst formation.

mosaic, in general terms, an animal consisting of genetically different cells, regardless of how it originated. However, the term has been more specifically used to distinguish an animal, such as one with a somatic mutation that is derived from a single zygote, from a chimera, which is derived from two or more different zygotes.

mutant allele, a variant allele that usually confers a phenotypically identifiable difference with reference to a "wild-type" phenotype. The term usually, but not necessarily, refers to an allele with a deleterious effect.

mutation, a new allele present in an organism that is not present in its parents. Spontaneous mutations occur without experimental intervention, whereas induced mutations are brought about by experimental means such as chemical mutagenesis or homologous recombination.

neo, neomycin resistance gene or neomycin resistance expression cassette. *neo* is a bacterial gene encoding aminoglycoside-3-phosphotransferase that confers resistance to the aminoglycoside G418.

neomorph, mutant protein with a new activity in comparison to wild type.

null, referring to a mutation that eliminates gene function.

OCT, optical coherence tomography.

OCT, Tissue-Tek Optimal Cutting Temperature Compound, a liquid embedding compound to freeze tissues for frozen sectioning.

oligonucleotide, short single strand of synthetic DNA used with CRISPR–Cas to introduce genetic modifications.

oogenesis, the process of oocyte formation from the female primordial germ cells; female gametogenesis.

oogonia, diploid female germ cells that give rise to oocytes.

OPT, optical projection tomography.

ORF, open reading frame.

organogenesis, the process of organ formation.

ortholog/orthologous, genes in two species that are related by evolution from a single common ancestral gene. Several genes in one species may have a single ortholog in another species and vice versa. (*See also homolog.* Note that all orthologs are homologs but not all homologs are orthologs.)

orthotopic, in the normal or usual position.

outbred or **noninbred,** also called random-bred, although the breeding scheme is not random but carefully controlled to avoid inbreeding. Outbred mice are usually bred in large closed colonies. These stocks are segregating for many alleles, some of which may be defined, and are generally robust mice. Different outbred stocks will vary in the range of alleles they contain.

outcross, a mating between genetically unrelated animals.

PACT, photoacoustic computed tomography.

PAM sequence, protospacer-adjacent motif, a 2- to 6-bp DNA sequence adjacent to the DNA region recognized by a guide RNA for CRISPR–Cas gene editing.

parietal yolk sac, extraembryonic membrane composed of trophoblast giant cells and parietal endoderm with the acellular Reichert's membrane; it lines the yolk sac cavity.

parturition, the process of birth.

PCNA, proliferating cell nuclear antigen, marks cells in mitosis and is used in an immunocytochemical assay for cell proliferation.

penetrance, percentage of genetically mutant individuals that exhibit the mutant phenotype. The mutant phenotype is fully penetrant when all mutant individuals exhibit the mutant phenotype.

pericardial effusion, edema resulting in a fluid-filled pericardial sac.

perinatal, time period soon after birth.

PFA, paraformaldehyde, a tissue fixative often used for embryos.

Pgk1, phosphoglycerate kinase 1 gene.

phenotype, the result of interaction between genotype and the environment.

placenta, in the mouse, typically used to refer to the chorioallantoic placenta.

PNS, peripheral nervous system.

polymorphic, more than one allele for a specific gene or DNA segment.

postnatal, after birth.

pronucleus (plural: **pronuclei**), nuclear structures derived from the sperm nucleus and the oocyte nucleus that form in the zygote after fertilization.

provirus, viral DNA that has integrated into the host genome.

pseudopregnant, a physiological state brought about in an estrous female by the stimulus of mating. Pseudopregnancy parallels the hormonal stages of early pregnancy even though there are no embryos. It can be induced by mating with vasectomized males to provide females as recipients for embryo transfer.

RBCs, red blood cells, erythrocytes.

recessive, the nature of inheritance of phenotypes. In relation to another allele at the same locus, a recessive phenotype is one that is not detectable unless both alleles have a particular mutation. A single mutation can be dominant for some phenotypic traits and recessive for others.

redundancy, overlap in function of different genes such that loss of function of one gene product may be insufficient to cause an overt mutant phenotype, but concomitant loss of function of the product of a second, related gene leads to a mutant phenotype.

regulatory mutation, a mutation altering transcription, translation, RNA splicing, RNA stability, etc.

rescuing transgene, a transgene randomly integrated into the genome that can provide wild-type function to suppress or "rescue" a mutant phenotype.

resorption or **resorption site,** the remains of a dead conceptus and its implantation site in the uterus.

retrovirus, RNA virus whose life cycle includes reverse transcription to generate DNA that integrates into the host genome as a provirus. This serves as a template to make more virus.

Rosa26, reverse orientation splice acceptor 26, a locus that was discovered in a gene trap experiment that has ubiquitous expression in the embryo proper, but patchy expression in the extraembryonic tissues.

semidominant, the nature of inheritance of phenotypes. A semidominant phenotype occurs when the phenotype in the heterozygote is intermediate between the wild-type and homozygous mutant.

site-specific recombinase, *see DNA recombinase systems.*

speed congenic, in transferring a mutation onto an inbred strain, this method makes use of genetic markers to select animals for breeding in the next generation that have undergone specific chromosomal recombination events, thus increasing the genetic contribution of the inbred strain.

spermatogenesis, the process of gametogenesis in the male germline resulting in the production of spermatozoa.

spermatogonia, diploid male germ cells that give rise to spermatozoa.

spermiogenesis, the stage of spermatogenesis in which haploid spermatids are morphologically transformed into spermatozoa.

SPF, specific pathogen–free, a categorization of animal facilities maintained to exclude a specific group of bacterial and viral pathogens.

spongiotrophoblast, the layer of the chorioallantoic placenta, derived from the trophoblast and surrounded by trophoblast giant cells.

Sry, sex determining region on the Y chromosome.

stock, referring to outbred mice.

substrain, an inbred strain maintained as a separate breeding population from the parent strain for 18 generations or more. Substrains may differ because of the fixation of spontaneous mutations or because of genetic contamination.

superovulation, hormonal induction of ovulation of a large number of oocytes using timed injections of pregnant mare serum gonadotropin and human chorionic gonadotropin.

syncytiotrophoblast, a syncytium of multinucleated trophoblast cells that makes up the labyrinth of the placenta.

syntenic, loci in the same linkage group. Conserved synteny between species refers to the situation in which linked loci in one species have orthologs that are linked in another species.

targeted mutagenesis, the technology that allows for directed mutagenesis of a specific locus.

TEM, transmission electron microscopy.

teratocarcinoma, malignant tumor similar in origin to a teratoma, but one in which a permanent population of stem cells renders the tumor transplantable. Embryonal carcinoma cells are derived from teratocarcinomas.

teratoma, a benign tumor with a variety of differentiated cell types. Teratomas are derived from the transplantation of embryos, genital ridges, or other fragments of embryos. The range of differentiated cell types is a reflection of the developmental potential of the transplanted tissue.

term or **full term,** full duration of gestation, ~19–20 days in mice.

tetraploid, containing two diploid sets of chromosomes, or 4N. Tetraploids are typically induced by electrofusion of blastomeres at the two-cell stage.

totipotency, the potential to differentiate into all cell types including embryonic and extraembryonic tissues.

transgene, foreign DNA that has been introduced and incorporated into the genome.

transgenic mice, mice containing foreign DNA that has been introduced into their genome. Technically, this includes mice made by viral infection or injection of DNA into preimplantation embryos (which results in random insertion of the foreign DNA) as well as mice produced by targeted mutagenesis. However, the term is usually reserved for mice made by random insertion of DNA following zygote injection.

transposase, enzyme that can excise and insert a transposon into DNA.

transposon, mobile DNA element.

trophectoderm, the outermost layer of epithelial cells of the blastocyst of the preimplantation embryo; also called trophoblast.

trophoblast giant cells, the nondividing, polyploid cells of the chorioallantoic placenta that are formed by endoreduplication and derived from the trophectoderm layer of the blastocyst.

trophoblast, the outer layer of epithelial cells of the preimplantation blastocyst embryo and all of its derivatives at later stages, comprising the giant cells and the majority of the fetal component of the placenta.

TS cells, trophoblast stem cells, stem cells derived from trophoblast of preimplantation or early postimplantation embryos that can be perpetuated in vitro as undifferentiated cells but have a developmental potential similar to trophectoderm (i.e., they can contribute to trophoblast derivatives in chimeras).

TUNEL, TDT-mediated UTP nick end labeling, method to visualize cells undergoing programmed cell death.

umbilicus, umbilical, cord-like structure, containing fetal veins and arteries, that connects the fetus to the placenta.

urogenital ridges, paired longitudinal ridges that form in the dorsal body wall; contain the mesonephroi and gonads.

visceral yolk sac, extraembryonic membrane composed of extraembryonic mesoderm and visceral endoderm.

vitelline, referring to the yolk sac. Vitelline circulation refers to the blood vessels in the yolk sac.

weaning, removal of pups from a lactating mother, usually done in mouse husbandry between 3 and 4 weeks of age.

wild type, an animal or allele that is considered normal and is usually the common type found in natural populations.

X-chromosome linked or X linked, a locus on the X chromosome.

XEN cells, extraembryonic endoderm–derived stem cells

X-gal, 5-bromo-4-chloro-3-indolyl-β-D-galactoside.

XO, referring to a mouse with only a single X chromosome and no Y chromosome; these animals are phenotypically female and fertile.

XX, referring to a mouse with two X chromosomes, normally a female.

XY, referring to a mouse with an X and a Y chromosome; normally a male.

ZP, zona pellucida, the thin, noncellular membrane, deposited around an oocyte during oogenesis, that the sperm penetrates during fertilization. It surrounds the preimplantation embryo until close to the time of implantation.

zygote, the fertilized oocyte.

APPENDIX 4

General Safety and Hazardous Material Information

Cold Spring Harbor Laboratory Manuals should be used by laboratory personnel with experience in laboratory and chemical safety or students under the supervision of trained personnel. The procedures, chemicals, and equipment referenced in these manuals may be hazardous and could cause serious injury unless performed, handled, and used with care and in a manner consistent with safe laboratory practices. Students and researchers using the procedures in these manuals do so at their own risk. It is essential for your safety that you consult the appropriate Material Safety Data Sheets, the manufacturers' manuals provided with the relevant equipment, and your institution's Environmental Health and Safety Office (hereafter referred to as safety office), as well as the General Safety and Disposal Cautions in this appendix for proper handling of hazardous materials. Cold Spring Harbor Laboratory makes no representations or warranties with respect to the material set forth in its manuals and has no liability in connection with the use of these materials.

All registered trademarks, trade names, and brand names mentioned in these manuals are the property of the respective owners. Readers should consult individual manufacturers and other resources for current and specific product information.

Appropriate sources for obtaining safety information and general guidelines for laboratory safety are provided in the General Safety and Hazardous Material Information Appendix below.

Users should always consult individual manufacturers, the manufacturers' safety guidelines, and other resources, including local safety offices, for current and specific product information and for guidance regarding the use and disposal of hazardous materials.

THE PRIMARY SAFETY INFORMATION RESOURCES FOR LABORATORY PERSONNEL ARE THE FOLLOWING

Institutional Safety Office. The best source of toxicity, hazard, storage, and disposal information is your institutional safety office, which maintains and makes available the most current information. Always consult this office for proper use and disposal procedures.

Post the phone numbers for your local safety office, security office, poison control center, and laboratory emergency personnel in an obvious place in your laboratory.

Material Safety Data Sheets (MSDSs). The Occupational Safety and Health Administration (OSHA) requires that MSDSs accompany all hazardous products that are shipped. These data sheets contain detailed safety information. MSDSs should be filed in the laboratory in a central location as a reference guide.

GENERAL SAFETY AND DISPOSAL CAUTIONS

The guidance offered here is intended to be generally applicable. However, proper waste disposal procedures vary among institutions; therefore, always consult your local safety office for specific instructions. All chemical waste must be disposed of in a suitable container clearly labeled with the type of material it contains and the date the waste was initiated.

It is essential for laboratory workers to be familiar with the potential hazards of materials used in laboratory experiments, and to follow recommended procedures for their use, handling, storage, and disposal.

The following general cautions should always be observed.

- **Before beginning the procedure,** become completely familiar with the properties of substances to be used.

- **The absence of a warning** does not necessarily mean that the material is safe, because information may not always be complete or available.

- **If exposed** to toxic substances, contact your local safety office immediately for instructions.

- **Use proper disposal procedures** for all chemical, biological, and radioactive waste.

- **For specific guidelines on appropriate gloves to use,** consult your local safety office.

- **Handle concentrated acids and bases** with great care. Wear goggles and appropriate gloves. A face shield should be worn when handling large quantities.

    Do not mix strong acids with organic solvents because they may react. Sulfuric acid and nitric acid especially may react highly exothermically and cause fires and explosions.

    Do not mix strong bases with halogenated solvents because they may form reactive carbenes that can lead to explosions.

- **Handle and store pressurized gas containers** with caution because they may contain flammable, toxic, or corrosive gases; asphyxiants; or oxidizers. For proper procedures, consult the Material Safety Data Sheet that is required to be provided by your vendor.

- **Never pipette** solutions using mouth suction. This method is not sterile and can be dangerous. Always use a pipette aid or bulb.

- **Keep halogenated and nonhalogenated** solvents separately (e.g., mixing chloroform and acetone can cause unexpected reactions in the presence of bases). Halogenated solvents are organic solvents such as chloroform, dichloromethane, trichlorotrifluoroethane, and dichloroethane. Nonhalogenated solvents include pentane, heptane, ethanol, methanol, benzene, toluene, N,N-dimethylformamide (DMF), dimethylsulfoxide (DMSO), and acetonitrile.

- **Laser radiation,** visible or invisible, can cause severe damage to the eyes and skin. Take proper precautions to prevent exposure to direct and reflected beams. Always follow the manufacturer's safety guidelines and consult your local safety office. See caution below for more detailed information.

- **Flash lamps,** because of their light intensity, can be harmful to the eyes. They also may explode on occasion. Wear appropriate eye protection and follow the manufacturer's guidelines.

- **Photographic fixatives, developers, and photoresists** contain chemicals that can be harmful. Handle them with care and follow the manufacturer's directions.

- **Power supplies and electrophoresis equipment** pose serious fire hazard and electrical shock hazards if not used properly.

- **Microwave ovens and autoclaves** in the laboratory require certain precautions. Accidents have occurred involving their use (e.g., when melting agar or Bacto Agar stored in bottles or when sterilizing). If the screw top is not completely removed and there is inadequate space for the steam to vent, the bottles can explode and cause severe injury when the containers are removed from the microwave or autoclave. Always completely remove bottle caps before microwaving or autoclaving. An alternative method for routine agarose gels that do not require sterile agar is to weigh out the agar and place the solution in a flask.

- **Ultrasonicators** use high-frequency sound waves (16–100 kHz) for cell disruption and other purposes. This "ultrasound," conducted through air, does not pose a direct hazard to humans, but the associated high volumes of audible sound can cause a variety of effects, including headache, nausea, and tinnitus. Direct contact of the body with high-intensity ultrasound (not medical imaging equipment) should be avoided. Use appropriate ear protection and display signs on the door(s) of laboratories where the units are used.

- **Use extreme caution when handling cutting devices**, such as microtome blades, scalpels, razor blades, or needles. Microtome blades are extremely sharp. Use care when sectioning. If unfamiliar with their use, have an experienced user demonstrate proper procedures. For proper disposal, use the "sharps" disposal container in your laboratory. Discard used needles *unshielded*, with the syringe still attached. This prevents injuries and possible infections when manipulating used needles because many accidents occur while trying to replace the needle shield. Injuries may also be caused by broken Pasteur pipettes, coverslips, or slides.

- **Procedures for the humane treatment of animals** must be observed at all times. Consult your local animal facility for guidelines. Animals such as rats are known to induce allergies that can increase in intensity with repeated exposure. Always wear a lab coat and gloves when handling these animals. If allergies to dander or saliva are known, wear a mask.

DISPOSAL OF LABORATORY WASTE

There are specific regulatory requirements for the disposal of all medical waste and biological samples. In the United States, these are mandated by the U.S. Environmental Protection Agency (see http://www.epa.gov/hw) and regulated by the individual states and territories (see http://www.epa.gov/hw/state-universal-waste-programs-united-states). Medical and biological samples that require special handling and disposal are generally termed Medical Pathological Waste (MPW), and medical, veterinary, and biological facilities will have programs for the collection of MPW and its disposal. Restrictions on how radioactive waste can be disposed of as regulated in the United States by the U.S. Nuclear Regulatory Commission can be found in 10 CFR 20.2001, General requirements for waste disposal (see http://www.nrc.gov/reading-rm/doc-collections/cfr/part020/part020-2001.html) or the individual **Agreement States**. The preferred method for the disposal of radioactively contaminated MPW is decay-in-storage (see http://www.nrc.gov/reading-rm/doc-collections/cfr/part035/part035-0092.html). Contact your institution's safety office and local regulatory agencies for instructions and requirements.

Waste and any materials contaminated with biohazardous materials must be decontaminated and disposed of as regulated medical waste. No harmful substances should be released into the environment in an uncontrolled manner. This includes all tissue samples, needles, syringes, scalpels, etc. Be sure to contact your institution's safety office concerning the proper practices associated with the handling and disposal of biohazardous waste.

Some basic rules are outlined below. For treatment of radioactive and biological waste, see sections on Radioactive Safety Procedures and Biological Safety Procedures.

- In practice, only **neutral aqueous solutions** without heavy metal ions and without organic solvents can be poured down the drain (e.g., most buffers). Acid and basic aqueous solutions need to be neutralized cautiously before their disposal by this method.

- For proper disposal of **strong acids and bases**, dilute them by placing the acid or base onto ice and neutralize them. Do not pour water into them. If the solution does not contain any other toxic compound, the salts can be flushed down the drain.

- For disposal of **other liquid waste**, similar chemicals can be collected and disposed of together, whereas chemically different wastes should be collected separately. This avoids chemical reactions between components of the mixture (see above). Collect at least inorganic aqueous waste, non-halogenated solvents, and halogenated solvents separately.

- Waste **from photo processing and automatic developers** should be collected separately to recycle the silver traces found in it.

RADIOACTIVE SAFETY PROCEDURES

In the United States and other countries, the access to radioactive substances is strictly controlled. You may be required to become a registered user (e.g., by attending a mandatory seminar and receiving a personal dosimeter). A convenient calculator to perform routine radioactivity calculations can also be found at http://www.graphpad.com/quickcalcs/ChemMenu.cfm.

If you have never worked with radioactivity before, follow the steps below.

- **Avoid it when possible.** Many experiments that are traditionally performed with the help of radioactivity can now be done using alternatives based on fluorescence or chemiluminescence and colorimetric assays, including, for example, DNA sequencing, Southern and northern blots, and protein kinase assays. However, in other cases (e.g., metabolic labeling of cells), use of radioactivity cannot be avoided.

- **Be informed.** While planning an experiment that involves the use of radioactivity, consider the physicochemical properties of the isotope (half-life, emission type, and energy), the chemical form of the radioactivity, its radioactive concentration (specific activity), total amount, and its chemical concentration. Order and use only as much as is really needed.

- **Familiarize yourself** with the designated working area. Perform a mental and practical dry run (replacing radioactivity with a colored solution) to make sure that all equipment needed is available and to get used to working behind a shield. Handle your samples as if sterility would be required to avoid contamination.

- **Always wear appropriate gloves**, lab coat, and safety goggles when handling radioactive material.

- **Check the work area** for contamination before, during, and after your experiment (including your lab coat, hands, and shoes).

- **Localize your radioactivity.** Avoid formation of aerosols or contamination of large volumes of buffers.

- **Liquid scintillation cocktails** are often used to quantitate radioactivity. They contain organic solvents and small amounts of organic compounds. Try to avoid contact with the skin. After use, they should be regarded as radioactive waste; the filled vials are usually collected in designated containers, separate from other (aqueous) liquid radioactive waste.

- **Dispose of radioactive waste** only into designated, shielded containers (separated by isotope, physical form [dry/liquid], and chemical form [aqueous/organic solvent phase]). Always consult your safety office for further guidance in the appropriate disposal of radioactive materials.

- Among the experiments requiring **special precautions** are those that use [^{35}S]methionine and ^{125}I, because of the dangers of airborne radioactivity. [^{35}S]methionine decomposes during storage into sulfoxide gases, which are released when the vial is opened. The isotope ^{125}I accumulates in the thyroid and is a potential health hazard. ^{125}I is used for the preparation of Bolton–Hunter reagent to radioiodinate proteins. Consult your local safety office for further guidance in the appropriate use and disposal of these radioactive materials before initiating any experiments. Wear appropriate gloves when handling potentially volatile radioactive substances, and work only in a radioiodine fume hood.

BIOLOGICAL SAFETY PROCEDURES

Biological safety fulfills three purposes: to avoid contamination of your biological sample with other species; to avoid exposure of the researcher to the sample; and to avoid release of living material into the environment. Biological safety begins with the receipt of the living sample; continues with its storage, handling, and propagation; and ends only with the proper disposal of all contaminated materials. A catalog of operations known as "sterile handling" is usually employed in manipulating living matter. However, the correct procedures largely depend on the samples, which can be quite diverse: *Escherichia coli* and other bacterial strains, yeasts, tissues of animal or plant origin, cultures of mammalian cells, or even derivatives from human blood are routinely handled in a biological laboratory. Two of these, bacteria and human blood products, are discussed in more detail below.

The U.S. Department of Health, Education, and Welfare (HEW) has classified various bacteria into different categories with regard to shipping requirements (see Sanderson and Zeigler, *Methods Enzymol* 204: 248–264 [1991]). Nonpathogenic strains of *E. coli* (such as K12) and *Bacillus subtilis* are in Class 1 and are considered to present no or minimal hazard under normal shipping conditions. However, *Salmonella, Haemophilus*, and certain strains of *Streptomyces* and *Pseudomonas* are in Class 2. Class 2 bacteria are "Agents of ordinary potential hazard: agents which produce disease of varying degrees of severity... but which are contained by ordinary laboratory techniques." Contact your institution's safety office concerning shipping biological material.

Human blood, blood products, and tissues may contain occult infectious materials such as hepatitis B virus and human immunodeficiency virus (HIV) that may result in laboratory-acquired infections. Investigators working with lymphoblast cell lines transformed by Epstein–Barr virus (EBV) are also at risk of EBV infection. Any human blood, blood products, or tissues should be considered a biohazard and should be handled accordingly until proved otherwise. Wear appropriate disposable gloves, use mechanical pipetting devices, work in a biological safety cabinet, protect against the possibility of aerosol generation, and disinfect all waste materials before disposal. Autoclave contaminated plasticware before disposal; autoclave contaminated liquids or treat with bleach (10% [v/v] final concentration) for at least 30 minutes before disposal (this is valid also for used bacterial media).

Always consult your local institutional safety officer about specific handling and disposal procedures for samples. Further information can be found in the Frequently Asked Questions of the ATCC homepage (http://www.atcc.org) and is also available from the U.S. National Institute of Environmental Health and Human Services, Biological Safety (http://www.niehs.nih.gov/about/stewardship).

GENERAL PROPERTIES OF COMMON HAZARDOUS CHEMICALS

The hazardous materials list can be summarized in the following categories:

- Inorganic acids, such as hydrochloric, sulfuric, nitric, or phosphoric, are colorless liquids with stinging vapors. Avoid spills on skin or clothing. Spills should be diluted with large amounts of water. The concentrated forms of these acids can destroy paper, textiles, and skin and cause serious injury to the eyes.

- Inorganic bases, such as sodium hydroxide, are white solids that dissolve in water and under heat development. Concentrated solutions will slowly dissolve skin and even fingernails.

- Salts of heavy metals are usually colored, powdered solids that dissolve in water. Many of them are potent enzyme inhibitors and, therefore, toxic to humans and the environment (e.g., fish and algae).

- Most organic solvents are flammable volatile liquids. Avoid breathing the vapors, which can cause nausea or dizziness. Also avoid skin contact.

- Other organic compounds, including organosulfur compounds such as mercaptoethanol or organic amines, can have very unpleasant odors. Others are highly reactive and should be handled with appropriate care.

- If improperly handled, dyes and their solutions can stain not only your sample but also your skin and clothing. Some are also mutagenic (e.g., ethidium bromide), carcinogenic, and toxic.

- Nearly all names ending with "ase" (e.g., catalase, β-glucuronidase, or zymolyase) refer to enzymes. There are also other enzymes with nonsystematic names such as pepsin. Many of them are provided by manufacturers in preparations containing buffering substances, etc. Be aware of the individual properties of materials contained in these substances.

- Toxic compounds are often used to manipulate cells. They can be dangerous and should be handled appropriately.

- Be aware that several of the compounds listed have not been thoroughly studied with respect to their toxicological properties. Handle each chemical with appropriate care. Although the toxic effects of a compound can be quantified (e.g., LD_{50} values), this is not possible for carcinogens or mutagens, where one single exposure can have an effect. Also realize that dangers related to a given compound may also depend on its physical state (fine powder vs. large crystals/diethyl ether vs. glycerol/dry ice vs. carbon dioxide under pressure in a gas bomb). Anticipate under which circumstances during an experiment exposure is most likely to occur and how best to protect yourself and your environment.

Cold Spring Harbor Laboratory Press (CSHLP) has used its best efforts in collecting and preparing the material contained herein but does not assume, and hereby disclaims, any liability for any loss or damage caused by errors and omissions in the publication, whether such errors and omissions result from negligence, accident, or any other cause. CSHLP does not assume responsibility for the user's failure to consult more complete information regarding the hazardous substances listed in this publication.

REFERENCE

Sanderson KE, Zeigler DR. 1991. Storing, shipping, and maintaining records on bacterial strains. *Methods Enzymol* **204:** 248–264. doi:10.1016/0076-6879(91)04012-D

WWW RESOURCES

ATCC Home page. http://www.atcc.org
ATCC, for Sample Handling (in Frequently Asked Questions). http://www .atcc.org/CulturesandProducts/TechnicalSupport/FrequentlyAskedQu estions/tabid/469/Default.aspx
GraphPad Software, Radioactivity Calculations. http://www.graphpad.com/ quickcalcs/ChemMenu.cfm
National Institute of Environmental Health and Human Services, Biological Safety (NIEHS). http://www.niehs.nih.gov/about/stewardship
U.S. Environmental Protection Agency (EPA), Federal waste disposal regulations, Laboratory. http://www.epa.gov/epawaste/hazard/tsd/index.htm

U.S. Environmental Protection Agency (EPA), Individual States and Territories. http://www.epa.gov/epawaste/wyl/stateprograms.htm
U.S. Nuclear Regulatory Commission (NRC), Medical Pathological Radioactively Contaminated Waste (Decay-in-Storage). http://www.nrc.gov/ reading-rm/doc-collections/cfr/part035/part035-0092.html
U.S. Nuclear Regulatory Commission (NRC), Radioactive Waste Disposal Regulations: General Requirements. http://www.nrc.gov/reading-rm/ doc-collections/cfr/part020/part020-2001.html

Index